The Pandas Workshop

A comprehensive guide to using Python for data analysis with real-world case studies

Blaine Bateman

Saikat Basak

Thomas V. Joseph

William So

BIRMINGHAM—MUMBAI

The Pandas Workshop

Publishing Product Manager: Heramb Bhavsar
Senior Editor: David Sugarman
Content Development Editor: Joseph Sunil
Technical Editor: Devanshi Ayare
Copy Editor: Safis Editing
Project Coordinator: Aparna Ravikumar Nair
Proofreader: Safis Editing
Indexer: Manju Arasan
Production Designer: Ponraj Dhandapani
Marketing Coordinator: Nivedita Singh

First published: June 2022

Production reference: 1270522

Published by Packt Publishing Ltd.
Livery Place
35 Livery Street
Birmingham
B3 2PB, UK.

ISBN 978-1-80020-893-3

`www.packt.com`

To my wife Cynthia, who steadfastly supports me in these efforts and is a constant source of inspiration.

-Blaine Bateman

"To all my friends, who couldn't believe I wrote a book about panda(s)."

-William So

To my mother Marykutty, and to the memory of my father V. T. Joseph, for laying the foundation of what I am. To my wife Anu, for being the pillar of support in all my endeavors. My children Joe and Tess, for reminding me that life is not all about Data Science.

-Thomas V. Joseph

Contributors

About the authors

Blaine Bateman has more than 35 years of experience working with various industries, from government R&D to start-ups to $1 billion public companies. His experience focuses on analytics, including machine learning and forecasting. His hands-on abilities include Python and R coding, Keras/TensorFlow, and AWS and Azure machine learning services. As a machine learning consultant, he has developed and deployed actual machine learning models in industry.

Saikat Basak is a data scientist and a passionate programmer. Having worked with multiple industry leaders, he has a good understanding of problem areas that can potentially be solved using data. Apart from being a data guy, he is also a science geek and loves to explore new ideas on the frontiers of science and technology.

Thomas V. Joseph is a data science practitioner, researcher, trainer, mentor, and writer with more than 19 years of experience. He has extensive experience in solving business problems using machine learning toolsets across multiple industry segments.

William So is a Data Scientist with both a strong academic background and extensive professional experience. He is currently the Head of Data Science at Douugh and also a Lecturer for Master of Data Science and Innovation at the University of Technology Sydney.

During his career, he successfully covered the end-end spectrum of data analytics from ML to Business Intelligence helping stakeholders derive valuable insights and achieve amazing results that benefits the business.

William So is a co-author of the "The Applied Artificial Intelligence Workshop" published by Packt.

About the reviewer

Vishwesh Ravi Shrimali graduated from BITS Pilani, where he studied mechanical engineering, in 2018. He also completed a masters in machine learning and AI at LJMU in 2021. He authored *Machine Learning for OpenCV* (*2nd edition*) and *Computer Vision Workshop and Data Science for Marketing Analytics* (*2nd edition*), both available from Packt. When he is not writing blogs or working on projects, he likes to go on long walks or play his acoustic guitar.

Table of Contents

3
Data I/O

4
Pandas Data Types

Part 2 – Working with Data

5
Data Selection – DataFrames

6
Data Selection – Series

7

Data Exploration and Transformation

8

Understanding Data Visualization

Part 3 – Data Modeling

9
Data Modeling – Preprocessing

10
Data Modeling – Modeling Basics

11
Data Modeling – Regression Modeling

Part 4 – Additional Use Cases for pandas

12

Using Time in pandas

13

Exploring Time Series

14

Applying pandas Data Processing for Case Studies

15

Appendix

Index

Other Books You May Enjoy

Preface

The Pandas Workshop will teach you how to be more productive with data and generate real business insights to inform your decision-making. You will be guided through real-world data science problems and shown how to apply key techniques in the context of realistic examples and exercises. Engaging activities will then challenge you to apply your new skills in a way that prepares you for real data science projects.

You'll see how experienced data scientists tackle a wide range of problems using data analysis with pandas. Unlike other Python books, which focus on the theory and spend too long on dry, technical explanations, this workshop is designed to quickly get you writing clean code and building your understanding through hands-on practice.

As you work through this Python pandas book, you'll tackle various real-world scenarios, such as using an air quality dataset to understand the pattern of nitrogen dioxide emissions in a city, as well as analyzing transportation data to improve bus transportation services.

By the end of this data analytics book, you'll have the knowledge, skills, and confidence to solve your own challenging data science problems with pandas.

Who this book is for

This data analysis book is for anyone with prior experience working with the Python programming language who wants to learn the fundamentals of data analysis with pandas. Previous knowledge of pandas is not necessary.

What this book covers

Chapter 1, Introduction, shows how pandas is one of the most versatile applications for data processing today and why it is the most sought-after tool for any data scientist. This chapter gives a brief introduction to many of the versatile features of pandas. It also takes a tour through all the topics that will be covered in this book, along with some introductory exercises using pandas.

Chapter 2, Data Structures, covers a key benefit of pandas, which is that it provides intuitive data structures that align to a wide range of data analysis tasks. The focus here is on learning about the important data structures in pandas, especially DataFrames, Series, and pandas index structures.

Chapter 3, Data I/O, explores the built-in functions that pandas provides to read data from a large variety of sources, as well as write data back to them, or to new files. In this chapter, you will learn all the important supported I/O methods.

Chapter 4, Data Types, explains why, when doing data analysis with pandas, it is critical to use the correct data type, otherwise, unexpected results or errors might appear. In this chapter, you will learn about pandas data types and how to use them.

Chapter 5, Data Selection – DataFrames, does a deep dive into using DataFrames now that you are well versed in the available data structures and methods in pandas.

Chapter 6, Data Selection – Series, highlights some of the important differences when working with pandas Series and is a companion to *Chapter 5, Data Selection – DataFrames*.

Chapter 7, Data Transformation, talks about how any dataset comes with challenges to its quality. In this chapter, you will learn how to use pandas to solve these challenges and make them ready for your analysis.

Chapter 8, Data Visualization, discusses how pandas offers in-built data visualization methods to accelerate your data analysis. In this chapter, you will learn how to build data visualizations from a DataFrame and how to further customize them with matplotlib.

Chapter 9, Data Modeling – Preprocessing, helps you to understand how to do some preliminary data review and analysis in pandas as a preparatory step to modeling, as well as some transformations important to successful modeling.

Chapter 10, Data Modeling – Modeling Basics, introduces you to some powerful pandas methods for resampling and smoothing data to find patterns and gain insights that can be used in more complex modeling tasks.

Chapter 11, Data Modeling – Regression Modeling, focuses on a workhorse method, regression modeling, as the next step toward using models to understand data and make predictions. By the end of the chapter, you will be tackling complex multi-variate datasets with regression models.

Chapter 12, Using Time in pandas, describes another type of data supported by pandas, time series data. It also looks at how pandas provides a wide range of methods to handle data organized by dates and/or times. You will learn how to do operations on time stamps, and see all the additional time-related attributes provided by pandas.

Chapter 13, Exploring Time Series, focuses on how to use a time series index to perform operations on time series data to gain insights. By the end of the chapter, you will apply regression modeling to time series data.

Chapter 14, Case Studies/Mini Projects, enables you to apply your knowledge to data analytics problems, as you will have learned a great deal about pandas throughout this book. This chapter will cover three case studies where you will apply all the skill sets you have gained through this book.

To get the most out of this book

This book assumes a good working knowledge of Python and, in particular, using Jupyter Notebook to create code. You need a Python environment set up on your local computer including Jupyter Notebook, and of course, pandas. There are additional dependencies you may need to install depending on how you created your local environment. The full list is available in the `requirements.txt` file in the GitHub repository for this workshop. The main tools used are given here:

Software/hardware covered in the book	Operating system requirements
Python 9.x	Windows, macOS, or Linux
Jupyter 1.0.0	
pandas 1.3	
matplotlib 3.3	

If you are using the digital version of this book, we advise you to type the code yourself or access the code from the book's GitHub repository (a link is available in the next section). Doing so will help you avoid any potential errors related to the copying and pasting of code.

Download the example code files

You can download the example code files for this book from GitHub at `https://github.com/PacktWorkshops/The-Pandas-Workshop`. If there's an update to the code, it will be updated in the GitHub repository.

We also have other code bundles from our rich catalog of books and videos available at `https://github.com/PacktPublishing/`. Check them out!

Download the color images

We also provide a PDF file that has color images of the screenshots and diagrams used in this book. You can download it here: `https://static.packt-cdn.com/downloads/9781800208933_ColorImages.pdf`.

Conventions used

There are a number of text conventions used throughout this book.

`Code in text`: Indicates code words in text, database table names, folder names, filenames, file extensions, pathnames, dummy URLs, user input, and Twitter handles. Here is an example: "`pandas` offers us the `.corr()` method to use with `DataFrames` as follows."

A block of code is set as follows:

```
lin_model = sm.OLS(metal_data['alloy_hardness'], X)
my_model = lin_model.fit()
print(my_model.summary())
```

When we wish to draw your attention to a particular part of a code block, the relevant lines or items are set in bold:

```
import pandas as pd
my_data = pd.read_csv('Datasets/auto-mpg.data.csv')
my_data.head()x1 and x2:   -0.9335045017430936
```

Bold: Indicates a new term, an important word, or words that you see onscreen. For instance, words in menus or dialog boxes appear in **bold**. Here is an example: "Select **System info** from the **Administration** panel."

> **Tips or Important Notes**
> Appear like this.

Get in touch

Feedback from our readers is always welcome.

General feedback: If you have questions about any aspect of this book, email us at `customercare@packtpub.com` and mention the book title in the subject of your message.

Errata: Although we have taken every care to ensure the accuracy of our content, mistakes do happen. If you have found a mistake in this book, we would be grateful if you would report this to us. Please visit www.packtpub.com/support/errata and fill in the form.

Piracy: If you come across any illegal copies of our works in any form on the internet, we would be grateful if you would provide us with the location address or website name. Please contact us at `copyright@packt.com` with a link to the material.

If you are interested in becoming an author: If there is a topic that you have expertise in and you are interested in either writing or contributing to a book, please visit authors.packtpub.com.

Share Your Thoughts

Once you've read *The Pandas Workshop*, we'd love to hear your thoughts! Scan the QR code below to go straight to the Amazon review page for this book and share your feedback.

`https://packt.link/r/1-800-20893-6`

Your review is important to us and the tech community and will help us make sure we're delivering excellent quality content.

Part 1 – Introduction to pandas

This section will serve as a brief introduction to the world of pandas, its functionalities, and its history. It also covers the various data structures that are used in pandas and how they can be used for data analysis and machine learning. We will also see how to efficiently access data from various sources, as well as the various data types that pandas uses for its various operations.

This section contains the following chapters:

- *Chapter 1, Introduction to pandas*
- *Chapter 2, Data Structures*
- *Chapter 3, Data I/O*
- *Chapter 4, pandas Data Types*

1
Introduction to pandas

From creating basic DataFrames to optimizing your code, this chapter will serve as a crash course that will help you realize, through code examples and practical exercises, the prowess of pandas in data wrangling and analytics. By the end of this chapter, you will have gained rudimentary experience in reading and writing data, performing DataFrame aggregation, working with time series, preprocessing before modeling, data visualization, and more. You will also be able to create datasets and perform preprocessing tasks on them. You will implement most of the core features of the library, which we will cover in more depth in the chapters that follow. The final chapter of this book will help you consolidate what you've learned through a series of activities.

In this chapter, we will cover the following topics:

- Introduction to the world of pandas
- Exploring the history and evolution of pandas
- Components and applications of pandas
- Understanding the basic concepts of pandas
- Activity 1.01 – comparing sales data for two stores

Introduction to the world of pandas

Tess's latest project has turned out to be much more time-consuming than she initially anticipated. Her client, who develops and provides content for schools, wants her to find insights into their students' needs by analyzing data that's been collected through various sources. Things would have been much easier had this data been in a single format, but unfortunately, that's not the case. The client has sent her data in multiple formats, including HTML, JSON, Excel, and CSV. She has to extract the relevant information from all these files. These are not the only data sources she'll be working with, though. She also has to access the records of the top-performing and struggling students from a SQLite database so that she can analyze their performance patterns. All these disparate data elements differ in their data types, velocities, frequencies, and volumes. She must now extract different elements from these data sources by slicing, subsetting, grouping, merging, and reshaping the data to get a comprehensive list of features for further analysis. Since the volumes are large, she must also optimize her methods for efficient processing.

Does this scenario sound familiar to you? Are you overwhelmed by the data wrangling tasks that must be performed before the analytics processes? Well, you do not have to struggle anymore. pandas is a Python library that is capable of carrying out all these tasks and more. Over the years, pandas has become the go-to tool for all the preprocessing tasks involved in the life cycle of data analytics.

In this chapter, you will begin to explore and have fun with pandas, an amazing library that's used extensively by the data science and machine learning community. As you work through the exercises and activities in this chapter and the ones that follow, you will understand why pandas is considered the de facto standard when working with data. But first, let's take a short trip through time to understand the evolution of the library and get a glimpse into all the functionalities you will be learning about in this chapter.

Exploring the history and evolution of pandas

pandas, in its basic version, was open sourced in 2009 by Wes McKinney, an MIT graduate with experience in quantitative finance. He was unhappy with the tools available at the time, so he started building a tool that was intuitive and elegant and required minimal code. pandas went on to become one of the most popular tools in the data science community, so much so that it even helped increase Python's popularity to a great extent.

One of the primary reasons for the popularity of pandas is its ability to handle different types of data. pandas is well suited for handling the following:

- Tabular data with columns that are capable of storing different types of data (such as numerical data and text data)

- Ordered and unordered series data (an arbitrary sequence of numbers in a list, such as *[2,4,8,9,10]*)
- Multi-dimensional matrix data (three-dimensional, four-dimensional, and so on)
- Any other form of observational/statistical data (such as SQL data and R data)

Besides this, a large repertoire of intuitive and easy-to-use functions/methods makes pandas the go-to tool for data analytics. In the next section, we'll cover the components of pandas and their main applications.

Components and applications of pandas

An introduction to the `pandas` library would be incomplete without a glimpse into its architecture. The `pandas` library is comprised of the following components:

- `pandas/core`: This contains the implementations of the basic data structures of pandas, such as Series and DataFrames. Series and DataFrames are basic toolsets that are very handy for data manipulation and are used extensively by data scientists. They will be covered in *Chapter 2, Data Structures*.
- `pandas/src`: This consists of algorithms that provide the basic functionalities of pandas. These functionalities are part of the architecture of pandas, which you will not be using explicitly. This layer is written in C or Cython.
- `pandas/io`: This comprises toolsets for the input and output of files and data. These toolsets facilitate data input from sources such as CSV and text and allow you to write data to formats such as text and CSV. They will be covered in detail in *Chapter 3, Data I/O*.
- `pandas/tools`: This layer contains all the code and algorithms for pandas functions and methods, such as `merge`, `join`, and `concat`.
- `pandas/sparse`: This contains the functionalities for handling missing values within its data structures, such as DataFrames and Series.
- `pandas/stats`: This contains a set of tools for handling statistical functions such as regression and classification.
- `pandas/util`: This contains all the utilities for debugging the library.
- `pandas/rpy`: This is the interface for connecting to R.

The versatility of its different architectural components makes pandas useful in many real-world applications. Various data-wrangling functionalities in pandas (such as merge, join, and concatenation) save time when building real-world applications. Some notable applications where the pandas library can come in handy are as follows:

- Recommendation systems
- Advertising
- Stock predictions
- Neuroscience
- **Natural language processing** (NLP)

The list goes on. What's more important to note is that these are applications that have an impact on people's daily lives. For this reason, learning pandas has the potential to give a fillip to your analytics career. Benjamin Franklin, one of the founding fathers of the United States, once said the following:

"*An investment in knowledge pays the best interest.*"

Throughout this book, you are going to invest your time in a tool that can have a profound impact on your analytics career. Do make the best use of this opportunity.

Understanding the basic concepts of pandas

This section will take you through a quick tour of the fundamental concepts of pandas. You will use Jupyter Notebooks to run the code snippets in this book. In the *Preface*, you learned how to install Anaconda and the required libraries. If you created and installed a separate virtual environment (and kernel) for this book, as shown in the *Preface*, you'll need to open the Jupyter Notebook, click **New** in the top-right corner of the Jupyter Notebook navigator, and select the **Pandas_Workshop** kernel:

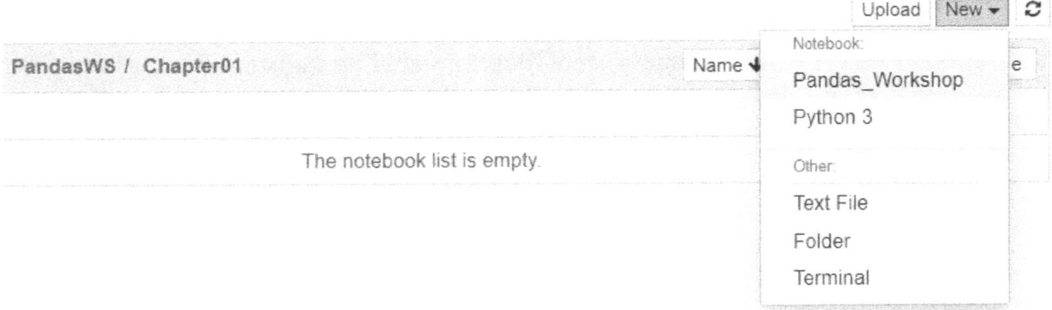

Figure 1.1 – Selecting the Pandas_Workshop kernel

A new untitled Jupyter notebook will open that should look something like this:

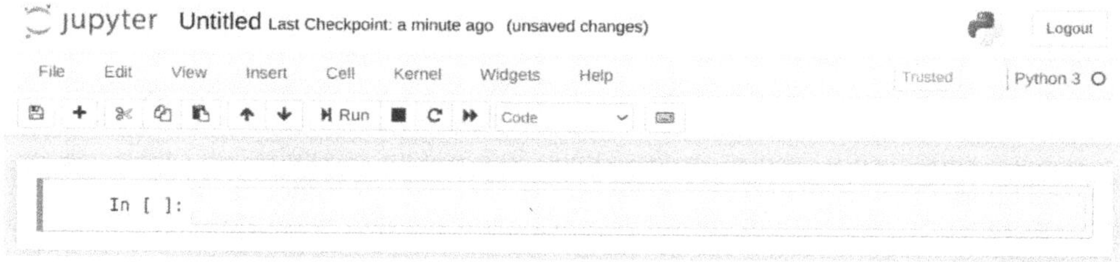

Figure 1.2 – Jupyter Notebook interface

Installing the `conda` environment will have also installed pandas. All you need to do now is import the library. You can do so by typing or pasting the following command in a fresh notebook cell:

```
import pandas as pd
```

Press the *Shift* + *Enter* keyboard shortcut (or click the **Run** button in the toolbar) to execute the command you just entered:

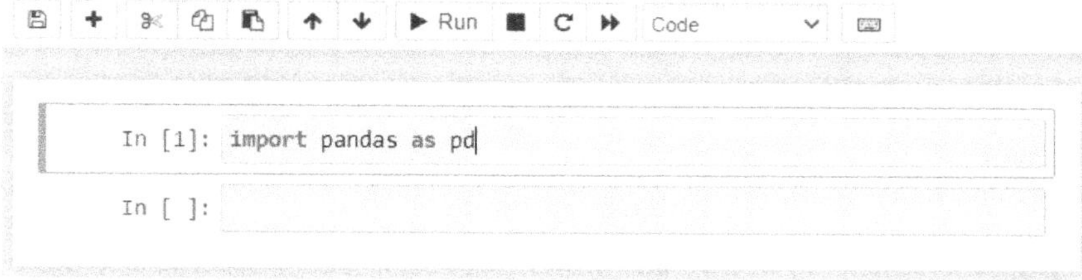

Figure 1.3 – Importing the pandas library

If you don't see any errors, this means that the library was imported successfully. In the preceding code, the `import` statement provides all the functionalities of pandas for you to work with. The term `pd` is a commonly used alias for referring to pandas. You can use any other alias, but `pd` is used the most.

Now that pandas has been imported, you are ready to roll. Before you start getting your hands dirty and learning about the library in more depth, it is worth taking a quick look at some of the key components and functionalities that pandas offers. At this stage, you don't need to know the details of how they work as you'll be learning about them as you progress through this book.

The Series object

To learn about data wrangling with pandas, you'll need to start from the very beginning, and that is with one-dimensional data. In pandas, one-dimensional data is represented as Series objects. Series objects are initialized using the `pd.Series()` constructor.

The following code shows how the `pd.Series()` constructor is used to create a new Series called `ser1`. Then, simply calling the new Series by its assigned name will display its contents:

> **Note**
>
> If you're running the following code example in a new Jupyter notebook, don't forget to run the `import` statement shown in the previous section. You'll need to import the pandas library, in the same manner, every time you create a fresh Jupyter notebook.

```
# Creating a Series
ser1 = pd.Series([10,20,30,40])
# Displaying the Series
ser1
```

Running the preceding code in a fresh Jupyter notebook cell will produce the following output:

```
0    10
1    20
2    30
3    40
dtype: int64
```

Figure 1.4 – Series object

From the output, you can see that the one-dimensional list is represented as a Series. The numbers to the left of the Series (`0`, `1`, `2`, `3`) are its indices.

You can represent different types of data in a Series. For example, consider the following snippet:

```
ser2 = pd.Series([[10, 20],\
                  [30, 40.5,'series'],\
                  [50, 55],\
                  {'Name':'Tess','Org':'Packt'}])
ser2
```

Running this snippet will result in the following output:

```
0                              [10, 20]
1                    [30, 40.5, series]
2                              [50, 55]
3    {'Name': 'Tess', 'Org': 'Packt'}
dtype: object
```

Figure 1.5 – Series output

In the preceding example, a Series was created with multiple data types. You can see that there are lists with numeric and text data (lines 0, 1, and 2) along with a dictionary (line 3). From the output, you can see how these varied data types are represented neatly.

Series help you deal with one-dimensional data. But what about multi-dimensional data? That's where DataFrames come in handy. In the next section, we'll provide a quick overview of DataFrames, one of the most used data structures in pandas.

The DataFrame object

One of the basic building blocks of pandas is the DataFrame structure. A DataFrame is a two-dimensional representation of data in rows and columns, which can be initialized in pandas using the `DataFrame()` constructor. In the following code, a simple list object is being converted into a one-dimensional DataFrame:

```
# Create a DataFrame using the constructor
df = pd.DataFrame([30,50,20])
# Display the DataFrame
df
```

You should get the following output:

	0
0	30
1	50
2	20

Figure 1.6 – DataFrame for list data

This is the simplest representation of a DataFrame. In the preceding code, a list of three elements was converted into a DataFrame using the `DataFrame()` constructor. The shape of a DataFrame can be visualized using the `df.shape()` command, as follows:

```
df.shape
```

The output will be as follows:

```
(3, 1)
```

The output is a DataFrame with a shape of `(3,1)`. Here, the first element (`3`) is the number of rows, while the second (`1`) is the number of columns.

If you look at the DataFrame in *Figure 1.6*, you will see `0` at the top of the column. This is the default name that will be assigned to the column when the DataFrame is created. You can also see the numbers `0`, `1`, and `2` along the rows. These are called **indices**.

To display the column names of the DataFrame, you can use the following command:

```
df.columns
```

This will result in the following output:

```
RangeIndex(start=0, stop=1, step=1)
```

The output shows that it is a range of indices starting at `0` and stopping at `1` with a step size of `1`. So, in effect, there is just one column with the name `0`. You can also display the names of the indices for the rows using the following command:

```
df.index
```

You will see the following output:

```
RangeIndex(start=0, stop=3, step=1)
```

As you can see, the indices start from `0` and end at `3` with a step value of `1`. This will give the indices `0`, `1`, and `2`.

There are many instances where you would want to use the column names and row indices for further processing. For such purposes, you can convert them into a list using the `list()` command. The following snippet converts the column names and row indices into lists and then prints those values:

```
print("These are the names of the columns",list(df.columns))

print("These are the row indices",list(df.index))
```

You should get the following output:

```
These are the names of the columns [0]
These are the row indices [0, 1, 2]
```

From the output, you can see that the column names and the row indices are represented as a list.

You can also rename the columns and row indices by assigning them to any list of values. The command for renaming a column is `df.columns`, as shown in the following snippet:

```
# Renaming the columns
df.columns = ['V1']
df
```

You should see the following output:

	V1
0	30
1	50
2	20

Figure 1.7 – Renamed column

Here, you can see that the column has been renamed `V1`. The command for renaming an index is `df.index`, as shown in the following snippet:

```
# Renaming the indices
df.index = ['R1','R2','R3']
df
```

Running the preceding snippet produces the following output:

	V1
R1	30
R2	50
R3	20

Figure 1.8 – Renamed indices

These were examples of DataFrames with just one column. But what if you need to create a DataFrame that contains multiple columns from the list data? This can easily be achieved using a nested list of lists, as follows:

```
# Creating DataFrame with multiple columns
df1 = pd.DataFrame([[10,15,20],[100,200,300]])

print("Shape of new data frame",df1.shape)

df1
```

You should get the following output:

Shape of new data frame (2, 3)

	0	1	2
0	10	15	20
1	100	200	300

Figure 1.9 – Multi-dimensional DataFrame

From the new output, you can see that the new DataFrame has two rows and three columns. The first list forms the first row and each of its elements gets mapped to the three columns. The second list becomes the second row.

You can also assign the column names and row names while creating the DataFrame. To do that for the preceding DataFrame, the following command must be executed:

```
df1 = pd.DataFrame([[10,15,20],[100,200,300]],\
                   columns=['V1','V2','V3'],\
                   index=['R1','R2'])

df1
```

The output will be as follows:

	V1	V2	V3
R1	10	15	20
R2	100	200	300

Figure 1.10 – Renamed columns and indices

From the output, you see that the column names (`V1`, `V2`, and `V3`) and index names (`R1` and `R2`) have been initialized with the user-provided values.

These examples have been provided to initiate you into the world of pandas DataFrames. You can do a lot more complex data analysis using DataFrames, as you will see in the upcoming chapters.

Working with local files

Working with pandas entails importing data from different source files and writing back the outputs in different formats. These operations are indispensable processes when working with data. In the following exercise, you will perform some preliminary operations with a CSV file.

First, you need to download a dataset, *Student Performance Data*, sourced from the UCI Machine Learning library. This dataset details student achievement in secondary education in two Portuguese schools. Some of the key variables of the dataset include student grades, demographic information, and other social and school-related features, such as hours of study time and prior failures.

> **Note**
>
> You can download the dataset (`student-por.csv`) from this book's GitHub repository at `https://github.com/PacktWorkshops/The-Pandas-Workshop/tree/master/Chapter01/Datasets`.
>
> The dataset has been sourced from *P. Cortez and A. Silva. Using Data Mining to Predict Secondary School Student Performance. In A. Brito and J. Teixeira Eds., Proceedings of 5th Future Business Technology Conference (FUBUTEC 2008) pp. 5-12, Porto, Portugal, April 2008, EUROSIS, ISBN 978-9077381-39-7.*
>
> The following is the link to the original dataset: `https://archive.ics.uci.edu/ml/datasets/Student+Performance`.

Once the dataset has been downloaded locally, you will be able to read data from it and then display that data using DataFrames. In the next section, we'll briefly cover the pseudocode that will help you read the file you just downloaded. Later, in *Exercise 1.01 – reading and writing data using pandas*, you will implement the pseudocode to read the CSV file's data into a DataFrame.

Reading a CSV file

To read a CSV file, you can use the following command:

```
pd.read_csv(filename, delimiter)
```

The first step of importing data is to define the path to the CSV file. In the preceding code example, the path is defined separately and stored in a variable called `filename`. Then, this variable is called inside the `pd.read_csv()` constructor. Alternatively, you can also provide the path directly inside the constructor; for example, `'Datasets/student-por.csv'`.

The second parameter, which is the **delimiter**, specifies how different columns of data are separated within the file. If you open the `student-por.csv` file in a program such as Excel, you will notice that all the columns in this file are separated using just semicolons (`;`), as shown in the following screenshot. Therefore, in this case, the delimiter would be a semicolon (`;`). In the `pd.read_csv()` constructor, you can represent this as `delimiter=';'`:

```
school;sex;age;address;famsize;Pstatus;Medu;Fedu;Mjob;Fjob;reason;guardian;traveltime;studyti
GP;"F";18;"U";"GT3";"A";4;4;"at_home";"teacher";"course";"mother";2;2;0;"yes";"no";"no";"no";"
GP;"F";17;"U";"GT3";"T";1;1;"at_home";"other";"course";"father";1;2;0;"no";"yes";"no";"no";"no";
GP;"F";15;"U";"LE3";"T";1;1;"at_home";"other";"other";"mother";1;2;0;"yes";"no";"no";"no";"yes";
```

Figure 1.11 – Snapshot of the source CSV

Displaying a snapshot of the data

After reading the data from external sources, it is important to ensure that it is loaded correctly. When you have a dataset that contains thousands of rows, it's not a good idea to print out all the rows. In such cases, you can get a snapshot of the dataset by printing only the first few rows. This is where the `head()` and `tail()` functions come in handy.

To see the first few rows of the data, you can use the `df.head()` command, where `df` is the name of the DataFrame. Similarly, to see the last few rows, you can use the `df.tail()` command. In both these cases, the first (or last) five rows will be displayed by default. If you want to see more (or fewer) rows, you can specify the number as an argument in the `head()` or `tail()` functions. For example, to see the first 10 rows, you can use the `df.head(10)` function. To see the last seven rows, you can use the `df.tail(7)` function.

Writing data to a file

There are many instances where you need to write data to a file and store it in a disk for future use. In the next exercise, you will be writing data to a CSV file. You will use the `df.to_csv(outpath)` command to write data to the file. Here, `df` is the name of the DataFrame and the `outpath` parameter is the path where the data must be written.

With that, it is time to put everything we've covered so far into action. The exercise that follows will help you do just that.

Exercise 1.01 – reading and writing data using pandas

In this exercise, you will be working with the student performance dataset you downloaded earlier. The goal of this exercise is to read the data in that file into a DataFrame, display the top rows of the data, and then store a small sample of this data in a new file.

> **Note**
>
> If you haven't downloaded the file yet, the `student-por.csv` file can be downloaded from `https://github.com/PacktWorkshops/The-Pandas-Workshop/tree/master/Chapter01/Datasets`.

The following steps will help you complete this exercise:

1. Open a new Jupyter notebook and select the **Pandas_Workshop** kernel, as shown in the following screenshot:

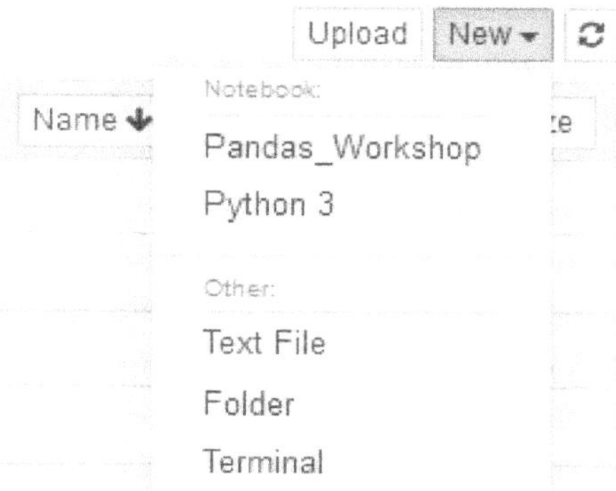

Figure 1.12 – Selecting the Pandas_Workshop kernel

2. Import the `pandas` library by typing or pasting the following command in a fresh Jupyter Notebook cell. Press the *Shift* + *Enter* keyboard combination to run the command:

```
import pandas as pd
```

3. Define the path to the downloaded file. Specify the path where the data has been downloaded in a variable named `filename`:

```
filename = '../Datasets/student-por.csv'
```

> **Note**
>
> The preceding code assumes that the CSV file is stored in a directory called `datasets` outside of the directory where the code is run. Based on where you have downloaded and stored the file, you'll have to make changes to the pathname (highlighted in the preceding line of code). If your file is stored in the same directory as your Jupyter notebook, the `filename` variable would just contain the `student-por.csv` value.

4. Read the file using the `pd.read_csv()` function, as follows:

```
studentData = pd.read_csv(filename, delimiter=';')
```

Here, you have stored the data in a variable called `studentData`, which will be the name of your DataFrame.

5. Display the first five rows of data, as follows:

```
studentData.head()
```

You should get the following output:

	school	sex	age	address	famsize	Pstatus	Medu	Fedu	Mjob	Fjob	...	famrel	freetime	goout	Dalc	Walc	health	absences	G1	G2	G3
0	GP	F	18	U	GT3	A	4	4	at_home	teacher	...	4	3	4	1	1	3	4	0	11	11
1	GP	F	17	U	GT3	T	1	1	at_home	other	...	5	3	3	1	1	3	2	9	11	11
2	GP	F	15	U	LE3	T	1	1	at_home	other	...	4	3	2	2	3	3	6	12	13	12
3	GP	F	15	U	GT3	T	4	2	health	services	...	3	2	2	1	1	5	0	14	14	14
4	GP	F	16	U	GT3	T	3	3	other	other	...	4	3	2	1	2	5	0	11	13	13

5 rows × 33 columns

Figure 1.13 – The first five rows of the studentData DataFrame

After displaying the top five rows of data, you can see that the data is spread across 33 columns.

6. Make a new DataFrame consisting of only the first five rows. Take the first five rows of the data and then store it in another variable called `studentSmall`:

```
studentSmall = studentData.head()
```

You'll be writing this small dataset to a new file in the subsequent steps.

7. Define the output path of the file to write to using the following command:

```
outpath = '../Datasets/studentSmall.csv'
```

> **Note**
> The path provided in the preceding command (highlighted) specifies the output path and the filename. You can change these values based on where you want the output files to be saved.

8. Run the following command to create a CSV file called `studentSmall.csv` in the output path you have specified:

```
# Write the data to disk
studentSmall.to_csv(outpath)
```

9. Using File Explorer (on Windows), Finder (on macOS), or even the command line, you can check whether the file has been saved to the `Datasets` folder (or whichever folder you chose). Upon opening the newly saved file in Excel (or any other compatible program), you'll notice that its contents are the same as the DataFrame you created in *Step 6*:

A	B	C	D	E	F	G	H	I	J	K
	school	sex	age	address	famsize	Pstatus	Medu	Fedu	Mjob	Fjob
0	GP	F	18	U	GT3	A	4	4	at_home	teacher
1	GP	F	17	U	GT3	T	1	1	at_home	other
2	GP	F	15	U	LE3	T	1	1	at_home	other
3	GP	F	15	U	GT3	T	4	2	health	services
4	GP	F	16	U	GT3	T	3	3	other	other

Figure 1.14 – Contents of the studentSmall.csv file

In this exercise, you read data from a CSV file, created a small sample of the dataset, and wrote it to disk. This exercise was meant to demonstrate one facet of pandas' input/output capabilities. pandas can also read and write data in multiple formats, such as Excel files, JSON files, and HTML, to name a few. We will explore how to input and output different data sources in more detail in *Chapter 3, Data I/O*.

Data types in pandas

pandas supports different data types, such as int64, float64, date, time, and Boolean. There will be innumerable instances in your data analysis life cycle where you will need to convert data from one type into the other. In such instances, it is imperative to understand these different data types. The operations you'll learn about in this section will help you understand these types.

For this section, you'll need to continue with the same Jupyter notebook that you used to implement *Exercise 1.01 – reading and writing data using pandas*, as the same dataset will be used in this section too.

The first aspect to know when working with any data is the different data types involved. This can be done using the `df.dtypes` method, where `df` is the name of the DataFrame. If you tried the same command on the `studentData` DataFrame you created in *Step 4* of the first exercise, the command would look something like this:

```
studentData.dtypes
```

The following is a small part of the output you will get:

```
Out[7]: school      object
        sex         object
        age          int64
        address     object
        famsize     object
        Pstatus     object
        Medu         int64
        Fedu         int64
        Mjob        object
        Fjob        object
        reason      object
```

Figure 1.15 – Data types in the studentData DataFrame

From the output, you can see that the DataFrame has two data types – `object` and `int64`. The `object` data type refers to string/text data or a combination of numeric and non-numeric data. The `int64` data type pertains to integer values.

Additionally, you can get information about the data types by using the `df.info()` method:

```
studentData.info()
```

The following is a small part of the output you will get:

```
<class 'pandas.core.frame.DataFrame'>
RangeIndex: 649 entries, 0 to 648
Data columns (total 33 columns):
 #   Column      Non-Null Count  Dtype
---  ------      --------------  -----
 0   school      649 non-null    object
 1   sex         649 non-null    object
 2   age         649 non-null    int64
 3   address     649 non-null    object
 4   famsize     649 non-null    object
 5   Pstatus     649 non-null    object
 6   Medu        649 non-null    int64
 7   Fedu        649 non-null    int64
 8   Mjob        649 non-null    object
```

Figure 1.16 – Information about the data types

Here, you can see that this method provides information about the number of null/non-null values, along with the number of rows, in the dataset. In this case, all `649` rows contain non-null data.

pandas also allows you to easily convert data from one type into another using the `astype()` function. Suppose you want to convert one of the `int64` data types into `float64`. You can do this for the `Medu` feature (in other words, the column), as follows:

```
# Converting 'Medu' to data type float
studentData['Medu'] = studentData['Medu'].astype('float')

studentData.dtypes
```

The following is a small part of the output you will get:

```
Out[13]: school       object
         sex          object
         age           int64
         address      object
         famsize      object
         Pstatus      object
         Medu        float64
         Fedu          int64
         Mjob         object
```

Figure 1.17 – Data type after conversion

When you converted the data type using the `astype()` function, you had to specify the target data type. Then, you stored the changes in the same variable. The output shows that the `int64` data type has changed to `float64`.

You can display the head of the changed DataFrame and see the changes in the values of the variable by using the following command:

```
studentData.head()
```

You should get the following output:

	school	sex	age	address	famsize	Pstatus	Medu	Fedu	Mjob	Fjob	...	famrel	freetime	goout	Dalc	Walc	health	absences	G1	G2	G3
0	GP	F	18	U	GT3	A	4.0	4	at_home	teacher	...	4	3	4	1	1	3	4	0	11	11
1	GP	F	17	U	GT3	T	1.0	1	at_home	other	...	5	3	3	1	1	3	2	9	11	11
2	GP	F	15	U	LE3	T	1.0	1	at_home	other	...	4	3	2	2	3	3	6	12	13	12
3	GP	F	15	U	GT3	T	4.0	2	health	services	...	3	2	2	1	1	5	0	14	14	14
4	GP	F	16	U	GT3	T	3.0	3	other	other	...	4	3	2	1	2	5	0	11	13	13

5 rows × 33 columns

Figure 1.18 – DataFrame after type conversion

From the output, you can see that the values have been converted into float values.

What you have seen so far are only the basic operations for converting data types. There are some interesting transformations you can do on data types, which you will learn about in *Chapter 4, Data Types*.

In the next section, we'll cover data selection methods. You will need to use the same Jupyter notebook that you've used so far in the next section as well.

Data selection

So far, you have seen operations that let you import data from external files and create data objects such as DataFrames from that imported data. Once a data object such as a DataFrame has been initialized, it is possible to extract relevant data from that data object using some of the intuitive functionalities pandas provides. One such functionality is **indexing**.

Indexing (also known as **subsetting**) is a method of extracting a cross-section of data from a DataFrame. First, you will learn how to index some specific columns of data from the studentData DataFrame.

To extract the age column from the DataFrame, you can use the following command:

```
ageDf = studentData['age']
ageDf
```

You should see the following output:

```
0      18
1      17
2      15
3      15
4      16
       ..
644    19
645    18
646    18
647    17
648    18
Name: age, Length: 649, dtype: int64
```

Figure 1.19 – The age variable

From the output, you can see how the age column is saved as a separate DataFrame. Similarly, you can subset multiple columns from the original DataFrame, as follows:

```
# Extracting multiple columns from DataFrame
studentSubset1 = studentData[['age','address','famsize']]
studentSubset1
```

You should get the following output:

	age	address	famsize
0	18	U	GT3
1	17	U	GT3
2	15	U	LE3
3	15	U	GT3
4	16	U	GT3
...	...	...	...
644	19	R	GT3
645	18	U	LE3
646	18	U	GT3
647	17	U	LE3
648	18	R	LE3

649 rows × 3 columns

Figure 1.20 – Extracting multiple features

Here, you can see how multiple columns (`age`, `address`, and `famsize`) have been extracted into a new DataFrame. When multiple columns must be subsetted, all of them are represented as a list, as shown in the preceding example.

You can also subset specific rows from a DataFrame, as follows:

```
studentSubset2 = studentData.
loc[:25,['age','address','famsize']]
studentSubset2.shape
```

You will see the following output for the preceding snippet:

```
(26, 3)
```

To subset specific rows, you can use a special operand called `.loc()`, along with the label indices of the rows you want. In the previous example, the subsetting was done until the 25th row of the dataset. From the output, you can see that the data has 26 rows and that the 25th row is also included in the subsetting.

Subsetting/indexing is a critical part of the data analytics process. pandas has some versatile functions for extracting cross-sections of data. These will be covered in detail in *Chapter 5, Data Selection – DataFrames*, and *Chapter 6, Data Selection – Series*.

In the next section, you will look at some data transformation methods. Again, you will continue using the same Jupyter notebook you've used so far.

Data transformation

Once you have created a DataFrame from source files or subset data to a new form, further transformations, such as cleaning up variables or treating for missing data, will be required. There may also be instances where you must group the data based on some variables to analyze data from a different perspective. This section will cover examples of some useful methods for data transformation.

One scenario where you may need to group data would be when you want to verify the number of students under each category of family size. In the family size feature, there are two categories: `GT3` (greater than three members) and `LE3` (less than three members). Let's say you want to know how many students there are under each of these categories.

To do that, you must take the family size column (`famsize`), identify all the unique family sizes within this column, and then find the number of students under each size category. This can be achieved by using two simple functions called `groupby` and `agg`:

```
studentData.groupby(['famsize'])['famsize'].agg('count')
```

```
famsize
GT3    457
LE3    192
Name: famsize, dtype: int64
```

Figure 1.21 – Data aggregation

Here, you grouped the DataFrame by the `famsize` variable using the `groupby` function. After this, you used the `count` aggregate function to count the number of records under each `famsize` category. From the output, you can see that there are two categories of family size, `LE3` and `GT3`, and that the majority fall under the `GT3` category.

As you can see from this example, all the different steps listed earlier can be achieved using a single line of code. Just like these, different types of transformations are possible with pandas.

In the next section, you will learn about some data visualization methods.

Data visualization

There is an old saying: "*A picture is worth a thousand words.*" This adage has a profound place in data analytics. Data analytics would be incomplete and would look hollow without proper visualization. matplotlib is a popular library for data visualization that works quite well with pandas. In this section, you will create some simple visualizations of the aggregated data you created in the previous section.

Consider that you want to visualize the number of students under each family size category:

```
aggData = studentData.groupby(['famsize'])['famsize'].
agg('count')
aggData
```

You should get the following output:

```
famsize
GT3    457
LE3    192
Name: famsize, dtype: int64
```

Figure 1.22 – Data aggregation

Here, the `groupby` function is used to aggregate the family size (`famsize`) column.

Before plotting this data, you need to create the *x*- and *y*-axis values. To define them, with the help of unique indices of grouped data, you can add the following code:

```
x = list(aggData.index)
x
```

You will see the following output:

```
['GT3', 'LE3']
```

In the preceding code, you took the index values of the aggregated data as the *x*-axis values. The next step is to create the *y*-axis values:

```
y = aggData.values
y
```

You should see the following output:

```
array([457, 192], dtype=int64)
```

The aggregated values for each category in the family size column are on the *y* axis.

Having obtained the *x*- and *y*-axis values, you can plot them using `matplotlib`, as follows:

```
import matplotlib.pyplot as plt
%matplotlib inline
plt.style.use('ggplot')

# Plotting the data
plt.bar(x, y, color='gray')
plt.xlabel("Family Sizes")
plt.ylabel("Count of Students ")
plt.title("Distribution of students against family sizes")
plt.show()
```

> **Note**
>
> In the preceding snippet, the `gray` value for the `color` attribute (emboldened) was used to generate graphs in grayscale. You can use other colors, such as `darkgreen` or `maroon`, as values of `color` parameters to get colored graphs.

You will get the following output:

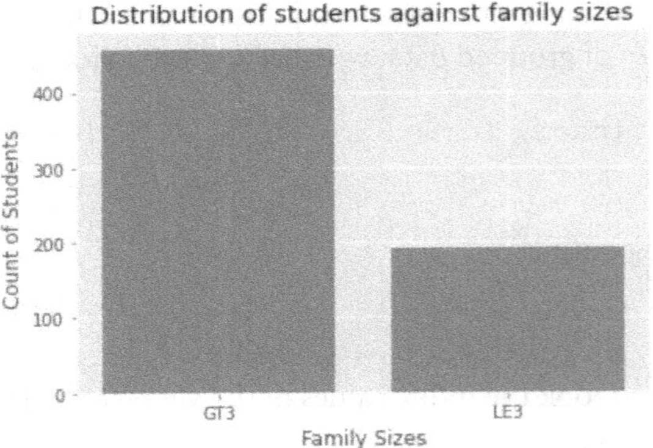

Figure 1.23 – Plots of family size

In the preceding snippet, the first line imports the `matplotlib` package. The `% inline` command ensures that the visualization appears in a cell in the same notebook rather than having it pop up as a separate window. This plotting uses a styling called `ggplot`, which is a popular plotting library that's used in R.

The `plt.bar()` method plots the data in the form of a bar chart. Inside this method, you can define the *x*-axis and *y*-axis values and also define the color. The rest of the lines in the snippet define the labels of the *x* and *y* axes, along with the title of the chart. Finally, the plot is displayed with the `plt.show()` line.

Matplotlib works well with pandas to create impressive visualizations. This topic will be explored in detail in *Chapter 8, Data Visualization.*

In the next section, you will see how pandas provides utilities to manipulate date objects.

Time series data

Time series data can be seen everywhere in your day-to-day life. Data emitted from social media, sensor data, browsing patterns from e-commerce sites, and log stream data from data centers are all examples of time series data. Preprocessing time series data necessitates performing various transformation operations on time components such as month, year, hours, minutes, and seconds.

In this section, we'll briefly cover some transformation functions with date and time objects. Suppose that you want to convert a string into a date object in pandas. This can be achieved using a handy function called `pd.to_datetime()`:

```
date = pd.to_datetime('15th of January, 2021')
print(date)
```

You should get the following output:

```
2021-01-15 00:00
```

From the output, you can see that the string object has been converted into a date. Next, suppose you want to calculate the date 25 days from the date you got in the preceding output. To do so, you can run the following code:

```
# date after 25 days
newdate = date+pd.to_timedelta(25,unit='D')
print(newdate)
```

You will get the following output:

```
2021-02-09 00:00:00
```

As you can see, the `pd.to_timedelta()` function provides an intuitive way of calculating the date after a prescribed number of days. The `unit='D'` parameter is used to define that the conversion must be in days.

Next, suppose you want to get all the dates from a start date to a certain period of days. This can be achieved as follows:

```
# Get all dates within the next 7 days
futureDate = pd.date_range(start=newdate, periods=7, freq='D')
futureDate
```

You should get the following output:

```
DatetimeIndex(['2021-02-09', '2021-02-10', '2021-02-11', '2021-02-12',
               '2021-02-13', '2021-02-14', '2021-02-15'],
              dtype='datetime64[ns]', freq='D')
```

Figure 1.24 – Date range output

In this example, the `date_range()` function was used to get a list of all future dates. The `start` parameter defines the date where you need to start calculating the ranges. The `period` parameter indicates the number of days it has to calculate from the start date, while `freq = 'D'` indicates that the unit is in days.

There are many more transformation functions available in pandas for working with time objects. These functions will be covered in greater depth in *Chapter 11, Time Series Data*.

Code optimization

If datasets get larger or if speed becomes a concern, there are numerous optimization techniques available to increase performance or improve the memory footprint. These range from idiomatic loops over appropriate data types and methods to custom extensions.

One of the many optimization techniques available in pandas is called **vectorization**. Vectorization is the process of applying an operation over an entire array. pandas has some efficient vectorization processes that enable faster data processing, such as `apply` and `lambda`, which will be briefly covered in this section.

As you might know, the `apply` method is used to apply a function to every element of a Series and the `lambda` function is a way to create anonymous functions in an operation. Very often, the `apply` and `lambda` functions are used in conjunction for efficient implementation.

In the following example, you will learn how to create a new dataset by applying a function to each of the grade variables (`G1`, `G2`, `G3`). The function takes each element of the grade feature and increments it by 5. First, you will only use the `apply` function; after, you will use the `lambda` function.

To implement the `apply` function, you must create a simple function. This function takes an input and adds 5 to it:

```
# Defining the function
def add5(x):
    return  x + 5
```

Now, this function will be used on each element of the grade feature using the `apply` function:

```
# Using apply method
df = studentData[['G1','G2','G3']].apply(add5)
df.head()
```

You should get the following output:

	G1	G2	G3
0	5	16	16
1	14	16	16
2	17	18	17
3	19	19	19
4	16	18	18

Figure 1.25 – Output after using the apply function

This can also be achieved without defining a separate function using the `lambda` method, as follows:

```
df = studentData[['G1','G2','G3']].apply(lambda x:x+5)
df.head()
```

You should get the following output:

	G1	G2	G3
0	5	16	16
1	14	16	16
2	17	18	17
3	19	19	19
4	16	18	18

Figure 1.26 – Output after using the lambda function

Here, the `lambda()` method lets you execute the operation in just one line of code. There is no need to create a separate function.

The example you have implemented is a relatively simple one in terms of using vectorization. Many more complex tasks can be optimized using vectorization. Moreover, vectorization is one of the many methods of code optimization that are used in pandas. You will learn more about it in *Chapter 12, Code Optimization.*

So far, you have got a bird's-eye view of the topics that will be covered in this book. All these topics will be dealt with in more depth. Now, it's time to apply the concepts you've learned so far in a couple of exercises. Then, you can test what you've learned by completing this chapter's activity.

Utility functions

In the exercises that follow, you will be using some utility functions. These are as follows:

- Generating random numbers using `random()`:

  There will be many instances in the data analytics life cycle where you will have to generate data randomly. numpy is a library in Python that contains some good functions for generating random numbers. Consider the following example:

  ```
  # Generating random numbers
  import numpy as np
  np.random.normal(2.0, 1, 10)
  ```

You should get the following output:

```
array([3.38456318, 1.76608323, 2.14843901, 2.95586157, 2.4149523 ,
       2.3740889 , 1.50526992, 1.91944094, 0.03591338, 3.45030228])
```

Figure 1.27 – Output of using the random function

In this example, you imported the numpy library and then used the `np.random.normal()` function to generate 10 numbers from a normal distribution with a mean of `2` and a standard deviation of `1`. You will be utilizing this function in the upcoming exercises.

- Concatenating multiple data series using `pd.concat()`:

  Another commonly used function is `concat`, which is used to concatenate multiple data elements together. This can be seen in the following code:

```
# Concatenating three series
pd.concat([ser1,ser2,ser3], axis=1)
```

  In this example, you are concatenating three series using the `pd.concat()` function. The `axis=1` parameter specifies that the concatenation is to be done along the columns. Alternatively, `axis=0` means concatenation along the rows, as shown in the following screenshot:

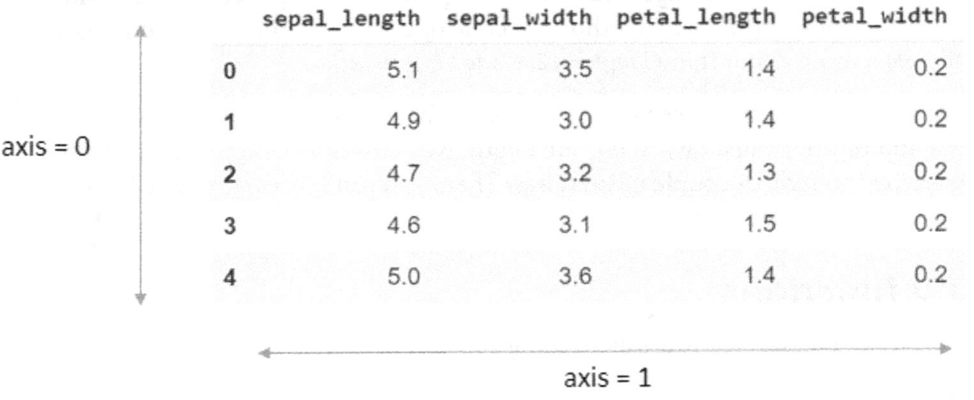

	sepal_length	sepal_width	petal_length	petal_width
0	5.1	3.5	1.4	0.2
1	4.9	3.0	1.4	0.2
2	4.7	3.2	1.3	0.2
3	4.6	3.1	1.5	0.2
4	5.0	3.6	1.4	0.2

Figure 1.28 – Representation of the axis argument

- The `df.sum`, `df.mean`, and `divmod` numeric functions:

  You will be using these numeric functions in the upcoming exercises. The first one, `df.sum`, is used to calculate the sum of the values in a DataFrame. If you want to find the sum of the DataFrame elements in the columns, you can use the `axis=0` argument; for example, `df.sum(axis=0)`.

For example, for the DataFrame shown in the preceding screenshot, `df.sum(axis=0)` would give you the following output:

```
sepal_length    24.3
sepal_width     16.4
petal_length       7
petal_width        1
```

Figure 1.29 – Example of axis summation

Defining `axis=1`, on the other hand, sums up the values along the rows. For example, for the same dataset, `df.sum(axis=1)` will give the following output:

```
0    10.2
1     9.5
2     9.4
3     9.4
4    10.2
```

Figure 1.30 – Output of axis summation

Similarly, `df.mean` calculates the mean values of the data. The direction of calculation can also be defined using the `axis` parameter. Finally, `divmod` is used to calculate the quotient and remainder of a division operation.

You will also use two data manipulation methods, `apply` and `applymap`, in the upcoming exercises. The first method, `apply`, can be used for both Series and DataFrame objects. The second method, `applymap`, can only be used for DataFrames. The output you get after applying these methods depends on the functions that were used inside these methods as arguments.

- The `to_numeric` and `to_numpy` data type conversion functions:

  Two commonly used data type conversion functions are `to_numeric` and `to_numpy`. The first one, `to_numeric`, is used to change the data type of any data object to the numeric data type. The second function, `to_numpy`, is used to convert a pandas DataFrame into a `numpy` array. Both of these functions are used extensively for data manipulation.

- The `df.eq`, `df.gt`, and `df.lt` DataFrame comparison functions:

  There will be many instances where you will need to compare one DataFrame to another. pandas provides some good comparison functions to do this. Some of the common ones are `df.eq`, `df.gt`, and `df.lt`. The first method displays all the elements of one DataFrame that are equal to the one in the second DataFrame. Similarly, `df.gt` finds those elements that are greater in one DataFrame, while the `df.lt` method finds those elements that are less than those in the other one.

- List comprehension:

  List comprehension is a syntax that's used in Python to create lists based on existing lists. Using list comprehension helps avoid the use of `for` loops. For example, suppose you want to extract every letter from the word `Pandas`. One way to do this is to use a `for` loop. Consider the following snippet:

  ```
  # Define an empty list
  letters = []

  for letter in 'Pandas':
      letters.append(letter)

  print(letters)
  ```

  You should see the following output, as expected:

  ```
  ['P', 'a', 'n', 'd', 'a', 's']
  ```

  However, using a list comprehension is much easier and more compact, as follows:

  ```
  letters = [ letter for letter in 'Pandas' ]
  print(letters)
  ```

  Again, you should get the following output:

  ```
  ['P', 'a', 'n', 'd', 'a', 's']
  ```

  As shown in the preceding example, using a list comprehension has optimized the code into just one line.

- The `df.iloc()` data selection method:

  The `df.loc()` method is used to select specific rows of a DataFrame. Similar to this, there is another function, `df.iloc()`, that selects specific indices of the data. Let's understand this with a simple example.

  Let's create a DataFrame, `df`:

  ```
  # Create a Data Frame
  lst = [['C', 45], ['A', 60],
         ['A', 26], ['C', 57], ['C', 81]]

  df = pd.DataFrame(lst, columns =['Product', 'Sales'])
  df
  ```

You should see the following output:

	Product	Sales
0	C	45
1	A	60
2	A	26
3	C	57
4	C	81

Figure 1.31 – Creating a DataFrame

Suppose you want to subset the first three rows for the `Sales` column. Here, you can use the `df.loc()` method, as follows:

```
df.loc[:2,'Sales']
```

Running this snippet should result in the following output:

```
0    45
1    60
2    26
Name: Sales, dtype: int64
```

Figure 1.32 – Selecting the first three rows

In the preceding code, `:2` indicates rows up to index 2. The second parameter, `Sales`, indicates the columns that need to be selected.

If you want to select a particular index in the DataFrame, you can use the `df.iloc()` method, as shown here:

```
df.iloc[3]
```

You should see the following output:

```
Product     C
Sales      57
Name: 3, dtype: object
```

Figure 1.33 – Selecting an index

- Boolean indexing:

  Another important concept in pandas is Boolean indexing. Consider the same sales data you created in the previous example:

	Product	Sales
0	C	45
1	A	60
2	A	26
3	C	57
4	C	81

Figure 1.34 – Sales data

Suppose you want to find out all the `Sales` values that equal `45`. To do this, you can subset the data, as follows:

```
df['Sales'].eq(45)
```

Running this will produce the following output:

```
0    True
1    False
2    False
3    False
4    False
```

Figure 1.35 – Finding which Sales values are equal to 45

Here, you get a Boolean output (`True` or `False`), depending on whether the row satisfies the condition – that is, the value of `Sales` is equal to `45`. However, if you want the actual row and its value, rather than the Boolean output, you should apply another type of subsetting, as follows:

```
df[df['Sales'].eq(45)]
```

With the additional subsetting, you will get the following output:

	Product	Sales
0	C	45

Figure 1.36 – Finding which row has a Sales value equal to 45

Here, you can see that the first row satisfies the condition. Now that you have learned about the utility functions, it is time to dive into the second exercise.

Exercise 1.02 – basic numerical operations with pandas

Tess is running a training session on pandas for a few apprentice data analysts who will be working with her. As part of their training, they need to generate some data and perform numerical operations on it, such as summation, mean, and modulus operations. The goal of this exercise is to help Tess conduct her training by running and validating all the code she will be using.

To achieve this, you will need to do the following:

1. Create DataFrames from the Series objects.
2. Find summary statistics of the DataFrame using the mean, sum, and modulus operations.
3. Use the `applymap()` method along with lambda functions.

4. Create new features using list comprehension and concatenate them with an existing DataFrame.
5. Change the data type to `int64`.
6. Convert the DataFrame into a numpy array.

Follow these steps to complete this exercise:

1. Open a new Jupyter notebook and select the **Pandas_Workshop** kernel.
2. Import `pandas`, `numpy`, and `random` into your notebook:

```
import pandas as pd
import numpy as np
import random
```

3. Create three Series by generating some random data points. To do this, start by sampling 100 data points from three different normal distributions. The first one should have a mean of `3.0` and a standard deviation of `1`. The second and third Series should have the mean and standard values at `5.0` and `3` and `1.0` and `0.5`, respectively. The sampled data must be converted into a pandas Series:

```
# Initialize a random seed
np.random.seed(123)

# Create three series
ser1 = pd.Series(np.random.normal(3.0, 1, 100))
ser2 = pd.Series(np.random.normal(5.0, 3, 100))
ser3 = pd.Series(np.random.normal(1.0, 0.5, 100))
```

Here, you used the `random.seed()` method. This method ensures that you get the same results that were mentioned in this example.

4. Concatenate all three Series into a DataFrame. Name the columns and display the first few rows of the newly created DataFrame:

```
Df = pd.concat([ser1,ser2,ser3], axis=1)

# Name the columns
Df.columns=['V1','V2','V3']

# Display the head of the column
Df.head()
```

You should get the following output:

	V1	V2	V3
0	1.914369	6.926164	1.351655
1	3.997345	-0.933664	0.700947
2	3.282978	7.136794	2.100351
3	1.493705	12.794912	1.344148
4	2.421400	4.926122	0.996846

Figure 1.37 – Concatenating the Series into a DataFrame

5. Next, find the sum of the DataFrame along the columns using the `df.sum` function:

```
Df.sum(axis=0)
```

You should get the following output:

```
V1    302.710907
V2    494.139331
V3     95.243434
dtype: float64
```

Figure 1.38 – Result of summation along the columns

6. The `axis=0` parameter defines the operation across the rows. Here, you can see three mean values corresponding to each of the columns, `V1`, `V2`, and `V3`. Alternatively, `axis=1` carries out the operation across the columns. Here, you will be able to see 100 values, where there will be one value corresponding to each row.

7. Calculate the mean of the values across both the columns and rows using the `df.mean` function:

```
# Find the mean of the columns
Df.mean(axis=0)
```

You will get the following output:

```
V1    3.027109
V2    4.941393
V3    0.952434
dtype: float64
```

Figure 1.39 – Mean of the columns

Now, find the mean across the rows:

```
# Find the mean across the rows
Df.mean(axis=1)
```

You should get the following output:

```
0       3.397396
1       1.254876
2       4.173374
3       5.210922
4       2.781456
          ...
95      0.228614
96      2.681817
97      2.503941
98      2.811963
99      2.792288
Length: 100, dtype: float64
```

Figure 1.40 – Mean across the rows

8. Next, find the modulus of each column by using the `divmod` function. Use this function in conjunction with the `lambda` and `apply` functions:

```
# Apply the divmod function to each of the series
Df.apply(lambda x: divmod(x,3))
```

You should get the following output:

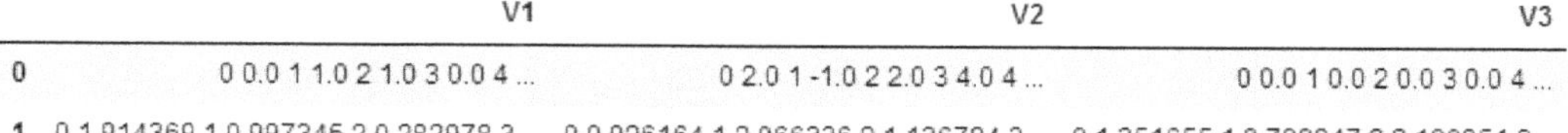

	V1	V2	V3
0	0 0.0 1 1.0 2 1.0 3 0.0 4 ...	0 2.0 1 -1.0 2 2.0 3 4.0 4 ...	0 0.0 1 0.0 2 0.0 3 0.0 4 ...
1	0 1.914369 1 0.997345 2 0.282978 3...	0 0.926164 1 2.066336 2 1.136794 3...	0 1.351655 1 0.700947 2 2.100351 3...

Figure 1.41 – Output of the divmod function

From the code, you can see how the `apply` function applies the `divmod` function to each of the Series. The `divmod` function generates the quotients and remainders for each of the values in the Series. The first row corresponds to the quotients, while the second row corresponds to the remainders.

9. In the preceding step, when you implemented the `apply` method, `divmod` was applied column-wise. So, you can see that the quotients for all the rows under the `V1` column are aggregated in row `0`, column `V1`, while the remainders for column `V1` are aggregated in row `1`, column `V1`. However, if you want to find the quotient and remainder for each cell of the DataFrame, you can do so using `applymap`, as follows:

```
Df.applymap(lambda x: divmod(x,3))
```

You should get the following output:

	V1	V2	V3
0	(0.0, 1.9143693966994388)	(2.0, 0.9261640678154937)	(0.0, 1.3516550589033651)
1	(1.0, 0.9973454465835858)	(-1.0, 2.0663362054386534)	(0.0, 0.7009473343295873)
2	(1.0, 0.28297849805199204)	(2.0, 1.1367939064115546)	(0.0, 2.1003510496085642)
3	(0.0, 1.493705286081908)	(4.0, 0.7949117818079436)	(0.0, 1.3441484651110442)
4	(0.0, 2.4213997480314635)	(1.0, 1.9261220557055578)	(0.0, 0.9968463745430639)
...	...	...	...
95	(1.0, 1.031114458921742)	(-2.0, 1.3068349762420635)	(0.0, 1.347893659569804)
96	(0.0, 1.9154320879942335)	(1.0, 1.1921195307477372)	(0.0, 1.9379002733568176)
97	(0.0, 1.6365284553814157)	(1.0, 1.6674478368409789)	(0.0, 1.207847269946217)
98	(1.0, 0.37940061207813613)	(1.0, 0.9762148513802424)	(0.0, 1.080272210739859)
99	(0.0, 2.6208235654274477)	(1.0, 1.3461612136911745)	(0.0, 1.4098803048050945)

100 rows × 3 columns

Figure 1.42 – Output after using divmod with applymap

When you use `applymap()` for the `divmod()` function in this way, the quotient along with the remainder is generated as a tuple in the output.

10. Now, create another character Series and then use the `to_numeric` function to convert the Series into a numeric format. After converting it into a numeric format, concatenate the Series with the existing DataFrame. As a first step, create a list of characters and display its length, as follows:

```
# Create a list of characters
list1 = [['20']*10,['35']*15,['40']*10,['10']*25,['15']*4
0]
# Convert them into a single list using list
comprehension
charlist = [x for sublist in list1 for x in sublist]
# Display the output
len(charlist)
```

You should get the following output:

```
100
```

In the first line, you created a nested list of character numbers. Then, this nested list was converted into a single list using list comprehension. Finally, the length of the list was displayed, using the `len()` function, to ensure that there are 100 elements in the list.

11. Next, convert the preceding list into a Series object using the `pd.Series()` function. Randomly shuffle the list using the `random.shuffle()` function before creating the Series:

```
# Randomly shuffle the character list
random.seed(123)
random.shuffle(charlist)

# Convert the list to a series
ser4 = pd.Series(charlist)
ser4
```

You should get the following output:

```
0     15
1     40
2     15
3     10
4     15
      ..
95    15
96    10
97    35
98    40
99    20
Length: 100, dtype: object
```

Figure 1.43 – The new Series

Here, you can see that the data type is `object`, indicating that this is a Series of character data types.

12. Now, convert the data type into numeric using the `to_numeric` function:

```
ser4 = pd.to_numeric(ser4)
ser4
```

You will get the following output:

```
0      15
1      40
2      15
3      10
4      15
       ..
95     15
96     10
97     35
98     40
99     20
Length: 100, dtype: int64
```

Figure 1.44 – The Series after converting the data type into numeric

Note that the data type has now changed to `int64` from `object`.

13. Concatenate the fourth Series that you created in the preceding step, store it in a new DataFrame, and then display its contents by converting it into a `numpy` array:

```
Df = pd.concat([Df,ser4],axis=1)

# Renaming the data frame
Df.rename(columns={0:'V4'}, inplace=True)

# Displaying the data frame
Df
```

You should get the following output:

	V1	V2	V3	V4
0	1.914369	6.926164	1.351655	15
1	3.997345	-0.933664	0.700947	40
2	3.282978	7.136794	2.100351	15
3	1.493705	12.794912	1.344148	10
4	2.421400	4.926122	0.996846	15
...	...	...	...	...
95	4.031114	-4.693165	1.347894	15
96	1.915432	4.192120	1.937900	10
97	1.636528	4.667448	1.207847	35
98	3.379401	3.976215	1.080272	40
99	2.620824	4.346161	1.409880	20

100 rows × 4 columns

Figure 1.45 – New DataFrame after the addition of a new Series

Here, the `inplace=True` parameter means that a new DataFrame is created after a new variable has been added.

14. Next, convert the DataFrame into a `numpy` array:

```
numpArray = Df.to_numpy()
```

Now, display the array:

```
numpArray
```

You should get the following output. Please note that this has been truncated:

```
array([[ 1.91436940e+00,  6.92616407e+00,  1.35165506e+00,
         1.50000000e+01],
       [ 3.99734545e+00, -9.33663795e-01,  7.00947334e-01,
         4.00000000e+01],
       [ 3.28297850e+00,  7.13679391e+00,  2.10035105e+00,
         1.50000000e+01],
       [ 1.49370529e+00,  1.27949118e+01,  1.34414847e+00,
         1.00000000e+01],
       [ 2.42139975e+00,  4.92612206e+00,  9.96846375e-01,
         1.50000000e+01],
       [ 4.65143654e+00,  5.10242639e+00,  8.96668848e-01,
         1.50000000e+01],
       [ 5.73320757e-01,  5.53864845e+00,  9.56738857e-01,
         2.00000000e+01],
       [ 2.57108737e+00, -5.85927132e-01,  5.42346465e-01,
         1.50000000e+01],
       [ 4.26593626e+00,  6.27843992e+00,  9.52398730e-01,
         1.50000000e+01],
       [ 2.13325960e+00,  1.83770768e-01,  1.13934176e+00,
```

Figure 1.46 – DataFrame converted into a numpy array

Here, you can see that the DataFrame has been converted into a `numpy` array.

In this exercise, you implemented several numerical functions, such as `df.sum`, `df.mean`, and `divmod`. You also reinforced your understanding of other important functions, such as `apply` and `lambda`.

Having learned about some utility functions and applied them in this exercise, you can now move on to the last topic in this chapter. There, you will briefly learn how pandas can be used for building models. You will be using the same Jupyter notebook you've used so far in the next section as well.

Data modeling

A big factor behind the popularity of pandas is its utility in the data science life cycle. pandas has become a tool for most of the data science preprocessing steps, such as data imputation, scaling, and normalization. These are important steps for building machine learning models.

Another important process in the data science life cycle where pandas can be very useful is in creating train and test sets. Training datasets are used to create machine learning models, while test datasets are used to evaluate the performance of the machine learning models that have been built after using the training set. pandas is a go-to tool for creating train and test sets.

> **Note**
> If you do not understand the concepts of modeling at this point, that's okay. Just run the code as-is. These concepts will be explained in detail in *Chapter 10, Data Modeling.*

Consider that a given dataset has been split into train and test sets using pandas. You can use the student dataset you used earlier and split it into two parts, as follows:

```
# Sampling 80% of data as train data
train=studentData.sample(frac=0.8,random_state=123)

# Rest of the data is taken as test data
test=studentData.drop(train.index)
```

First, you sampled 80% of the data randomly. The fraction of train data is specified using the `frac=0.8` parameter. The data sampling for the train and test sets happens randomly. However, to get the same train and test sets, you can use a parameter called `random_state` and define a seed number for this (in this case, `123`). Every time you use the same seed number, you will get a similar dataset. You can change the dataset by changing the seed number. This process is called reproducing results using a pseudorandom number. The `random_state=123` parameter has been set so that you also get results that are similar to the ones shown here.

Once the training data has been sampled, the next task is to drop those samples from the original dataset to get the test data. Using the following code, you can look at the shapes of the train and test datasets:

```
print('Shape of the training data',train.shape)
print('Shape of the test data',test.shape)
```

You will get the following output:

```
Shape of the training data (519, 33)
Shape of the test data (130, 33)
```

The preceding output shows that the train data contains `519` rows, which is almost 80% of the data, and that the rest of the data is with the `test` set. To see the head (top rows) of the training dataset, you can type in the following command:

```
train.head()
```

You should get the following output:

	school	sex	age	address	famsize	Pstatus	Medu	Fedu	Mjob	Fjob	...	famrel	freetime	goout	Dalc	Walc	health	absences	G1	G2	G3
376	GP	F	18	U	GT3	T	1.0	1	other	other	...	4	5	5	1	2	2	0	14	14	14
142	GP	M	18	U	LE3	T	3.0	1	services	services	...	3	3	4	4	5	4	2	11	11	12
43	GP	M	15	U	GT3	T	2.0	2	services	services	...	5	4	1	1	1	1	0	9	10	10
162	GP	M	15	U	LE3	A	2.0	1	services	other	...	4	5	5	2	5	5	0	12	11	11
351	GP	M	20	U	GT3	A	3.0	2	services	other	...	5	5	3	1	1	5	0	14	15	15

5 rows × 33 columns

Figure 1.47 – Top rows of the training set

The output shows that the dataset is shuffled when the train and test sets are generated, as evident from the indices that have been shuffled here. Data shuffling is important when you're generating train and test sets.

In the next exercise, you will compare two DataFrames using the `merge` operation.

Exercise 1.03 – comparing data from two DataFrames

In her training, Tess must demonstrate to the apprentice data analysts how two DataFrames can be compared. To do that, she needs to create sales datasets for two imaginary stores and compare their sales figures side by side.

In this exercise, you will create these sales datasets using a data series. Each DataFrame will contain two columns. One column will be the list of products in the store, while the second and the third columns will enlist the sales of these products, as shown in the following table:

Product	SALES 1	SALES 2
A	0	0
B	1	0
C	1	0

Figure 1.48 – Sample data format

You will use DataFrame comparison techniques to compare these DataFrames. You will also perform a `merge` operation on these DataFrames for easier comparison.

Specifically, you will do the following in this exercise:

1. Create Series data and concatenate it to create two DataFrames.
2. Apply comparison methods such as `eq()`, `lt()`, and `gt()` to compare the DataFrames.
3. Use the `groupby()` and `agg` methods to consolidate the DataFrames.
4. Merge the DataFrames so that it's easy to compare them.

Follow these steps to complete this exercise:

1. Open a new Jupyter notebook and select the **Pandas_Workshop** kernel.
2. Import the `pandas` and `random` libraries into your notebook:

```
import pandas as pd
import random
```

3. Create a pandas Series for product lists. You have three different products, `A`, `B`, and `C`, with a varying number of transactions. You will use a random seed of `123`. Once the list has been generated, shuffle it and convert it into a Series:

```
# Create a list of characters and convert it to a series
random.seed(123)

list1 = [['A']*3,['B']*5,['C']*7]
charlist = [x for sublist in list1 for x in sublist]
random.shuffle(charlist)
# Creating a series from the list
ser1 = pd.Series(charlist)
ser1
```

You should get the following output:

```
0     C
1     B
2     C
3     B
4     A
5     B
6     C
7     C
8     C
9     C
10    C
11    B
12    A
13    B
14    A
dtype: object
```

Figure 1.49 – Series for product categories

4. The next Series you will create is a numeric one, which will be the sales values of the products. You will randomly select 15 integer values between `10` and `100` to get the list of sales figures:

```
# Create a series of numerical elements by random sampling
random.seed(123)
ser2 = pd.Series(random.sample(range(10, 100), 15))
ser2
```

You will see the following output:

```
0     16
1     44
2     21
3     62
4     23
5     14
6     58
7     78
8     81
9     52
10    53
11    30
12    27
13    99
14    41
dtype: int64
```

Figure 1.50 – Sales data figures

5. Next, concatenate the product and data Series you created into a pandas DataFrame:

```
# Creating a products data frame
prodDf1 = pd.concat([ser1,ser2],axis=1)
prodDf1.columns=['Product','Sales']
prodDf1
```

You should see the following output:

	Product	Sales
0	C	16
1	B	44
2	C	21
3	B	62
4	A	23
5	B	14
6	C	58
7	C	78
8	C	81
9	C	52
10	C	53
11	B	30
12	A	27
13	B	99
14	A	41

Figure 1.51 – First DataFrame

6. Create the second DataFrame similar to how you created the first one. First, create the product lists, as follows:

```
# Create the second series of products
random.seed(321)
list1 = [['A']*2,['B']*8,['C']*5]
charlist = [x for sublist in list1 for x in sublist]
random.shuffle(charlist)
ser3 = pd.Series(charlist)
ser3
```

Running this will result in the following output:

```
0     C
1     A
2     A
3     C
4     C
5     B
6     C
7     B
8     C
9     B
10    B
11    B
12    B
13    B
14    B
dtype: object
```

Figure 1.52 – Second DataFrame for products

7. Next, create the sales figures:

```
# Creating sales figures
random.seed(321)
ser4 = pd.Series(random.sample(range(10, 100), 15))
ser4
```

You should get an output similar to the following:

```
0     45
1     60
2     26
3     57
4     81
5     66
6     53
7     41
8     87
9     68
10    64
11    95
12    38
13    11
14    75
dtype: int64
```

Figure 1.53 – Series for sales figures

8. Finally, create the DataFrame by concatenating both Series (`ser3` and `ser4`):

```
# Creating a products data frame
prodDf2 = pd.concat([ser3,ser4],axis=1)
prodDf2.columns=['Product','Sales']
prodDf2
```

The output should be as follows:

	Product	Sales
0	C	45
1	A	60
2	A	26
3	C	57
4	C	81
5	B	66
6	C	53
7	B	41
8	C	87
9	B	68
10	B	64
11	B	95
12	B	38
13	B	11
14	B	75

Figure 1.54 – DataFrame created by concatenating ser3 and ser4

9. Now, find out how many sales values in the second DataFrame equal `45`. This value has been arbitrarily chosen – you may select another value. Do this using the `df.eq()` function:

```
prodDf2['Sales'].eq(45)
```

You should get an output similar to the following:

```
0     True
1    False
2    False
3    False
4    False
5    False
6    False
7    False
8    False
9    False
10   False
11   False
12   False
13   False
14   False
Name: Sales, dtype: bool
```

Figure 1.55 – Snapshot of the compared dataset

The output is a Boolean data type. Here, you can see that the first record is where the sales value equals `45`. To get only the actual values where the condition has been met and not the Boolean output, subset the DataFrame with the equality comparison within brackets, as follows:

```
# Comparing values
prodDf2[prodDf2['Sales'].eq(45)]
```

You should get an output similar to the following:

	Product	Sales
0	C	45

Figure 1.56 – Records after comparison

From the output, you can see that only the relevant record has been generated.

10. Now, verify the number of records where the sales value in the second DataFrame is greater than that in the first one. You can use the `df.gt` function for this:

```
prodDf2['Sales'].gt(prodDf1['Sales'])
```

The output will look as follows:

```
0      True
1      True
2      True
3     False
4      True
5      True
6     False
7     False
8      True
9      True
10     True
11     True
12     True
13    False
14     True
Name: Sales, dtype: bool
```

Figure 1.57 – Subset of the DataFrame

11. Subset this and find the actual values:

```
prodDf2[prodDf2['Sales'].gt(prodDf1['Sales'])]
```

You should get the following output:

	Product	Sales
0	C	45
1	A	60
2	A	26
4	C	81
5	B	66
8	C	87
9	B	68
10	B	64
11	B	95
12	B	38
14	B	75

Figure 1.58 – Records after comparison

12. Now, implement the `lt` function to get those rows where `prodDf2` is less than the corresponding row of `prodDf1`:

```
prodDf2[prodDf2['Sales'].lt(prodDf1['Sales'])]
```

The output for this should be as follows:

	Product	Sales
3	C	57
6	C	53
7	B	41
13	B	11

Figure 1.59 – Records after comparison

13. Select specific data points from the DataFrame. Use the `df.iloc()` method for this, as follows:

```
prodDf2.iloc[[2,5,6,8]]
```

You should get an output similar to the following:

	Product	Sales
2	A	26
5	B	66
6	C	53
8	C	87

Figure 1.60 – Accessing records based on their index values

14. Now, find the total sales for each product. Then, compare the first DataFrame with the second one by merging both DataFrames based on an overlapping column. Group each DataFrame based on the `Product` column and find the total sum of all the values in each group using the aggregate function (`.agg()`):

```
tab1 = prodDf1.groupby(['Product']).agg('sum')
tab2 = prodDf2.groupby(['Product']).agg('sum')
print(tab1)
print(tab2)
```

You should get the following output:

	Sales
Product	
A	91
B	249
C	359

	Sales
Product	
A	86
B	458
C	323

Figure 1.61 – Aggregation of products

From the output, you can see that the `Sales` values for each product have been aggregated.

15. Now, merge both DataFrames based on the `Product` column:

```
tab3 = pd.DataFrame(pd.merge(tab1,tab2,on=['Product']))
tab3.columns = ['Sales1','Sales2']
tab3
```

You should get the following output:

Product	Sales1	Sales2
A	91	86
B	249	458
C	359	323

Figure 1.62 – Merged DataFrame

The preceding output shows how the sales values, when placed side by side against the corresponding products, help you compare them.

In this exercise, you used some interesting methods to compare two DataFrames and combine two DataFrames based on some overlapping columns. Having learned about these data manipulation methods, it is time to test what you've learned by solving an interesting test case.

Activity 1.01 – comparing sales data for two stores

ABC Corporation is a retail company with two big stores for grocery and stationery products. The company is planning to create an ambitious marketing campaign next year. As a data analyst, your task is to derive the following insights from the data and relay those insights to the sales team so that they can plan the campaign effectively:

- *Which store has greater sales for the quarter?*
- *Which store has the highest sales for grocery products?*
- *Which store has the highest sales for March?*
- *For how many days were the sales of stationery products greater in store 1 than in store 2?*

In this activity, you will create the datasets for the two stores and use all the methods you have learned so far to answer the preceding questions. The following steps will help you complete this activity:

1. Open a new Jupyter notebook.
2. Load the data that corresponds to the two stores (`Store1.csv` and `Store2.csv`). These datasets are available in this book's GitHub repository at `https://github.com/PacktWorkshops/The-Pandas-Workshop/tree/master/Chapter01/Datasets`.
3. Use the different methods you have learned about in this chapter to answer the questions.
4. Print the resulting DataFrames. Note that the DataFrames you create should be in the following format:

	Months	Grocery_sales	Stationary_sales
0	Jan	16	57
1	Jan	44	139
2	Jan	15	85
3	Jan	59	8
4	Jan	36	106

Figure 1.63 – Final output

With that, we have covered everything you need to know to get started with pandas.

> **Note**
> You can find the solution for this activity in the *Appendix*.

Summary

This chapter gave you a high-level walkthrough of the key features that make pandas a vital tool in the data analytics life cycle. You started by learning briefly about the library's architecture and the topics that are going to be covered in this book. You discovered the library's capabilities with the help of hands-on examples. Then, you learned about data objects such as Series and DataFrames, data types such as `int64`, `float`, and `object`, and different methods you can use to input data from external sources and also write data to formats such as CSV. After that, you implemented different methods to manipulate data, such as data selection and indexing. Later, you performed data transformation using aggregation and grouping methods and implemented various data visualization techniques. You also worked with time series data and discovered ways to optimize code in pandas. Finally, you learned how pandas can be used for preparing data for modeling.

In the next chapter, you will learn about the main data structures in pandas: Series and DataFrames.

2
Working with Data Structures

This chapter introduces you to the core pandas data structures—DataFrames and Series. First, you will create both these data structures from scratch, and then learn how to store them as CSV files. Then, you'll discover how to load the same data structures from CSV files. You will learn how to manipulate row indexes and columns in pandas DataFrames and Series. Furthermore, you will also discover how to convert a column into a new index. By the end of this chapter, you will be adept at manipulating pandas Series and DataFrames in Python.

This chapter covers the following topics:

- The need for data structures
- Exploring indexes and columns
- Working with pandas Series

Introduction to data structures

Data structures are fundamental to computer programming languages. In Python, the core data structures are lists, sets, tuples, and dictionaries. When working in a programming environment, data structures are an abstraction that helps keep track of data, manipulate it, or change it. They also help pass large collections of data as single objects, such as sending an entire Python dictionary to a function. However, organized collections of data can be much more complex, often comprising numerous rows and columns. In this chapter, you will learn about the data structures in pandas that help you deal with such collections of data more effectively. You will dive deeper into the inner workings of these structures and discover how you can use them to accomplish your goals efficiently in Python.

In *Chapter 1, Introduction to pandas*, you were introduced to the ideas that led to the creation of pandas and basic concepts, such as DataFrames and Series. There, you learned about some basic I/O operations, data selection methods, and the transformations supported by pandas.

In this chapter, you will learn about the basic operations involving DataFrames and Series, including the `pandas.read_csv()` method and the corresponding `.to_csv()` methods. You will also be introduced to some ideas about the index. An index is used to keep track of the rows and columns in DataFrames (and items in Series). In fact, columns are another form of an index. Later in the book, as you work through some exercises and activities, you will build on these ideas to become proficient in the highly efficient data manipulations supported by pandas.

The need for data structures

Consider that you are working with quarterly **gross domestic product** (**GDP**) data for the US. A natural way to think about the data and work with it would be to use it in a table. An example might be viewing the data in spreadsheet software, as shown here:

	A	B
1	date	GDP
2	2017-03-31	19190.4
3	2017-06-30	19356.6
4	2017-09-30	19611.7
5	2017-12-31	19918.9
6	2018-03-31	20163.2
7	2018-06-30	20510.2
8	2018-09-30	20749.8
9	2018-12-31	20897.8
10	2019-03-31	21098.8
11	2019-06-30	21340.3
12	2019-09-30	21542.5
13	2019-12-31	21729.1

Figure 2.1 – Tabular data

In *Figure 2.1*, you see two columns of data. The spreadsheet software has labeled the columns with letters and the rows with numbers. In addition, the column names representing the data (`date`, `GDP`) are present in the first row.

The table shown in *Figure 2.1* is a data structure. Having this data in two columns makes it easier to understand and work with. However, in the spreadsheet, it's complicated to work with the data as a single object (a table). This is where pandas gives you an edge over the core Python data structures (and over spreadsheets). As you saw in *Chapter 1, Introduction to pandas*, in pandas you can refer to the entire dataset as an object, for example, a DataFrame called `GDP_by_quarter`. Without such a structure, you would have to keep track of, say, two lists: one for the dates and one for the GDP values. Another approach would be to put the data into a **dictionary**, but that would make simple operations such as summing over a time range more difficult.

As an aside, pandas is built in part "on top of" modules such as NumPy. In many ways, a NumPy array is similar to a pandas DataFrame. So, why not just keep using that? Well, for one thing, in a NumPy array, you don't have row or column names, only numeric indexing (zero-indexed). So, you would have to keep track of the fact that column `0` refers to the dates and column `1` refers to the GDP values. This is similar to the limitations in the spreadsheet; you would have to refer to the `GDP` column as `B` if you were using the spreadsheet shown in *Figure 2.1*.

When working with data for data science and analytics, among other areas, tabular data is ubiquitous. Through DataFrames, pandas provides a natural structure for storing tabular data. As DataFrames comprise Series as well, Series is the other key data structure concept in pandas, and the two often go hand in hand.

Data structures

If you load the data present in *Figure 2.1* into a Jupyter notebook using pandas, it looks as follows:

	date	GDP
0	2017-03-31	19190.4
1	2017-06-30	19356.6
2	2017-09-30	19611.7
3	2017-12-31	19918.9
4	2018-03-31	20163.2
5	2018-06-30	20510.2
6	2018-09-30	20749.8
7	2018-12-31	20897.8
8	2019-03-31	21098.8
9	2019-06-30	21340.3
10	2019-09-30	21542.5
11	2019-12-31	21729.1

Figure 2.2 – Data in Jupyter Notebook

You see the same data in the same tabular structure, but there are a few key differences. In the spreadsheet in *Figure 2.1*, the first row includes the columns as part of the data, with the spreadsheet's header names being `A`, `B`, and `C`. In pandas, on the other hand, the first row in *Figure 2.2* contains the `date` and `GDP` labels, and they are not part of the rows that comprise GDP data, which start on the next row.

In a pandas DataFrame, the row numbers start with `0` by default, as with all Python indexing. The column names are not part of the data; they are an **index**. In fact, both the rows and columns have indices, with which you can refer to the data, select it, modify it, and more, in much the same way that you would with a spreadsheet. These operations will be covered in detail later on; first, you will learn how to create a DataFrame from scratch.

> **Note**
> All the code for the examples used in this chapter is contained in a single Jupyter notebook in a file called `Examples` in the GitHub repository for this chapter, at `https://github.com/PacktWorkshops/The-Pandas-Workshop/tree/master/Chapter02`.

Creating DataFrames in pandas

Suppose you want to create a simple DataFrame in a Python script. The first step is to import pandas:

```
import pandas as pd
```

As a reminder, you can run any single cell in a Jupyter notebook with the *Shift* + *Enter* keyboard combination.

The next step is to use a method to create a DataFrame. Since you have imported the `pandas` library as `pd`, the method to create a DataFrame is a method attached to `pd`, and in this case, it's nicely named `DataFrame`. In a new Jupyter notebook, if you type and run the following code, it will create a pandas DataFrame called `sample_df_construction`:

```
sample_df_construction = pd.DataFrame({'col1' : range(0, 100),\
                                       'col2' : range(1, 200, 
2)})
```

In the method, you provide a Python dictionary to pandas and use the core Python `range()` function to generate a numeric series. In the first case, you use the default increment of `1`, and in the second case, you set the increment to `2`. Note how you can specify the column names as well as the individual data points using this approach.

The next step is to inspect the data to ensure it's what you wanted. For this, you can use the pandas `.head()` and `.tail()` methods, as follows:

```
sample_df_construction.head()
```

The preceding code produces the following output:

Out[4]:

	col1	col2
0	0	1
1	1	3
2	2	5
3	3	7
4	4	9

Figure 2.3 – First five rows of a simple pandas DataFrame

Similarly, you can use the `tail()` method, as follows:

```
sample_df_construction.tail()
```

This produces the following output:

Out[5]:

	col1	col2
95	95	191
96	96	193
97	97	195
98	98	197
99	99	199

Figure 2.4 – Last five rows of a simple pandas DataFrame

Let's explore in detail what was done in the `DataFrame` **constructor**. First, note that `DataFrame` is a **class** from the pandas library, so it can be instantiated by calling `pd.DataFrame()` (assuming pandas is imported as `pd`, which is the commonly used convention). Each time you do this, a new **instance** of the `DataFrame` class is generated. You can check the data type of the `sample_df_construction` variable using the following code, by calling the core Python `type()` function, as follows:

```
type(sample_df_construction)
```

This will result in the following output:

```
pandas.core.frame.DataFrame
```

A good habit to form if you want to understand methods, classes, and functions in Python is to look at the help documentation readily available within Jupyter. You can view the documentation for `DataFrame` using the following code, where `?` is a shortcut for `help()`:

```
?pd.DataFrame
```

Running this line in a Jupyter notebook cell should produce the following (truncated here for brevity):

```
In [6]:  ?pd.DataFrame
```

```
Init signature:
pd.DataFrame(
    data=None,
    index: Union[Collection, NoneType] = None,
    columns: Union[Collection, NoneType] = None,
    dtype: Union[str, numpy.dtype, ForwardRef('ExtensionDtype'), NoneType] = None,
    copy: bool = False,
)
Docstring:
Two-dimensional, size-mutable, potentially heterogeneous tabular data.

Data structure also contains labeled axes (rows and columns).
Arithmetic operations align on both row and column labels. Can be
thought of as a dict-like container for Series objects. The primary
pandas data structure.

Parameters
----------
data : ndarray (structured or homogeneous), Iterable, dict, or DataFrame
    Dict can contain Series, arrays, constants, or list-like objects.
```

Figure 2.5 – pandas DataFrame help documentation

You can see that in addition to providing data, you can specify the index and the columns and force a particular data type (`dtype`). In the description for the `data` parameter, note that data can be a NumPy array, an iterable, `dict` (dictionary), or a DataFrame.

In the preceding example, the `dictionary` method is used, which is easy to read. For each column, there is a `key : value` pair, where the key becomes the column name and the value becomes the data in the column. Since a dictionary can handle any type of data, this form allows complex structures to be added to a column.

The next example shows the difference between using a dictionary (as in the previous example) and using other data types. In the following snippet, you first load the numpy library and create a couple of arrays using `np.array()`. Note that `np.array()` is a function that creates a NumPy array, and an array is a structure with one or more dimensions to hold numeric data. Here, you are creating two one-dimensional arrays:

```
import numpy as np
sample_np_array_1 = np.array(range(0, 100))
sample_np_array_2 = np.array(range(1, 200, 2))
```

Now, you can combine those one-dimensional arrays into a two-dimensional array, called `sample_np_2D`, using a NumPy function called `column_stack()`:

```
sample_np_2D = np.column_stack((sample_np_array_1,\
                                sample_np_array_2))
print(sample_np_2D[0:5])
```

This generates the following output:

```
[[0 1]
 [1 3]
 [2 5]
 [3 7]
 [4 9]]
```

You can see the same two columns that were present in *Figure 2.3*, where you used the `.head()` method to inspect the DataFrame. But, notice that there are no row numbers or column names. NumPy is great for efficient numerical operations. It also provides a structure for higher-dimensional data; however, it's hard to keep track of things in this form. That is part of the motivation for building pandas on top of NumPy. You can convert the NumPy array to a pandas DataFrame, using the constructor as follows, where you create a new DataFrame called `sample_df_from_np`:

```
sample_df_from_np = pd.DataFrame(sample_np_2D,
                                 columns = ['col1', 'col2'])
print(sample_df_from_np.head())
```

This creates the following output, which looks exactly like *Figure 2.3*:

```
   col1  col2
0     0     1
1     1     3
2     2     5
3     3     7
4     4     9
    col1  col2
95    95   191
96    96   193
97    97   195
98    98   197
99    99   199
```

Figure 2.6 – A new pandas DataFrame created using the NumPy two-dimensional array

This example was kind of a roundabout method, but it demonstrates just how easy it is to get data into a DataFrame, getting the benefits of the row index and the column names.

As another example, suppose you had data that started as lists. In the following snippet, you use two **list comprehensions** to create two lists with the data:

```
list_1 = [i for i in range(100)]
list_2 = [i for i in range(1, 200, 2)]
print(list_1, list_2)
```

This produces the following output. Notice this is not a very user-friendly format:

```
[0, 1, 2, 3, 4, 5, 6, 7, 8, 9, 10, 11, 12, 13, 14, 15, 16, 17, 18, 19, 20, 21, 22, 23, 24, 25, 26, 27, 28,
29, 30, 31, 32, 33, 34, 35, 36, 37, 38, 39, 40, 41, 42, 43, 44, 45, 46, 47, 48, 49, 50, 51, 52, 53, 54, 5
5, 56, 57, 58, 59, 60, 61, 62, 63, 64, 65, 66, 67, 68, 69, 70, 71, 72, 73, 74, 75, 76, 77, 78, 79, 80, 81,
82, 83, 84, 85, 86, 87, 88, 89, 90, 91, 92, 93, 94, 95, 96, 97, 98, 99] [1, 3, 5, 7, 9, 11, 13, 15, 17, 1
9, 21, 23, 25, 27, 29, 31, 33, 35, 37, 39, 41, 43, 45, 47, 49, 51, 53, 55, 57, 59, 61, 63, 65, 67, 69, 71,
73, 75, 77, 79, 81, 83, 85, 87, 89, 91, 93, 95, 97, 99, 101, 103, 105, 107, 109, 111, 113, 115, 117, 119,
121, 123, 125, 127, 129, 131, 133, 135, 137, 139, 141, 143, 145, 147, 149, 151, 153, 155, 157, 159, 161, 1
63, 165, 167, 169, 171, 173, 175, 177, 179, 181, 183, 185, 187, 189, 191, 193, 195, 197, 199]
```

Figure 2.7 – Printing the two lists

You can use the constructor again to combine the lists into a DataFrame. Note that earlier, it was mentioned that the data can be iterable. Lists are iterables in Python, but you want both lists to be stored in the DataFrame. Python's `zip()` method produces an iterator by pairing up two iterables, and you can iterate using a `zip` iterator. Therefore, you can use this method to easily combine the lists into one DataFrame, as follows:

```
sample_df_from_iterable = pd.DataFrame(zip(list_1, list_2),
                                        columns = ['col1',
'col2'])
sample_df_from_iterable
```

This code snippet produces the following output:

```
   col1  col2
0     0     1
1     1     3
2     2     5
3     3     7
4     4     9
    col1  col2
95    95   191
96    96   193
97    97   195
98    98   197
99    99   199
```

Figure 2.8 – A two-column DataFrame created from two iterables

Imagine how cumbersome it would be if you had lots of lists and had to manage them without pandas.

You can also use the dictionary format for the data, so you can easily match the lists with the respective column names:

```
sample_df_from_lists = pd.DataFrame({'col1' : list_1,
                                     'col2' : list_2})
print(sample_df_from_lists.head())
print(sample_df_from_lists.tail())
```

The output from the preceding code snippet is as follows:

```
   col1  col2
0     0     1
1     1     3
2     2     5
3     3     7
4     4     9
    col1  col2
95    95   191
96    96   193
97    97   195
98    98   197
99    99   199
```

Figure 2.9 – The new sample_df_from_lists DataFrame

You have returned to the now-familiar pandas-formatted DataFrame. You might have noticed in *Figure 2.5* that `data` is optional and defaults to `None`. `None` indicates that there is a data structure ready to hold some data, but there is no actual data yet. In the next example, a DataFrame called `empty_data` is created by calling the `.DataFrame()` method with no arguments. You can use the `shape` method to get the dimensions of the DataFrame created. The `shape` method returns a tuple containing the shape of the DataFrame in the `(rows, columns)` form:

```
empty_data = pd.DataFrame()
empty_data.shape
```

You will see the following output:

```
(0,0)
```

So, you can create a DataFrame with `0` rows and `0` columns (the output of `shape`). You might want to do that because, in Python, you cannot reference an object until it is created, so you can create an empty DataFrame that can be used later.

The following snippet shows how to use the empty DataFrame created in the previous code snippet, and add more data to it in later stages using a NumPy array and the `.concat` method:

```
import numpy as np
my_array = np.array([[0, 1, 2, 3, 4],\
                     [2, 3, 4, 5, 6]])
```

Now, suppose you want to concatenate the NumPy array onto the empty DataFrame. For this, you can use the pandas `.concat` method, which can combine Series or DataFrames. In the following code, you specify `axis = 0` to mean that you are adding data in the row direction. The `axis` parameter tells pandas how the new data will be appended to the original DataFrame. When `axis` is set to `0`, data is appended in a row-wise manner; when it is set to `1`, data is appended in a column-wise manner. Since, in this case, there is no data in the original DataFrame, it does not make any difference whether you set the `axis` parameter to `1` or `0`. Note that `pd.DataFrame()` is called on the NumPy array, as pandas can only concatenate pandas objects:

```
filled_dataframe=pd.concat([empty_data,pd.DataFrame(my_
array)],\
                            axis=0)
filled_dataframe
```

This produces the following output. You can see that the two rows of the NumPy array are now rows of the new DataFrame:

Out[35]:

	0	1	2	3	4
0	0	1	2	3	4
1	2	3	4	5	6

Figure 2.10 – Result of concatenating array data to an initially empty DataFrame

Note that this example is a bit contrived; you could have achieved the same result in a simpler way as done before, using the NumPy array as the data:

```
filled_data_from_np = pd.DataFrame(my_array)
filled_data_from_np
```

However, this example was given to illustrate that you can create DataFrames directly using the constructor, create and then fill empty DataFrames, and convert appropriate objects, such as a two-dimensional NumPy array, into a DataFrame. You can also combine `Pandas.Series()` objects to make DataFrames, but let's spend a little more time on the DataFrame index and columns first.

Exercise 2.01 – Creating a DataFrame

You are developing a few data modeling methods and you need some sample data that will help you test and debug your models before using those methods on real data. In this exercise, you will use the `DataFrame` constructor to create a synthetic dataset where one column contains the time (measured in seconds) and another column contains some fictitious measurements. These measurements need to be collected at an interval of 0.1 seconds.

> **Note**
>
> You can find the code for this exercise here: https://github.com/PacktWorkshops/The-Pandas-Workshop/tree/master/Chapter02/Exercise2_01.

Perform the following steps to complete the exercise:

1. Open a new Jupyter notebook and select the **Pandas_Workshop** kernel.
2. For this exercise, all you will need is the `pandas` library and the `numpy` library. Load them in the first cell of the notebook:

```
import pandas as pd
import numpy as np
```

3. You know the data you plan to analyze is collected at intervals of 0.1 seconds. Also, in most cases, the data is periodic. You decide to make 1,000 samples and use the `np.sin()` function for the measurement portion. For the time, use `range(1000)` to generate the samples, and wrap that in `pd.Series()` to get all the values, then divide by `10`. For the data, use the same values inside of `np.sin()` and multiply by `2` and `np.pi`, because `np.sin()` expects values in radians:

```
test_data = pd.DataFrame({'time' :
pd.Series(range(1000))/10,\
'measurement' : np.sin(2 * np.pi *
pd.Series(range(1000))/10/1)})
```

4. Now, use `test_data.head(11)` to list out the first 11 rows. Confirm that the measurement returns to `0` at `0.5` and `1.0` seconds:

```
test_data.head(11)
```

The output should now look like the following:

Out[19]:

	time	measurement
0	0.0	0.000000e+00
1	0.1	5.877853e-01
2	0.2	9.510565e-01
3	0.3	9.510565e-01
4	0.4	5.877853e-01
5	0.5	1.224647e-16
6	0.6	-5.877853e-01
7	0.7	-9.510565e-01
8	0.8	-9.510565e-01
9	0.9	-5.877853e-01
10	1.0	-2.449294e-16

Figure 2.12 – The test_data DataFrame

You see that the values at `0.5` and `1.0` are less than 10-15, which is effectively `0`. Of course, since you have already seen the `pd.DataFrame.to_csv()` method, you could save your synthetic data for later use.

In this exercise, you used the `DataFrame` constructor with a dictionary of labels (column names) and values to create the `test_data` DataFrame. The next section goes deeper into the topics of row index and column names.

Indexes and columns

We have already referred to **indexes** and **columns** without fully defining them. An **index** contains references to the rows of a DataFrame. The index of a pandas DataFrame is analogous to the row numbers you might see in a spreadsheet. In spreadsheets, it's common to use the so-called A1 notation, where *A* refers to the columns, which usually begin with `A`, and *1* refers to the rows, which usually begin with `1`.

We will start by looking at the index, and continue with the `sample_df_from_lists` DataFrame created earlier. You can use the `.index` method to display information about the index, as follows:

```
sample_df_from_lists.index
```

This line of code produces the following output:

```
RangeIndex(start=0, stop=100, step=1)
```

You may recall that **ranges** in Python are inclusive of the start value and exclusive of the end value. You see that the index for `sample_df_from_lists` runs from `0` to `99`, which matches the rows. As you will learn in detail in *Chapter 5*, *Data Selection – DataFrames*, pandas separates the ideas of integer row or column numbers from the index. In fact, you can use nearly anything for the index.

In the next example, you will see how there are several methods to set, reset, and change the index in pandas. First, you'll need to create some data that can be used for the new index. To do that, you can once again use the `range()` function along with the Python `map()` function to apply the `chr()` function to the range, as follows:

```
letters = pd.Series(map(chr, range(97, 122)))
letters = pd.DataFrame({'letter' :\
                        (list(letters) + list(letters * 2)\
                       + list(letters * 3) + list(letters*4))})
```

The `map()` function "maps" a given function (in this case, `chr()`) to the arguments given (in this case, `range(97, 122)`). The `chr()` function returns the *ASCII* character for a given integer value. Note that when using `map()`, you need to pass the name of the function without parentheses (that is, `chr`). This is because you are not calling the function, but sending the function to map as an argument.

The first line of the preceding code generates the lowercase alphabet up to the letter `y`, and contains 25 characters. Then, the `DataFrame` constructor is used with data in a dictionary form again, and the Series created is extended by appending it with each character doubled, tripled, and quadrupled using Python's operator overloading, which allows the use of `*` with string data to replicate values.

In the DataFrame constructor, the first list (`letters`) puts 25 items in the `letter` column, then the next list (`letters * 2`) puts another 25 elements in with the items `aa`, `bb`, and so on; the third adds `aaa`, `bbb`, and so forth, and the fourth adds `aaaa`, and `bbbb`. The result is a DataFrame with one column named `letter` and 100 rows.

Now, take a look at some of the data. The following code uses the **bracket notation** for a pandas DataFrame (with index values enclosed in brackets, `[]`):

```
letters[21:29]
```

The preceding line of code provides the rows matching the range from the first value up to 28. This produces the following output:

Out[33]:

	letter
21	v
22	w
23	x
24	y
25	aa
26	bb
27	cc
28	dd

Figure 2.13 – Eight rows of data in the new letters DataFrame

> **Note**
>
> One of the many benefits of pandas is the expansion of the so-called bracket notation. This will be covered in depth in *Chapter 5, Data Selection – DataFrames*, but in general, the pandas bracket notation lets you use column names instead of numbers. It also lets you use string values (labels) for the row index. Furthermore, it comes with a host of useful features, such as subsetting. For DataFrames, the general notation is `some_dataframe.iloc[rows, columns]` if using all integers, or logical (Boolean) expressions, and `some_dataframe.loc[rows, columns]` if using labels.
>
> When using integers and the `.iloc` method, pandas follows the Python convention of including the first value and excluding the last value, but when using labels, both the first and last values are included.

Now for the fun part—you will replace the index of `sample_df_from_lists` with the `letter` column from the `letters` DataFrame. You will do that with another method named `.set_index()`, which will be explored more later; for now, you just need to know that the `.set_index()` method of the `DataFrame` class allows you to set the index of a DataFrame to new values.

In the following code snippet, you first print the current index of `sample_df_from_lists`, then replace it with the `letter` column. Then, you print the index again. Note that, instead of making an assignment to the `sample_df_from_lists` object, you use the `inplace = True` option that tells pandas to make the change to the existing DataFrame without having to reassign the result to it. Also, note that you are using the `.iloc` method and bracket notation to select just the `letter` column. Bracket notation lets you use a `':'` shorthand for the rows (or columns), which simply means "all," so `[:, 0]` tells pandas "all rows and the first column":

```
print(sample_df_from_lists.index)
sample_df_from_lists.set_index(keys = letters.iloc[:, 0],\
                               inplace = True)
print(sample_df_from_lists.index)
```

Running this creates the following output:

```
RangeIndex(start=0, stop=100, step=1)
Index(['a', 'b', 'c', 'd', 'e', 'f', 'g', 'h', 'i', 'j', 'k', 'l', 'm', 'n',
       'o', 'p', 'q', 'r', 's', 't', 'u', 'v', 'w', 'x', 'y', 'aa', 'bb', 'cc',
       'dd', 'ee', 'ff', 'gg', 'hh', 'ii', 'jj', 'kk', 'll', 'mm', 'nn', 'oo',
       'pp', 'qq', 'rr', 'ss', 'tt', 'uu', 'vv', 'ww', 'xx', 'yy', 'aaa',
       'bbb', 'ccc', 'ddd', 'eee', 'fff', 'ggg', 'hhh', 'iii', 'jjj', 'kkk',
       'lll', 'mmm', 'nnn', 'ooo', 'ppp', 'qqq', 'rrr', 'sss', 'ttt', 'uuu',
       'vvv', 'www', 'xxx', 'yyy', 'aaaa', 'bbbb', 'cccc', 'dddd', 'eeee',
       'ffff', 'gggg', 'hhhh', 'iiii', 'jjjj', 'kkkk', 'llll', 'mmmm', 'nnnn',
       'oooo', 'pppp', 'qqqq', 'rrrr', 'ssss', 'tttt', 'uuuu', 'vvvv', 'wwww',
       'xxxx', 'yyyy'],
      dtype='object', name='letter')
```

Figure 2.14 – The index of sample_df_from_lists before and after the change

Note

In many cases in Python and when using pandas, the application of a method returns a new object. If you want to change the existing object, you have to assign the result back to the object, like so:

`object = object.method()`

In many methods, such as the pandas `.set_index()` method that was used in the preceding code, the option is provided to use the `inplace = True` parameter, which changes the object directly. This makes the code more compact and can reduce memory usage.

You can see that the index has been changed from a range of integers to the expanding alphabet. Now, consider a quick example of what you can do with this new index.

In the following snippet, you simply list out a range of rows. Note that the colon (`:`) notation is used, just as you did with integers, with the new string index values. The `.loc` method is used to subset from uuuu to yyyy:

```
sample_df_from_lists.loc['uuuu':'yyyy', :]
```

Running this code line produces the following output:

Out[48]:

	col1	col2
letter		
uuuu	95	191
vvvv	96	193
wwww	97	195
xxxx	98	197
yyyy	99	199

Figure 2.15 – The new string-based index of sample_df_from_lists

Using the index in this way can be very useful for natural, readable data, as well as code.

You've now seen that it is possible to use text (strings) for the index, and pandas understands the order, so you can refer to ranges just like you do with other objects, such as lists. In the next example, you'll see that there are times when making labels for the index is very useful—such as when the labels represent categories you are interested in. Keep in mind that you can use integer indexing (with `.iloc`) anytime, but that requires, in many cases, for you to know the rows or columns you want, by number. Using label indexing (with `.loc`) allows you to use natural groups to work with your data.

The pandas `.set_index()` method can also take column names as an argument, supporting the common pattern of setting the index to the values in an existing column. In the next example, you will make a list to contain 100 values, of which half are `cat` and half are dog, and then print out the first and last five values:

```
animal_type = ['cat'] * 50 + ['dog'] * 50
print(animal_type[:5], animal_type[-5:])
```

Running this code produces the following output:

```
['cat', 'cat', 'cat', 'cat', 'cat'] ['dog', 'dog', 'dog',
'dog', 'dog']
```

Now, you will use `animal_type` to add a column to `sample_df_from_lists`. Notice here that the pandas bracket notation for the DataFrame, when given a string value, will either result in replacing that column (if it exists) or adding it to the DataFrame. As you can see again, pandas makes the code very readable. After adding the column, you can list the DataFrame for verification:

```
sample_df_from_lists['animal_type'] = animal_type
sample_df_from_lists
```

This produces the following output:

Out[129]:

	col1	col2	animal_type
letter			
a	0	1	cat
b	1	3	cat
c	2	5	cat
d	3	7	cat
e	4	9	cat
...	...	...	...
uuuu	95	191	dog
vvvv	96	193	dog
wwww	97	195	dog
xxxx	98	197	dog
yyyy	99	199	dog

100 rows × 3 columns

Figure 2.16 – The sample_df_from_lists DataFrame updated to have animal_type as a column

You can now replace the index with the new column. Here, again, you will use the pandas `set_index()` method, where you'll specify the column name (`animal_type`). And, just like you saw earlier, `inplace = True` can be used to make the change directly to the DataFrame:

```
sample_df_from_lists.set_index('animal_type', inplace = True)
sample_df_from_lists
```

This produces the following output:

```
Out[133]:
```

animal_type	col1	col2
cat	0	1
cat	1	3
cat	2	5
cat	3	7
cat	4	9
...	...	...
dog	95	191
dog	96	193
dog	97	195
dog	98	197
dog	99	199

100 rows × 2 columns

Figure 2.17 – The sample_df_from_lists DataFrame with the new animal_type row index

Note that it's perfectly fine to have duplicate values in an index of this type. You'll learn more about working with the row and column indices later, but as a preview, try changing the generic `col1` and `col2` to `good` and `bad`, respectively. First, you can see the structure of the column names index by using the `.columns` attribute of the DataFrame, as follows:

```
print(sample_df_from_lists.columns)
```

Running this code results in the following output:

```
Index(['col1', 'col2'], dtype='object')
```

You can see that the column index values are stored in a Python list as strings. You can simply assign a list of new values directly to the `.columns` attribute.

> **Note**
>
> In general, since an attribute contains data associated with a Python object, you can assign new values freely to attributes. This pattern of assigning new values to column names is fairly common.

The new list is simply `['good', 'bad']`, and you can assign it as follows:

```
sample_df_from_lists.columns = ['good', 'bad']
sample_df_from_lists.head()
```

The following output shows the changed names:

Out[34]:

	good	bad
animal_type		
cat	0	1
cat	1	3
cat	2	5
cat	3	7
cat	4	9

Figure 2.18 – The sample_df_from_lists DataFrame with new column names

With the new index, you can select just the `cat` rows, and list out 10 of them. The following line of code uses bracket notation twice; first, you use the pandas `.loc` method (using labels) and select all the `'cat'` rows and all the columns (after `'cat'`, the colon (`:`) means *all columns*), and then you can use the bracket notation again to get just the first `10` results:

```
sample_df_from_lists.loc['cat', :][:10]
```

This code will produce the following output:

Out[35]:

	good	bad
animal_type		
cat	0	1
cat	1	3
cat	2	5
cat	3	7
cat	4	9
cat	5	11
cat	6	13
cat	7	15
cat	8	17
cat	9	19

Figure 2.19 – sample_df_from_lists selecting only cat rows and then listing the first 10

In addition to simply listing values, you can use the indexing methods to subset if you want to apply some logic to parts of the data. For example, suppose you want to sum up how many *good cats* and *good dogs* there are; pandas provides many math methods, such as `.sum()`, for these purposes. So, you can again use `.loc` to index by labels, and use both the animal type and the column name to select the data you want, before applying the `.sum()` method. Here, you apply this approach to cats and dogs separately:

```
print('good cats', sample_df_from_lists.loc['cat', 'good'].
sum())
print('good dogs', sample_df_from_lists.loc['dog', 'good'].
sum())
```

This gives the desired result, as follows:

```
good cats 1225
good dogs 3725
```

In addition to the fact that there seem to be three times more good dogs than cats, you can see that the pandas notation to accomplish this task and answer the question is easily understood just by looking at the code.

Exercise 2.02 - Reading DataFrames and manipulating the index

In this exercise, you will read data from two CSV files, each containing data on pet food sales. You will combine them into a single DataFrame and replace the index with the corresponding animal type. Then, using the new index, you will find the total amount due for the cat food and dog food shipments.

> **Note**
>
> You can find the code for this exercise here: `https://github.com/PacktWorkshops/The-Pandas-Workshop/tree/master/Chapter02/Exercise2_02`.

Perform the following steps to complete the exercise:

1. Open a new Jupyter notebook and select the **Pandas_Workshop** kernel.
2. For this exercise, all you will need is the `pandas` library. Load it in the first cell of the notebook:

    ```
    import pandas as pd
    ```

3. Suppose you run a small pet supplies store, and your supplier of dog and cat food sends you CSV files with the details of each order. You have received two new files for the latest orders. Read the `dog_food_orders.csv` and `cat_food_orders.csv` files from the `Datasets` subdirectory, then print the first three lines of each DataFrame. Use the `read_csv()` method prepended by the `Datasets` path, as follows:

```
dog_food_orders = pd.read_csv('../Datasets/dog_food_
orders.csv')
cat_food_orders = pd.read_csv('../Datasets/cat_food_
orders.csv')
print(dog_food_orders.head(3))
print(cat_food_orders.head(3))
```

> **Note**
>
> Please change the path of the dataset file (highlighted) based on where you have downloaded it on your system. You can download `dog_food_orders.csv` from `https://github.com/PacktWorkshops/The-Pandas-Workshop/blob/master/Chapter02/Datasets/dog_food_orders.csv`, and `cat_food_orders.csv` from `https://github.com/PacktWorkshops/The-Pandas-Workshop/blob/master/Chapter02/Datasets/cat_food_orders.csv`.

The output should look like the following:

```
            product  wholesale_price   msrp  qty_ordered  qty_shipped
0     skippys_dream             8.99  18.38          100          100
1     just_the_beef             4.99  10.43          200          195
2  potatos_and_lamb             5.19  11.43           50           50
          product  wholesale_price   msrp  qty_ordered  qty_shipped
0     cat_delight             4.95   9.98           50            0
1   tuna_surprise             7.17  15.27          100          100
2  hint_of_catnip             3.99   8.23           25           25
```

Figure 2.20 – The dog_food_orders and cat_food_orders DataFrames

> **Note**
>
> The pandas `.read_csv()` method is amazingly versatile. Use `?pd.read_csv` in your notebook to explore all the options available.

4. Now, add an `animal` column to each DataFrame, using the `['animal']` bracket notation, and print the first three lines of each DataFrame again:

```
dog_food_orders['animal'] = 'dog'
cat_food_orders['animal'] = 'cat'
print(dog_food_orders.head(3))
print(cat_food_orders.head(3))
```

The output should now look like the following:

```
           product  wholesale_price   msrp  qty_ordered  qty_shipped animal
0     skippys_dream             8.99  18.38          100          100    dog
1     just_the_beef             4.99  10.43          200          195    dog
2  potatos_and_lamb             5.19  11.43           50           50    dog
         product  wholesale_price   msrp  qty_ordered  qty_shipped animal
0    cat_delight             4.95   9.98           50            0    cat
1  tuna_surprise             7.17  15.27          100          100    cat
2  hint_of_catnip            3.99   8.23           25           25    cat
```

Figure 2.21 – The updated DataFrames with the animal column

5. Combine the two DataFrames into a new DataFrame called `orders`, using the pandas `.concat()` method, setting the `axis` parameter to `0` so that they are combined as rows. List out the `orders` DataFrame:

```
orders = pd.concat([dog_food_orders,\
                    cat_food_orders],\
                    axis = 0)
orders
```

The result should be as follows:

Out[8]:

	product	wholesale_price	msrp	qty_ordered	qty_shipped	animal
0	skippys_dream	8.99	18.38	100	100	dog
1	just_the_beef	4.99	10.43	200	195	dog
2	potatos_and_lamb	5.19	11.43	50	50	dog
3	turkey_and_cranberries	5.98	12.00	50	50	dog
4	roasted_duck	9.59	17.48	15	15	dog
0	cat_delight	4.95	9.98	50	0	cat
1	tuna_surprise	7.17	15.27	100	100	cat
2	hint_of_catnip	3.99	8.23	25	25	cat
3	roast_chicken	5.57	12.08	30	30	cat
4	lamb_w_rice	5.83	11.68	30	30	cat

Figure 2.22 – The new orders DataFrame

6. Replace the index with the `animal` column, using the pandas `.set_index()` method. Use `inplace = True` to make the change directly to the DataFrame. List the result again:

```
orders.set_index('animal', inplace = True)
orders
```

The result should be as follows:

Out[9]:

animal	product	wholesale_price	msrp	qty_ordered	qty_shipped
dog	skippys_dream	8.99	18.38	100	100
dog	just_the_beef	4.99	10.43	200	195
dog	potatos_and_lamb	5.19	11.43	50	50
dog	turkey_and_cranberries	5.98	12.00	50	50
dog	roasted_duck	9.59	17.48	15	15
cat	cat_delight	4.95	9.98	50	0
cat	tuna_surprise	7.17	15.27	100	100
cat	hint_of_catnip	3.99	8.23	25	25
cat	roast_chicken	5.57	12.08	30	30
cat	lamb_w_rice	5.83	11.68	30	30

Figure 2.23 – The orders DataFrame with the updated row index

7. Now, you need the total amount due for these shipments. For this, first create a column with the subtotal by row, multiplying the wholesale price by the quantity actually shipped, and save that in a new column called `net_due`. You can then get the result in a single line; simply use `.loc` to select by labels and select each animal type along with the new `net due` column. You can then use `.sum()` to get the total:

```
orders['net_due'] = \
(orders['wholesale_price'] * orders['qty_shipped'])
print('cat food orders due: ',\
      orders.loc['cat', 'net_due'].sum(),
      '\ndog food orders due: ',\
      orders.loc['dog', 'net_due'].sum())
```

The result should be as follows:

```
cat food orders due:  1158.75
dog food orders due:  2574.4
```

In this exercise, you began by implementing the `.read_csv()` method to read two CSV files. After storing the data from those files into two new DataFrames, you combined those DataFrames with the `.concat()` method. Next, you used bracket notation to add a column. With `.set_index()`, you set the newly created column as the index. After performing a calculation, you were able to store its result in a new column using the bracket notation once again. Finally, you used `.sum()` to total the costs and present a sum for each animal food type.

Working with columns

You have learned about the index; however, most of what you saw applies in some ways to columns too. There are some specific methods for columns, though. For example, suppose you have data on the kinds of foods consumed and how people rate the taste, and you get this data in an unlabeled form. Here, you will use the DataFrame constructor again, passing a small array as the data and giving the columns generic names, such as `col1` and `col2`:

```
food_taste = pd.DataFrame(data = np.array([[60, 3.5],\
                                          [40, 8]]),\
                          columns = ['col1', 'col2'])
```

Instead of the `.index` method that you have been using on the rows, you can use the `.columns` method to see the column names (which are the index labels). This was shown earlier in the *Indexes and columns* section without fully explaining the relationship of columns to an index. The following example uses the `.columns` method on the `food_taste` DataFrame:

```
food_taste.columns
```

The result is as follows:

```
Index(['col1', 'col2'], dtype='object')
```

The result shows that you can get the column names (labels) using the `.columns` method and that the result is an index that includes a list object with the column names. Something that may be confusing at first is that if you assign the result of the `.columns` method to a variable, the result is not a list, but an index object, as the following code snippet shows:

```
food_columns = food_taste.columns
food_columns
```

Running this snippet will produce the following output:

```
Index(['col1', 'col2'], dtype='object')
```

To get the column names as a list, you can use the `list()` constructor. You can then reassign the result of the `list()` constructor called on the column index to the same variable, then list it out:

```
food_columns = list(food_columns)
food_columns
```

This produces a list, as expected, as shown in the following:

```
['col1', 'col2']
```

Unlike a row index, there is no `set` or `reset` method for the column index. Instead, you can assign new values directly, or use the `.rename()` method. Suppose you know that the first column of the raw data represents consumption values, and the second column represents an aggregate taste index (a rating). Here, the pandas `.rename()` method can be used to reassign the column names to meaningful values, as done in the following snippet. Notice the use of the dictionary structure to provide the mapping of old names to new names:

```
food_taste.rename({'col1' : 'food_consumption',\
                   'col2' : 'taste_index'},\
                   axis = 1,\
                   inplace = True)
food_taste
```

Running the code produces the following output:

Out[44]:

	food_consumption	taste_index
0	60.0	3.5
1	40.0	8.0

Figure 2.24 – The food_taste DataFrame with renamed columns

To replace the names, you can simply assign a new list of names using the .columns method. The choice of using .rename() or direct assignment is up to you; in some cases, one or the other will be shorter—if there are many columns and you only need to rename a few, then .rename() is likely the best choice. In contrast, if there are only a few columns, using direct assignment can be easier. Additionally, if you happen to have the names in an existing list object, then direct assignment makes sense there. This can come in handy in certain cases. Suppose that while accessing data without column names from an external source, you find that the names are provided in a separate file (often called a header). In such cases, you would need to use direct assignment to assign a new list of names, and then list out the result, as follows:

```
food_taste.columns = ['food_cons', 'taste']
food_taste
```

The output of this snippet is as follows:

Out[168]:

	food_cons	taste
0	60.0	3.5
1	40.0	8.0

Figure 2.25 – Renaming the columns of food_taste by direct assignment to the column names

So far, you have seen that pandas provides a natural tabular structure for data, the DataFrame, and provides several methods to manage the indices (rows and columns) of the data. For example, you have learned how to create a column index that labels the rows with text instead of integers, and how to name and rename the columns. Using these methods, you have seen how easy it is to manipulate data naturally, such as easily getting all the dog food and cat food orders. However, while the Series data structure was introduced in the preceding examples that used pd.Series(), it was never fully defined. The next section explores pandas Series.

Series

The Series is the other fundamental pandas data structure. You can consider a DataFrame to be an organized collection of series, where each column is, in fact, a Series. Looking at the food_cons column of the food_taste DataFrame, you can see this relationship. The following line of code calls the type() method on the food_cons column of food_taste:

```
type(food_taste['food_cons'])
```

This generates the following output:

```
pandas.core.series.Series
```

So, every DataFrame column is a pandas Series, once separated and on its own. This would also be the case if you separated a single row from a DataFrame. Recall that you can use `?` in Jupyter to get the help documentation. Try to do that and look at the first part of the Series documentation. You can use the following code to get the documentation:

```
?pd.Series
```

This provides the following output (truncated for brevity):

```
Init signature:
pd.Series(
    data=None,
    index=None,
    dtype=None,
    name=None,
    copy=False,
    fastpath=False,
)
Docstring:
One-dimensional ndarray with axis labels (including time series).
```

Figure 2.26 – The first portion of the help documentation for pandas Series

Note the similarities with `pd.DataFrame()`. You see the same `data`, `index`, `dtype`, and `copy` parameters. A Series is a one-dimensional structure, so, unlike what you saw in a `DataFrame` constructor, you won't find a `columns` parameter here. `fastpath` is an internal parameter related to behind-the-scenes data operations, so you don't need to be concerned about it. So, much of what you have already learned about using the DataFrame constructor will naturally apply to the `Series` constructor. You can easily extract Series from DataFrames (by selecting a column) and add Series to DataFrames as new columns.

Since the emphasis has been on tabular data so much until now, you might wonder about the importance of Series other than just being a precursor to DataFrames. There are several use cases where it is natural to have a series of values, including cases where values are naturally ordered. One such example is when the data is ordered by time, such as when a purchase is made through e-commerce applications. pandas time series will be covered in detail in *Chapter 9, Time Series*. However, since Series also has an index, and since the index can hold time information, you can work with **time series** using Series without needing a DataFrame.

Another natural use of Series is in data collection. Suppose you are monitoring a process in a chemical plant, and one of the things to monitor is the flow rate. In such cases, it is common that measurements would be taken periodically and, therefore, collected naturally in a series.

However, Series are much more flexible and can be used for a lot more than just holding ordered data. You can also envision cases where you might have a series of responses to a question (such as a survey). In such cases, time is irrelevant, but having an index is still useful. Also, objects in a Series don't have to be of the same type (this is a big difference compared, to, say, NumPy arrays). The next section looks into pandas Series in more detail and shows how the Series index can be used.

The Series index

You have already seen the importance of Series in *Chapter 1, Introduction to pandas*. Though their structure is similar to that of an array, Series have the additional advantage of having an index that can be either an integer or labels, unlike arrays, which can have only integers as the index.

You'll begin by using `.read_csv()` to read some data from a file. Suppose you have some data representing a measurement over time, say, percent improvement in fuel mileage of new cars. Here, you read a file called `noisy_series.csv` (this file can be downloaded from `https://github.com/PacktWorkshops/The-Pandas-Workshop/blob/master/Chapter02/Datasets/noisy_series.csv`). By default, pandas will read data from a file into a DataFrame; since you want the data to be a Series, you can use the `squeeze = True` parameter to tell pandas to produce a Series, if possible. Note that without writing `squeeze = True`, the result will be a DataFrame with a single column, instead of a Series:

```
noisy_series = pd.read_csv('noisy_series.csv', squeeze = True)
```

Now, take a look at the data. In *Chapter 8, Data Visualization*, you will learn about many methods of looking at data in pandas, but for now, note that pandas provides the `.plot()` method, which works on Series. You can use `noisy_series.plot()` to visualize the data, as follows:

```
noisy_series.plot()
```

This produces the following basic plot:

```
Out[8]: <matplotlib.axes._subplots.AxesSubplot at 0x2c0ab0d5248>
```

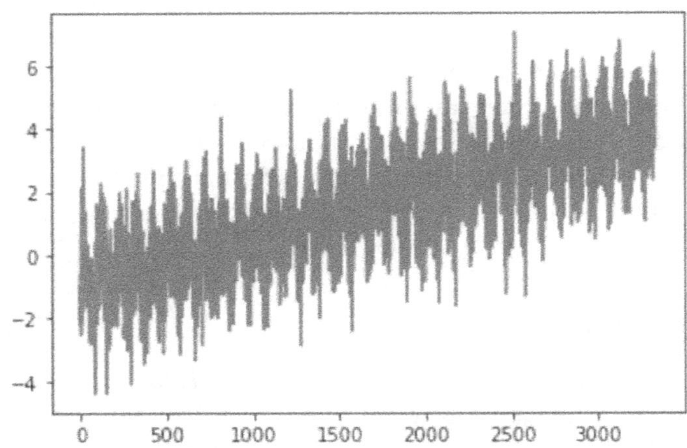

Figure 2.27 – A simple plot of the data in noisy_series

The output shows that there are more than 3,000 data points, ranging from around `-4` to `7`. Is this consistent with the Series index? To find out, you can inspect both the shape and the index of the data, as follows:

```
print(noisy_series.shape)
print(noisy_series.index)
```

You will see the following result:

```
(3330,)
RangeIndex(start=0, stop=3330, step=1)
```

As expected, the shape, `(3330, )`, matches the plot, as does the index.

Recall earlier when you created a DataFrame with the constructor approach, and used the `range()` method to fill it with data. Try to do that again, but this time, using just one range and the `Series` constructor. The following code will create a Series named `my_series` using `range(26)`. Then, the `head()` method is used to list some of the values:

```
my_series = pd.Series(range(26))
my_series.head()
```

This produces the following output:

```
Out[26]: 0    0
         1    1
         2    2
         3    3
         4    4
         dtype: int64
```

Figure 2.28 – The first five elements of your my_series Series

Now, as you did with DataFrames, try to generate some letters. You will again use `map` and pass the `chr` function with a range to generate all the lowercase letters of the alphabet. Then, the `.head()` and `.tail()` methods will help verify that you obtained all the letters:

```
letters = pd.Series(map(chr, range(97, 123)))
print(letters.head(), '\n', letters.tail())
```

Running this produces the following output:

```
0    a
1    b
2    c
3    d
4    e
dtype: object
 21    v
22    w
23    x
24    y
25    z
dtype: object
```

Figure 2.29 – A series of letters to be used as an index

Now, suppose you want to replace the index of `my_series` with `letters`. Here, it is important to note one difference between handling the index for Series and for DataFrames. In the case of DataFrames, the `.set_index()` method was used to replace the index. However, Series do not have a `.set_index()` method; instead, you can assign new values directly to the existing index. In the following snippet, you assign `letters` to `my_series.index` and inspect the results using bracket notation:

```
my_series.index = letters
my_series[13:27]
```

You should see the following output:

```
Out[33]: n     13
         o     14
         p     15
         q     16
         r     17
         s     18
         t     19
         u     20
         v     21
         w     22
         x     23
         y     24
         z     25
         dtype: int64
```

Figure 2.30 – my_series with the index replaced with letters

Now, let's explore what can be stored in a series, by way of some examples. Here, you will do something different from what you've done so far—you will store multiple object types in a single series, including objects with multiple values, such as lists and ranges. In the next code, the `Series` constructor is called and passed an integer, a string, another string, a list of integers, and a range, in that order:

```
my_mixed_series = pd.Series([1, 'cat', 'yesterday',\
                             [1, 2, 3], range(5)])
```

The following code lists out the Series just created:

```
my_mixed_series
```

You should see an output as follows:

```
Out[35]: 0                  1
         1                cat
         2          yesterday
         3          [1, 2, 3]
         4    (0, 1, 2, 3, 4)
         dtype: object
```

Figure 2.31 – A Series with multiple data types

You can see from the output that it preserved all the objects in the appropriate locations. This example illustrates the point that pandas does not remove any of the flexibility of storing data in Python; it only adds a lot of convenience.

Next, you will explicitly look at the data types of each of the elements in the Series to verify they are just as they were passed to the constructor. The following code loops over the index of the Series, and prints out each value in the Series along with the type. Note the use of the index as the loop iterable. As with DataFrames, the Series index is an iterable, so you can iterate over it using a `for` loop:

```
for i in my_mixed_series.index:
    print(my_mixed_series[i],' is type ',type(my_mixed_
series[i]))
```

Running this snippet results in the following output:

```
1  is type  <class 'int'>
cat  is type  <class 'str'>
yesterday  is type  <class 'str'>
[1, 2, 3]  is type  <class 'list'>
range(0, 5)  is type  <class 'range'>
```

Figure 2.32 – The different types of data stored in my_mixed_series

You might notice that the Series itself shows a data type (`dtype`) of `object`. This is expected with a single Series or column of *mixed* types; since the Series is made up of multiple types, pandas reports it as the `object` type as a whole, while the individual elements retain their respective types.

> **Note**
>
> As of pandas version 1.2.3, when reading data from a file, pandas will also report a Series or a DataFrame column that is the string (`str`) type as `object`. This behavior could change in the future.

You are now equipped to do a couple of exercises illustrating what has been covered so far.

Exercise 2.03 – Series to DataFrames

In this exercise, you will read data from a CSV file into a Series object. The dataset to be used contains noisy data of some sensor measurements. These measurements appear to have a periodic behavior, for which you may eventually want to construct a model to predict future measurements. You are told that there is a natural period in the data—that is, after 92 instances, the data seems to show a repeating pattern. You will make a second Series that captures this periodic behavior by operating (using a `sine` function) on the index, then combine the two Series into a DataFrame. After creating the DataFrame, you will save it into a new CSV file.

> **Note**
>
> You can find the code for this exercise here: `https://github.com/PacktWorkshops/The-Pandas-Workshop/tree/master/Chapter02/Exercise2_03`.

Perform the following steps to complete the exercise:

1. Open a new Jupyter notebook and select the **Pandas_Workshop** kernel.
2. For this exercise, all you will need is the `pandas` library. Load it in the first cell of the notebook:

```
import pandas as pd
```

Read the `test_series.csv` file from the `Datasets` subdirectory (available at `https://github.com/PacktWorkshops/The-Pandas-Workshop/blob/master/Chapter02/Datasets/test_series.csv`). Here, you will store the path in the `fname` variable, then provide that to the `pd.read_csv()` method. pandas defaults to loading the result of `.read_csv()` to a DataFrame. Since you want this data to be a Series, supply the `squeeze = True` option, which tells pandas to put the data in a Series, if possible. Then, list out the result:

```
fname = '../Datasets/test_series.csv'
my_series = pd.read_csv(fname, squeeze = True)
my_series
```

> **Note**
>
> Please change the path of the dataset file (highlighted) based on where you have downloaded it in your system.

The output should look like the following:

```
Out[2]: 0      -0.161666
        1      -0.487261
        2      -0.392687
        3      -0.489594
        4      -0.334385
                  ...
        360    -0.386292
        361    -0.311027
        362    -0.437957
        363    -0.569164
        364    -0.727658
        Name: data, Length: 365, dtype: float64
```

Figure 2.33 – The result of reading my_series from the test_series.csv file

3. To be sure that you have a `Series` object, inspect the type:

```
type(my_series)
```

This produces the following output, confirming you have stored the data in a Series:

```
pandas.core.series.Series
```

A few things to note so far. If you open the CSV file directly (for example, in a text editor), it looks like this:

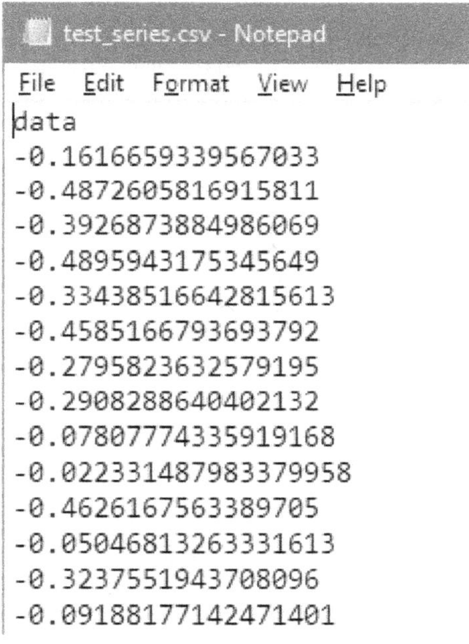

Figure 2.34 – Top rows of test_series.csv

You can see that the first row has a string, `data`. Note that pandas automatically reads that and stores it as the name of the data in the Series. The remainder of the rows are real numbers.

> **Note**
>
> By default, pandas tries to *infer* names (the name of the Series, or of one or more columns for a DataFrame) from the data in the file, so it correctly finds the `data` name for the values in the Series.

4. As you know, the sensor measurements have a period of `92`. You want to build a feature that would let you predict the sensor measurements later. As a first step, construct a second Series that contains the `sine` function of the index (the index is equivalent to the time points in this case). To make a `sine` function with a period of `92`, the general formula is `sine(2 * pi * time / 92)`, which you can apply using the NumPy `sin()` method and the `pi` NumPy constant. Note that you need to import NumPy, as you have not yet done so. After creating the Series, list it out, as follows:

```
import numpy as np
new_series = (pd.Series(np.sin((my_series.index) * 2 *
np.pi / 92)))
new_series
```

You will see the following output:

```
Out[3]: 0        0.000000
        1        0.068242
        2        0.136167
        3        0.203456
        4        0.269797
                   ...
        360     -0.519584
        361     -0.460065
        362     -0.398401
        363     -0.334880
        364     -0.269797
        Length: 365, dtype: float64
```

Figure 2.35 – new_series created from my_series

5. Construct a DataFrame named `model_data` using the DataFrame constructor syntax, with `my_series` as `col1`, and `new_series` as `col2`. Inspect the new DataFrame with the `head()` method:

```
model_data = pd.DataFrame({'col1' : my_series,\
                           'col2' : new_series})
model_data.head()
```

You will get the output as follows:

Out[4]:

	col1	col2
0	-0.161666	0.000000
1	-0.487261	0.068242
2	-0.392687	0.136167
3	-0.489594	0.203456
4	-0.334385	0.269797

Figure 2.36 – First five rows of the new model_data DataFrame

6. Save the new DataFrame to a new CSV file using the `to_csv()` method. Use the `index = None` option, because, by default, pandas would write the index to the first column. In this case, you don't need the index stored with the data, since it is just integers. Note that in the `.to_csv()` method, you specify the path and filename you want:

```
model_data.to_csv('../Datasets/model_data.csv', index = None)
```

> **Note**
> Please change the path of the dataset file (highlighted) based on where you have downloaded it in your system.

By completing this exercise, you have implemented the `read_csv()` and `to_csv()` methods and combined two Series into a new DataFrame. The next section looks at another aspect of pandas data structures, which is the ability to index by time values.

Using time as the index

As you will explore in detail in *Chapter 9*, *Time Series*, pandas makes use of objects of the `timestamp` type as the index, to work with data that is ordered by time. In this section, you will learn how to convert a date stored as a string to the `timestamp` type, and how to use that as the index.

You will start by creating a simple DataFrame that comprises some dates as strings, along with some numeric data. You will use the `Series` constructor, the `DataFrame` constructor, and the Python `range()` method. First, you need to create a series with six consecutive dates in the *YYYY-MM-DD* string format, and then another series of six integers using `range()`. Then, you need to combine them into a DataFrame with the `date` and `data` column names. Finally, the result should be listed out, as follows:

```
dates = pd.Series(['2017-01-01', '2017-01-02', '2017-01-03',
                   '2017-01-04', '2017-01-05', '2017-01-06'])
time_series = pd.Series(range(6))
time_series_df = pd.DataFrame({'date' : dates,\
                               'data' : time_series})
time_series_df
```

This produces the following output:

Out[19]:

	date	data
0	2017-01-01	0
1	2017-01-02	1
2	2017-01-03	2
3	2017-01-04	3
4	2017-01-05	4
5	2017-01-06	5

Figure 2.37 – The time_series_df DataFrame

Now, using the `type()` method, inspect the first element of the `date` column to determine its type. You can use the bracket notation to select the `date` column and the `[0]` element:

```
type(time_series_df['date'][0])
```

This shows that the data in the column is of the `str` type:

```
str
```

pandas provides the `.to_datetime()` method to convert strings to timestamps. You will use this method to convert the `date` column to a timestamp, then inspect the first date element with `type()` as done before:

```
time_series_df['date'] = pd.to_datetime(time_series_df['date'])
type(time_series_df['date'][0])
```

This produces the expected result:

```
pandas._libs.tslibs.timestamps.Timestamp
```

Now, again, you will replace the index, in this case with the `date` column. For this, you can use the `set_index()` method, telling pandas to perform the operation directly on the existing DataFrame (with `inplace = True`). After making this change, the DataFrame should be listed out again, as follows:

```
time_series_df.set_index('date', inplace = True)
time_series_df
```

This produces the following output:

Out[31]:

date	data
2017-01-01	0
2017-01-02	1
2017-01-03	2
2017-01-04	3
2017-01-05	4
2017-01-06	5

Figure 2.38 – The updated time_series_df, now with the timestamps as the index instead of a column

Now, you can learn about one of pandas' time series methods, the `resample` method. pandas' `resample` takes an argument that is the new time step you want for the data. The method is called *resample* because it can either *upsample* or *downsample* the data. **Upsampling** refers to the case where you have a Series, and you decrease the spacing (in this case, the time) between the points, thereby increasing the number of total data points. Similarly, **downsampling** refers to increasing the spacing, thereby decreasing the number of total points. pandas uses the index and the first argument of `resample` to determine the new timestamps.

To adjust the data, pandas requires the addition of an aggregation function after the `resample` operation. Aggregation functions include functions such as `mean`. In this next example, you will use the `interpolate` method, which linearly interpolates between the existing points. While this may sound complicated, it is actually a very common pattern in data manipulation.

Let's say you want to resample some data that is currently on a spacing of 1 day to 12 hours. pandas' `resample` can take intuitive strings as the argument, so you just pass `12h` to it, followed by `interpolate`. You can then list out the results to see the change, as follows:

```
time_series_df = time_series_df.resample('12h').interpolate()
time_series_df
```

This produces the following output:

Out[25]:

date	data
2017-01-01 00:00:00	0.0
2017-01-01 12:00:00	0.5
2017-01-02 00:00:00	1.0
2017-01-02 12:00:00	1.5
2017-01-03 00:00:00	2.0
2017-01-03 12:00:00	2.5
2017-01-04 00:00:00	3.0
2017-01-04 12:00:00	3.5
2017-01-05 00:00:00	4.0
2017-01-05 12:00:00	4.5
2017-01-06 00:00:00	5.0

Figure 2.39 – The resampled index on 12-hour intervals with the interpolated data

You can see how powerful pandas can be, combining the availability of the index along with some methods designed to leverage that. You are now well prepared for an exercise to work with the index.

Exercise 2.04 – DataFrame indices

In this exercise, you will use the `read_csv()` method to read data directly into a DataFrame. The DataFrame, as read from the drive, will have a `time` column and several other columns with the column names being dates. The data is some economic data for the year 2019, with rows representing the data every 15 minutes on the last day of each month, and the columns representing the month. You will convert the times or dates into an index and use the index to resample the data to a smaller time step. Finally, you will finish this exercise by saving the updated data into a new file using the `to_csv()` function.

> **Note**
>
> You can find the code for this exercise here: `https://github.com/PacktWorkshops/The-Pandas-Workshop/tree/master/Chapter02/Exercise2_04`.

Perform the following steps to complete the exercise:

1. Open a new Jupyter notebook and select the **Pandas_Workshop** kernel.
2. For this exercise, all you will need is the `pandas` library. Load it in the first cell of the notebook:

```
import pandas as pd
```

3. Load a DataFrame from a CSV file, using the `test_data_frame.csv` file, and name it `economic_data`:

```
fname = '../Datasets/test_data_frame.csv'
economic_data = pd.read_csv(fname)
economic_data.head()
```

> **Note**
>
> Please change the path of the dataset file (highlighted) based on where you have downloaded it on your system. You can download the file from `The-Pandas-Workshop/test_data_frame.csv at master · PacktWorkshops/The-Pandas-Workshop · GitHub`.

The output should be as follows:

Out[28]:

	time	1/31/2019	2/28/2019	3/31/2019	4/30/2019	5/31/2019	6/30/2019	7/31/2019	8/31/2019	9/30/2019	10/31/2019	11/30/2019	12/31/2019
0	12:00:00 AM	2312.22	2403.93	2285.59	1841.71	1144.73	579.97	184.34	217.88	609.83	1098.53	1832.15	2409.02
1	12:15:00 AM	2357.01	2503.56	2319.69	1863.97	1183.33	511.77	225.56	158.63	531.24	1132.16	1797.98	2354.98
2	12:30:00 AM	2298.20	2475.26	2386.27	1875.62	1259.22	555.14	167.05	199.51	536.58	1126.26	1725.46	2336.46
3	12:45:00 AM	2359.41	2615.92	2368.70	1825.99	1139.68	525.37	117.55	149.68	482.08	1087.29	1816.17	2374.96
4	1:00:00 AM	2328.82	2565.09	2298.29	1802.28	1178.65	586.78	212.88	129.09	551.16	1145.26	1802.78	2318.55

Figure 2.40 – First five lines of economic_data

4. Now, inspect the data types of the DataFrame, using `.dtypes`:

```
economic_data.dtypes
```

The output should be as follows:

```
Out[29]: time            object
         1/31/2019      float64
         2/28/2019      float64
         3/31/2019      float64
         4/30/2019      float64
         5/31/2019      float64
         6/30/2019      float64
         7/31/2019      float64
         8/31/2019      float64
         9/30/2019      float64
         10/31/2019     float64
         11/30/2019     float64
         12/31/2019     float64
         dtype: object
```

Figure 2.41 – Data types of the economic_data DataFrame

5. Now, convert the `time` column into a pandas `timestamp` type, using the `to_datetime()` method. Since the values are in a particular string format, you will need to provide a string to the `format` option to specify the format to pandas. In this case, use `%I` to mean hours in a 12-hour format, `%M` to mean minutes in a numeric two-digit format, `%S` to mean seconds in a numeric two-digit format, and `%p` to mean A.M./P.M. The colons (`:`) tell pandas that the values are separated by colons in the strings, and similarly, the space between `%S` and `%p` represents the space before A.M./P.M. After converting to a timestamp, subtract the first time from every time value, so that the values start at `0` days and `0` hours and increase over the time of the day. List the DataFrame to confirm that the required changes have been made:

```
economic_data['time'] = pd.to_datetime(economic_
data['time'],\
                                       format = '%I:%M:%S
%p')
economic_data['time'] = economic_data['time']\
                      - economic_data['time'][0]
economic_data
```

The output should be as follows:

Out[22]:

	time	1/31/2019	2/28/2019	3/31/2019	4/30/2019	5/31/2019	6/30/2019	7/31/2019	8/31/2019	9/30/2019	10/31/2019	11/30/2019	12/31/2019
0	0 days 00:00:00	2312.22	2403.93	2285.59	1841.71	1144.73	579.97	184.34	217.88	609.83	1098.53	1832.15	2409.02
1	0 days 00:15:00	2357.01	2503.56	2319.69	1863.97	1183.33	511.77	225.56	158.63	531.24	1132.16	1797.98	2354.98
2	0 days 00:30:00	2298.20	2475.26	2386.27	1875.62	1259.22	555.14	167.05	199.51	536.58	1126.26	1725.46	2336.46
3	0 days 00:45:00	2359.41	2615.92	2368.70	1825.99	1139.68	525.37	117.55	149.68	482.08	1087.29	1816.17	2374.96
4	0 days 01:00:00	2328.82	2565.09	2298.29	1802.28	1178.65	586.78	212.88	129.09	551.16	1145.26	1802.78	2318.55
...	...	...	...	...	...	...	...	...	...	...	...	...	...
91	0 days 22:45:00	2347.70	2549.58	2351.71	1850.14	1064.03	534.39	208.35	176.65	580.45	1100.67	1821.87	2263.76
92	0 days 23:00:00	2234.47	2570.41	2296.87	1778.81	1180.03	584.29	108.65	243.64	477.31	1214.94	1816.56	2231.82
93	0 days 23:15:00	2302.04	2469.22	2273.06	1865.61	1146.12	535.60	112.78	137.46	554.88	1131.31	1894.77	2360.27
94	0 days 23:30:00	2276.66	2401.19	2326.91	1801.10	1125.49	535.32	156.38	242.52	585.82	1121.86	1786.20	2293.01
95	0 days 23:45:00	2338.81	2521.09	2317.67	1825.98	1164.15	561.30	177.96	106.36	574.28	1116.48	1834.12	2305.17

Figure 2.42 – The time column is now converted to a timestamp

6. You can now replace the index with the `timestamp` values. Use `set_index` and provide the column name (`time`) and tell pandas to drop the column (`drop = True`). Use the `head` method to verify the result, as follows:

```
economic_data.set_index('time', inplace = True)
economic_data.head()
```

The output should be as follows:

Out[23]:

	1/31/2019	2/28/2019	3/31/2019	4/30/2019	5/31/2019	6/30/2019	7/31/2019	8/31/2019	9/30/2019	10/31/2019	11/30/2019	12/31/2019
time												
0 days 00:00:00	2312.22	2403.93	2285.59	1841.71	1144.73	579.97	184.34	217.88	609.83	1098.53	1832.15	2409.02
0 days 00:15:00	2357.01	2503.56	2319.69	1863.97	1183.33	511.77	225.56	158.63	531.24	1132.16	1797.98	2354.98
0 days 00:30:00	2298.20	2475.26	2386.27	1875.62	1259.22	555.14	167.05	199.51	536.58	1126.26	1725.46	2336.46
0 days 00:45:00	2359.41	2615.92	2368.70	1825.99	1139.68	525.37	117.55	149.68	482.08	1087.29	1816.17	2374.96
0 days 01:00:00	2328.82	2565.09	2298.29	1802.28	1178.65	586.78	212.88	129.09	551.16	1145.26	1802.78	2318.55

Figure 2.43 – economic_data with the index converted to timestamps

7. You can now use the time-based index with other pandas methods. For this, use the `resample` and `interpolate` methods. pandas' `resample` understands intuitive strings for the time step, so here you use `5min`. Use `head` to verify the result, as follows:

```
economic_data = economic_data.resample('5min').
interpolate()
economic_data.head()
```

The output should be as follows:

Out[6]:

time	1/31/2019	2/28/2019	3/31/2019	4/30/2019	5/31/2019	6/30/2019	7/31/2019	8/31/2019	9/30/2019	10/31/2019	11/30/2019	12/31
00:00:00	2312.220000	2403.930000	2285.590000	1841.710000	1144.730000	579.970000	184.340000	217.880000	609.830000	1098.530000	1832.150000	2409.0
00:05:00	2327.150000	2437.140000	2296.956667	1849.130000	1157.596667	557.236667	198.080000	198.130000	583.633333	1109.740000	1820.760000	2391.0
00:10:00	2342.080000	2470.350000	2308.323333	1856.550000	1170.463333	534.503333	211.820000	178.380000	557.436667	1120.950000	1809.370000	2372.9
00:15:00	2357.010000	2503.560000	2319.690000	1863.970000	1183.330000	511.770000	225.560000	158.630000	531.240000	1132.160000	1797.980000	2354.9
00:20:00	2337.406667	2494.126667	2341.883333	1867.853333	1208.626667	526.226667	206.056667	172.256667	533.020000	1130.193333	1773.806667	2348.8

Figure 2.44 – economic_data resampled to a 5-minute period from 15 minutes

8. Now, use `to_csv()` to save the data to a new file, `economic_data.csv`:

```
economic_data.to_csv('../Datasets/economic_data.csv')
```

In this exercise, you have used `read_csv()` and `to_csv()` to read and store data from a drive. You have seen how data in string format can be converted to timestamps, and how to use it as an index that then supports advanced methods such as `resample`. This pattern is very common in data analysis where time is involved. Now, you are ready to test your new knowledge on some US GDP data.

Activity 2.01 – Working with pandas data structures

In this activity, you will read a DataFrame from the `US_GDP.csv` file, which contains information about the GDP of the US, from the first financial quarter of 2017 to the last financial quarter of 2019. The data is stored in two columns, `date` and `GDP`, and the date is read in (by default) as the `object` type. The goal of this activity is to first convert the `date` column into a timestamp and then set this column as the index. Finally, you'll save the updated dataset to a new file:

> **Note**
>
> You can download the file from `https://github.com/PacktWorkshops/The-Pandas-Workshop/blob/master/Chapter02/Datasets/US_GDP.csv`

1. Import the `pandas` library.
2. Read the `US_GDP.csv` file from the `Datasets` directory into a DataFrame named `GDP_data`. The data is stored as dates and values, and you wish to use the dates as the index, so that in future work you may apply pandas time series methods to this data.

3. Display the head of `GDP_data` so that you can see the formats of the data in the file.
4. Inspect the object types of `GDP_data`, in particular the `date` column.
5. Use the `pd.to_datetime()` method to convert the `date` column into a timestamp.
6. Use the `.set_index()` method to replace the index with the `date` column. Be sure to use `inplace = True` so the result is applied to the existing DataFrame, and `drop = True` to remove the `date` column after it is used for the index. Use `.head()` to confirm the result. The output should be as follows:

Out[4]:

date	GDP
2017-03-31	19190.4
2017-06-30	19356.6
2017-09-30	19611.7
2017-12-31	19918.9
2018-03-31	20163.2

Figure 2.45 – The GDP_data DataFrame after using the date for the index

7. Use the `.to_csv()` method to save the file to a new `.csv` file named `US_GDP_date_index.csv`.

> **Note**
> The solution for this activity can be found in the *Appendix*.

Summary

In this chapter, you were introduced to the two fundamental pandas data structures, DataFrames and Series, along with the basic concepts of the pandas index. With the help of some basic I/O functions, such as `read_csv()` and `to_csv()`, you saw how pandas makes it easy to read from, or write data directly into, DataFrames and Series. To illustrate the ideas, a few pandas methods were introduced in the chapter. You also learned about methods such as `set_index()` and the use of `timestamp` as an index, and used `resample()`, a pandas time series method that can change the time interval of data, as well as `concat()`, which is used to combine pandas data structures into other structures.

By now, you should be comfortable with the concept of a DataFrame and Series. The rest of the chapters in this book will build upon these concepts. In the next chapter, you will learn about data I/O using pandas.

3
Data I/O

In the previous chapter, you learned how pandas stores information in Series and DataFrames, and were introduced to the `.read_csv()` and `.to_csv()` methods, which are used to read data from a storage drive into DataFrames or Series and save data in a CSV file. In this chapter, you will learn about a range of other data sources, including different file formats and data from web pages and **application programming interfaces** (**APIs**). By the end of this chapter, you will be comfortable with getting data into pandas from a wide range of common sources, enabling you to work with different teams in your organization.

In this chapter, we will cover the following topics:

- The world of data
- Exploring data sources
- Fundamental formats
- Additional text formats
- Manipulating SQL data
- Activity 3.01 – using SQL data for pandas analytics

The world of data

In today's digitally driven world, data is being generated at an ever-increasing pace. The *World Economic Forum* reports that, by 2025, 463 exabytes of data will be created each day globally. An exabyte is a 1 followed by 18 zeros. That is just a little less than 5,359 TB per second, or 5.3 million GB per second. Unsurprisingly, not all of this data is in the form of simple text files. While CSV files are very common and highly useful, they are just one of the many possible data formats we may want to work with while using pandas. In this chapter, we will explore more options to bring data into pandas DataFrames and Series, and store data back in memory. Such data operations are called **input/output** or **I/O** operations. For example, data from your finance team might be from software such as SAS or Stata. Furthermore, if you work with "big data," you may need to access Parquet or HDF data. Depending on your business requirements and the complexity of the task, the data formats you'll need to use will vary. Sometimes, you'll be limited to the formats the business chooses to use due to its policies or cost constraints, and in other cases where such constraints are absent, you'll need to be wise in picking the most efficient format. In either of these cases, pandas' ability to seamlessly work with numerous formats will prove invaluable.

For example, imagine that you've been tasked with setting up a database where your company will stream customer order data into the database continuously. As an additional requirement, you need to run data science analyses on the data in Python, reading portions of the data into pandas at any given time. Traditionally, the company has used CSV format and the data is expected to be about 1 TB in size as a `.csv` file since it's been collected over the last 6 months. Rather than store it in `.csv` format, you work with the Data Engineering team to store the data in Parquet format, and instead of 1 TB, now you only need about 130 GB of storage space. The beauty of the pandas library is that you will be able to run the required analyses on the data in the new format without any difference in functionality.

In cloud computing, storage, and data streaming, other binary formats often come into play. Examples include proprietary software formats such as SAS or Stata and big data open formats such as HDF5. HDF5, for example, stores "metadata" with the data; thus, this type of data is "self-describing." It is designed for fast I/O operations regarding big data and heterogeneous data structures such as those with multiple related tables. Regarding the latter point, you've probably heard of SQL (often pronounced "sequel"), which is a relational database format that's used extensively in the industry. Python supports a version of SQL directly called `sqlite`.

When it comes to data involving the web or from data service APIs, the most commonly used formats are JSON, XML, and HTML. Additionally, the ubiquity of spreadsheets means that we often need to read data into pandas from Excel or similar formats, where the data is binary – binary formats are usually more compact in that they require less memory and provide a richer structure. As a result, the relationships among the different data formats can also be easily stored.

The following table shows the data types, file/system types, and the corresponding pandas methods for reading and writing to them. In this chapter, we will cover the bold items in the table as they are commonly encountered. Once you have hands-on experience with them, you will be prepared to use other formats as the need arises.

The right-hand column of the table, labeled **Dependencies**, refers to some additional requirements:

Type of Data	File / system	Input	Output	dependencies
text	**CSV**	**read_csv**	**to_csv**	
text	**JSON**	**read_json**	**to_json**	
text	**HTML**	**read_html**	**to_html**	**lxml or bs4/html5lib**
text	**XML**			**pandas-read-xml**
text	Local clipboard	read_clipboard	to_clipboard	
text	Fixed-Width Text File	read_fwf		
binary	**Matlab / Octave**			**scipy.io**
binary	**Excel**	**read_excel**	**to_excel**	**xlrd or openpyxl**
binary	**HDF5**	**read_hdf**	**to_hdf**	**zlib, lzo, etc.**
binary	**Stata**	**read_stata**	**to_stata**	**pyreadstat**
binary	**SAS**	**read_sas**		
binary	OpenDocument	read_excel		
binary	Feather	read_feather	to_feather	
binary	Parquet	read_parquet	to_parquet	
binary	ORC	read_orc		
binary	Msgpack	read_msgpack	to_msgpack	
binary	**SPSS**	**read_spss**		
binary	Pickle	read_pickle	to_pickle	
SQL	**SQL**	**read_sql**	**to_sql**	**sqlite3**
SQL	**BigQuery (Google)**	read_gbq	to_gbq	**pandas-gbq** **google-cloud-bigquery**

Figure 3.1 – pandas I/O methods

> **Note**
>
> There may be other options/choices available regarding the dependencies listed in the preceding table, and you can refer to the pandas documentation for more details. In particular, to convert to and from various formats, pandas relies on "engines," which are installed separately from pandas, and in some cases, "compression libraries," which are also separate from pandas. In most cases, they only have to be installed in your environment; you don't need to import something into your code. However, depending on other modules and libraries you may have installed, some of them may be missing, causing an error message. Those messages will inform you of what additional components you need to install in your environment.

In the case of `pyreadstat` and `pandas-gbq`, for example, those libraries need to be installed in your environment separately (using `pip` or `conda`) for the pandas methods to function. In the case of `sqlite3`, it needs to be imported in Python for you to use it, even though it is part of the standard Python distribution. You'll see examples of some of these later in this chapter.

Several of the formats listed in the preceding table may be new to you, and if you are faced with something new, it may feel like it does not appear in the pandas I/O documentation. A common example is Matlab (or the open source version, Octave), which is used a lot in the industry and academia. It can be seen in the preceding table but it doesn't have a pandas method listed for it. In cases where a format is not built into pandas, it's a good idea to do a quick internet search to see if there is a Python package available to read those files. For Matlab, such a search produces various links, many of which teach you how to use the `.loadmat()` function from `scipy.io`. For example, suppose you are helping a colleague from the engineering department who is new to pandas. They bring you a USB drive that contains Matlab files they generated in a product test. The data is from a laboratory data collection system that uses Matlab to collect and store data and is stored in a file called `matlab.mat`. Here, you can use the `scipy` method (contained in the `scipy.io` module) to read the data in, as follows:

```
import scipy.io
mat = scipy.io.loadmat('datasets/matlab.mat')
mat
```

> **Note**
>
> Please change the path of the dataset file (highlighted) based on where you have downloaded it on your system.

This produces the following output:

```
Out[18]: {'__header__': b'MATLAB 5.0 MAT-file Platform: nt, Created on: Tue Feb  2 14:21:02 2021',
          '__version__': '1.0',
          '__globals__': [],
          'storage': array([[0.00000000e+00],
                 [3.60020368e-04],
                 [7.26299303e-04],
                 ...,
                 [1.36616373e-05],
                 [1.35810556e-05],
                 [1.36134929e-05]]),
          'T1': array([[475.5],
                 [475.5],
                 [475.4],
                 ...,
                 [476.8],
                 [476.8],
                 [476.8]]),
          'time': array([[10256548.8],
                 [10256549. ],
                 [10256549.2],
                 ...,
                 [10273672.4],
                 [10273672.6],
                 [10273672.8]]),
          'value': array([[10256548.8       ],
                 [10256550.09106825],
                 [10256550.31226313],
                 ...,
                 [10273670.63541315],
                 [10273672.1572869 ],
                 [10273672.87393071]])}
```

Figure 3.2 – Reading a Matlab data file into Python using the scipy.io loadmat() method

The result is `dictionary`, which we could process into a DataFrame or some other useful form, as we saw in *Chapter 2, Data Structures*.

> **Note**
>
> All the examples for this chapter can be found in the `examples.ipynb` notebook in the `Chapter03` folder of this book's GitHub repository, while the data files can be found in the `datafiles` folder. To ensure the examples run correctly, you need to run the notebook from start to finish in order.

Exploring data sources

Data can be obtained from a variety of sources, such as files on your computer, files on your company network, files in the cloud (such as Amazon AWS S3 storage), and web sources. You saw CSV files that contain data as text in the previous chapter. Now, let's consider different types of data that may appear in files.

Text files and binary files

You are already familiar with text files. A simple definition, albeit a bit circular, is that if you can open, read, and understand data in a text editor (such as Notepad on Windows, Notepad++, or other similar applications), you are dealing with text data. For example, in *Chapter 2, Data Structures*, you worked with small files that contained sales records of pet foods. If you open `dog_food_orders.csv` (located in the `Chapter02/Datasets` folder) in a text editor (here, Notepad in Windows), you will see the following:

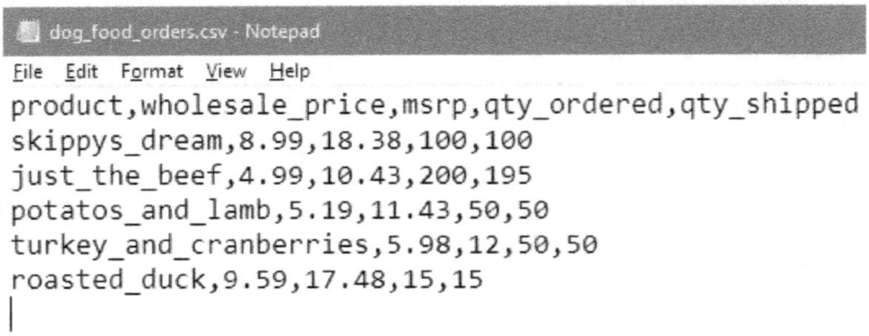

dog_food_orders.csv - Notepad

File Edit Format View Help

```
product,wholesale_price,msrp,qty_ordered,qty_shipped
skippys_dream,8.99,18.38,100,100
just_the_beef,4.99,10.43,200,195
potatos_and_lamb,5.19,11.43,50,50
turkey_and_cranberries,5.98,12,50,50
roasted_duck,9.59,17.48,15,15
```

Figure 3.3 – The dog_food_orders.csv file as a text file

Here, you can see all the values. They are separated by commas, which makes sense as the file is a CSV file. You'll learn about some other common separators later in this chapter.

Now, suppose you saved the same data in Excel, then viewed it again in Notepad. You would see something like this:

dog_food_orders.xlsx - Notepad

File Edit Format View Help

PK ! bîh^ [Content_Types].xml ¢ (

¬"ËNÃ0E÷HüCä-JÜ²@5í,Ç*Q>ÀÄ"Æªc[ž
˜q·ÃR4DáAJ¬hÚ>€ãÚÇVßÆ¹ªZ¨9ÈÛÁàNVÞ8Ê©ÓãÑÔji){^óã-I<"{Üv^¥P!XS)b
Pk<´-2nÏ}Ä?£LËðÂ Ýû%áÄßdºždN"m,à¥ÇžDO97*,~§ÈÉ¸8ÀOíc|n¦ÑäEøÿöéº
_rels/.rels ¢ (

¬'MOÃ0†ïHü‡È÷ÕÝBKwAH»!T~€IÜµ£$Ý¿'TƒG¼~üÊÛÝ<êÈ!öâ4¬<;#¶w-†-
†°·7 ê"Ï>x-¦é
?ˆ9LìÒ™Èsbgù®|Èl!õùUSh9i°bžr:"y_dlÀóD>¿ý|-NœÈR"4ø2ÏGÇ% õZ´4ñËyÄ7
xl/_rels/workbook.xml.rels ¢(

¬RMKÄ0¼þ‡Øw›vÙt/"iUëÉ´)Û&!3~õß*º]XÖK/
£¦"Dô¹Ò†4jÎ0u2jsÐÊMYÞË´ä€ú„Sì-,´·· š)fåÿ¹CÛöŸ,yÑó I<

Figure 3.4 – The dog_food_orders data, saved in .xlsx format and reopened in Notepad

Although you can see hints of legible text, this does not look intelligible at all. We call such data files binary data. The term "binary" comes from the 0s and 1s associated with raw computer data – data that is intended to be interpreted by machines, but not directly by humans. Although you can't read this file, pandas can read it, along with many of the other formats that were listed in *Figure 3.1*. In many cases, if you are given data in a binary format, you should know that it came from some other software, such as Excel, or a data software program such as SPSS or SAS. If you don't have that software, pandas lets you read many of these files directly. In the following snippet, the `.read_excel()` method is being used to read the `.xlsx` file version of `dog_food_orders`:

```
import pandas as pd
dog_food_orders = \
    pd.read_excel('datasets/dog_food_orders.xlsx', engine =
'openpyxl')
dog_food_orders
```

> **Note**
>
> Please change the path of the dataset file (highlighted) based on where you have downloaded it on your system.

The preceding code produces the following output:

Out[19]:

	product	wholesale_price	msrp	qty_ordered	qty_shipped
0	skippys_dream	8.99	18.38	100	100
1	just_the_beef	4.99	10.43	200	195
2	potatos_and_lamb	5.19	11.43	50	50
3	turkey_and_cranberries	5.98	12.00	50	50
4	roasted_duck	9.59	17.48	15	15

Figure 3.5 – The dog food orders data, read directly from the .xlsx file by pandas

You obtained the desired result here, in part because pandas used the default values in the `.read_excel()` method that worked. We will explore these parameters in more detail in the Excel section.

Online data sources

Data exists online in many forms. As an example, a web page may contain a table, like the Wikipedia *Wind power* page does (`https://en.wikipedia.org/wiki/Wind_power`).

If you scroll down the page, you will see the following table:

Large onshore wind farms

Wind farm	Capacity (MW)	Country	Refs
Gansu Wind Farm	7,965	China	[18][19]
Muppandal wind farm	1,500	India	[20]
Alta (Oak Creek-Mojave)	1,320	United States	[21]
Jaisalmer Wind Park	1,064	India	[22]
Shepherds Flat Wind Farm	845	United States	[23]
Roscoe Wind Farm	782	United States	
Horse Hollow Wind Energy Center	736	United States	[24][25]
Capricorn Ridge Wind Farm	662	United States	[24][25]
Fântânele-Cogealac Wind Farm	600	Romania	[26]
Fowler Ridge Wind Farm	600	United States	[27]
Whitelee Wind Farm	539	United Kingdom	[28]

Figure 3.6 – A Wikipedia page with a table of data we would like to get into pandas

In addition to the table, there is a lot of other information on the page. There are a wide variety of modules and packages available for you to obtain data from web pages (often called **scraping**), but pandas provides the `read_html()` method, which attempts to collect groups of things together into a list of Python objects, including pandas DataFrames. In the following code, we are using `read_html()` to get the data and take a quick look at it:

```
import pandas as pd
data_url = 'https://en.wikipedia.org/wiki/Wind_power'
data = pd.read_html(data_url)
data
```

Running this code produces the following output. Note that the output has been truncated for brevity:

```
Out[6]: [                                                     0
         0                                    Part of a series about
         1                                        Sustainable energy
         2                                                  Overview
         3                      Carbon-neutral fuel Fossil fuel phase-out
         4                                       Energy conservation
         5    Cogeneration Efficient energy use Energy stora...
         6                                          Renewable energy
         7    Hydroelectricity Solar Wind Bioenergy Geotherm...
         8                                     Sustainable transport
         9        Electric vehicle Green vehicle Plug-in hybrid
         10          Renewable energy portal  Environment portal
         11   .mw-parser-output .navbar{display:inline;font-...,
                                    Wind farm  Capacity(MW)         Country      Refs
         0                        Gansu Wind Farm          7965           China  [18][19]
         1                     Muppandal wind farm          1500           India      [20]
         2                  Alta (Oak Creek-Mojave)         1320   United States      [21]
         3                     Jaisalmer Wind Park          1064           India      [22]
         4                  Shepherds Flat Wind Farm          845   United States      [23]
         5                       Roscoe Wind Farm          782   United States       NaN
         6    Horse Hollow Wind Energy Center          736   United States  [24][25]
         7             Capricorn Ridge Wind Farm          662   United States  [24][25]
         8        Fântânele-Cogealac Wind Farm          600         Romania      [26]
         9                 Fowler Ridge Wind Farm          600   United States      [27]
         10                   Whitelee Wind Farm          539  United Kingdom     [28],
```

Figure 3.7 – The result of using pandas .read_html() on a web page

This is a bit messy, but the first thing to note is that pandas returns all the information in a list. The first item in the list appears to be information about the article. Then, there is a comma on line 11, followed by the data you want from the table. There are more tables on the page, and pandas tries to load everything it can. So, you'll need to extract just the second element from the list, which you can do as follows, selecting element `[1]` (the second item) in the list:

```
data[1]
```

> **Note**
>
> Since Wikipedia pages are constantly updated, the outputs you get for the commands in this section may differ slightly from ours. As a consequence, you may need to modify the preceding command slightly based on the output you got in *Figure 3.7*.

This produces the following output:

```
Out[9]:
```

	Wind farm	Capacity(MW)	Country	Refs
0	Gansu Wind Farm	7965	China	[18][19]
1	Muppandal wind farm	1500	India	[20]
2	Alta (Oak Creek-Mojave)	1320	United States	[21]
3	Jaisalmer Wind Park	1064	India	[22]
4	Shepherds Flat Wind Farm	845	United States	[23]
5	Roscoe Wind Farm	782	United States	NaN
6	Horse Hollow Wind Energy Center	736	United States	[24][25]
7	Capricorn Ridge Wind Farm	662	United States	[24][25]
8	Fântânele-Cogealac Wind Farm	600	Romania	[26]
9	Fowler Ridge Wind Farm	600	United States	[27]
10	Whitelee Wind Farm	539	United Kingdom	[28]

Figure 3.8 – The first item in the list produced by pd.read_html() on the web page

That was easy and demonstrates the power of using pandas for a wide range of data I/O needs.

Exercise 3.01 – reading data from web pages

You are working on a project where you're studying the market for renewable energy sources. After obtaining the wind power data (from the previous example), you decide to gather similar data for solar power stations. You find a suitable web page on Wikipedia, *Solar power*, as the source (`https://en.wikipedia.org/wiki/Solar_power`). The goal of this exercise is to read this newly found data into a new pandas DataFrame.

> **Note**
> You can find the code for this exercise at `https://github.com/PacktWorkshops/The-Pandas-Workshop/tree/master/Chapter03/Exercise03_01`.

Follow these steps to complete this exercise:

1. Open a new Jupyter notebook and select the **Pandas_Workshop** kernel.
2. For this exercise, all you will need is the `pandas` library. Load it in the first cell of the notebook:

```
import pandas as pd
```

3. Read the web page into pandas using `pd.read_html`:

```
page_url = \
    ('https://en.wikipedia.org/w/index.php?' +
     'title=Solar_power&oldid=1022764142')
data = pd.read_html(page_url)
data
```

This produces the following output:

```
Out[3]: [                                                   0
         0                          Part of a series about
         1                              Sustainable energy
         ...
         10       Renewable energy portal  Environment portal
         11  .mw-parser-output .navbar{display:inline;font-...,
             Solar Electricity Generation                          \
                                     Year              Energy (TWh)
         0                           2004                       2.6
         1                           2005                       3.7
         2                           2006                       5.0
         ...
         12                         1.31%
         13                         1.73%
         14                         2.68%
         15  Sources:[32][33][34][35][36]  ,
                                       Name        Country  CapacityMWp  \
         0               Pavagada Solar Park          India         2050
         1         Tengger Desert Solar Park          China         1547
         2                 Bhadla Solar Park          India         1515
         3       Kurnool Ultra Mega Solar Park        India         1000
         4  Datong Solar Power Top Runner Base        China         1000
         5         Longyangxia Dam Solar Park         China          850
         6              Rewa Ultra Mega Solar         India          750
         7         Kamuthi Solar Power Project        India          648
         8              Solar Star (I and II)  United States          579
         9                  Topaz Solar Farm  United States          550

            GenerationGWh p.a.  Sizekm2  Year                  Ref
         0                 NaN       53  2017          [2][52][53]
         1                 NaN       43  2016             [54][55]
         2                 NaN       40  2017         [56][57][58]
         3                 NaN       24  2017                 [59]
         4                 NaN      NaN  2016         [60][61][62]
         5                 NaN       23  2015 [63][64][65][66][67]
         6                 NaN      NaN  2018                 [68]
         7                 NaN     10.1  2016             [69][70]
         8              1664.0       13  2015             [71][72]
         9              1301.0 24.6[73]  2014         [74][75][76]  ,
```

Figure 3.9 – Large PV solar power plants data

> **Note**
>
> Again, depending on whether the Wikipedia page has been updated after this book has been published, your output will be different and you may need to modify your code accordingly in the step that follows.

4. Scrolling through the output (truncated in *Figure 3.10* for brevity), you can see that the desired table is the third item in the list produced by `pd.read_html()`. Use a list slice to select only that element, and assign it to a new variable named `solar_PV_data`:

```
solar_PV_data = data[2]
solar_PV_data
```

The output should be as follows:

Out[7]:

	Name	Country	CapacityMWp	GenerationGWh p.a.	Sizekm2	Year	Ref
0	Pavagada Solar Park	India	2050	NaN	53	2017	[2][52][53]
1	Tengger Desert Solar Park	China	1547	NaN	43	2016	[54][55]
2	Bhadla Solar Park	India	1515	NaN	40	2017	[56][57][58]
3	Kurnool Ultra Mega Solar Park	India	1000	NaN	24	2017	[59]
4	Datong Solar Power Top Runner Base	China	1000	NaN	NaN	2016	[60][61][62]
5	Longyangxia Dam Solar Park	China	850	NaN	23	2015	[63][64][65][66][67]
6	Rewa Ultra Mega Solar	India	750	NaN	NaN	2018	[68]
7	Kamuthi Solar Power Project	India	648	NaN	10.1	2016	[69][70]
8	Solar Star (I and II)	United States	579	1664.0	13	2015	[71][72]
9	Topaz Solar Farm	United States	550	1301.0	24.6[73]	2014	[74][75][76]

Figure 3.10 – The solar_PV_data DataFrame

With that, you have successfully obtained the data you needed in just a few lines of code and can compare it to the wind power data in Python. With these basic examples in hand, let's work through some important formats in more detail.

Fundamental formats

We have already learned about the basics of text data and binary data. In this section, we'll look at these formats in a bit more detail and introduce some additional important data structures.

Text data

Earlier, we mentioned that, in general, text data can be viewed in a text editor. Text files can often be recognized by their file extensions; common ones include `.csv` (comma separated), `.txt` (plain text), `.sql` (SQL database script files), and others. Note that the extension is only a convention and does not guarantee the format of the contents. For example, it's not unusual to receive files with `.txt` extensions that are in `.csv` format.

However, there is an additional complexity that may arise, depending on how the data was created and stored. Text data may appear the same but be stored in different binary versions of each character. These binary representations are called encodings, and in most cases, you will find data encoded in UTF-8 format.

Encodings have names, such as ASCII or UTF-8, and each encoding essentially defines which character is matched to a numeric value. You may have seen **ASCII** before; it stands for **American Standard Code for Information Interchange** and was developed in the 1960s, evolving from teletype machines. The ASCII encoding allows you to use the uppercase and lowercase alphabet, digits, and some punctuation characters, making 128 in all. The following screenshot shows a version of the ASCII table (`https://simple.wikipedia.org/wiki/ASCII`):

Decimal	Hex	Char	Decimal	Hex	Char	Decimal	Hex	Char	Decimal	Hex	Char
0	0	[NULL]	32	20	[SPACE]	64	40	@	96	60	`
1	1	[START OF HEADING]	33	21	!	65	41	A	97	61	a
2	2	[START OF TEXT]	34	22	"	66	42	B	98	62	b
3	3	[END OF TEXT]	35	23	#	67	43	C	99	63	c
4	4	[END OF TRANSMISSION]	36	24	$	68	44	D	100	64	d
5	5	[ENQUIRY]	37	25	%	69	45	E	101	65	e
6	6	[ACKNOWLEDGE]	38	26	&	70	46	F	102	66	f
7	7	[BELL]	39	27	'	71	47	G	103	67	g
8	8	[BACKSPACE]	40	28	(	72	48	H	104	68	h
9	9	[HORIZONTAL TAB]	41	29	)	73	49	I	105	69	i
10	A	[LINE FEED]	42	2A	*	74	4A	J	106	6A	j
11	B	[VERTICAL TAB]	43	2B	+	75	4B	K	107	6B	k
12	C	[FORM FEED]	44	2C	,	76	4C	L	108	6C	l
13	D	[CARRIAGE RETURN]	45	2D	-	77	4D	M	109	6D	m
14	E	[SHIFT OUT]	46	2E	.	78	4E	N	110	6E	n
15	F	[SHIFT IN]	47	2F	/	79	4F	O	111	6F	o
16	10	[DATA LINK ESCAPE]	48	30	0	80	50	P	112	70	p
17	11	[DEVICE CONTROL 1]	49	31	1	81	51	Q	113	71	q
18	12	[DEVICE CONTROL 2]	50	32	2	82	52	R	114	72	r
19	13	[DEVICE CONTROL 3]	51	33	3	83	53	S	115	73	s
20	14	[DEVICE CONTROL 4]	52	34	4	84	54	T	116	74	t
21	15	[NEGATIVE ACKNOWLEDGE]	53	35	5	85	55	U	117	75	u
22	16	[SYNCHRONOUS IDLE]	54	36	6	86	56	V	118	76	v
23	17	[ENG OF TRANS. BLOCK]	55	37	7	87	57	W	119	77	w
24	18	[CANCEL]	56	38	8	88	58	X	120	78	x
25	19	[END OF MEDIUM]	57	39	9	89	59	Y	121	79	y
26	1A	[SUBSTITUTE]	58	3A	:	90	5A	Z	122	7A	z
27	1B	[ESCAPE]	59	3B	;	91	5B	[	123	7B	{
28	1C	[FILE SEPARATOR]	60	3C	<	92	5C	\	124	7C	|
29	1D	[GROUP SEPARATOR]	61	3D	=	93	5D	]	125	7D	}
30	1E	[RECORD SEPARATOR]	62	3E	>	94	5E	^	126	7E	~
31	1F	[UNIT SEPARATOR]	63	3F	?	95	5F	_	127	7F	[DEL]

Figure 3.11 – The ASCII character encoding

Notice that the numeric values for ASCII run from 0 to 127. ASCII is encoded in 7 binary bits, so the values begin at 0 and go to 2^7 – 1 (127). When a file is read in Python, the data that's stored in the file must be interpreted according to the encoding to get the correct results. As an example, in *Exercise 3.03 – working with SQL*, you will use a file named `new_customers.csv`, located in `Chapter03/datasets/new_customers.csv`. Here, however, we have used a Windows PowerShell utility called `format-hex` to view this file as raw hex codes:

```
          00 01 02 03 04 05 06 07 08 09 0A 0B 0C 0D 0E 0F

00000000  43 75 73 74 6F 6D 65 72 5F 4E 75 6D 62 65 72 2C  Customer_Number,
00000010  43 6F 6D 70 61 6E 79 2C 43 69 74 79 2C 53 74 61  Company,City,Sta
00000020  74 65 0D 0A 31 39 38 32 38 2C 52 65 70 74 69 6C  te..19828,Reptil
00000030  65 20 44 65 73 65 72 74 2C 42 61 6C 74 69 6D 6F  e Desert,Baltimo
00000040  72 65 2C 4D 44 0D 0A 31 39 31 38 36 2C 41 71 75  re,MD..19186,Aqu
00000050  61 74 69 63 20 46 72 69 65 6E 64 73 2C 53 61 6E  atic Friends,San
00000060  20 42 65 72 6E 61 64 69 6E 6F 2C 43 41 0D 0A 31   Bernadino,CA..1
00000070  39 39 34 38 2C 41 72 61 63 68 6E 61 70 68 69 6C  9948,Arachnaphil
00000080  69 61 2C 4E 65 77 61 72 6B 2C 4E 4A 0D 0A 31 39  ia,Newark,NJ..19
00000090  36 39 37 2C 53 6F 6E 67 62 69 72 64 20 4D 75 73  697,Songbird Mus
000000A0  69 63 20 53 74 6F 72 65 2C 4D 65 6D 70 68 69 73  ic Store,Memphis
000000B0  2C 54 58 0D 0A 31 39 37 38 38 2C 45 71 75 65 73  ,TX..19788,Eques
000000C0  74 72 69 61 6E 20 50 61 6C 61 63 65 2C 43 6F 6C  trian Palace,Col
000000D0  6F 72 61 64 6F 20 53 70 72 69 6E 67 73 2C 43 4F  orado Springs,CO
000000E0  0D 0A 31 39 31 31 35 2C 4A 75 73 74 20 53 68 6F  ..19115,Just Sho
000000F0  77 20 44 6F 67 73 2C 42 61 74 6F 6E 20 52 6F 75  w Dogs,Baton Rou
00000100  67 65 2C 4C 41 0D 0A 31 39 36 37 38 2C 4D 79 20  ge,LA..19678,My
00000110  46 61 76 6F 72 69 74 65 20 42 75 74 74 65 72 66  Favorite Butterf
00000120  6C 79 2C 4C 75 62 62 6F 63 6B 2C 54 58 0D 0A     ly,Lubbock,TX..
```

Figure 3.12 – The contents of the new_customers.csv file as hexadecimal codes

Notice that the first word, `Customer`, is coded as `43 75 73 74 6F 6D 65 72`.

With the widespread use of computers, it has become common to refer to raw data as bits (binary) and bytes (8 bits of binary), and 8-bit character encodings appeared. However, for many years, they were not standard until the 1990s, when so-called Unicode encodings were defined using 8 bits, and different ones were defined to accommodate languages such as Cyrillic or Hebrew. The UTF-8 encoding is the most widely used today, and since it has more bits than ASCII, it can encode more characters. Also, utf-8 encoding can use more than one byte (more than 8 bits) per character to encode even more alphabets and special characters.

Depending on your editor, you can usually see the encoding. The following screenshot shows the `bike_share.csv` data (located in `Chapter03/datasets/bike_share.csv`) open in Notepad ++. Notice the encoding in the lower right-hand corner:

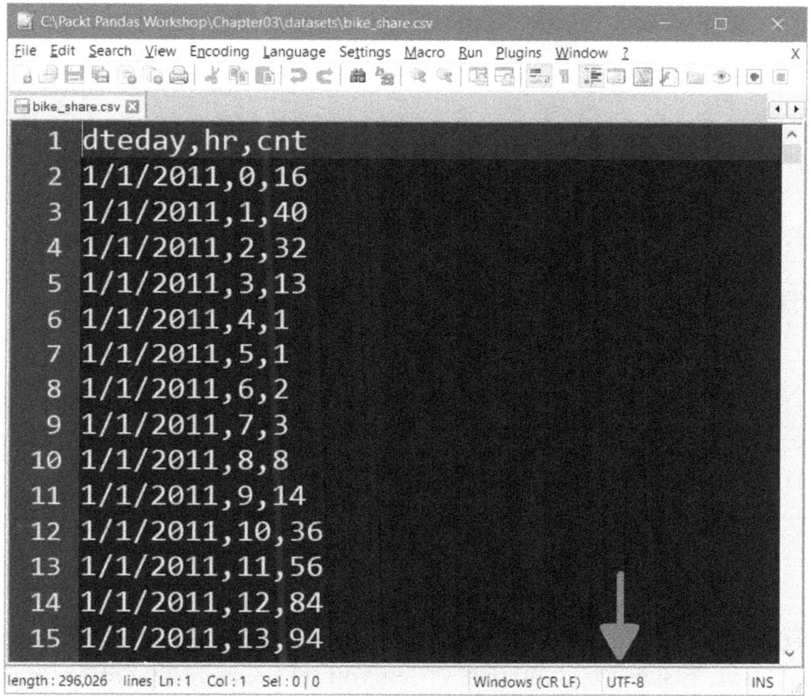

Figure 3.13 – The bike_share.csv file open in Notepad ++

You can read this file without any special parameters in pandas by using `.read_csv()`:

```
import pandas as pd
pd.read_csv('datasets/bike_share.csv')
```

> **Note**
>
> Please change the path of the dataset file (highlighted) based on where you have downloaded it on your system.

This produces the following output:

Out[2]:

	dteday	hr	cnt
0	1/1/2011	0	16
1	1/1/2011	1	40
2	1/1/2011	2	32
3	1/1/2011	3	13
4	1/1/2011	4	1
...	...	...	...
17374	12/31/2012	19	119
17375	12/31/2012	20	89
17376	12/31/2012	21	90
17377	12/31/2012	22	61
17378	12/31/2012	23	49

17379 rows × 3 columns

Figure 3.14 – The bike_share.csv file read using read_csv()

In most text editors, not only can you see the encoding, but you can also choose the encoding that the file is saved in. Notepad ++ lets you set the encoding of `bike_share.csv` to UCS-2 LE BOM and save it as `bike_share_UCS_2_LE_BOM.csv`. If you try to read this file as follows, you may run into an issue:

```
import pandas as pd
pd.read_csv('datasets/bike_share_UCS_2_LE_BOM.csv')
```

> **Note**
> Please change the path of the dataset file (highlighted) based on where you have downloaded it on your system.

This produces the following output (only the last line has been shown to save space):

```
UnicodeDecodeError: 'utf-8' codec can't decode byte 0xff in position 0: invalid start byte
```

Figure 3.15 – Error trying to read the bike_share_UCS_2_LE_BOM.csv file

pandas defaults to utf-8 encoding, but the data here is in a different encoding, so an error occurs. Fortunately, pandas can handle this sort of thing easily if you specify one more parameter. Among the many possible parameters for `.read_csv()`, there is `encoding`. The following snippet uses the `encoding = 'utf_16_le'` parameter to read the file:

```
import pandas as pd
pd.read_csv('datasets/bike_share_UCS_2_LE_BOM.csv',\
          encoding = 'utf_16_le')
```

This produces the desired result:

Out[8]:

	dteday	hr	cnt
0	1/1/2011	0	16
1	1/1/2011	1	40
2	1/1/2011	2	32
3	1/1/2011	3	13
4	1/1/2011	4	1
...	...	...	...
17374	12/31/2012	19	119
17375	12/31/2012	20	89
17376	12/31/2012	21	90
17377	12/31/2012	22	61
17378	12/31/2012	23	49

17379 rows × 3 columns

Figure 3.16 – The result of using a specific encoding with pd.read_csv()

As you can see, the exact name of the encoding that was used in `.read_csv()` and `utf_16_le` isn't identical to the name of the encoding that you saved the file with, `UCS-2 LE BOM`. This can be confusing in some cases but trying encodings that are similar to what you think is correct or searching the internet can usually give you the information you need.

> **Note**
>
> The encoding names that can be used in `.read_csv()` are given on the Python documentation page at `https://docs.python.org/3/library/codecs.html#standard-encodings`.

Another common variation in text data files is that they are separated by some character other than a comma. They can even be separated by "invisible" characters such as the tab character. In the following example, the `bike_share_UCS_2_LE_BOM.csv` file has been saved with tabs instead of commas – that is, as `bike_share_US_2_LE_BOM.tsv`. It looks like this:

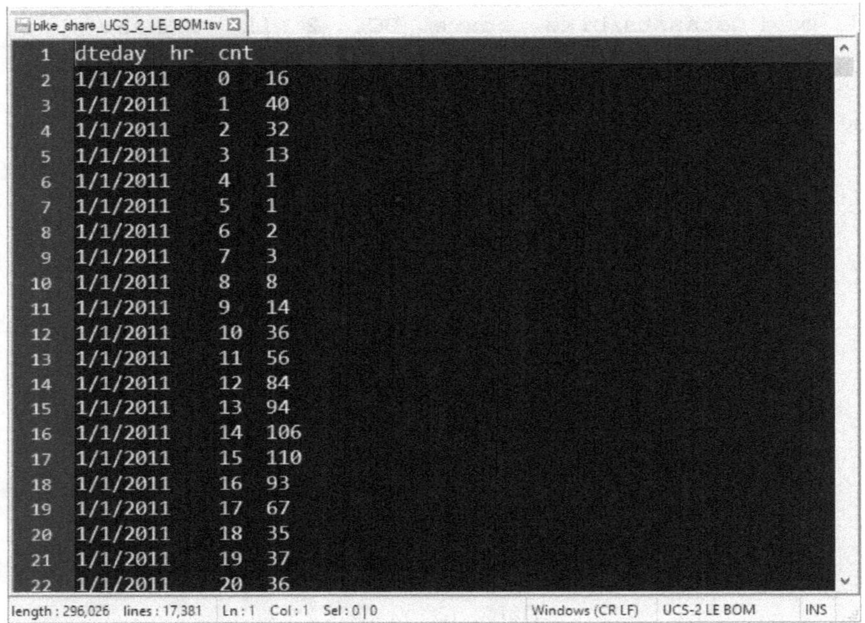

Figure 3.17 – The bike share data converted into tab separators instead of commas

All you need to do is explicitly specify the separator in `.read_csv()`. The following code shows how you can read the tab-separated file. Note that `'\t'` is the representation of a tab character in Python string data:

```
pd.read_csv('datasets//bike_share_UCS_2_LE_BOM.tsv',\
            encoding = 'utf_16_le',
            sep = '\t')
```

This produces the desired result:

Out[10]:

	dteday	hr	cnt
0	1/1/2011	0	16
1	1/1/2011	1	40
2	1/1/2011	2	32
3	1/1/2011	3	13
4	1/1/2011	4	1
...	...	...	...
17374	12/31/2012	19	119
17375	12/31/2012	20	89
17376	12/31/2012	21	90
17377	12/31/2012	22	61
17378	12/31/2012	23	49

17379 rows × 3 columns

Figure 3.18 – The bike_share_UCS_2_LE_BOM.tsv file read correctly

As you can see, with the correct separator specified, `pd.read_csv()` works, as you have seen already with CSV files. There are no explicit rules for what the separator character can be, although commas are the most common. There are good use cases for other separators. For example, when you're handling a lot of text data where the original text may contain commas (for example, user comments on products on the web), you cannot structure the file properly while using commas as separators.

The following exercise will reinforce what we've covered in this section.

Exercise 3.02 – text character encoding and data separators

As part of a research hospital data science team, you have been given a dataset that contains various metabolic and other measurements on patients. You plan to eventually make a new predictive model to detect thyroid-based ailments, but for now, you need to get the data into pandas. You know the data is in text format, but you don't know the encoding. Your goal is to read it into a DataFrame.

> **Note**
>
> You can find the code for this exercise at `https://github.com/PacktWorkshops/The-Pandas-Workshop/tree/master/Chapter03/Exercise03_02`.

Follow these steps to complete this exercise:

1. Open a new Jupyter notebook and select the **Pandas_Workshop** kernel.
2. For this exercise, all you will need is the `pandas` library. Load it into the first cell of the notebook:

```
import pandas as pd
```

3. Now, read the `thyroid.tsv` data file into pandas:

```
data = pd.read_csv('../datasets/thyroid.tsv')
data.head()
```

> **Note**
>
> In the preceding code and the code in the subsequent steps, please change the path of the dataset file (highlighted) based on where you have downloaded it on your system.

This produces the following result (only the last line has been shown to save space):

```
UnicodeDecodeError: 'utf-8' codec can't decode byte 0xff in position 0: invalid start byte
```

Figure 3.19 – An error while attempting to read the thyroid data

4. Looking at the output of the previous step, you realize that the data must be encoded in a different format than utf-8. Consulting the source of the data (a colleague in the oncology group), you determine it is `utf_16_le`. Change the `.pd.read_csv()` file accordingly:

```
data = pd.read_csv('../datasets/thyroid.tsv',\
                   encoding = 'utf_16_le')
data
```

The output should be as follows:

Out[3]:

	age\tsex\ton_thyroxine\tquery_on_thyroxine\ton_antithyroid_medicatio ... 4U\tFTI_measured\tFTI\tTBG_measured\tTBG\treferral_source\tresult\tvalue
0	41\tF\tf\tf\tf\tf\tf\tf\tf\tf\tf\tf\tf\t...
1	23\tF\tf\tf\tf\tf\tf\tf\tf\tf\tf\tf\tf\t...
2	46\tM\tf\tf\tf\tf\tf\tf\tf\tf\tf\tf\tf\t...
3	70\tF\tt\tf\tf\tf\tf\tf\tf\tf\tf\tf\tf\t...
4	70\tF\tf\tf\tf\tf\tf\tf\tf\tf\tf\tf\tf\t...
...	...
2795	70\tM\tf\tf\tf\tf\tf\tf\tf\tf\tf\tf\tf\t...
2796	73\tM\tf\tt\tf\tf\tf\tf\tf\tf\tf\tf\tf\t...
2797	75\tM\tf\tf\tf\tf\tf\tf\tf\tt\tf\tf\tf\t...
2798	60\tF\tf\tf\tf\tf\tf\tf\tf\tf\tf\tf\tf\t...
2799	81\tF\tf\tf\tf\tf\tf\tf\tf\tf\tf\tf\tf\t...

2800 rows × 1 columns

Figure 3.20 – Correcting the encoding allows you to read the file, but it only reads as one column

Here, you can see that there is another issue –the data is in one very wide column. As shown in *Figure 3.20*, you need to scroll to the right to see the values. There are also lots of `\t` values. Checking the extension on the file, you realize it is tab-delimited instead of comma-separated since in the data, as well as the column names, `\t` (the tab character) is present after each column name and the data.

5. Change `.read_csv()` one more time to specify the separator as `'\t'`:

```
data = pd.read_csv('../datasets/thyroid.tsv',\
                   encoding = 'utf_16_le',\
                   sep = '\t')
data
```

This should produce the following output:

Out[4]:

	age	sex	on_thyroxine	query_on_thyroxine	on_antithyroid_medication	sick	pregnant	thy
0	41	F	f	f	f	f	f	
1	23	F	f	f	f	f	f	
2	46	M	f	f	f	f	f	
3	70	F	t	f	f	f	f	
4	70	F	f	f	f	f	f	
...	...	...	...	...	...	...	...	
2795	70	M	f	f	f	f	f	
2796	73	M	f	t	f	f	f	
2797	75	M	f	f	f	f	f	
2798	60	F	f	f	f	f	f	
2799	81	F	f	f	f	f	f	

2800 rows × 31 columns

Figure 3.21 – The correctly read thyroid data

As you can see, the data has been successfully read into pandas. In this exercise, you used the `.read_csv` file with parameters for character encoding and the data separator to read a text file into a DataFrame. With that, you have learned how to recognize and address issues related to non-default character encodings and value separators. You should be able to recognize these issues in the future and correct them when they arise.

Binary data

You were introduced to the idea of binary data earlier regarding Matlab data and Excel. Unlike text data, there are no universal methods to read binary data. The reason for this is that for binary data, the details of the encoding are up to the designer of the format. To help with this, pandas provides many software/format-specific methods for reading such data. In general, most proprietary data systems use binary formats that are specific to that system. As noted for text formats, the file extension is usually a good indication, but not a guarantee, of the format. Aside from these points, the other thing to keep in mind is that pandas cannot write in every binary format. Refer to *Figure 3.1* and note that SAS and SPSS are two important examples where you can read data into pandas, but not write it back in the native format.

Databases – SQL data

A particular type of binary data is databases. Technically, a database can store any type of data, but here, we are referring to **relational databases**. The "relational" adjective arises from the fact that such databases store data in more than one table, and the relationship among the tables is defined by one or more keys. The following screenshot shows a simple database schematically. It contains two tables – `Customers` and `Orders`:

Customers		
Customer_ID	Address	Credit_Limit
02349	1324 S. My Way	10000
13795	2987 West St.	13000
93298	3756 East Ave.	9500
39873	12 North Gary Ln.	13500
...		

Orders			
Order_ID	Customer_ID	Item	Qty
347991	02349	23-0495	1000
347991	02349	17-0311	200
269981	13795	99-0000	1
459812	93298	45-2391	237
...			

Figure 3.22 – An example of a relational database

Here, you can see that we can match `Customer_ID` in each table to relate orders to customers, to generate shipping information or invoices, for example.

Suppose you want to find all the orders that have been sent to `1324 S. My Way`. Here, you could look in the `Orders` table by using `Customer_ID 02349`. The key takeaway here is that unlike some of the simple examples we've seen so far, binary data for a relational database contains more information that allows multiple tables and its related keys to be stored with the data. There are ways to discover, or query, the structure of a database if you don't know the structure ahead of time, but knowing the complete structure is very useful if you want to use the data in Python.

Possibly the most commonly encountered database type is **SQL**, which stands for **Structured Query Language** and is usually pronounced like the word *sequel*. SQL is a programming language as well as a database type and, as mentioned earlier, is supported in Python using `sqlite3`, a module that is part of the Python distribution but must be imported in order to be used.

> **Note**
>
> Python is usually distributed in what is called the "standard distribution," which comprises the **Python Standard Library**. This, in turn, provides all the base functionalities that more advanced modules are built on, such as NumPy and pandas. However, many modules that must be imported for you to use them are included in the base distribution, such as the `math` module, the `os` module, and more. The `sqlite3` module is listed in the **Data Persistence** group. You can find the complete list at `https://docs.python.org/3.7/library/`.

Note that there are many variations of SQL languages, such as Microsoft SQL Server, PostgreSQL, MySQL, SQLite, and more. Although much of the syntax is similar, each variation has some uniqueness and requires knowledge of the language's syntax.

Using `sqlite3`, you can load SQL data (notably, data that has the `.db`, `.sqlite`, or `.sqlite3` extension) into pandas, as well as execute SQL language commands from within Python programs, so long as the data is consistent with the SQLite syntax. Let's look at a simple example of using `sqlite3`.

sqlite3

An important concept in working with databases as opposed to files is that you need a "connection" to the database before you can work with it. In the following code, we are importing `sqlite3`, then using it to connect to a database called `bike_share.db`. Note that if this database does not already exist, it will be created in this step.

Then, we must create a `cursor` object with the connection to execute commands on the database. Following that, we must load the same data that we did previously and create a table in the database named `RENTALS` with three columns (corresponding to `dteday`, `hr`, and `cnt` from the `.tsv` file). Finally, we must use the `.to_sql()` method to write our DataFrame into the table, which will update the database. In this case, we must tell pandas to replace the table if it is already there (there are other options too, such as append):

```
import pandas as pd
import sqlite3
conn = sqlite3.connect('datasets/bike_share.db')
c = conn.cursor()
data = pd.read_csv('datasets/bike_share_UCS_2_LE_BOM.tsv',
                   encoding = 'utf_16_le',
                   sep = '\t')
c.execute('CREATE TABLE IF NOT EXISTS RENTALS (Date, Hour, Qty)')
conn.commit()
data.to_sql("RENTALS", conn, if_exists = 'replace')
```

The preceding code will create the `bike_share.db` file in the `datasets` directory, which contains the same data as `bike_share.csv`. If you have a database tool, you can connect to it and view the data.

We will use the popular open source database tool Dbeaver (`https://dbeaver.io/`) to connect to the database and view the table:

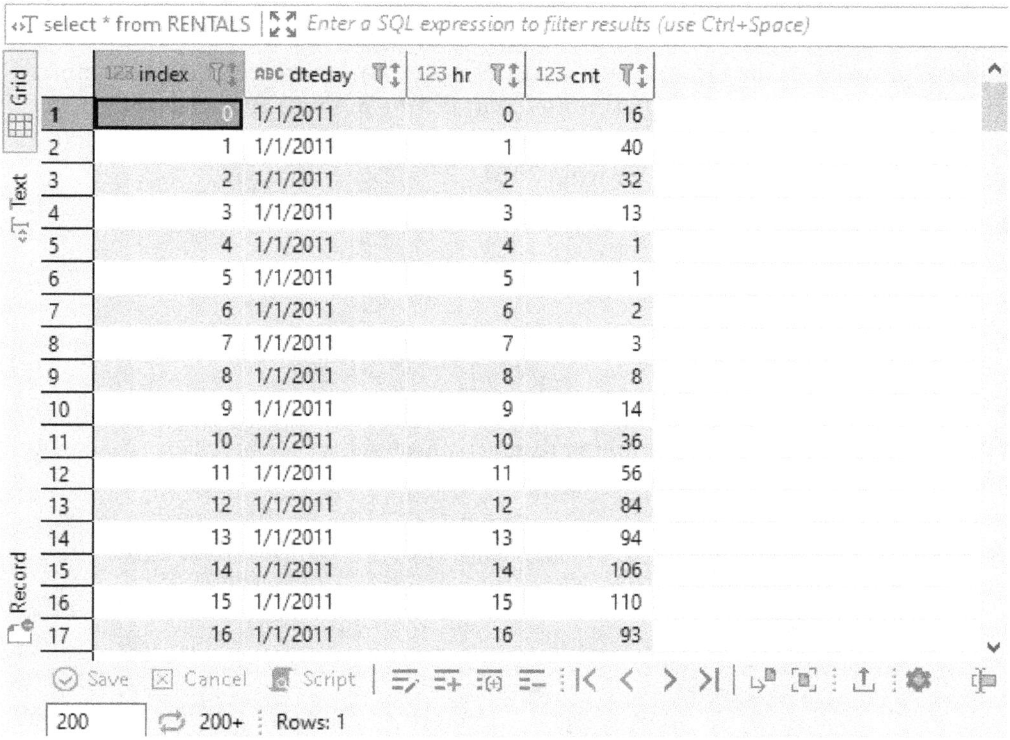

select * from RENTALS | Enter a SQL expression to filter results (use Ctrl+Space)

	index	dteday	hr	cnt
1	0	1/1/2011	0	16
2	1	1/1/2011	1	40
3	2	1/1/2011	2	32
4	3	1/1/2011	3	13
5	4	1/1/2011	4	1
6	5	1/1/2011	5	1
7	6	1/1/2011	6	2
8	7	1/1/2011	7	3
9	8	1/1/2011	8	8
10	9	1/1/2011	9	14
11	10	1/1/2011	10	36
12	11	1/1/2011	11	56
13	12	1/1/2011	12	84
14	13	1/1/2011	13	94
15	14	1/1/2011	14	106
16	15	1/1/2011	15	110
17	16	1/1/2011	16	93

Save Cancel Script

200 200+ Rows: 1

Figure 3.23 – Viewing the bike_share.db RENTALS table using Dbeaver, after creating it in Python

Here, you can see that the same values have been stored in the table that were stored in the database.

Additional text formats

Although text data can be viewed and read in a text editor, that doesn't mean it always contains plain text or simple columns of data. Two formats are encountered so often in today's projects that we need to spend some additional time studying them: JSON and HTML/XML. The JSON format is plain text but is structured much like a Python `dictionary`. Because it is plain text, it's easy to send and receive over internet connections, and because it has structure, it can encode complex table structures, including hierarchical or tree-like tables and other forms. You will find that many APIs use JSON by default, so you will likely encounter this format at some point. If you are reading data from a website, then it is likely encoded as HTML or XML data. In *Exercise 3.01 – reading data from web pages*, you saw a simple example of scraping a web page using `.read_html()`.

Let's look at these formats in more detail.

Working with JSON

Let's access some JSON data from a public API. Suppose you are researching economic information and need a list of all US counties as a starting point to build a dataset. Here, we can use the `requests` library to call an API provided by the US Census Bureau to obtain a list of every county in the US from 2010:

```
import requests
US_counties_query = \
requests.get\
('https://api.census.gov/data/2010/dec/sf1?get=NAME&for=county:*')
US_counties_query.text
```

This produces a somewhat messy-looking result (truncated here for brevity):


```
Out[5]: '[["NAME","state","county"],\n["Sebastian County, Arkansas","05","131"],\n["Sevier County, Arkansas","0
5","133"],\n["Sharp County, Arkansas","05","135"],\n["Stone County, Arkansas","05","137"],\n["Union Coun
ty, Arkansas","05","139"],\n["Van Buren County, Arkansas","05","141"],\n["Washington County, Arkansa
s","05","143"],\n["White County, Arkansas","05","145"],\n["Yell County, Arkansas","05","149"],\n["Colusa
County, California","06","011"],\n["Butte County, California","06","007"],\n["Alameda County, Californi
a","06","001"],\n["Alpine County, California","06","003"],\n["Amador County, California","06","005"],\n
["Calaveras County, California","06","009"],\n["Contra Costa County, California","06","013"],\n["Del Nor
te County, California","06","015"],\n["Kings County, California","06","031"],\n["Glenn County, Californi
a","06","021"],\n["Humboldt County, California","06","023"],\n["Imperial County, California","06","02
5"],\n["El Dorado County, California","06","017"],\n["Fresno County, California","06","019"],\n["Inyo Co
unty, California","06","027"],\n["Kern County, California","06","029"],\n["Mariposa County, Californi
a","06","043"],\n["Lake County, California","06","033"],\n["Lassen County, California","06","035"],\n["L
os Angeles County, California","06","037"],\n["Madera County, California","06","039"],\n["Marin County,
```


Figure 3.24 – The raw JSON text returned by the API

The result is a string (it begins with a single quotation mark), and since the first character inside the quotes is an opening square bracket, the content is a list of some sort. If you were to scroll to the end of this output, you would find a closing bracket. You can also see additional bracketed text in-between, which contains a word for each county. The next thing you'd probably want to do is get this data into a DataFrame.

Before we look at how pandas can help here, let's extract this data into Python.

You will need to use a module called `ast` (which stands for **Abstract Syntax Trees**). This module provides the `.literal_eval()` method, which can be used to extract data stored as a string. The `.literal_eval()` method takes the string and *evaluates* it as Python code. In this case, since the string is a representation of a list, using `.literal_eval()`, you will get a list object. After that, you can use list comprehension to step through the individual lists, getting a new list with one list for each county:

```
import ast
US_counties_data = ast.literal_eval(US_counties_query.text)
[US_counties_data[i] for i in range(len(US_counties_data))]
```

This produces the following output:

```
Out[10]: [['NAME', 'state', 'county'],
          ['Sebastian County, Arkansas', '05', '131'],
          ['Sevier County, Arkansas', '05', '133'],
          ['Sharp County, Arkansas', '05', '135'],
          ['Stone County, Arkansas', '05', '137'],
          ['Union County, Arkansas', '05', '139'],
          ['Van Buren County, Arkansas', '05', '141'],
          ['Washington County, Arkansas', '05', '143'],
          ['White County, Arkansas', '05', '145'],
          ['Yell County, Arkansas', '05', '149'],
          ['Colusa County, California', '06', '011'],
          ['Butte County, California', '06', '007'],
          ['Alameda County, California', '06', '001'],
          ['Alpine County, California', '06', '003'],
          ['Amador County, California', '06', '005'],
          ['Calaveras County, California', '06', '009'],
          ['Contra Costa County, California', '06', '013'],
          ['Del Norte County, California', '06', '015'],
```

Figure 3.25 – A list of lists obtained from the API JSON data

This output looks understandable, albeit not yet in pandas. However, we intentionally did this conversion "the hard way" to show you more of the power of pandas. You can simply pass the JSON text that results from the `requests()` call (`US_counties_query.text`) to the `pandas.read_json()` method and get a DataFrame, as follows:

```
import pandas as pd
US_counties_data = pd.read_json(US_counties_query.text)
US_counties_data
```

This produces the following output:

Out[11]:

	0	1	2
0	NAME	state	county
1	Sebastian County, Arkansas	05	131
2	Sevier County, Arkansas	05	133
3	Sharp County, Arkansas	05	135
4	Stone County, Arkansas	05	137
...	...	...	...
3217	Eau Claire County, Wisconsin	55	035
3218	Florence County, Wisconsin	55	037
3219	Fond du Lac County, Wisconsin	55	039
3220	Forest County, Wisconsin	55	041
3221	Jefferson County, Wisconsin	55	055

3222 rows × 3 columns

Figure 3.26 – The US counties data in a pandas DataFrame

The only issue remaining is that the column names are in the first row.

It's impressive what pandas can do in these situations, dramatically simplifying the work of ingesting JSON data. If you knew beforehand that the data from your request was in JSON format, you could have accomplished this task much more easily by just passing the URL to the `.read_json()` method of pandas, then fixing the column names. The following code shows how you can use the `.read_json()` method, drop the first row, and then add column names:

```
URL =\
'https://api.census.gov/data/2010/dec/sf1?get=NAME&for=county:*'
US_counties_data = pd.read_json(URL).loc[1:, :]
US_counties_data.columns = ['County',
                            'state_code',
                            'county_code']
US_counties_data
```

This produces the following output:

Out[6]:

	County	state_code	county_code
1	Sebastian County, Arkansas	05	131
2	Sevier County, Arkansas	05	133
3	Sharp County, Arkansas	05	135
4	Stone County, Arkansas	05	137
5	Union County, Arkansas	05	139
...	...	...	...
3217	Eau Claire County, Wisconsin	55	035
3218	Florence County, Wisconsin	55	037
3219	Fond du Lac County, Wisconsin	55	039
3220	Forest County, Wisconsin	55	041
3221	Jefferson County, Wisconsin	55	055

3221 rows × 3 columns

Figure 3.27 – The US counties DataFrame

Working with HTML/XML

Sometimes, you find the data you need on a web page, but there isn't an API or a file to download. You can check out *Figure 3.7* for an example.

Here, you can see the data you want, but how do you get it into pandas? Earlier, you learned how to use `.read_html()` to accomplish this. Before you put it to use again, you'll need to understand how pandas makes sense of the underlying data on a web page. In the following screenshot, we have used the browser to show the raw HTML that would be read from this page by pandas. This is done by right-clicking on the page and selecting **View Page Source** in Chrome:

```
<!DOCTYPE html>
<html class="client-nojs" lang="en" dir="ltr">
<head>
<meta charset="UTF-8"/>
<title>Wind power - Wikipedia</title>
<script>document.documentElement.className="client-js";RLCONF={"wgBreakFrames":!1,"wgSeparatorTransformTable":["",""],"wgDigitTransformTable":["",
"Articles with permanently dead external links","CS1 maint: location","Articles with Spanish-language sources (es)","Articles with short descripti
"wgIsProbablyEditable":!1,"wgRelevantPageIsProbablyEditable":!1,"wgRestrictionEdit":["autoconfirmed"],"wgRestrictionMove":["sysop"],"wgFlaggedRevs
"#87ceeb","#a4a1a2"]}],"version":2,"marks":[{"type":"line","properties":{"hover":{"stroke":{"value":"red"}},"update":{"stroke":{"scale":"color","f
"format":{"parse":{"y":"number","x":"date"},"type":"json"},"name":"chart","values":[{"y":6.1,"series":"","x":"1996"},{"y":7.6,"series":"","x":"199
"wgULSPosition":"interlanguage","wgGENewcomerTasksGuidanceEnabled":!0,"wgGEAskQuestionEnabled":!1,"wgGELinkRecommendationsFrontendEnabled":!1,"wgW
"jquery.makeCollapsible","mediawiki.toc","skins.vector.legacy.js","ext.gadget.ReferenceTooltips","ext.gadget.charinsert","ext.gadget.extra-toolbar
<script>(RLQ=window.RLQ||[]).push(function(){mw.loader.implement("user.options@1hzgi",function($,jQuery,require,module){/*@nomin*/mw.user.tokens.s
});});</script>
<link rel="stylesheet" href="/w/load.php?lang=en&amp;modules=ext.cite.styles%7Cext.graph.styles%7Cext.math.styles%7Cext.timeline.styles%7Cext.tmh.
<script async="" src="/w/load.php?lang=en&amp;modules=startup&amp;only=scripts&amp;raw=1&amp;skin=vector"></script>
<meta name="ResourceLoaderDynamicStyles" content=""/>
<link rel="stylesheet" href="/w/load.php?lang=en&amp;modules=site.styles&amp;only=styles&amp;skin=vector"/>
<meta name="generator" content="MediaWiki 1.37.0-wmf.4"/>
<meta name="referrer" content="origin"/>
<meta name="referrer" content="origin-when-crossorigin"/>
<meta name="referrer" content="origin-when-cross-origin"/>
<meta property="og:image" content="https://upload.wikimedia.org/wikipedia/commons/thumb/e/e0/Wind_power_plants_in_Xinjiang%2C_China.jpg/1200px-Win
<meta property="og:title" content="Wind power - Wikipedia"/>
<meta property="og:type" content="website"/>
<link rel="preconnect" href="//upload.wikimedia.org"/>
<link rel="alternate" media="only screen and (max-width: 720px)" href="//en.m.wikipedia.org/wiki/Wind_power"/>
<link rel="apple-touch-icon" href="/static/apple-touch/wikipedia.png"/>
<link rel="shortcut icon" href="/static/favicon/wikipedia.ico"/>
<link rel="search" type="application/opensearchdescription+xml" href="/w/opensearch_desc.php" title="Wikipedia (en)"/>
<link rel="EditURI" type="application/rsd+xml" href="//en.wikipedia.org/w/api.php?action=rsd"/>
<link rel="license" href="//creativecommons.org/licenses/by-sa/3.0/"/>
<link rel="canonical" href="https://en.wikipedia.org/wiki/Wind_power"/>
<link rel="dns-prefetch" href="//login.wikimedia.org"/>
<link rel="dns-prefetch" href="//meta.wikimedia.org" />
</head>
<body class="mediawiki ltr sitedir-ltr mw-hide-empty-elt ns-0 ns-subject page-Wind_power rootpage-Wind_power skin-vector action-view skin-vector-l
<div id="mw-head-base" class="noprint"></div>
```

Figure 3.28 – The raw HTML of the Wikipedia page

Earlier, you obtained the first table that results from using `.read_html()` on this page by selecting the second item in the list you got. However, there are other tables on this page, such as one for offshore wind farms:

World's largest offshore wind farms

Wind farm	Capacity (MW)	Country	Turbines and model	Commissioned	Refs
Walney Extension	659	United Kingdom	47 x Vestas 8MW 40 x Siemens Gamesa 7MW	2018	[48]
London Array	630	United Kingdom	175 × Siemens SWT-3.6	2012	[49][50][51]
Gemini Wind Farm	600	The Netherlands	150 × Siemens SWT-4.0	2017	[52]
Gwynt y Môr	576	United Kingdom	160 × Siemens SWT-3.6 107	2015	[53]
Greater Gabbard	504	United Kingdom	140 × Siemens SWT-3.6	2012	[54]
Anholt	400	Denmark	111 × Siemens SWT-3.6–120	2013	[55]
BARD Offshore 1	400	Germany	80 BARD 5.0 turbines	2013	[56]

Figure 3.29 – Offshore wind farms table on the Wikipedia page

If you inspect the data you obtained earlier, you will find it has a list that's `27` in length:

```
import pandas as pd
data_url = 'https://en.wikipedia.org/wiki/Wind_power'
data = pd.read_html(data_url)
len(data)
```

This produces the following output:

```
27
```

As you did earlier when you extracted element `[1]` of this list, use the following command to look at the next element, `[2]`:

```
data[2]
```

You will see the following output:

Out[11]:

	Wind farm	Capacity (MW)	Country	Turbines and model	Commissioned	Refs
0	Walney Extension	659	United Kingdom	47 x Vestas 8MW 40 x Siemens Gamesa 7MW	2018	[48]
1	London Array	630	United Kingdom	175 × Siemens SWT-3.6	2012	[49][50][51]
2	Gemini Wind Farm	600	The Netherlands	150 × Siemens SWT-4.0	2017	[52]
3	Gwynt y Môr	576	United Kingdom	160 × Siemens SWT-3.6 107	2015	[53]
4	Greater Gabbard	504	United Kingdom	140 × Siemens SWT-3.6	2012	[54]
5	Anholt	400	Denmark	111 × Siemens SWT-3.6–120	2013	[55]
6	BARD Offshore 1	400	Germany	80 BARD 5.0 turbines	2013	[56]

Figure 3.30 – The third element of the list that's returned by .read_html() is a DataFrame

Similarly, you can inspect other parts of the page and obtain additional tables there, if they're needed for your project.

One consideration to keep in mind when you're collecting HTML data this way is that there is no guarantee that the page(s) you access will remain the same over time. So, using sources like these within a larger data system could incur some risk that it would break in the future.

Working with XML data

XML is similar to HTML and is frequently used to store data in a structured way so that it can be accessed via the internet. For example, the US government provides a site that contains over 35,000 datasets that can be accessed in XML format. pandas doesn't provide a `.read_xml()` method, but you can use an installable module to help. The following example shows how to use the `pandas-read-xml` module. It can be installed using `pip`, as follows:

```
pip install pandas-read-xml
```

> **Note**
>
> On Linux or macOS, use `pip3` instead of `pip`. Alternatively, if you're running the `pip` command in a Jupyter Notebook, use `!pip` instead.

Suppose you are interested to model student success based on some factors, and you have chosen a URL that provides math testing scores for the city of New York (`data.cityofnewyork.us/api/views/825b-niea/rows.xml`). Let's inspect the page directly:

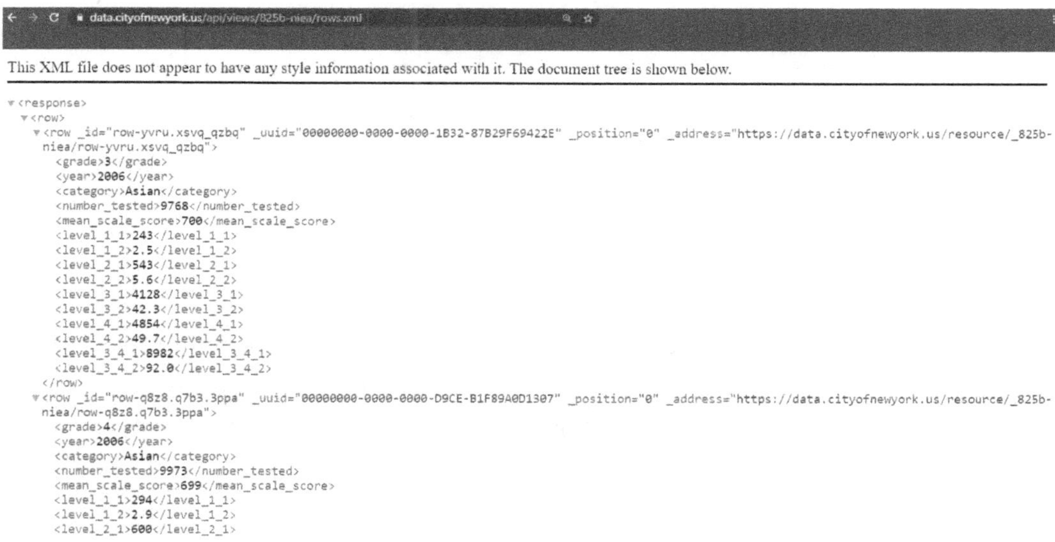

Figure 3.31 – Raw XML for the City of New York student data

XML-formatted data is often very complex and decoding it manually can be extremely challenging.

The following code shows how to load the `pandas_read_xml` package, which provides the `.read_xml()` method. You'll need to pass the URL to that method, along with a list that defines the "tree" or the hierarchy of the data in the XML data – in this case, `['response', 'row', 'row']`:

```
import pandas as pd
import pandas _ read _ xml as pdx
pdx.read _ xml\
('https://data.cityofnewyork.us/api/views/825b-niea/rows.
xml?accessType=DOWNLOAD',
                   ['response', 'row', 'row'],
                   root _ is _ rows = False)
```

Running this snippet produces the following output:

Out[8]:

	@_id	@_uuid	@_position	@_address	grade	year	category	number_tested	mean_scale_score	le
0	row-yvru.xsvq_qzbq	00000000-0000-0000-1B32-87B29F69422E	0	https://data.cityofnewyork.us/resource/_825b-n...	3	2006	Asian	9768	700	
1	row-q8z8.q7b3.3ppa	00000000-0000-0000-D9CE-B1F89A0D1307	0	https://data.cityofnewyork.us/resource/_825b-n...	4	2006	Asian	9973	699	
2	row-i23x-4prc-46fj	00000000-0000-0000-C9EE-2418870B5F93	0	https://data.cityofnewyork.us/resource/_825b-n...	5	2006	Asian	9852	691	
3	row-7u9v-dwwy.fhw3	00000000-0000-0000-17FD-7D50A499A0E1	0	https://data.cityofnewyork.us/resource/_825b-n...	6	2006	Asian	9606	682	
4	row-64kf_k4ma_4zgq	00000000-0000-0000-6A3C-917EFD40527E	0	https://data.cityofnewyork.us/resource/_825b-n...	7	2006	Asian	9433	671	
...	...	...	...	...	...	...	...	...	...	
163	row-i6yz_wbge_khnu	00000000-0000-0000-11E2-D5CA802D0782	0	https://data.cityofnewyork.us/resource/_825b-n...	5	2011	White	10808	699	

Figure 3.32 – The data that was retrieved by using .read_xml() from the pandas-read-xml module

The question, then, is how did you determine the correct list for the data hierarchy? If you look at the raw XML in the browser again, you will see the following structure:

```
▼<response>
  ▼<row>
    ▼<row _id="row-yvru.xsvq_qzbq" _uuid="00000000-0000-0000-1B32-87B29F69422E" _position="0" _address="https://data.cit
        <grade>3</grade>
        <year>2006</year>
        <category>Asian</category>
        <number_tested>9768</number_tested>
        <mean_scale_score>700</mean_scale_score>
        <level_1_1>243</level_1_1>
        <level_1_2>2.5</level_1_2>
        <level_2_1>543</level_2_1>
        <level_2_2>5.6</level_2_2>
        <level_3_1>4128</level_3_1>
        <level_3_2>42.3</level_3_2>
        <level_4_1>4854</level_4_1>
        <level_4_2>49.7</level_4_2>
        <level_3_4_1>8982</level_3_4_1>
        <level_3_4_2>92.0</level_3_4_2>
      </row>
    ▼<row _id="row-q8z8.q7b3.3ppa" _uuid="00000000-0000-0000-D9CE-B1F89A0D1307" _position="0" _address="https://data.cit
        <grade>4</grade>
        <year>2006</year>
        <category>Asian</category>
        <number_tested>9973</number_tested>
        <mean_scale_score>699</mean_scale_score>
        <level_1_1>294</level_1_1>
```


Figure 3.33 – The raw XML data for the test scores

The data begins three levels down into the tree, and those are the levels that are passed to the method (`response`, `row`, and `row`). In many cases, this could be more complex.

Working with Excel

When you read Excel data in the *Text files and binary files* section, it appeared to work just like `.read_csv()`. However, as you saw back then, this was mainly because the defaults that were used in `.read_excel()` were correct for the simple file you were using. If you look at the possible parameters for `.read_excel()`, you will find the following, among others:

> **Note**
>
> We have only listed a few here; when you're using pandas I/O functions, always consult the official pandas documentation for a full list and the meanings of the parameters.

parameter (= default value)	meaning
io	the object containing the Excel data--can be a path etc.
sheet_name = 0	defaults to the first sheet
header = 0	what row, if any, contains the column names?
usecols = None	what columns to read, if not all; can be a list of letters or integers etc.

Figure 3.34 – Some parameters for the read_excel() method

As you can see, pandas will automatically take the first row of the Excel file for the column names of the resulting DataFrame, use the first sheet, and parse all the columns. Often, Excel files contain many objects besides data, and the data may not always start in the first row or column. The following screenshot shows a partial view of an Excel file we want to load into a DataFrame, which contains some sensor data with times and three values from three sensors:

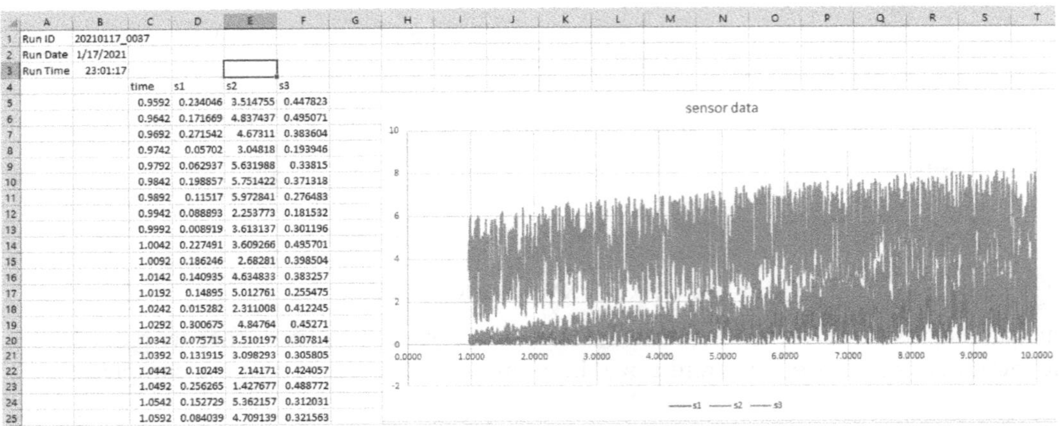

	A	B	C	D	E	F
1	Run ID	20210117_0037				
2	Run Date	1/17/2021				
3	Run Time	23:01:17				
4			time	s1	s2	s3
5			0.9592	0.234046	3.514755	0.447823
6			0.9642	0.171669	4.837437	0.495071
7			0.9692	0.271542	4.67311	0.383604
8			0.9742	0.05702	3.04818	0.193946
9			0.9792	0.062937	5.631988	0.33815
10			0.9842	0.198857	5.751422	0.371318
11			0.9892	0.11517	5.972841	0.276483
12			0.9942	0.088893	2.253773	0.181532
13			0.9992	0.008919	3.613137	0.301196
14			1.0042	0.227491	3.609266	0.495701
15			1.0092	0.186246	2.68281	0.398504
16			1.0142	0.140935	4.634833	0.383257
17			1.0192	0.14895	5.012761	0.255475
18			1.0242	0.015282	2.311008	0.412245
19			1.0292	0.300675	4.84764	0.45271
20			1.0342	0.075715	3.510197	0.307814
21			1.0392	0.131915	3.098293	0.305805
22			1.0442	0.10249	2.14171	0.424057
23			1.0492	0.256265	1.427677	0.488772
24			1.0542	0.152729	5.362157	0.312031
25			1.0592	0.084039	4.709139	0.321563

Figure 3.35 – A typical Excel file that contains information in addition to the target data

Using the appropriate parameters, it's possible to load only the data table portion of this file. The following code shows how to use `.read_excel()` and how to specify the columns to read (`C` through `F` in Excel, which correspond to `2` through `5` in Python). You'll also need to specify a row for the header, which is used for the column names. Finally, you'll need to specify the sheet's name:

```
sensor_data = pd.read_excel('datasets/sensor_data.xlsx',\
                            usecols = [2, 3, 4, 5],\
                            header = 3,\
                            sheet_name = '20210117_0037',
                            engine = 'openpyxl')
sensor_data
```

This results in the following output:

Out[20]:

	time	s1	s2	s3
0	0.95924	0.234046	3.514755	0.447823
1	0.96424	0.171669	4.837437	0.495071
2	0.96924	0.271542	4.673110	0.383604
3	0.97424	0.057020	3.048180	0.193946
4	0.97924	0.062937	5.631988	0.338150
...	...	...	...	...
10669	54.30424	15.066911	7.506722	29.028388
10670	54.30924	17.264761	10.195260	24.272862
10671	54.31424	9.744161	7.956116	10.244286
10672	54.31924	1.722525	10.254374	2.513277
10673	54.32424	10.190016	11.267764	0.942601

10674 rows × 4 columns

Figure 3.36 – Using the appropriate parameters, we only extract the required data from an Excel file

Thus, you can easily control what data is extracted from an Excel file. Note that you could do the same thing for multiple sheets in a workbook. However, you would need to know the location of the data and column names in a sheet to read them.

Suppose that, for compatibility with another group, you want to save the data in a new Excel file, now that you have done the work to extract the required information. pandas provides the `.to_excel()` method for this purpose. In the following code snippet, we are saving to a new file, `sensor_data_clean.xlsx`, naming the sheet `sensor_data`, and specifying `index = None` so that we don't get an extra index column (since Excel has its own row numbers):

```
sensor_data.to_excel('datasets/sensor_data_clean.xlsx',\
                     sheet_name = 'sensor_data',\
                     index = None,
                     engine = 'openpyxl')
```

The resulting Excel file will look as follows:

	A	B	C	D
1	time	s1	s2	s3
2	0.95924	0.23405	3.51476	0.44782
3	0.96424	0.17167	4.83744	0.49507
4	0.96924	0.27154	4.67311	0.3836
5	0.97424	0.05702	3.04818	0.19395
6	0.97924	0.06294	5.63199	0.33815
7	0.98424	0.19886	5.75142	0.37132
8	0.98924	0.11517	5.97284	0.27648
9	0.99424	0.08889	2.25377	0.18153
10	0.99924	0.00892	3.61314	0.3012
11	1.00424	0.22749	3.60927	0.4957
12	1.00924	0.18625	2.68281	0.3985
13	1.01424	0.14093	4.63483	0.38326
14	1.01924	0.14895	5.01276	0.25547

Figure 3.37 – The sensor data stored in a new Excel file

SAS data

SAS is a leading data analytics platform that's used in many industries, so you will likely encounter SAS data at some point. SAS data is usually binary and stored in a proprietary format and pandas can read SAS data files using the `.read_sas()` method. Suppose the finance team uses SAS and has provided you with some data on the airline industry that you want to use for market analysis. You can use this method to read the SAS dataset directly into a DataFrame, as follows:

```
import pandas as pd
data = pd.read_sas('datasets//airline.sas7bdat')
data.head()
```

This produces the following output:

Out[10]:

	YEAR	Y	W	R	L	K
0	1948.0	1.214	0.243	0.1454	1.415	0.612
1	1949.0	1.354	0.260	0.2181	1.384	0.559
2	1950.0	1.569	0.278	0.3157	1.388	0.573
3	1951.0	1.948	0.297	0.3940	1.550	0.564
4	1952.0	2.265	0.310	0.3559	1.802	0.574

Figure 3.38 – The result of using .read_sas() on the airline.sas7bdat file

The finance department has told you that column `Y` is the total industry revenue and that columns `W` through `K` are the costs, so you can now calculate profit for your analysis. Since you cannot write to a SAS data file with pandas, if you were to modify the table, you would need to store it as a CSV or some other suitable format.

SPSS data

SPSS is a statistical software platform that has been in use since 1968 and is still widely used in industries such as pharmaceuticals and many others. **SPSS** is an acronym for **Statistical Product and Service Solutions**.

As a data scientist supporting a study to estimate body fat from skinfold measurements, you have been given data in SPSS. The data is a table with the actual body fat determined from laboratory measurements, plus values for the skinfold thickness at the triceps, thigh, and mid-arm. Reading this data into pandas is easy – you can use the `.read_spss()` method:

```
import pandas as pd
data = pd.read_spss('datasets/bodyfat.sav')
data.head()
```

This produces the following output:

Out[11]:

	y	x1	x2	x3
0	19.5	43.1	29.1	11.9
1	24.7	49.8	28.2	22.8
2	30.7	51.9	37.0	18.7
3	29.8	54.3	31.1	20.1
4	19.1	42.2	30.9	12.9

Figure 3.39 – The result of reading the bodyfat.sav SPSS data file into a pandas DataFrame

With the data now in a DataFrame, you are ready to test various predictive models. As with SAS data, pandas cannot write to SPSS data files, so changes would need to be stored in another format.

Stata data

Stata is another statistical analysis platform that has been in use since 1985. The following is an example that uses the `.read_stata()` method, using data from the US Federal Reserve that contains the Survey of Consumer Finances. Unlike the SAS and SPSS methods, pandas can write to Stata files, so we must add the steps to output to a Stata data file, and then reread and compare the two DataFrames to ensure they are the same:

```
import pandas as pd
data = pd.read_stata('datasets//rscfp2019.dta')
print('data:\n', data.head(2))
data.to_stata('datasets//rscfp2019_write.dta',write_index=False)
data2 = pd.read_stata('datasets//rscfp2019_write.dta')
print('data2:\n', data2.head(2))
print('differences between rscfp2019 and rscfp2019_write:\n',\
      data.compare(data2))
```

Running this code results in the following output:

```
data:
   yy1  y1          wgt  hhsex  age  agecl  educ  edcl  married  kids  ...  \
0    1  11  6119.779308      2   75      6    12     4        2     0  ...
1    1  12  4712.374912      2   75      6    12     4        2     0  ...

   nwcat  inccat  assetcat  ninccat  ninc2cat  nwpctlecat  incpctlecat  \
0      5       3         6        3         2          10            6
1      5       3         6        3         1          10            5

   nincpctlecat  incqrtcat  nincqrtcat
0             6          3           3
1             5          2           2

[2 rows x 351 columns]
data2:
   yy1  y1          wgt  hhsex  age  agecl  educ  edcl  married  kids  ...  \
0    1  11  6119.779308      2   75      6    12     4        2     0  ...
1    1  12  4712.374912      2   75      6    12     4        2     0  ...

   nwcat  inccat  assetcat  ninccat  ninc2cat  nwpctlecat  incpctlecat  \
0      5       3         6        3         2          10            6
1      5       3         6        3         1          10            5

   nincpctlecat  incqrtcat  nincqrtcat
0             6          3           3
1             5          2           2

[2 rows x 351 columns]
differences between rscfp2019 and rscfp2019_write:
 Empty DataFrame
Columns: []
Index: []
```

Figure 3.40 – Results from.read_stata(), .to_stata(), using the economic data

The last part, using pandas `DataFrame.compare()`, will show any rows or columns that differ in the two DataFrames. In this case, there are no differences (as expected), so the results are empty. By inspecting the first two rows of each DataFrame, you can see that they appear to be the same but using `compare()` ensures this.

HDF5 data

HDF data refers to **Hierarchical Data Format**, which was originally developed by the National Center for Supercomputing Applications. Compared to the proprietary binary formats we've discussed so far, HDF5 is an open format. HDF5 is intended to support very large data applications and is available on many cloud systems. Because pandas can read and write to HDF5 file formats, here, we will create some data to simulate some time-based data collection process. We will store it in HDF5 format, then read it again and view it:

```
import pandas as pd
import numpy as np
time = np.arange(0, 100, 0.01)
values = np.sin(2 * np.pi * time / 17)
data = pd.DataFrame({'time': time, 'data': values})
data.to_hdf('datasets/store_data_h5.h5', 'table', append =
True)
data_reread = pd.read_hdf('datasets/store_data_h5.h5',
'table',\
                            where = ['index > 9'])
data_reread.head()
```

We will see the following output:

Out[21]:

	time	data
10	0.10	0.036951
11	0.11	0.040645
12	0.12	0.044337
13	0.13	0.048029
14	0.14	0.051721

Figure 3.41 – The first five rows of the synthetic data for an index greater than 9, read back in from an HDF5 file

Notice the additional parameter that was used in the `.read_hdf()` method, `where = ['index > 9']`. This option supports loading an arbitrary part of the data, which can be useful for very large files if not all the rows are needed.

Manipulating SQL data

Earlier, when we introduced SQL, we showed you a simple example where we took our existing `bike_share` data, created a database, and stored the data as a table in the new database. We mentioned that you can execute SQL commands by using the `cursor.execute()` method. Here, you will see another example, this time using a database that contains multiple tables.

Suppose you have a database for a pet supply company. You can use `pd.read_sql()` to issue a command to the database and return the names of all the tables. SQL databases can contain multiple tables, so there is a "master" table, called `sqlite_master` here, that can be queried so that you can see every other table in the database. In the following snippet, the `SELECT` statement says to get a variable called `name` and return its values where the variable called type is `'table'` – in other words, a list of tables. The `sqlite3.connect()` statement opens a connection to the database so that we can read from it. This creates the connection and completes the command in one statement:

```
import pandas as pd
import sqlite3
tables = \
    pd.read_sql(
    "SELECT name FROM sqlite_master WHERE type = 'table' ORDER 
BY name ASC",
    sqlite3.connect('datasets/pet_stores.db'))
tables
```

This produces the following output:

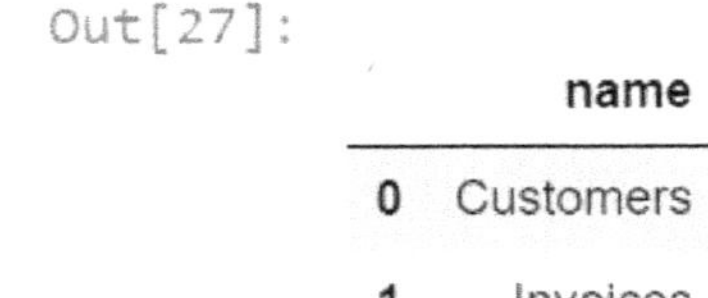

	name
0	Customers
1	Invoices

Figure 3.42 – The tables in the pet_stores.db database

Note that in this case, pandas automatically makes the connection to the database for you when we use `read_sql()`, by passing the `sqlite3.connect()` statement within the `.read_sql()` method. Here, you can see that there are two tables. Before you perform some operations on the data, a copy of the database has been made so that the original will be available in the future. In most business situations, you would likely not need to do this. Here, using the two table names, you can read each table into a pandas DataFrame using `pd.read_sql()`, then write them back to a copy, `pet_stores_2.db`, using `.to_sql()`:

```
stores = pd.read_sql("SELECT * FROM Customers",
                     sqlite3.connect('datasets/pet_stores.db'))
invoices = pd.read_sql("SELECT * FROM Invoices",
                     sqlite3.connect('datasets/pet_stores.db'))
stores.to_sql("Customers",
              sqlite3.connect('datasets/pet_stores_2.db'),
              if_exists = 'replace',
              index = True)
invoices.to_sql("Invoices",
                sqlite3.connect('datasets/pet_stores_2.db'),
                if_exists = 'replace',
                index = True)
```

Let's read all the data in the `Customers` table and view it. You can do that with `.read_sql()` as well:

```
customers = \
    pd.read_sql(
    'select Customer_Number, Company, City, State from
Customers',
    sqlite3.connect('datasets/pet_stores_2.db'))customers
```

This produces the following output:

Out[10]:

	Customer_Number	Company	City	State
0	15846	Pet Radio	Minneapolis	MN
1	13197	Just Pets	Columbus	OH
2	11154	Love Strays	Pittsburgh	PA
3	15540	WebPet	Mesa	AZ
4	18397	Pet-ng-Zoo	San Antonio	TX
5	17293	Pet Fud	St. Paul	MN
6	19977	Canine Cravings	Henderson	NV
7	15238	Stock Ur Pet	Stockton	CA
8	15217	Kittie Lullaby	New Orleans	LA
9	17114	Big Dogs Only	Anchorage	AK
10	18448	K9s4Ever	Dallas	TX
11	13388	Bird Sanctuary	Newark	NJ
12	11485	GrrrtoPurr	Plano	TX

Figure 3.43 – The customers for our pet supply company

The `select * from Customers` statement simply says to return all the rows and columns from the table. Suppose you wanted to only look at customers in Texas (`TX`). Here, you can retrieve only those rows by adding a `WHERE` clause to the SQL statement:

```
TX _ customers = \
    pd.read _ sql(
    "select Customer _ Number, Company, City, State from
Customers " +
    "WHERE State = 'TX'",
    sqlite3.connect('datasets/pet _ stores _ 2.db'))     sqlite3.
connect('datasets/pet _ stores _ 2.db'))TX _ customers
```

This produces the following output:

Out[13]:

	Customer_Number	Company	City	State
0	18397	Pet-ng-Zoo	San Antonio	TX
1	18448	K9s4Ever	Dallas	TX
2	11485	GrrrtoPurr	Plano	TX

Figure 3.44 – Listing only TX customers' details

Alternatively, you could accomplish the same thing by slicing the full `Customers` DataFrame you obtained earlier:

```
customers.loc[customers['State'] == 'TX', :]
```

This produces a result that's identical to the preceding one, except for the pandas index:

```
Out[14]:
```

	Customer_Number	Company	City	State
4	18397	Pet-ng-Zoo	San Antonio	TX
10	18448	K9s4Ever	Dallas	TX
12	11485	GrrrtoPurr	Plano	TX

Figure 3.45 – Obtaining the TX customers via slicing in pandas

If the SQL database were extremely large, you may prefer the SQL method, since it only loads the data you want into memory. Contrastingly, if you need the original index numbers, or you have many manipulations to do on different customer groups, then reading the data into a DataFrame and slicing might be preferred.

pandas supports more than simply reading from SQL databases. Suppose you were asked to add some new invoices to the `Invoices` table. First, you would read the table to inspect it:

```
invoices = pd.read_sql("select * from Invoices",\
sqlite3.connect('datasets/pet_stores_2.db'))
print(invoices.head(3), '\n', invoices.tail(3))
```

This produces the following output:

```
   index       Date  Customer_Number        Invoice   Amount
0      0  2/20/2020            18397  2020022018397  1038.95
1      1  2/25/2020            17114  2020022517114  1523.97
2      2  2/25/2020            15846  2020022515846  1535.56
    index       Date  Customer_Number        Invoice   Amount
35     35  3/19/2020            17114  2020031917114  1041.22
36     36  3/19/2020            13388  2020031913388  1043.63
37     37  3/24/2020            15217  2020032415217  1542.85
```

Figure 3.46 – The first and last rows of the Invoices table

Here, you can see there are 38 rows, and unlike the `Customers` table, this table is indexed, as shown by the fact that you get values in the `index` column from the query.

You are provided a `.csv` file with new invoices to add to the database. Here, you can read that into a pandas DataFrame, as follows:

```
new_invoices = pd.read_csv('datasets/new_invoices.csv')
new_invoices
```

This produces the following output:

Out[5]:

	Date	Customer_Number	Invoice	Amount
0	3/24/2020	15846	2020032415846	1355.73
1	3/24/2020	17293	2020032417293	1375.67
2	3/24/2020	18448	2020032418448	1415.38
3	3/24/2020	11485	2020032411485	1025.46
4	3/25/2020	11154	2020032511154	1245.01
5	3/25/2020	13388	2020032513388	1055.32
6	3/25/2020	13197	2020032513197	1105.15
7	3/25/2020	15217	2020032515217	1495.33
8	3/26/2020	17114	2020032617114	1185.30
9	3/26/2020	13197	2020032613197	1290.44
10	3/26/2020	15238	2020032615238	1170.75
11	3/26/2020	18397	2020032618397	1330.36

Figure 3.47 – The new invoices DataFrame

In this case, you should preserve the SQL index and extend it. You can do that in `pandas` by setting the index on the DataFrame to the desired values before you use SQL to append it to the `Invoices` table. Here, you can use the maximum of the existing index plus 1 as a start and create a list using the `range()` function and the size of the `new_invoices` DataFrame:

```
new_invoices.index = list(range(invoices['index'].max() + 1,\
                                invoices['index'].max() +\
                                new_invoices.shape[0] + 1))
new_invoices
```

The updated DataFrame will look as follows:

Out[6]:

	Date	Customer_Number	Invoice	Amount
38	3/24/2020	15846	2020032415846	1355.73
39	3/24/2020	17293	2020032417293	1375.67
40	3/24/2020	18448	2020032418448	1415.38
41	3/24/2020	11485	2020032411485	1025.46
42	3/25/2020	11154	2020032511154	1245.01
43	3/25/2020	13388	2020032513388	1055.32
44	3/25/2020	13197	2020032513197	1105.15
45	3/25/2020	15217	2020032515217	1495.33
46	3/26/2020	17114	2020032617114	1185.30
47	3/26/2020	13197	2020032613197	1290.44
48	3/26/2020	15238	2020032615238	1170.75
49	3/26/2020	18397	2020032618397	1330.36

Figure 3.48 – The new_invoices DataFrame with an updated index

`pandas` provides the `.to_sql()` method, which you can use to add this data to the existing `Invoices` table. You need to specify the table (`Invoices`) and the database file, and since we want to add this data and keep the existing data, you must specify `if_exists = 'append'` and use `index = True`. The `index = True` parameter tells `pandas` to use the `DataFrame` index as the values in the SQL table index:

```
new_invoices.to_sql("Invoices",
                    sqlite3.connect('datasets/pet_store_2.db'),
                    if_exists = 'append',
                    index = True)
```

Now, you can read the `Invoices` table again to see the result:

```
invoices = pd.read_sql("select * from Invoices",
sqlite3.connect('datasets/pet_stores_2.db'))
print(invoices.head(), '\n', invoices.tail())
```

This shows that the new values were added and that the index has been updated properly:

```
   index       Date  Customer_Number        Invoice   Amount
0      0  2/20/2020            18397  2020022018397  1038.95
1      1  2/25/2020            17114  2020022517114  1523.97
2      2  2/25/2020            15846  2020022515846  1535.56
3      3  2/25/2020            15540  2020022515540  1568.95
4      4  2/26/2020            18448  2020022618448  1509.51
    index       Date  Customer_Number        Invoice   Amount
45     45  3/25/2020            15217  2020032515217  1495.33
46     46  3/26/2020            17114  2020032617114  1185.30
47     47  3/26/2020            13197  2020032613197  1290.44
48     48  3/26/2020            15238  2020032615238  1170.75
49     49  3/26/2020            18397  2020032618397  1330.36
```

Figure 3.49 – The updated Invoices table

You are now equipped to work with SQL data, which is an important tool in most business settings. Note that while you can perform many SQL operations using `pandas`, it isn't intended to be a database management system. As you saw in this example, just keeping track of the index adds extra work. But to access SQL data and make some database changes, `pandas` makes this easy to do.

Exercise 3.03 – working with SQL

In this exercise, you will reuse the database from the previous example. You have been given a `.csv` file that contains a few new customers and asked to add it to the database. To accomplish this, you will need to open the existing database, open the `.csv` file, add SQL actions to the `Customers` table, and store the updated SQL database back on disk.

> **Note**
>
> You can find the code for this exercise at `https://github.com/PacktWorkshops/The-Pandas-Workshop/tree/master/Chapter03/Exercise03_03`.

Follow these steps to complete this exercise:

1. Open a new Jupyter notebook and select the **Pandas_Workshop** kernel.
2. For this exercise, all you will need is the `pandas` library and `sqlite3`. Load them into the first cell of the notebook:

```
import pandas as pd
import sqlite3
```

3. Read the names of all the tables in the database and list the tables:

```
tables = \
    pd.read_sql(
    "SELECT name FROM sqlite_master WHERE type = 'table'
ORDER BY name ASC",
    sqlite3.connect('../datasets/company_database.db'))
tables
```

This should product the following output:

```
       name
0 Customers
1  Invoices
```

4. Make a copy of the database by reading both tables and writing them to a new database:

```
customers = pd.read_sql("SELECT * FROM Customers",
                        sqlite3.connect('../datasets/
company_database.db'))
invoices = pd.read_sql("SELECT * FROM Invoices",
                     sqlite3.connect('../datasets/company_
database.db'))
customers.to_sql("Customers",
                 sqlite3.connect('../datasets/company_
database_2.db'),
                 if_exists = 'replace',
                 index = False)
invoices.to_sql("Invoices",
                sqlite3.connect('../datasets/company_
database_2.db'),
                if_exists = 'replace',
                index = False)
```

5. Read in the existing `Customers` table from the `company_database_2.db` SQL database:

```
customers =\
pd.read_sql("select * from Customers",\
sqlite3.connect("../datasets/company_database_2.db"))
customers
```

> **Note**
> Please change the path of the dataset file (highlighted) based on where you have downloaded it on your system.

This will produce the following output:

```
Out[2]:
```

	index	Customer_Number	Company	City	State
0	None	15846	Pet Radio	Minneapolis	MN
1	None	13197	Just Pets	Columbus	OH
2	None	11154	Love Strays	Pittsburgh	PA
3	None	15540	WebPet	Mesa	AZ
4	None	18397	Pet-ng-Zoo	San Antonio	TX
5	None	17293	Pet Fud	St. Paul	MN
6	None	19977	Canine Cravings	Henderson	NV
7	None	15238	Stock Ur Pet	Stockton	CA
8	None	15217	Kittle Lullaby	New Orleans	LA
9	None	17114	Big Dogs Only	Anchorage	AK
10	None	18448	K9s4Ever	Dallas	TX
11	None	13388	Bird Sanctuary	Newark	NJ
12	None	11485	GrrrtoPurr	Plano	TX

Figure 3.50 – The existing Customers table

Note that there are 13 customers in the database.

6. Now, read the `new_customers.csv` file into `pandas`:

```
new_customers = pd.read_csv('../datasets/new_customers.
csv')
new_customers
```

This will produce the following output:

Out[2]:

	Customer_Number	Company	City	State
0	19828	Reptile Desert	Baltimore	MD
1	19186	Aquatic Friends	San Bernadino	CA
2	19948	Arachnaphilia	Newark	NJ
3	19697	Songbird Music Store	Memphis	TX
4	19788	Equestrian Palace	Colorado Springs	CO
5	19115	Just Show Dogs	Baton Rouge	LA
6	19678	My Favorite Butterfly	Lubbock	TX

Figure 3.51 – New customers we want to add to company_database.db

7. Add the data in the `new_customers` DataFrame to the `Customers` table. To append, instead of overwrite, the table, use `if_exists = 'append'`. Set `index = False` as this table hasn't been indexed:

```
new_customers.to_sql\
("Customers",\
sqlite3.connect('../datasets/company_database_2.db'),\
if_exists = 'append',\
index = False)
```

8. Read and display the entire `Customers` table:

```
customers = pd.read_sql("select * from Customers",\
sqlite3.connect("../datasets/company_database_2.db"))
customers
```

This should produce the following output:

Out[6]:

	index	Customer_Number	Company	City	State
0	None	15846	Pet Radio	Minneapolis	MN
1	None	13197	Just Pets	Columbus	OH
2	None	11154	Love Strays	Pittsburgh	PA
3	None	15540	WebPet	Mesa	AZ
4	None	18397	Pet-ng-Zoo	San Antonio	TX
5	None	17293	Pet Fud	St. Paul	MN
6	None	19977	Canine Cravings	Henderson	NV
7	None	15238	Stock Ur Pet	Stockton	CA
8	None	15217	Kittle Lullaby	New Orleans	LA
9	None	17114	Big Dogs Only	Anchorage	AK
10	None	18448	K9s4Ever	Dallas	TX
11	None	13388	Bird Sanctuary	Newark	NJ
12	None	11485	GrrrtoPurr	Plano	TX
13	None	19828	Reptile Desert	Baltimore	MD
14	None	19186	Aquatic Friends	San Bernadino	CA
15	None	19948	Arachnaphilia	Newark	NJ
16	None	19697	Songbird Music Store	Memphis	TX
17	None	19788	Equestrian Palace	Colorado Springs	CO
18	None	19115	Just Show Dogs	Baton Rouge	LA
19	None	19678	My Favorite Butterfly	Lubbock	TX

Figure 3.52 – The updated Customers table from company_database_2.db

The output shows that there are 20 customers in the database. Note that if you repeated the `.to_sql()` portion, you would add more, duplicate, rows to the database, which would likely cause problems. You can try this and see the results, but in a real business scenario, you should avoid doing so. If this were a routine operation in business, we would create more robust code to check for the existence of records before adding them again.

In this exercise, you used the `.to_sql()` method to update a SQL database table. SQL data is ubiquitous in analytics and data science and having a comfort level while working with SQL data is a valuable skill. pandas makes basic operations easy and can enable you to perform analytics in pandas from SQL data.

> **Google BigQuery**
>
> Google BigQuery is a Data Warehouse and SQL database that provides fast access to big data in the cloud using SQL. Although pandas provides `.read_gbq()` to read from BigQuery, it requires the `pandas-gbq` package. Google, on the other hand, has a `google-cloud-bigquery` library. Since the question of which to use arises often, Google has posted a comparison here: `https://cloud.google.com/bigquery/docs/pandas-gbq-migration`. It may be that the Google library is preferred in the future since presumably, Google would support any updates or changes to the BigQuery services.

Choosing a format for a project

You've seen a wide range of text and binary formats so far in this chapter, including databases. If you are creating a new project, you may need to decide on which data format is most suitable for it. The most important question to answer is, who else will use the data from the project, and how? If the data is provided by your project via an API to users, then JSON is a good choice. If the data is naturally organized into multiple tables, a relational database may be a good choice. However, even in that case, it may make sense to create individual tables and merge them into the database in another step.

This is a good segue to the topic of **data pipelines**. In larger production projects, there is ongoing data creation, ingestion, transformation, storage, and more. Upfront design effort and communication with stakeholders are required if you wish to design the best data pipeline. This is so important that job titles such as Data Engineer, Machine Learning Engineer, and similar titles are now common. Keep in mind that most large data-centric projects are team efforts and require a lot of constant communication among team members.

Packt offers several titles in the area of data engineering, including *Data Engineering with Python, Big Data Architect's Handbook, Architecting Data-Intensive Applications*, and *GCP: Google Cloud Platform: Data Engineer, Cloud Architect*. If you find yourself addressing data architecture, take a look at the Packt catalog.

> **Other Formats and Methods**
>
> As we have seen with `to_excel()`, `to_csv()`, `to_sql()`, `to_stata()`, and `to_hdf()`, pandas provides methods for creating data in various formats from DataFrames, such as JSON and HTML.
>
> In addition to the methods we've discussed in this chapter, there are additional methods in pandas that support common formats that you may encounter in data projects or other work. Examples include Parquet format, which is a common big data format, and Pickle format, which is used for a variety of tasks, such as saving trained machine learning models and complex data structures so that they can be reloaded into Python later. If you encounter something that's not listed, searching for a Python module may be fruitful. Refer to the pandas documentation for all the supported formats.

Activity 3.01 – using SQL data for pandas analytics

As the data analyst at a supply company, you have been provided with a list of customers and orders for Q4 2020. The data is present in some tables in a database named `supply_company.db`, and you have been asked by the sales team to identify the largest purchasing customer for Q4.

> **Note**
>
> You can find the code for this activity at `https://github.com/PacktWorkshops/The-Pandas-Workshop/tree/master/Chapter03/Activity03_01`.

Follow these steps to complete this activity:

1. For this activity, you will only need the `pandas` and `sqlite3` libraries. Load them into the first cell of the notebook.
2. Get the list of tables that are contained in the `supply_company.db` file. The database can be downloaded from `https://github.com/PacktWorkshops/The-Pandas-Workshop/blob/master/Chapter03/datasets/supply_company.db`.
3. Use a pandas SQL method to load the table that contains the orders into a DataFrame.
4. Determine the number of customers with the largest sales in the data you have been given.

In SQL, you could construct SQL code to answer this question. However, for this activity, you'll just use pandas and Python. The following code will retrieve the ID of the customer with the largest purchases in the data. Here, `.groupby()` aggregates the data by the customer ID, while `sum()` tells pandas to do the aggregation by summing up the values. This has the effect that if a customer has more than one order, the total amounts will be summed up. `['amount']` is simply indexing the `amount` column, while `.sort_values(ascending = False)` is sorting the largest value in `amount` to the first position. Finally, `index[0]` returns the customer ID since `.groupby()` makes the grouping argument the index:

```
Largest_cust= \
    orders.groupby('Customer_Number').sum()['amount'].\
                            sort_values(ascending = False).
index[0]
largest
```

This should return the ID number of the target customer.

5. Find and list the row for this customer in the table containing the list of customers.

> **Note**
> You can find the solution for this activity in the Appendix.

Summary

In this chapter, you saw how pandas supports data I/O to and from a wide variety of formats, both text and digital. You saw how pandas supports acting on multi-table databases in SQL directly from Python. You also explored the different character encodings that you may encounter in text data, as well as how to extract only the desired data columns from a more complex Excel file. Given the large amounts of data on the internet, you saw how pandas can extract tables from web pages and decode more complex web data in XML or JSON formats. You also learned how to use APIs to obtain data. In most cases, all you need is the pandas `.read_xxx()` and `.to_xxx()` methods. With what you have learned and practiced in this chapter, you are ready to handle most data sources you may encounter in your work.

Here, you've focused on getting data into and out of pandas DataFrames from a wide range of file types. In the next chapter, you will begin digging into the finer details and exploring the various data types that pandas supports. When you read data from a file, pandas will use certain data types (such as strings or numbers) by default, but pandas gives you a lot of control regarding this. In this regard, you will learn how to work with and manipulate data types.

Summary

4
Pandas Data Types

In this chapter, you will learn about the different pandas data types and how to convert them from one type into another. By the end of this chapter, you will know how to inspect the pandas data types of DataFrames, and how to manipulate them for your analysis.

In this chapter, we will cover the following topics:

- Introducing pandas dtypes
- Missing data types
- Subsetting by data types

Introducing pandas dtypes

When working with pandas, it is vital to make sure that you assign the correct data types to the values you're working with. Otherwise, you may end up getting unexpected results or errors when running certain operations or calculating aggregations. Having a good understanding of every data type in pandas will save you a lot of time and energy as you will considerably reduce the number of errors in your code.

Data types in pandas are internal labels that a programming language uses to understand how to store and manipulate data. For example, a program needs to understand that you can add two numbers together, such as *1* + *2*, to get *3*. Or, if you have two strings such as "**data**" and "**frame**," they can be concatenated to get "**DataFrame**."

Data types in pandas are called **dtypes** and should not be confused with Python's data type. We shall be using both data types and dtypes interchangeably throughout this chapter.

Obtaining the underlying data types

It is crucial to understand data types when you're analyzing data. The following table lists the data types in pandas, along with relevant examples:

pandas dtypes	Usage	Examples
`object`	Includes any Python object such as a list, dict, set, str, or a user-defined class object.	`"pandas"`, `"pandas 1.0.5"`
`int64`, `int32`, `int8`	Integer numbers – the postfixes 64, 32, and 8 indicate the number of bits that have been reserved for the respective data type in memory.	`100`, `5000`
`float64`, `float32`	Floating-point numbers – again, the postfixes 64 and 32 refer to the number of bits that have been reserved for these types in memory.	`1.05`, `0.0004`
`bool`	Boolean values.	`True`, `False`
`datetime64[ns]`	Date and time values.	`"2015-01-10"`, `"2020-11-15 00:00:00"`
`timedelta[ns]`	Differences between two datetimes.	`"-2736 days"`
category	List of unique text values.	`["pandas", "dataframe"]`

Figure 4.1 – List of pandas dtypes

Let's start with an example where we're analyzing some customer details:

1. We'll start by importing the pandas and numpy libraries:

```
import pandas as pd
import numpy as np
```

2. Next, we'll define the column names and fill in the values in the individual rows, as follows:

```
column_names = ["Customer ID", "Customer Name",\
                "2018 Revenue", "2019 Revenue",\
                "Growth", "Start Year", "Start Month",\
                "Start Day", "New Customer"]
```

```
row1 = list([1001.0, 'Pandas Banking',\
             '€235000', '€248000',\
             '5.5%', 2013,3,10, 0])
row2 = list([1002.0, 'Pandas Grocery', \
             '€196000', '€205000', \
             '4.5%', 2016,4,30, 0])
row3 = list([1003.0, 'Pandas Telecom', \
             '€167000', '€193000', '15.5%',\
             2010,11,24, 0])
row4 = list([1004.0, 'Pandas Transport',\
             '€79000', '€90000', '13.9%', \
             2018,1,15, 1])
row5 = list([1005.0, 'Pandas Insurance', \
             '€241000', '€264000', '9.5%',\
             2009,6,1, 0])
```

3. Finally, we'll create a DataFrame from the rows we have defined and display it, as follows:

```
data_frame = pd.DataFrame(data=[row1, row2, row3, row4, 
row5],\
                          columns=column_names)
data_frame
```

The output will be as follows:

	Customer ID	Customer Name	2018 Revenue	2019 Revenue	Growth	Start Year	Start Month	Start Day	New Customer
0	1001.0	Pandas Banking	€235000	€248000	5.5%	2013	3	10	0
1	1002.0	Pandas Grocery	€196000	€205000	4.5%	2016	4	30	0
2	1003.0	Pandas Telecom	€167000	€193000	15.5%	2010	11	24	0
3	1004.0	Pandas Transport	€79000	€90000	13.9%	2018	1	15	1
4	1005.0	Pandas Insurance	€241000	€264000	9.5%	2009	6	1	0

Figure 4.2 – Displaying the customer details DataFrame

4. With that, our DataFrame is ready. Now, we can try performing some operations to analyze the data. Let's try adding the values for the `2018 Revenue` and `2019 Revenue` columns, as follows:

```
data_frame['2018 Revenue'] + data_frame['2019 Revenue']
```

The output will be as follows:

```
0     €235000€248000
1     €196000€205000
2     €167000€193000
3       €79000€90000
4     €241000€264000
dtype: object
```

Figure 4.3 – Added values for 2018 and 2019 revenue

This result is not what we were expecting. What we expected was the sum after adding the revenues for 2018 and 2019. Instead, pandas concatenated the values to create a new *string*. This behavior can be explained by the last line, `dtype: object`, which tells us that pandas recognizes the data in the columns as being of the `object` type instead of recognizing them as numbers. A string data type is recognized as the `object` type in pandas, which causes string addition (that is, concatenation) rather than numerical addition.

This is a perfect example of why it's important to make sure that pandas is assigning the correct dtypes to our data before we can work with it.

To see what dtype has been assigned to each column, you can use the `dtypes` attribute of the DataFrame, as follows:

```
data_frame.dtypes
```

You will see the following output:

```
Customer ID       float64
Customer Name      object
2018 Revenue       object
2019 Revenue       object
Growth             object
Start Year          int64
Start Month         int64
Start Day           int64
New Customer        int64
dtype: object
```

Figure 4.4 – Viewing the data types that have been assigned to each column

We can also use the `info()` attribute of the DataFrame to get more details:

```
data_frame.info()
```

We'll see the following output after running this snippet:

```
<class 'pandas.core.frame.DataFrame'>
RangeIndex: 5 entries, 0 to 4
Data columns (total 9 columns):
 #   Column         Non-Null Count  Dtype
---  ------         --------------  -----
 0   Customer ID    5 non-null      float64
 1   Customer Name  5 non-null      object
 2   2018 Revenue   5 non-null      object
 3   2019 Revenue   5 non-null      object
 4   Growth         5 non-null      object
 5   Start Year     5 non-null      int64
 6   Start Month    5 non-null      int64
 7   Start Day      5 non-null      int64
 8   New Customer   5 non-null      int64
dtypes: float64(1), int64(4), object(4)
memory usage: 488.0+ bytes
```

Figure 4.5 – Complete details regarding the DataFrame

Looking at the assigned data types, we can see that there are several concerns:

- `Customer ID` has `float64` as its type. However, ideally, we would want it to be an integer.
- `2018 Revenue`, `2019 Revenue`, and `Growth` are stored as objects, but these should be numerical values.

```
Start Year, Start Month, and Start Day are stored as int64
values.
We want to have these as Datetime objects.
```

- `New Customer` is an `int64` but should be a `boolean` (this condition can either be `True` or `False`).
- `Customer Name` is stored as an `object` but should be a category.

Until we convert these data types, it is going to be very difficult to use the data productively. We'll learn how to do that in the next section.

Converting from one type into another

To convert data types in pandas, you have three main options:

- Use the `astype()` function to force an appropriate dtype.
- Use pandas functions such as `to_numeric()` or `to_datetime()`.
- Create a custom function to convert the data type.

Continuing with the example from the previous section, let's learn how to use the `astype()` function. In the following snippet, we're using it to convert `Customer ID` from a float into an integer. We'll also need to assign it back to the `Customer ID` column after the conversion since the `astype()` function returns a copy:

```
data_frame["Customer ID"] =data_frame['Customer ID'].
astype('int')
data_frame["Customer ID"]
```

The output will be as follows:

```
0    1001
1    1002
2    1003
3    1004
4    1005
Name: Customer ID, dtype: int32
```

Figure 4.6 – The Customer ID column after data type conversion

Now, let's try doing the same for `2018 Revenue`:

```
data_frame['2018 Revenue'] =data_frame['2018 Revenue'].
astype('int')
```

You'll see the following output:

```
ValueError:   invalid literal for int() with base 10:
'€235000'
```

The preceding code raises an error because pandas considers the mix of a number and the € symbol to be a string. In this case, we need to create a custom function to remove the € symbol and convert the remaining number into an integer, as follows:

```
def remove_currency(column):
    new_column = column.replace('€', '')
    return int(new_column)
```

This function will replace the € symbol with a blank using the `replace()` function, and then convert the result into an integer using the `int()` function.

Now, we can use this function on the `2018 Revenue` column by using the `apply()` function:

```
data_frame['2018 Revenue'] =\
data_frame['2018 Revenue'].apply(remove_currency)
data_frame["2018 Revenue"]
```

The output will be as follows:

```
0    235000
1    196000
2    167000
3     79000
4    241000
Name: 2018 Revenue, dtype: int64
```

Figure 4.7 – The 2018 Revenue column after the remove_currency function has been applied to it

This time, we got the expected output. We can repeat this process for `2019 Revenue`:

```
data_frame['2019 Revenue'] =\
data_frame['2019 Revenue'].apply(remove_currency)
data_frame["2019 Revenue"]
```

The output will be as follows:

```
0    248000
1    205000
2    193000
3     90000
4    264000
Name: 2019 Revenue, dtype: int64
```

Figure 4.8 – The 2019 Revenue column after the remove_currency function has been applied to it

We can create another function for the `Growth` column to remove the `%` symbol and convert it from an `object` into a `float`:

```
def remove_percentage(column):
    new_column = column.replace('%', '')
    return float(new_column)

data_frame['Growth'] = data_frame['Growth'].apply(remove_
percentage)
data_frame["Growth"]
```

The output will be as follows:

```
0     5.5
1     4.5
2    15.5
3    13.9
4     9.5
Name: Growth, dtype: float64
```

Figure 4.9 – The Growth column after the remove_percentage function has been applied

Now, we need to create a new column named `Starting Date` from `Start Year`, `Start Month`, and `Start Day`. We can use the `to_datetime()` function from pandas but it requires the column's name to include `year`, `month`, and `day`, so we'll change those first:

```
data_frame.rename(columns={'Start Year': 'year',\
                           'Start Month': 'month',\
                           'Start Day': 'day'},\
                  inplace=True)


data_frame['Starting Date'] =\
pd.to_datetime(data_frame[['day', 'month', 'year']])
data_frame['Starting Date']
```

Running this code produces the following output:

```
0    2013-03-10
1    2016-04-30
2    2010-11-24
3    2018-01-15
4    2009-06-01
Name: Starting Date, dtype: datetime64[ns]
```

Figure 4.10 – The Starting Date column after the .to_datetime function has been applied

You can verify the type of data conversion from the last line of output, which is `datetime64[ns]`. This indicates that the data is now a `datetime` object and that we can directly perform date and time-related operations with it.

In the following snippet, we are converting `New Customer` from an `int64` into a `bool`. This can easily be done using the `astype()` function:

```
data_frame["New Customer"] =\
data_frame['New Customer'].astype('bool')
data_frame["New Customer"]
```

You will see the following output:

```
0    False
1    False
2    False
3     True
4    False
Name: New Customer, dtype: bool
```

Figure 4.11 – The New Customer column after converting its data type into a bool

Finally, we must convert `Customer Name` from an `object` into a `category`. Having `Customer Name` as a `category` will optimize our memory usage, as we'll see in more detail later in this chapter. This conversion can also be done by using the `astype()` function, as follows:

```
data_frame["Customer Name"] =\
data_frame['Customer Name'].astype('category')
data_frame["Customer Name"]
```

Running this results in the following output:

```
0       Pandas Banking
1       Pandas Grocery
2       Pandas Telecom
3     Pandas Transport
4     Pandas Insurance
Name: Customer Name, dtype: category
Categories (5, object): [Pandas Banking, Pandas Grocery, Pandas Insurance, Pandas Telecom, Pandas Transport]
```

Figure 4.12 – The Customer Name column

Now, let's display our `dtypes` in our DataFrame again:

```
data_frame.dtypes
```

The output will be as follows:

```
Customer ID                int32
Customer Name           category
2018 Revenue               int64
2019 Revenue               int64
Growth                    object
year                       int64
month                      int64
day                        int64
New Customer                bool
Starting Date     datetime64[ns]
dtype: object
```

Figure 4.13 – Displaying the DataFrame

Now, we can try running a few operations to check that everything is working as desired. For example, let's see if we can add `2018 Revenue` and `2019 Revenue`:

```
data_frame['2018 Revenue'] + data_frame['2019 Revenue']
```

We'll get the following output:

```
0    483000
1    401000
2    360000
3    169000
4    505000
dtype: int64
```

Figure 4.14 – Adding the revenues for 2018 and 2019

This time, the addition operation works properly since we have converted the data in both columns into integers.

Next, let's try to find out how many days have passed since the starting date of each row from September 1, 2020:

```
data_frame['Starting Date'] - pd.to_datetime('2020-09-01')
```

The output will be as follows:

```
0   -2732 days
1   -1585 days
2   -3569 days
3    -960 days
4   -4110 days
Name: Starting Date, dtype: timedelta64[ns]
```

Figure 4.15 – Finding the number of days that have passed since the start date of each row from September 1, 2020

We had to convert `2020-09-01` into a `datetime` as pandas would assign it to a string data type. This operation would have been invalid in that case. The result of the conversion is the `timedelta64[ns]` data type, which we haven't seen yet. This dtype indicates that the values that have been stored show the difference between two datetimes.

Exercise 4.01 – underlying data types and conversion

In this exercise, you will read a dataset into a DataFrame and find out whether the columns have proper data types. The goal here is to convert the data types wherever necessary to make the DataFrame suitable for further analysis.

> **Note**
>
> The data for this exercise can be found in the `retail_purchase.csv` file, which you can find at `https://github.com/PacktWorkshops/The-Pandas-Workshop/blob/master/Chapter04/Data/retail_purchase.csv`.

Follow these steps to complete this exercise:

1. Open a new Jupyter notebook and select the **Pandas_Workshop** kernel.
2. Import the pandas library:

```
import pandas as pd
```

3. Next, load the CSV file as a DataFrame into the notebook and read the CSV file:

```
file_url = '..//Data//retail_purchase.csv'
data_frame = pd.read_csv(file_url)
```

> **Note**
> Please change the path of the file (highlighted) based on where you have downloaded the file on your machine.

4. Use the head() function to display the first five rows of the DataFrame:

```
data_frame.head()
```

The output will be as follows:

	Receipt Id	Date of Purchase	Product Name	Product Weight	Total Price	Retail shop name
0	10001	24/05/20	Wheat	4.8lb	€17	Fline Store
1	10002	05/05/20	Fruit Juice	3.1lb	€19	Dello Superstore
2	10003	27/04/20	Vegetables	1.2lb	€15	Javies Retail
3	10004	05/05/20	Oil	3.1lb	€17	Javies Retail
4	10005	27/04/20	Wheat	4.8lb	€13	Javies Retail

Figure 4.16 – Displaying the first five rows of the DataFrame

5. Use the tail() function to display the last five rows of the DataFrame:

```
data_frame.tail()
```

You should see the following output:

	Receipt Id	Date of Purchase	Product Name	Product Weight	Total Price	Retail shop name
99995	109996	24/05/20	Oil	4.8lb	€25	Visco Retail
99996	109997	20/04/20	Rice	3.1lb	€12	Kelly Superstore
99997	109998	08/01/20	Fruit Juice	2.7lb	€24	Dello Superstore
99998	109999	05/05/20	Butter	3.1lb	€22	Dello Superstore
99999	110000	17/04/20	Bread	4.4lb	€27	Visco Retail

Figure 4.17 – Displaying the last five rows of the DataFrame

After looking at the DataFrame, we can decide on the desired type for each column. The **Receipt Id** column should be `int`, **Date of Purchase** should be `datetime`, and **Total Price** and **Product Weight** need to be converted into `float`. Similarly, the **Product Name** and **Retail shop name** columns need to be converted into `category`.

6. Use the `info()` function to display the data types of each column in the DataFrame:

```
data_frame.info()
```

The output will be as follows:

```
<class 'pandas.core.frame.DataFrame'>
RangeIndex: 100000 entries, 0 to 99999
Data columns (total 6 columns):
 #   Column             Non-Null Count   Dtype
---  ------             --------------   -----
 0   Receipt Id         100000 non-null  int64
 1   Date of Purchase   100000 non-null  object
 2   Product Name       100000 non-null  object
 3   Product Weight     100000 non-null  object
 4   Total Price        100000 non-null  object
 5   Retail shop name   100000 non-null  object
dtypes: int64(1), object(5)
memory usage: 4.6+ MB
```

Figure 4.18 – Displaying the full details of the DataFrame

As shown in the preceding screenshot, it looks like all the columns except for `Receipt Id` have been loaded as a string (`Dtype = object`). You need to convert them into the desired type.

7. Convert `Date of Purchase` into `datetime` and display it:

```
data_frame['Date of Purchase'] =\
pd.to_datetime(data_frame['Date of Purchase'],\
               format='%d/%m/%y')
data_frame['Date of Purchase']
```

You will get the following result:

```
0        2020-05-24
1        2020-05-05
2        2020-04-27
3        2020-05-05
4        2020-04-27
            ...
99995    2020-05-24
99996    2020-04-20
99997    2020-01-08
99998    2020-05-05
99999    2020-04-17
Name: Date of Purchase, Length: 100000, dtype: datetime64[ns]
```

Figure 4.19 – The Date of Purchase column after its data type has been converted into a datetime

Here, we can see that the `Date of Purchase` column has been correctly converted into a `datetime`.

8. Remove the € symbol from `Total Price` and display it:

```
data_frame['Total Price'] =\
data_frame['Total Price'].str[1:]
data_frame['Total Price']
```

Since € is the first character in each row of `Total Price`, you can use the `str[1:]` method to only keep the characters after the first one. The output will be as follows:

```
0        17
1        19
2        15
3        17
4        13
         ..
99995    25
99996    12
99997    24
99998    22
99999    27
Name: Total Price, Length: 100000, dtype: object
```

Figure 4.20 – The Total Price column

9. Convert `Total Price` into a `float` and then display it:

```
data_frame['Total Price'] =\
data_frame['Total Price'].astype('float')
data_frame['Total Price']
```

The output will be as follows:

```
0        17.0
1        19.0
2        15.0
3        17.0
4        13.0
         ...
99995    25.0
99996    12.0
99997    24.0
99998    22.0
99999    27.0
Name: Total Price, Length: 100000, dtype: float64
```

Figure 4.21 – The Total Price column after its data type has been converted into a float64

The preceding screenshot shows that `Total Price` has been successfully converted into a `float64`.

Remove `lb` from `Product Weight`, then display it. Since `lb` represents the last two first characters in each row of `Product Weight`, you can use `str[:-2]` to only keep characters before the last two:

```
data_frame['Product Weight'] =\
data_frame['Product Weight'].str[:-2]
data_frame['Product Weight']
```

The output will be as follows:

```
0        4.8
1        3.1
2        1.2
3        3.1
4        4.8
        ...
99995    4.8
99996    3.1
99997    2.7
99998    3.1
99999    4.4
Name: Product Weight, Length: 100000, dtype: object
```

Figure 4.22 – The Product Weight column

10. Convert `Product Weight` into a `float` and display it:

```
data_frame['Product Weight'] =\
data_frame['Product Weight'].astype('float')
data_frame['Product_Weight']
```

You should see the following output:

```
0        4.8
1        3.1
2        1.2
3        3.1
4        4.8
        ...
99995    4.8
99996    3.1
99997    2.7
99998    3.1
99999    4.4
Name: Product Weight, Length: 100000, dtype: float64
```

Figure 4.23 – The Product Weight column after converting its data type into a float64

11. Use the `unique()` function to find every unique value from `Product Name`:

```
data_frame['Product Name'].unique()
```

The output will be as follows:

```
array(['Wheat', 'Fruit Juice', 'Vegetables', 'Oil',
'Butter', 'Fruits',
          'Cheese', 'Rice', 'Bread'], dtype=object)
```

12. Convert `Product Name` into a `category` and then display it:

```
data_frame['Product Name'] =\
data_frame['Product Name'].astype('category')
data_frame['Product Name']
```

Running this results in the following output:

```
0              Wheat
1        Fruit Juice
2         Vegetables
3                Oil
4              Wheat
            ...
99995            Oil
99996           Rice
99997    Fruit Juice
99998         Butter
99999          Bread
Name: Product Name, Length: 100000, dtype: category
Categories (9, object): [Bread, Butter, Cheese, Fruit Juice, ..., Oil, Rice, Vegetables, Wheat]
```

Figure 4.24 – The Product Name column after converting it into a category

13. Repeat the previous two steps for `Retail shop name`. First, use the `unique()` function to find every unique value in `Retail shop name`:

```
data_frame['Retail shop name'].unique()
```

You will get the following output:

```
array(['Fline Store', 'Dello Superstore', 'Javies
Retail',
           'Oldi Superstore', 'Kanes Store', 'Kelly
Superstore',
           'Visco Retail', 'Rotero Retail'],
dtype=object)
```

14. Now, convert `Retail shop name` into a `category` and display it:

```
data_frame['Retail shop name'] =\
data_frame['Retail shop name'].astype('category')
data_frame['Retail shop name']
```

The output will be as follows:

```
0           Fline Store
1       Dello Superstore
2          Javies Retail
3          Javies Retail
4          Javies Retail
              ...
99995      Visco Retail
99996   Kelly Superstore
99997   Dello Superstore
99998   Dello Superstore
99999      Visco Retail
Name: Retail shop name, Length: 100000, dtype: category
Categories (8, object): [Dello Superstore, Fline Store, Javies Retail, Kanes Store, Kelly Superstore, Oldi Superstor
e, Rotero Retail, Visco Retail]
```

Figure 4.25 – The Retail shop name column after converting it into a category

15. Now that the data types have been converted for the columns, use the `info()` function to display the data types of each column, as follows:

```
data_frame.info()
```

You will see the following output:

```
<class 'pandas.core.frame.DataFrame'>
RangeIndex: 100000 entries, 0 to 99999
Data columns (total 6 columns):
 #   Column            Non-Null Count   Dtype
---  ------            --------------   -----
 0   Receipt Id        100000 non-null  int64
 1   Date of Purchase  100000 non-null  datetime64[ns]
 2   Product Name      100000 non-null  category
 3   Product Weight    100000 non-null  float64
 4   Total Price       100000 non-null  float64
 5   Retail shop name  100000 non-null  category
dtypes: category(2), datetime64[ns](1), float64(2), int64(1)
memory usage: 3.2 MB
```

Figure 4.26 – Full details of the DataFrame

With that, all the columns have been converted into the desired data types and can be used for further analysis.

In this exercise, you learned how to inspect the data types of each column inside a DataFrame and convert them into a usable format if needed.

Now, you know how to identify incorrectly assigned data types and correct them. But how do you deal with situations where the data type itself is missing from your data? We'll look at this in the next section.

Missing data types

While working with real-world datasets, you are bound to encounter missing data quite frequently during data analysis. Understanding how pandas displays missing data for each dtype is crucial to ensure that your data analysis is correct.

The missing alphabet soup

In the previous section, we learned about the different data types and how to convert them if needed. Here, we will learn about how to represent missing data for each data type.

We will continue with our previous example. However, this time, we will replace some values with `None`, as follows:

```
data_frame.drop(['year','month','day'], axis = 1, inplace=True)

data_frame.iloc[0,0] = None
data_frame.iloc[4,1] = None
data_frame.iloc[2,2] = None
data_frame.iloc[3,3] = None
data_frame.iloc[3,4] = None
data_frame.iloc[1,5] = None
data_frame.iloc[2,6] = None

data_frame
```

Upon running this snippet, you should see the following output:

	Customer ID	Customer Name	2018 Revenue	2019 Revenue	Growth	New Customer	Starting Date
0	NaN	Pandas Banking	235000.0	248000.0	5.5	0.0	2013-03-10
1	1002.0	Pandas Grocery	196000.0	205000.0	4.5	NaN	2016-04-30
2	1003.0	Pandas Telecom	NaN	193000.0	15.5	0.0	NaT
3	1004.0	Pandas Transport	79000.0	NaN	NaN	1.0	2018-01-15
4	1005.0	NaN	241000.0	264000.0	9.5	0.0	2009-06-01

Figure 4.27 – Replacing some values with None

Now, let's use the `info()` method on the DataFrame to get more details:

```
data_frame.info()
```

The output will be as follows:

```
<class 'pandas.core.frame.DataFrame'>
RangeIndex: 5 entries, 0 to 4
Data columns (total 7 columns):
 #   Column         Non-Null Count  Dtype
---  ------         --------------  -----
 0   Customer ID    4 non-null      float64
 1   Customer Name  4 non-null      category
 2   2018 Revenue   4 non-null      float64
 3   2019 Revenue   4 non-null      float64
 4   Growth         4 non-null      float64
 5   New Customer   4 non-null      float64
 6   Starting Date  4 non-null      datetime64[ns]
dtypes: category(1), datetime64[ns](1), float64(5)
memory usage: 573.0 bytes
```

Figure 4.28 – Full details of the DataFrame

Here, we can see that `Non-Null Count` is not equal to `5` anymore for each column since we have null values in the DataFrame. There are three ways pandas displays null values:

1. `NaN` ("Not a Number")
2. `None`
3. `NaT` ("Not a Time")

An interesting observation is that by introducing null values, some of the dtypes have been changed. For example, `Customer ID`, `2018 Revenue`, and `2019 Revenue` were `int64` and now they have been changed to `float64`. The same happened with `New Customer`, which was `bool` previously and is now `float64` too. This means that some data types can be `NaN` while other data types can be `NaT`. We call such data types nullable data types. We'll learn about them in the next section.

Nullable types

As we saw in the previous example, some dtypes allow null values, whereas other dtypes will not allow null values and will force them to be converted into types that can have null values.

To find nullable data types, we will use our previous example and try to convert them into other data types. Let's start with `Customer ID` by trying to convert it back into an `int64`:

```
data_frame["Customer ID"] =\
data_frame['Customer ID'].astype('int')
data_frame["Customer ID"]
```

You will see the following output:

```
ValueError:  Cannot convert non-finite values (NA or inf) to
integer
```

It seems that `int64` is a dtype that doesn't allow null values, but `float64` is nullable.

Now, let's try converting `Customer Name` into an `object`:

```
data_frame["Customer Name"] =\
data_frame['Customer Name'].astype('object')
data_frame["Customer Name"]
```

The output will be as follows:

```
0      Pandas Banking
1      Pandas Grocery
2      Pandas Telecom
3    Pandas Transport
4                 NaN
Name: Customer Name, dtype: object
```

Figure 4.29 – Converting the data type of Customer Name into an object

This conversion was successful because both `category` and `object` are nullable.

Now, let's try converting `New Customer`, which is a floating-point integer, back into a `bool`:

```
data_frame["New Customer"] =\
data_frame['New Customer'].astype('bool')
data_frame["New Customer"]
```

The output will be as follows:

```
0    False
1     True
2    False
3     True
4    False
Name: New Customer, dtype: bool
```

Figure 4.30 – Converting the data type of New Customer back into a bool

This case required a bit of caution; `bool` is not nullable, but a float with a nullable value can be converted into a Boolean. The null value has been converted into `True`.

Now, let's try calculating the number of differences between `Starting Date` and `2020-09-01`:

```
data_frame['Starting Date'] - pd.to_datetime('2020-09-01')
```

The output will be as follows:

```
0   -2732 days
1   -1585 days
2          NaT
3    -960 days
4   -4110 days
Name: Starting Date, dtype: timedelta64[ns]
```

Figure 4.31 – Calculating the differences between Starting Date and 2020-09-01

Both `datetime64[ns]` and `timedelta64[ns]` are nullable.

Exercise 4.02 – missing data and converting into non-nullable dtypes

In this exercise, you will read a DataFrame using `pd.read_csv()`. Then, you will handle the missing data in each column. Finally, you will convert these columns into their right dtypes, including the non-nullable dtypes.

> **Note**
>
> The data that we'll be using in this exercise, can be found in the `retail_purchase_missing.csv` file, which you can find on GitHub at `https://raw.githubusercontent.com/PacktWorkshops/The-Pandas-Workshop/master/Chapter04/Data/retail_purchase_missing.csv`.

Follow these steps to complete this exercise:

1. Open a new Jupyter notebook and select the **Pandas_Workshop** kernel.
2. Import `pandas` using the following command:

```
import pandas as pd
```

3. Load the CSV file as a DataFrame into the notebook and read the CSV file:

```
file_url = 'https://github.com/PacktWorkshops/The-Pandas-Workshop/blob/master/Chapter04/Data/retail_purchase.csv?raw=true'
data_frame = pd.read_csv(file_url)
```

4. Use the `head()` function to display the first five rows of the DataFrame:

```
data_frame.head()
```

The output will be as follows:

	Receipt Id	Date of Purchase	Product Name	Product Weight	Total Price	Retail shop name
0	10001.0	24/05/20	Wheat	87.0	99.0	NaN
1	NaN	05/05/20	NaN	NaN	25.0	Dello Superstore
2	10003.0	27/04/20	Vegetables	19.0	37.0	Javies Retail
3	10004.0	05/05/20	Oil	99.0	44.0	Javies Retail
4	10005.0	NaN	Wheat	30.0	NaN	Javies Retail

Figure 4.32 – Displaying the first five rows of the DataFrame

You can already see some missing data (`NaN`) in every column.

5. Use the `info()` function to display the data types of each column in the DataFrame:

```
data_frame.info()
```

The output will be as follows:

```
<class 'pandas.core.frame.DataFrame'>
RangeIndex: 58 entries, 0 to 57
Data columns (total 6 columns):
 #   Column            Non-Null Count  Dtype
---  ------            --------------  -----
 0   Receipt Id        44 non-null     float64
 1   Date of Purchase  46 non-null     object
 2   Product Name      46 non-null     object
 3   Product Weight    51 non-null     float64
 4   Total Price       51 non-null     float64
 5   Retail shop name  45 non-null     object
dtypes: float64(3), object(3)
memory usage: 2.8+ KB
```

Figure 4.33 – Displaying the full details of the DataFrame

This confirms that there is some missing data as we don't have 58 (the total number of entries) non-null values for any column.

6. To handle this missing data, you can use the `fillna()` function, which replaces every missing value in a column with a chosen value. Replace the missing values by `0` in every numeric column (`Receipt Id`, `Product Weight`, and `Total Price`) and display the top five rows:

```
data_frame.fillna(value = {'Receipt Id': 0, \
                           'Product Weight': 0,\
```

```
                                'Total Price': 0},\
                        inplace= True)
data_frame.head()
```

The output will be as follows:

	Receipt Id	Date of Purchase	Product Name	Product Weight	Total Price	Retail shop name
0	10001.0	24/05/20	Wheat	87.0	99.0	NaN
1	0.0	05/05/20	NaN	0.0	25.0	Dello Superstore
2	10003.0	27/04/20	Vegetables	19.0	37.0	Javies Retail
3	10004.0	05/05/20	Oil	99.0	44.0	Javies Retail
4	10005.0	NaN	Wheat	30.0	0.0	Javies Retail

Figure 4.34 – The top five rows of the DataFrame

7. Now, replace the missing values with `01/01/99` in the `Date of Purchase` column by using `fillna()` and display the top five rows:

```
data_frame['Date of Purchase'].fillna('01/01/99',inplace
= True)
data_frame.head()
```

You will see the following output:

	Receipt Id	Date of Purchase	Product Name	Product Weight	Total Price	Retail shop name
0	10001.0	24/05/20	Wheat	87.0	99.0	NaN
1	0.0	05/05/20	NaN	0.0	25.0	Dello Superstore
2	10003.0	27/04/20	Vegetables	19.0	37.0	Javies Retail
3	10004.0	05/05/20	Oil	99.0	44.0	Javies Retail
4	10005.0	01/01/99	Wheat	30.0	0.0	Javies Retail

Figure 4.35 – Displaying the top five rows of the DataFrame

8. Replace the missing values with `Missing Name` in the remaining columns (`Product Name` and `Retail shop name`) and display the top five rows:

```
data_frame.fillna(value = {'Product Name': 'Missing
Name', \
                                'Retail shop name': 'Missing
Name'},\
                        inplace= True)
data_frame.head()
```

Running this will produce the following output:

	Receipt Id	Date of Purchase	Product Name	Product Weight	Total Price	Retail shop name
0	10001.0	24/05/20	Wheat	87.0	99.0	Missing Name
1	0.0	05/05/20	Missing Name	0.0	25.0	Dello Superstore
2	10003.0	27/04/20	Vegetables	19.0	37.0	Javies Retail
3	10004.0	05/05/20	Oil	99.0	44.0	Javies Retail
4	10005.0	01/01/99	Wheat	30.0	0.0	Javies Retail

Figure 4.36 – Displaying the top five rows after replacing the missing values with "Missing Name"

9. Finally, convert the data types of the columns as follows, and use the `info()` method to display the data type of every column:

   - `Date of Purchase` into a `datetime`
   - `Total Price` into an `int`
   - `Product Weight` into an `int`
   - `Product Name` into a `category`
   - `Retail shop name` into a `category`:

```
data_frame['Date of Purchase'] = \
pd.to_datetime(data_frame['Date of Purchase'], \
               format='%d/%m/%y')
data_frame['Receipt Id'] = \
data_frame['Receipt Id'].astype('int')
data_frame['Total Price'] = \
data_frame['Total Price'].astype('int')
data_frame['Product Weight'] = \
data_frame['Product Weight'].astype('int')
data_frame['Product Name'] = \
data_frame['Product Name'].astype('category')
data_frame['Retail shop name'] = \
data_frame['Retail shop name'].astype('category')

data_frame.info()
```

The output will be as follows:

```
<class 'pandas.core.frame.DataFrame'>
RangeIndex: 58 entries, 0 to 57
Data columns (total 6 columns):
 #   Column            Non-Null Count  Dtype
---  ------            --------------  -----
 0   Receipt Id        58 non-null     int32
 1   Date of Purchase  58 non-null     datetime64[ns]
 2   Product Name      58 non-null     category
 3   Product Weight    58 non-null     int32
 4   Total Price       58 non-null     int32
 5   Retail shop name  58 non-null     category
dtypes: category(2), datetime64[ns](1), int32(3)
memory usage: 2.1 KB
```

Figure 4.37 – Displaying the DataFrame's details after converting the data types

With that, all the columns have been converted into the desired data types, including non-nullable dtypes.

By completing this exercise, you have learned how to handle missing data and convert each column inside a DataFrame into a usable format, if needed.

Activity 4.01 – optimizing memory usage by converting into the appropriate dtypes

In this activity, you will optimize the memory usage of a DataFrame by handling the missing values and converting the initial data type into the appropriate dtype. You will work on the Car Evaluation dataset, which is available in this book's GitHub repository at `https://raw.githubusercontent.com/PacktWorkshops/The-Pandas-Workshop/master/Chapter04/Data/car.csv`.

> **Note**
>
> The original Car Evaluation dataset was sourced from `https://archive.ics.uci.edu/ml/datasets/Car+Evaluation`. This dataset has been slightly modified for this activity.

Follow these steps to complete this activity:

1. Open a Jupyter Notebook.
2. Import the `pandas` package.
3. Load the CSV file as a DataFrame.

4. Display the first 10 rows of the DataFrame.
5. Display the data types of each column in the DataFrame using the `info()` method.
6. Replace the missing values appropriately.
7. Count the number of distinct unique values for `buying`, `maint`, `doors`, `persons`, `lug_boot`, `safety`, and `class`.
8. Convert the `object` columns into `category` columns wherever appropriate.
9. Finally, display the data types of each column in the DataFrame. You should see the following output:

```
<class 'pandas.core.frame.DataFrame'>
RangeIndex: 1728 entries, 0 to 1727
Data columns (total 7 columns):
 #   Column    Non-Null Count  Dtype
---  ------    --------------  -----
 0   buying    1728 non-null   category
 1   maint     1728 non-null   category
 2   doors     1728 non-null   int64
 3   persons   1728 non-null   int32
 4   lug_boot  1728 non-null   category
 5   safety    1728 non-null   category
 6   class     1728 non-null   category
dtypes: category(5), int32(1), int64(1)
memory usage: 29.8 KB
```

Figure 4.38 – Final output

> **Note**
>
> The solution to this activity is available in the *Appendix*.

Now that you have good knowledge of missing data and dtypes in pandas, in the next section, you will learn about the specific methods, functions, and operations you must use for each dtype.

Subsetting by data types

How can we extract specific data within data using its type? Through subsetting the data with pandas, we can locate and manipulate datasets easily. Subsetting is used frequently during a real analysis as it gives you the power to change data dynamically.

For text data (`dtype = object` or `dtype = category`), pandas provides methods to perform string transformation that are referred to as "on the fly." These string transforming methods can be accessed through the `str` attribute when they're used in a pandas Series or Index.

Let's look at the most important strings methods by covering a few examples:

- We will start by defining a series with a string:

```
import pandas as pd
s = pd.Series(['pandas is awesome'])
s
```

The output will be as follows:

```
0      pandas is awesome
dtype:  object
```

- Now, let's use the `split()` method to split the series into a list of three strings:

```
s.str.split()
```

The output will be as follows:

```
0      [pandas is, awesome]
dtype:  object
```

- We can use the `replace()` method to replace the word `pandas` with `python`:

```
s.str.replace('pandas', 'python')
```

The output will be as follows:

```
0      python is awesome
dtype:  object
```

- We can use the `count()` method to count the number of times the letter `a` is in the series:

```
s.str.count('a')
```

The output will be as follows:

```
0    3
dtype:  int64
```

- We can use the `len()` method to count the number of characters in the series:

```
s.str.len()
```

The output will be as follows:

```
0    17
dtype:  int64
```

- We can use the `capitalize()` method to capitalize the string in our series:

  ```
  s.str.capitalize()
  ```

  The output will be as follows:

  ```
  0       Pandas is awesome
  dtype:  object
  ```

- We can use the `islower()` method to test if our series only contains lowercase characters:

  ```
  s.str.islower()
  ```

  The output will be as follows:

  ```
  0      True
  dtype:  bool
  ```

The following table shows the main available methods and their usage for text data in pandas:

Method	Description	Example

`cat()`	This is used to concatenate strings.	`string1 = pd.Series(['pandas is'])`
		`string2 = pd.Series([' awesome'])`
		`string1.str.cat(string2)`
		
		`OUTPUT:`
		`0    pandas is awesome`
		`dtype: object`

`split()`	This is used to split a string into a list. The separator and maximum split parameters can be specified here to restrict the delimiter inside a string.	`string = pd.Series(['pandas is awesome'])`
		`string.str.split()`
		
		`OUTPUT:`
		`0    [pandas, is, awesome]`
		`dtype: object`

`contains()`	This is used to test if certain patterns are contained within a string. It returns a boolean output based on the given pattern or if a regex is included inside a string.	`string = pd.Series('pandas is awesome')`
		`string.str.contains('pandas')`
		
		`OUTPUT:`
		`0    True`
		`dtype: bool`

`replace()`	This replaces a specified pattern with another specified value in a string.	`string = pd.Series('pandas is awesome')`
		`string.str.replace('pandas', 'python')`
		
		`OUTPUT:`
		`0    python is awesome`
		`dtype: object`

Method	Description	Example

`repeat()`	This is used to consecutively duplicate each element of the current Series a specified number of times.	`string = pd.Series('pandas')`
		`string.str.repeat(2)`
		
		`OUTPUT:`
		`0    pandaspandas`
		`dtype: object`

`count()`	This returns the number of elements with the specified pattern.	`string = pd.Series('pandas is awesome')`
		`string.str.count('a')`
		
		`OUTPUT:`
		`0    3`
		`dtype: int64`

`len()`	This returns the number of items in an object. When it is used on a string, it returns the number of characters inside the string.	`string = pd.Series('pandas is awesome')`
		`string.str.len()`
		
		`OUTPUT:`
		`0    17`
		`dtype: int64`

`lower()`	This is used to transform the string into lowercase characters.	`string = pd.Series('PANDAS is awesome')`
		`string.str.lower()`
		
		`OUTPUT:`
		`0    pandas is awesome`
		`dtype: object`

`upper()`	This is used to transform the string into uppercase characters.	`string = pd.Series('PANDAS is awesome')`
		`string.str.upper()`
		
		`OUTPUT:`
		`0    PANDAS IS AWESOME`
		`dtype: object`

`capitalize()`	This is used to transform the first character of a string into an uppercase character.	`string = pd.Series('pandas is awesome')`
		`string.str.capitalize()`
		
		`OUTPUT:`
		`0    Pandas is awesome`
		`dtype: object`

`swapcase()`	This converts the lowercase characters into uppercase and the uppercase characters into lowercase inside a string.	`string = pd.Series('PANDAS is awesome')`
		`string.str.swapcase()`
		
		`OUTPUT:`
		`0    pandas IS AWESOME`
		`dtype: object`

Figure 4.39 – List of methods for text data that can be applied to Series

Working with the dtype category

When you're working with categorical data (`dtype = category`), pandas comes with a set of methods that are only available to categorical data. These methods can be accessed through the `cat` attribute when they're used in a pandas series, similar to the `str` attribute when we were using string methods on series.

Let's look at the most important category methods by covering a few examples:

- **Defining categories**: We will start by defining a series with categories:

  ```
  s = pd.Categorical(["large", "small", "medium",
  "small"],\
                     categories=["large", "small",
  "medium"],\
                     ordered=False)
  s
  ```

  The output will be as follows:

  ```
  [large, small, medium, small]
  Categories (3, object): [large, small, medium]
  ```

- **Displaying the categories**: Now, let's use the `categories` method to display the list of categories:

  ```
  s.categories
  ```

  The output will be as follows:

  ```
  Index(['large', 'small', 'medium'], dtype='object')
  ```

- **Adding new categories**: We can use the `add_categories()` method to add a new category called `extra large`:

  ```
  s.add_categories(['extra large'])
  ```

  The output will be as follows:

  ```
  [large, small, medium, small]
  Categories (4, object): [large, small, medium,
  extra large]
  ```

- **Reordering categories**: We can use the `reorder_categories()` method to reorder the categories of the series:

  ```
  s = s.reorder_categories(["small", "medium","large"],
  ordered=True)
  s
  ```

  The output will be as follows:

  ```
  [large, small, medium, small]
  Categories (3, object): [small < medium < large]
  ```

- **Sorting values categorically**: We can use the `sort_values()` method to sort the values of the series according to the order of the categories:

```
s.sort_values()
```

The output will be as follows:

```
[small, small, medium, large]
Categories (3, object): [small < medium < large]
```

The following table shows the main available methods and their usage for categorical data in pandas:

Method	Description	Example

`categories`	This is used to show all the categories of the series.	`series = pd.Categorical(['a', 'b'], categories=['a', 'b'])`
		`series.categories`
		
		`OUTPUT:`
		`Index(['a', 'b'], dtype='object')`

`rename_categories()`	This is used to rename the elements of the categories.	`series = pd.Categorical(['a', 'b'], categories=['a', 'b'])`
		`series.rename_categories({'a': 'A'}, inplace=True)`
		`series.categories`
		
		`OUTPUT:`
		`Index(['A', 'b'], dtype='object')`

`reorder_categories()`	This is used to reorder the categories.	`series = pd.Categorical(['a', 'b'], categories=['a', 'b'])`
		`series.reorder_categories(['b', 'a'], inplace=True)`
		`series.categories`
		
		`OUTPUT:`
		`Index(['b', 'a'], dtype='object')`

`add_categories()`	This is used to add a new category to the list of categories.	`series = pd.Categorical(['a', 'b'], categories=['a', 'b'])`
		`series.add_categories(['c'], inplace=True)`
		`series.categories`
		
		`OUTPUT:`
		`Index(['a', 'b', 'c'], dtype='object')`

Figure 4.40 – List of methods for categorical data

Working with dtype = datetime64[ns]

When you're working with datetime data (`dtype = datetime64[ns]`), pandas comes with a set of methods that are only available to datetime data. These methods can be accessed through the `dt` attribute when used in a pandas series, similar to how we used `str` and `cat` for string and category-based methods, respectively.

Let's look at the most important datetime methods by covering a few examples.

- **Defining a series**: We will start by defining a series with datetimes:

```
s = pd.to_datetime(pd.Series(['1990-05-31 10:00',\
                              '1995-06-05 15:00',\
                              '2020-09-09 12:00']))
s
```

The output will be as follows:

```
0    1990-05-31 10:00:00
1    1995-06-05 15:00:00
2    2020-09-09 12:00:00
dtype: datetime64[ns]
```

Figure 4.41 – Defining the series

- **Displaying the date**: Now, let's use the `date` method to display the date part of each datetime in the series:

```
s.dt.date
```

The output will be as follows:

```
0     1990-05-31
1     1995-06-05
2     2020-09-09
dtype: object
```

Figure 4.42 – Displaying the date part of each datetime

- **Display the time**: We can use the `time` method to display the time part of each `datetime` in the series:

```
s.dt.time
```

The output will be as follows:

```
0     10:00:00
1     15:00:00
2     12:00:00
dtype: object
```

Figure 4.43 – Displaying the time part of each datetime

- **Displaying the year**: We can use the `year` method to display the year of each date in the series:

```
s.dt.year
```

The output will be as follows:

```
0    1990
1    1995
2    2020
dtype: int64
```

Figure 4.44 – Displaying the year of each date

- **Displaying the day name of each date**: We can use the `day_name()` method to display the day name of each date in the series:

```
s.dt.day_name()
```

The output will be as follows:

```
0     Thursday
1       Monday
2    Wednesday
dtype: object
```

Figure 4.45 – Displaying the day names

The following table shows the main available methods and their usage for `datetime` data in pandas:

Method	Description	Example

`date`	This is used to get the date part of a datetime without the timezone.	`df = pd.DataFrame({'datetime': ['2020-10-01 10:00:00']})`
		`df['datetime'] = pd.to_datetime(df['datetime'],format='%Y-%m-%d %H:%M:%S')`
		`df.datetime.dt.date`
		
		`OUTPUT:`
		`0    2020-10-01`
		`Name: datetime, dtype: object`

`time`	This is used to get the time part of the datetime without the timezone.	`df = pd.DataFrame({'datetime': ['2020-10-01 10:00:00']})`
		`df['datetime'] = pd.to_datetime(df['datetime'],format='%Y-%m-%d %H:%M:%S')`
		`df.datetime.dt.time`
		
		`OUTPUT:`
		`0    10:00:00`
		`Name: datetime, dtype: object`

`year`	This is used to get the year of the datetime.	`df = pd.DataFrame({'datetime': ['2020-10-01 10:00:00']})`
		`df['datetime'] = pd.to_datetime(df['datetime'],format='%Y-%m-%d %H:%M:%S')`
		`df.datetime.dt.year`
		
		`OUTPUT:`
		`0    2020`
		`Name: datetime, dtype: int64`

`month`	This is used to get the month of the datetime.	`df = pd.DataFrame({'datetime': ['2020-10-01 10:00:00']})`
		`df['datetime'] = pd.to_datetime(df['datetime'],format='%Y-%m-%d %H:%M:%S')`
		`df.datetime.dt.month`
		
		`OUTPUT:`
		`0    10`
		`Name: datetime, dtype: int64`

Method	Description	Example

`day`	This is used to get the day of the datetime.	`df = pd.DataFrame({'datetime': ['2020-10-01 10:00:00']})`
		`df['datetime'] = pd.to_datetime(df['datetime'],format='%Y-%m-%d %H:%M:%S')`
		`df.datetime.dt.day`
		
		`OUTPUT:`
		`0    1`
		`Name: datetime, dtype: int64`

`hour`	This is used to get the hours of the datetime.	`df = pd.DataFrame({'datetime': ['2020-10-01 10:00:00']})`
		`df['datetime'] = pd.to_datetime(df['datetime'],format='%Y-%m-%d %H:%M:%S')`
		`df.datetime.dt.hour`
		
		`OUTPUT:`
		`0    10`
		`Name: datetime, dtype: int64`

`minute`	This is used to get the minutes of the datetime.	`df = pd.DataFrame({'datetime': ['2020-10-01 10:37:00']})`
		`df['datetime'] = pd.to_datetime(df['datetime'],format='%Y-%m-%d %H:%M:%S')`
		`df.datetime.dt.minute`
		
		`OUTPUT:`
		`0    37`
		`Name: datetime, dtype: int64`

`second`	This is used to get the seconds of the datetime.	`df = pd.DataFrame({'datetime': ['2020-10-01 10:00:50']})`
		`df['datetime'] = pd.to_datetime(df['datetime'],format='%Y-%m-%d %H:%M:%S')`
		`df.datetime.dt.second`
		
		`OUTPUT:`
		`0    50`
		`Name: datetime, dtype: int64`

`week`	This is used to get the week of the year of the datetime.	`df = pd.DataFrame({'datetime': ['2020-10-01 10:00:00']})`
		`df['datetime'] = pd.to_datetime(df['datetime'],format='%Y-%m-%d %H:%M:%S')`
		`df.datetime.dt.week`
		
		`OUTPUT:`
		`0    40`
		`Name: datetime, dtype: int64`

`dayofweek`	This is used to get the day of the week (as a number) of the datetime.	`df = pd.DataFrame({'datetime': ['2020-10-01 10:00:00']})`
		`df['datetime'] = pd.to_datetime(df['datetime'],format='%Y-%m-%d %H:%M:%S')`
		`df.datetime.dt.dayofweek`
		
		`OUTPUT:`
		`0    3`
		`Name: datetime, dtype: int64`

Method	Description	Example

`dayofyear`	This is used to get the day of the year of the datetime.	`df = pd.DataFrame({'datetime': ['2020-10-01 10:00:00']})`
		`df['datetime'] = pd.to_datetime(df['datetime'],format='%Y-%m-%d %H:%M:%S')`
		`df.datetime.dt.dayofyear`
		
		`OUTPUT:`
		`0    275`
		`Name: datetime, dtype: int64`

`quarter`	This is used to get the quarter of the datetime.	`df = pd.DataFrame({'datetime': ['2020-10-01 10:00:00']})`
		`df['datetime'] = pd.to_datetime(df['datetime'],format='%Y-%m-%d %H:%M:%S')`
		`df.datetime.dt.quarter`
		
		`OUTPUT:`
		`0    4`
		`Name: datetime, dtype: int64`

`daysinmonth`	This is used to get the number of days in the month.	`df = pd.DataFrame({'datetime': ['2020-10-01 10:00:00']})`
		`df['datetime'] = pd.to_datetime(df['datetime'],format='%Y-%m-%d %H:%M:%S')`
		`df.datetime.dt.daysinmonth`
		
		`OUTPUT:`
		`0    31`
		`Name: datetime, dtype: int64`

`month_name()`	This is used to get the month names of the datetime with a specified locale language.	`df = pd.DataFrame({'datetime': ['2020-10-01 10:00:00']})`
		`df['datetime'] = pd.to_datetime(df['datetime'],format='%Y-%m-%d %H:%M:%S')`
		`df.datetime.dt.month_name()`
		
		`OUTPUT:`
		`0    October`
		`Name: datetime, dtype: object`

`day_name()`	This is used to get the day names of the datetime with a specified locale language.	`df = pd.DataFrame({'datetime': ['2020-10-01 10:00:00']})`
		`df['datetime'] = pd.to_datetime(df['datetime'],format='%Y-%m-%d %H:%M:%S')`
		`df.datetime.dt.day_name()`
		
		`OUTPUT:`
		`0    Thursday`
		`Name: datetime, dtype: object`

Figure 4.46 – List of methods for datetime data

Working with dtype = timedelta64[ns]

When you're working with time difference data (`dtype = timedelta64[ns]`), pandas come with a set of methods that are only available to different data. These methods can be accessed through the `dt` attribute when used in a pandas series.

Let's look at the most important `timedelta` methods by covering a few examples:

- **Defining a series with timedelta**: We will start by defining a series with timedelta:

```
s = pd.to_datetime(pd.Series(['1990-05-31',\
                              '1995-06-05',\
                              '2020-09-09']))\
   - pd.to_datetime('2020-01-01')
s
```

The output will be as follows:

```
0   -10807 days
1    -8976 days
2      252 days
dtype: timedelta64[ns]
```

Figure 4.47 – Series defined with timedelta

- **Displaying the number of seconds**: Now, let's use the `total_seconds()` method to display the number of seconds for each timedelta of the series:

```
s.dt.total_seconds()
```

The output will be as follows:

```
0   -933724800.0
1   -775526400.0
2     21772800.0
dtype: float64
```

Figure 4.48 – Number of seconds for each timedelta

- **Displaying the series as datetime.timdelta array**: We can use the `to_pytimedelta()` method, which converts a Series of pandas timedeltas into the `datetime.timedelta` format that's the same length as the original Series:

```
s.dt.to_pytimedelta()
```

The output will be as follows:

```
array([datetime.timedelta(days=-10807), datetime.
timedelta(days=-8976),
          datetime.timedelta(days=252)], dtype=object)
```

- **Displaying each time component**: We can use the `components` method to display a DataFrame with each time component of the series:

```
s.dt.components
```

The output will be as follows:

	days	hours	minutes	seconds	milliseconds	microseconds	nanoseconds
0	-10807	0	0	0	0	0	0
1	-8976	0	0	0	0	0	0
2	252	0	0	0	0	0	0

Figure 4.49 – Displaying each time component

The following table shows the main available methods and their usage for `timedelta` data in pandas:

Method	Description	Example
`components`	This is used to get a DataFrame of the component of a timedelta.	`timedelta = pd.to_datetime(pd.Series(['2020-09-09'])) - pd.to_datetime('2020-01-01')` `timedelta.dt.components`

`to_pytimedelta()`	This is used to convert the timedelta into a Python datetime. timedelta object.	`timedelta = pd.to_datetime(pd.Series(['2020-09-09'])) - pd.to_datetime('2020-01-01')`
		`timedelta.dt.to_pytimedelta()`
		OUTPUT:
		`array([datetime.timedelta(days=252)], dtype=object)`

`total_seconds()`	This is used to convert the timedelta into seconds.	`timedelta = pd.to_datetime(pd.Series(['2020-09-09'])) - pd.to_datetime('2020-01-01')`
		`timedelta.dt.to_pytimedelta()`
		OUTPUT:
		`array([datetime.timedelta(days=252)], dtype=object)`

Figure 4.50 – List of methods for timedelta data

Now, let's practice using them.

Exercise 4.03 – working with text data using string methods

Working with text data usually requires a lot of transformation to shape it into the desired state. In this exercise, you will use pandas-specific string methods to work with text data.

Follow these steps to complete this exercise:

1. Open a new Jupyter notebook file.
2. Import `pandas` using the following command:

```
import pandas as pd
```

3. Next, run the following code to create a series with text data:

```
s = pd.Series\
(['              Data Analysis using python with pandas is
great',
                 'pandas DataFrame and pandas series are
useful   ',
                 'PYTHON3 PANDAS'])
s
```

The output will be as follows:

```
0              Data Analysis using python with panda...
1       pandas DataFrame and pandas series are useful
2                                      PYTHON3 PANDAS
dtype: object
```

Figure 4.51 – Creating a series with text data

4. Use the `count()` function to count the number of times the word `pandas` appears in the text stored in `s`:

```
s.str.count('pandas')
```

The output will be as follows:

```
0    1
1    2
2    0
dtype: int64
```

Figure 4.52 – Counting the number of times "pandas" appears in the text

As you can see, Python is case-sensitive, which is why `PANDAS` wasn't picked up by the `count()` function.

5. Use the `len()` function to count the number of characters in each string:

```
s.str.len()
```

The output will be as follows:

```
0    56
1    47
2    14
dtype: int64
```

Figure 4.53 – Number of characters in each string

The number of characters seems a bit high for our first string. This is due to numerous leading whitespaces in it. These can be removed by using the `strip()` function, combined with the `len()` function.

6. Remove the leading whitespaces with the `strip()` function and combine it with the `len()` function to count the number of characters in each string, excluding leading whitespaces:

```
s.str.strip().str.len()
```

The output will be as follows:

```
0    47
1    45
2    14
dtype: int64
```

Figure 4.54 – Combining the strip() and len() functions

7. Use the `startswith()` function to check if each string starts with the letter `p`:

```
s.str.startswith('p')
```

The output will be as follows:

```
0    False
1     True
2    False
dtype: bool
```

Figure 4.55 – Checking if a string starts with the letter "p"

Now, you want to force all our strings to start with the letter `p`. For that, you must use a combination of functions to remove the leading whitespaces (`strip`), replace the word `Data` with `pandas` (`replace`), lowercase the strings (`lower`), and check if the strings start with the letter `p` (`startswith`).

8. Use a combination of functions to force all the strings to start with the letter `p`:

```
s.str.strip().str.replace('Data', 'pandas').str.lower().
str.startswith('p')
```

The output will be as follows:

```
0    True
1    True
2    True
dtype: bool
```

Figure 4.56 – Forcing strings to start with the letter "p"

9. Use the `get_dummies()` function to create a DataFrame of dummy variables:

```
s.str.get_dummies(' ')
```

The output will be as follows:

	Analysis	Data	DataFrame	PANDAS	PYTHON3	and	are	great	is	pandas	python	series	useful	using	with
0	1	1	0	0	0	0	0	1	1	1	1	0	0	1	1
1	0	0	1	0	0	1	1	0	0	1	0	1	1	0	0
2	0	0	0	1	1	0	0	0	0	0	0	0	0	0	0

Figure 4.57 – Using the get_dummies() function

Dummy variables are Booleans that take the True/1 value when the column name is contained in the row.

Now, you want to create a new DataFrame of dummy variables that contains the words `pandas` and `python`. For that, you need to remove all the irrelevant words (including `3`), lowercase the strings, and use `get_dummies()` again.

10. Use a combination of functions to create a DataFrame of dummy variables that contains only `pandas` and `python`:

```
s = s.str.strip().str.lower()
s.str.get_dummies(' ')[['python','pandas']]
```

The output will be as follows:

	python	pandas
0	1	1
1	0	1
2	0	1

Figure 4.58 – Creating a DataFrame that only contains "pandas" and "python"

By completing this exercise, you have learned how to handle text data by using pandas-specific string methods.

Now that you have good knowledge of the specific methods, functions, and operations for each dtype, let's learn how to select data in a DataFrame based on its dtype.

Selecting data in a DataFrame by its dtype

When you're performing data analysis, you may only want to select the numerical columns in a DataFrame. pandas provides the `select_dtypes()` method so that you select columns based on their dtype. This method allows you to group columns of the same data type and helps us apply some general transformation to the data, depending on their types, without the need to explicitly name the columns.

Let's start with an example:

```
import pandas as pd
import numpy as np

column_names = ["Customer ID", "Customer Name",\
                "2018 Revenue", "2019 Revenue",\
                "Growth", "Start Year", "Start Month",\
                "Start Day", "New Customer"]

row1 = list([1001.0, 'Pandas Banking', '235000',\
             '248000', '5.5', 2013,3,10, 0])
row2 = list([1002.0, 'Pandas Grocery', '196000', '205000',\
             '4.5', 2016,4,30, 0])
row3 = list([1003.0, 'Pandas Telecom', '167000',\
             '193000', '15.5', 2010,11,24, 0])
row4 = list([1004.0, 'Pandas Transport', '79000',\
             '90000', '13.9', 2018,1,15, 1])
row5 = list([1005.0, 'Pandas Insurance', '241000',\
             '264000', '9.5', 2009,6,1, 0])

data_frame = pd.DataFrame(data=[row1, row2, row3, row4, row5],\
                          columns=column_names)

data_frame
```

The output will be as follows:

	Customer ID	Customer Name	2018 Revenue	2019 Revenue	Growth	Start Year	Start Month	Start Day	New Customer
0	1001.0	Pandas Banking	235000	248000	5.5	2013	3	10	0
1	1002.0	Pandas Grocery	196000	205000	4.5	2016	4	30	0
2	1003.0	Pandas Telecom	167000	193000	15.5	2010	11	24	0
3	1004.0	Pandas Transport	79000	90000	13.9	2018	1	15	1
4	1005.0	Pandas Insurance	241000	264000	9.5	2009	6	1	0

Figure 4.59 – Output of the preceding snippet

Let's see which dtype has been assigned to each column:

```
data_frame.dtypes
```

The output will be as follows:

```
Customer ID         float64
Customer Name        object
2018 Revenue         object
2019 Revenue         object
Growth               object
Start Year            int64
Start Month           int64
Start Day             int64
New Customer          int64
dtype: object
```

Figure 4.60 – Checking the data type of each column

To only select the `object` dtype columns, you can use the `select_dtypes()` method:

```
data_frame.select_dtypes('object')
```

The output will be as follows:

	Customer Name	2018 Revenue	2019 Revenue	Growth
0	Pandas Banking	235000	248000	5.5
1	Pandas Grocery	196000	205000	4.5
2	Pandas Telecom	167000	193000	15.5
3	Pandas Transport	79000	90000	13.9
4	Pandas Insurance	241000	264000	9.5

Figure 4.61 – Selecting the columns that contain object data types

Now, you only want to select numeric columns:

```
data_frame.select_dtypes('number')
```

The output will be as follows:

	Customer ID	Start Year	Start Month	Start Day	New Customer
0	1001.0	2013	3	10	0
1	1002.0	2016	4	30	0
2	1003.0	2010	11	24	0
3	1004.0	2018	1	15	1
4	1005.0	2009	6	1	0

Figure 4.62 – Selecting only numeric columns

Now, let's say you need to select numeric columns, but you also want to exclude the `int64` dtype columns. You can use the `exclude` parameter of `select_dtypes()` here:

```
data_frame.select_dtypes('number', exclude='int64')
```

The output will be as follows:

	Customer ID
0	1001.0
1	1002.0
2	1003.0
3	1004.0
4	1005.0

Figure 4.63 – Excluding the int64 types

Now, you should be able to extract columns from a DataFrame based on their data types. Knowing what data type to include or exclude in a DataFrame will save you a lot of time when it comes to data processing.

Summary

In this chapter, you learned about the fundamentals of pandas data types and how to control them. After going through the range of data types in pandas, you learned how to inspect the underlying dtype of your data and how to convert it from one type into another. You also saw how converting your initial data into the appropriate dtypes can greatly reduce the memory usage of a DataFrame. Next, you learned about missing data, from how it's represented to its impact on dtypes. Then, you delved deeper into nullable and non-nullable dtypes and learned about the methods, functions, and operations that are specific to certain dtypes.

In the next chapter, you will learn about the different methods you can use to select data with pandas, be it with a DataFrame or a Series.

Part 2 – Working with Data

In this section, we will dive deeper into the key benefits and features of pandas. We will learn how to use DataFrames and Series to our advantage, alongside various advanced techniques surrounding them. We will also learn how to understand and transform data according to our needs and also visualize data as needed for analysis.

This section contains the following chapters:

- *Chapter 5, Data Selection – DataFrames*
- *Chapter 6, Data Selection – Series*
- *Chapter 7, Data Exploration and Transformation*
- *Chapter 8, Data Visualization*

5
Data Selection – DataFrames

In this chapter, you will develop an understanding of the different forms of the pandas **index**, and how the index is involved in **slicing**, which is one way to get a subset of a pandas data structure. You will learn how to manipulate the index itself, as well as the different notations pandas provides for selection.

By the end of this chapter, you will be able to select subsets of data and work with the index efficiently. You will also learn how to implement pandas dot, bracket, `.loc()`, and `.iloc()` notations to slice and **index**.

This chapter covers the following topics:

- Introduction to DataFrames
- Data selection in pandas DataFrames
- Activity 5.01 – Creating a multi-index from columns
- Bracket and dot notation
- Changing DataFrame values using bracket and dot notation

Introduction to DataFrames

Imagine that you're working on a dataset that contains hundreds of columns and thousands of rows, of which only a small subset – say a dozen rows and two or three columns – matter to you for a particular analysis. In such cases, it's better to isolate and focus on those rows and columns rather than working with the entire dataset. In data analysis and data science, you will constantly need to work with a subset of a larger dataset. Thankfully, pandas provides selection methods that make this process easy and efficient. You will learn about these methods in this chapter. We will start by revisiting DataFrames and then see how pandas selection methods apply to DataFrames.

So far in this book, you have learned about the basics of the pandas data structures (*Chapter 2, Data Structures*), how to get data in or out of pandas (*Chapter 3, Data I/O*), and the different data types in pandas (*Chapter 4, Data Types*). Now, it is time to integrate these concepts with data selection. By the end of this chapter, you will have most of the tools you'll need to efficiently obtain data, understand its structure, select what you need, and manipulate the data as needed. In *Chapter 2, Data Structures*, we introduced the pandas DataFrame and related it to tabular data, which you were likely already familiar with. Now, we are ready to teach you how to leverage pandas row and column indexes to access and extract data from DataFrames.

The need for data selection methods

Suppose we have been asked to analyze some industry detailed **gross domestic product** (**GDP**) data from the United States. We can read the data we have been given using the pandas `.read_csv()` method. After starting a Jupyter notebook, we must load `pandas`, and then read in the data:

```
import pandas as pd
GDP_data = pd.read_csv('Datasets/US_GDP_Industry.csv')
GDP_data
```

The preceding code produces the following output:

Out[2]:

	period	Industry	GDP
0	2015_Q1	All industries	31917.8
1	2015_Q2	All industries	32266.2
2	2015_Q3	All industries	32406.6
3	2015_Q4	All industries	32298.7
4	2016_Q1	All industries	32303.8
...	...	...	...
2129	2019_Q2	Government enterprises	371.4
2130	2019_Q3	Government enterprises	373.5
2131	2019_Q4	Government enterprises	375.1
2132	2020_Q1	Government enterprises	372.8
2133	2020_Q2	Government enterprises	346.0

2134 rows × 3 columns

Figure 5.1 – The GDP data we have been asked to analyze

Here, we can see that the data has been arranged by quarter (from 2015 Q1 to 2020 Q2) and some information about the industry. Now, suppose we have been asked to tabulate the GDP for 2018 and 2019 for only certain industries. How would we do that? That's where `pandas` data selection methods come into the picture. Let's learn about them in more detail.

> **Note**
>
> All the examples for this chapter can be found in the `Examples.ipynb` notebook in the `Chapter05` folder, while the data files can be found in the `Datafiles` folder. Both of these can be found in this book's GitHub repository. To ensure the examples run correctly, you need to run the notebook from start to finish in order.

Data selection in pandas DataFrames

In *Chapter 3*, *Data Structures*, we studied the two core pandas data structures, `DataFrames` and `Series`. There, we did some very basic data selection without digging into the details of how it works. In this section, we will do a deeper dive and explore the **index**, which is fundamental to many `pandas` operations.

As you may recall when we introduced the idea of DataFrames, we drew analogies to spreadsheets. Let's revisit that analogy. Here is the same figure from *Chapter 2, Data Structures* (which is the data from *Figure 5.1* but in a spreadsheet):

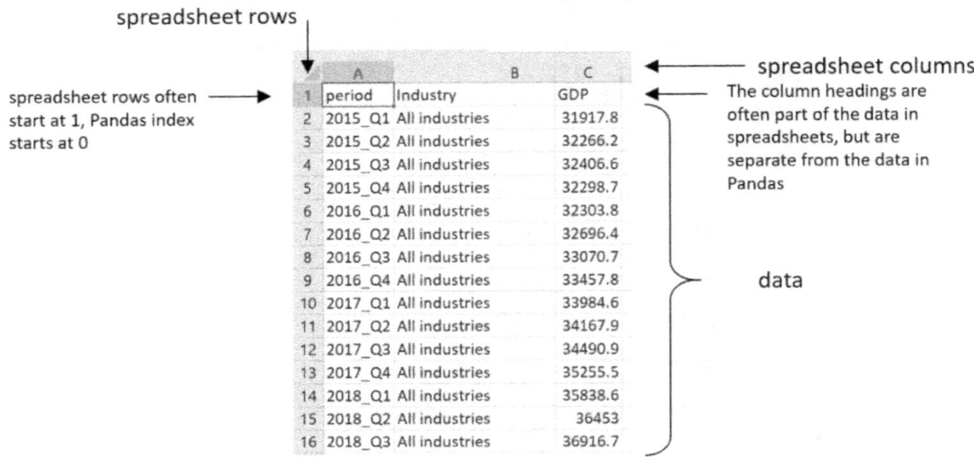

	A	B	C
1	period	Industry	GDP
2	2015_Q1	All industries	31917.8
3	2015_Q2	All industries	32266.2
4	2015_Q3	All industries	32406.6
5	2015_Q4	All industries	32298.7
6	2016_Q1	All industries	32303.8
7	2016_Q2	All industries	32696.4
8	2016_Q3	All industries	33070.7
9	2016_Q4	All industries	33457.8
10	2017_Q1	All industries	33984.6
11	2017_Q2	All industries	34167.9
12	2017_Q3	All industries	34490.9
13	2017_Q4	All industries	35255.5
14	2018_Q1	All industries	35838.6
15	2018_Q2	All industries	36453
16	2018_Q3	All industries	36916.7

Figure 5.2 – The industry GDP data in a spreadsheet

Here, we can see the same three columns of data that were shown in *Figure 5.1*, but we have annotated the key differences. In pandas, the standard row **index** starts at 0, while for most spreadsheets, it starts at row 1. This "0 indexing" is standard for Python. An **index** in pandas is a series of numbers or strings that have been assigned to the rows (and a separate series for the columns) that lets us refer to particular locations in the data. If we were to look inside the spreadsheet and where the data is stored, it may look something like this:

	A	B	C
1	period	Industry	GDP
2	(R2, C1)	(R2, C2)	(R2, C3)
3	(R3, C1)	(R3, C2)	(R3, C3)
4	(R4, C1)	(R4, C2)	(R4, C3)
5	(R5, C1)	(R5, C2)	(R5, C3)
6	(R6, C1)	(R6, C2)	(R6, C3)
7	(R7, C1)	(R7, C2)	(R7, C3)
8	(R8, C1)	(R8, C2)	(R8, C3)
9	(R9, C1)	(R9, C2)	(R9, C3)
10	(R10, C1)	(R10, C2)	(R10, C3)
11	(R11, C1)	(R11, C2)	(R11, C3)
12	(R12, C1)	(R12, C2)	(R12, C3)
13	(R13, C1)	(R13, C2)	(R13, C3)

Figure 5.3 – The row and column values for all the data locations in the spreadsheet

Here, we can see that each data location can be identified by a row number and a column number. We have the same thing in pandas, except for where the rows start. The row and column locations in pandas would look as follows:

period	Industry	GDP
(R0, C0)	(R0, C1)	(R0, C2)
(R1, C0)	(R1, C1)	(R1, C2)
(R2, C0)	(R2, C1)	(R2, C2)
(R3, C0)	(R3, C1)	(R3, C2)
(R4, C0)	(R4, C1)	(R4, C2)
(R5, C0)	(R5, C1)	(R5, C2)
(R6, C0)	(R6, C1)	(R6, C2)
(R7, C0)	(R7, C1)	(R7, C2)
(R8, C0)	(R8, C1)	(R8, C2)
(R9, C0)	(R9, C1)	(R9, C2)
(R10, C0)	(R10, C1)	(R10, C2)
(R11, C0)	(R11, C1)	(R11, C2)

Figure 5.4 – The row and column values for all the data locations in the pandas DataFrame

In both the spreadsheet and pandas, the row **index** is not part of the data itself, but is there to allow us to refer to particular rows, as we will see shortly. In most spreadsheets, the column names are generic letters or numbers, and if we want actual column names for the data, they must be part of the spreadsheet data. For example, in Excel, there might be data in cell E7 – E is the column name; generally, in spreadsheets, you cannot change that, but you can use a workaround by putting names in the first row. In pandas, the columns can have names specified in the column **index** and are not part of the data (refer to *Figure 5.1*). This provides us with the flexibility to use names and not worry about or even know the column order in pandas.

In *Figure 5.1*, where we read the GDP data into pandas in a Jupyter notebook, we can see three columns of data, with each column labeled with a name and each row labeled with a number. Recall from *Chapter 3*, *Data I/O*, that, by default, pandas will use the elements in the first row of the `.csv` data as the column names and will assign a row **index** as integers from 0. `pandas` provides a method to access the row **index** and a method to access the column **index**. We will learn about these in the next section.

The index and its forms

Considering the data we read from `US_GDP_Industry.csv` and the variable we created (`GDP_data`), we can now inspect its row index. The **index** can be accessed using the `.index` attribute of the DataFrame, as follows:

```
GDP_data.index
```

This will produce the following output:

```
Out[3]: RangeIndex(start=0, stop=2134, step=1)
```

This result might be confusing at first as you may expect to see a list of values, but instead, we get a description of a `RangeIndex()` object. What we can see is that the index is of the `RangeIndex` type and it starts at `0` and ends at `2134`, with a step of `1`. But why do we see this information instead of the values?

Recall that in Python, we have the notion of **iterables**. An **iterable** object represents a series of values that can be returned one at a time by some operation. Consider the `for` loop – we code it as `for` (temp variable) `in` (iterable). The `for` statement is the operation that steps through the possible values of the iterable, assigning each new value to the `temp` variable. Many object types in Python are also iterables, including ranges, lists, tuples, and series. However, ranges behave differently than, say, series as you must use a method such as `list` to get the values from the range. Let's look at the `range()` equivalent of this `index` by running the following line of code:

```
range(0, 2134, 1)
```

This will produce the following output:

```
Out[4]: range(0, 2134)
```

Here, we can see an output that's similar to using `.index`. Now, because `range` is an iterable, to access the values or see them, we need to iterate over `range`. One way to do this is with a `for` loop, while the other way is to use a method such as `list()`. The `for` loop iterates the temporary variable over the iterable, while the `list()` method iterates over the iterable and puts all the values into a list object. The following code iterates over `range` using a `for` loop:

```
for i in range(0, 2134, 1):
    if(i > 0 and i < 10):
        print(i)
```

This will produce the following output:

```
1
2
3
4
5
```

```
6
7
8
9
```

You can find the iteration by using the `list()` method, as follows:

```
print(list(range(0, 2134, 1))[:10])
```

This will produce the following output:

```
[0, 1, 2, 3, 4, 5, 6, 7, 8. 9]
```

Note that we used the `if` statement in the `for` loop and the list slice in the `print(list())` method to reduce the output to just the first 10 values in each case. You may recall from core Python that we can select a slice of a list using the bracket notation. In this case, `[:10]` specifies all the values up to, but not including, 10.

So, we can see that both methods give us the sequence of values in the range. Since we said that the result of using `.index` is an iterable, we should be able to use the same methods to see the values in the index. Here, we are using a `for` loop to iterate over the index:

```
for i in GDP_data.index:
    if(i > 0 and i < 10):
        print(i)
```

This code will produce the following output:

```
1
2
3
4
5
6
7
8
9
```

Now, we can see that the default row index of the `GDP_index` DataFrame is very similar to a `range()`, and it is an iterable we can access similarly to how we have already used `for` loops or created `lists` in the past with base Python.

We can also look at the column names. pandas provides the `.columns` attribute, which is analogous to the `.index` attribute for rows, except that it is for columns. Let's use it to look at the column index of `GDP_data`:

```
GDP_data.columns
```

You should see the following output:

```
Out[8]: Index(['period', 'Industry', 'GDP'], dtype='object')
```

Notice that similar to the row index, we can see that the column index is an `Index` type and that the values are in a list. Also, note that here, the columns have names, so the list is a list of names. If the columns had not been named, they would have been labeled with integers from 0 to the number of columns minus 1.

You may have guessed that this column index is an iterable, and you would be correct. This means we can do similar things as we did with the rows, such as iterating over the columns with a `for` loop or getting the values into a list with the `list()` method. We'll try to do that in the next exercise.

Exercise 5.01 – identifying the row and column indices in a dataset

You are the environmental compliance manager for a gas-fueled power plant and have received some data on the emissions and other measurements of a gas turbine in one of the generators for 2015. This data can be found in the `gt_2015.csv` file in the `Datasets` subdirectory. In this exercise, you will read data from the `.csv` file into a `DataFrame` object and inspect the row and column index values. You have been asked to find the largest value for NOx emissions, which will be a particular column in the data.

> **Note**
>
> You can find the code for this exercise in this book's GitHub repository at `https://github.com/PacktWorkshops/The-Pandas-Workshop/tree/master/Chapter05/Exercise05_01`.

Follow these steps to complete this exercise:

1. Create a `Chapter05` directory for all the exercises in this chapter. Inside the `Chapter05` directory, create the `Exercise05_01` directory.
2. Open your Terminal (macOS or Linux) or Command Prompt (Windows), navigate to the `Chapter05` directory, and type `jupyter notebook`. The Jupyter notebook should open.

3. Select the `Exercise05_01` directory to change the working Jupyter directory to that folder. Then, click **New** > **Python 3** to create a new **Python 3** notebook.
4. For this exercise, all you will need is the `pandas` library. Load it into the first cell of the notebook:

    ```
    import pandas as pd
    ```

5. Prepend the pandas `.read_csv()` method to the `Datasets` path, as follows:

    ```
    gas_turbine_data = pd.read_csv('../Datasets/gt_2015.csv')
    ```

    > **Note**
    >
    > Please change the path of the dataset file (highlighted) based on where you have downloaded it on your system.

6. Print a list of the column names by using the `.columns` attribute, as we did earlier in this chapter, as well as the length of the list of column names:

    ```
    gt_columns = list(gas_turbine_data.columns)
    print(gt_columns, '\n (', len(gt_columns), ' columns )')
    ```

    This will produce the following output:

    ```
    ['AT', 'AP', 'AH', 'AFDP', 'GTEP', 'TIT', 'TAT', 'TEY',
    'CDP', 'CO', 'NOX']
     ( 11  columns )
    ```

7. Inspect the row index using the `.index` attribute:

    ```
    gas_turbine_data.index
    ```

    The output should look as follows:

    ```
    Out[4]: RangeIndex(start=0, stop=7384, step=1)
    ```

    Here, we can see that we have 11 columns and 7,384 rows in the data.

8. Now that you have the column names, it is evident that the NOx emissions are in the column labeled `NOX`. There are several ways that we can find the largest value in that column. Here, it's best to use the `.max()` method after selecting the column using bracket notation. Bracket notation is a pandas shorthand that makes it easy to select columns and rows from a DataFrame; we will learn more about it later. Type in the following line:

    ```
    gas_turbine_data['NOX'].max()
    ```

The output should be as follows:

```
Out[7]: 119.68
```

In this exercise, we used the `.columns` and `.index` attributes of a DataFrame to see the column names and row index, respectively. This gives us all the information we need to access particular parts of the data for our analysis, which we will look at in more detail later in this chapter. By using what we have learned so far, we have completed the first step of our analysis – finding the highest `NOX` value.

Saving the index or columns

Since accessing the **index** or columns returns Python objects, we can also assign the indices to variables and manipulate them. This is a common pattern when it comes to exploring and manipulating data in Python. In the following example, where we're using the `GDP_Data` dataset, we're using the `.index` and `.columns` attributes that we used previously, but we are assigning the result to two variables (`GDP_data_index` and `GDP_data_cols`). Then, instead of manipulating the index itself, we are operating on the newly created variables:

```
GDP_data_index = GDP_data.index
GDP_data_cols = GDP_data.columns
print('the index is type ', type(GDP_data_index),
      '\nwhile the columns are type ', type(GDP_data_cols))
print('the second item in the index is ', GDP_data_index[1],
     '\nand the second column is ', GDP_data_cols[1])
```

The output will be as follows:

```
the index is type  <class 'pandas.core.indexes.range.RangeIndex'>
while the columns are type  <class 'pandas.core.indexes.base.Index'>
the second item in the index is  1
and the second column is  Industry
```

Figure 5.5– Storing the indexes in variables and accessing specific values

Here, we can access the items in either variable using the normal Python list indexing notation of `[]` with integer values. By default, the row index is created as a `RangeIndex`, which means it is an iterable that's equivalent to a `range()` with values from 0 to the length of the dataset less than 1 (due to 0 indexing). The column index is a `base.Index`, which simply means it is a list of values, but not a range, per se. However, it is also an iterable, as we saw earlier.

We have been using the pandas `.read_csv()` method with default parameters for the most part. One of those parameters is `header`, with its default value being `header = 0`. This means that pandas will try to get column names from the first row it reads from the `.csv` file. If we set `header = None`, we will find that the default column index is a list of integers. In the following example, we are reading a file called `bare_csv.csv`, where `bare` implies there are no column names in the file. We are reading it with `header = None` and looking at the column index:

```
bare_data = pd.read_csv('Datasets/bare_csv.csv', header = None)
print(bare_data.index, bare_data.columns)
```

This will produce the following output:

```
RangeIndex(start=0, stop=19, step=1) Int64Index([0, 1],
dtype='int64')
```

Here, we can see that the default column index is an `Int64Index()` that consists of integers beginning at 0, similar to the row index.

We've been simplifying the indexes and their values a little so far. Now, let's look at the pandas indexing and slicing methods in more detail.

Slicing and indexing methods

pandas provide several methods for selecting items from DataFrames, but we will start by learning how to index with two methods: `DataFrame.loc` and `DataFrame.iloc`. pandas lets us index a DataFrame by row and column integer values (0 indexed) or labels. Looking at the DataFrame in *Figure 5.1*, you may think that the rows are always integers and that the columns are always labeled, but that isn't always true. In the previous example, you saw that we could use integers for the columns. pandas provides `DataFrame.iloc` to let us use index by row and column *integer* values, and `DataFrame.loc` can be used if we want to use *labels*. The key idea is that the integer values automatically adjust to match the number of rows or columns, but labels are attached to the rows and columns they have been assigned to, so if we delete a row or column, it is removed from the sequence. Since this is an important distinction, let's explore it by looking at an example. First, we will create a DataFrame containing some information about how many and what kind of fruits we have by using the `DataFrame` constructor and the `Dictionary` data form, as follows:

```
labels_vs_integers =\
pd.DataFrame({'values' : [6, 1, 5, 2],
              'names' : ['oranges', 'apples',
```

```
                                'bananas', 'pears']})
labels_vs_integers
```

This will produce the following output:

Out[11]:

	values	names
0	6	oranges
1	1	apples
2	5	bananas
3	2	pears

Figure 5.6 – The labels_vs_integers DataFrame containing our fruit inventory

Here, we have four rows, numbered 0 to 3, and two columns called `values` and `names`. pandas provides a feature that lets us assign labels to the rows (or the columns). We'll look at more ways to do this later, but a simple way is to assign a new list of values directly to the index. In the following code, we are using the `.index` attribute on the left-hand side of the assignment and passing a list of labels on the right-hand side:

```
labels_vs_integers.index = ['citrus', 'non_citrus',\
                            'non_citrus', 'non_citrus']
labels_vs_integers
```

> **Note**
>
> All the examples for this chapter can be found in the `Examples.ipynb` notebook in the `Chapter05` folder, including the data files (in the `Datafiles` folder). To have the examples run correctly, you need to run the notebook from start to finish in order. If you were to start in the middle, you may need to run the following line of code for the examples to work properly:
>
> `import pandas as pd`

This will produce the following output:

Out[12]:

	values	names
citrus	6	oranges
non_citrus	1	apples
non_citrus	5	bananas
non_citrus	2	pears

Figure 5.7 – The updated labels_vs_integers DataFrame, with labels assigned to the rows

Earlier, we mentioned that `.loc` acts on labels, while `.iloc` acts on integer values. The syntax for both methods is `DataFrame.loc[rows, columns]` or `DataFrame.iloc[rows, columns]`, where `rows, columns` represents the row/column labels (for `.loc`) or the row/column integer values (for `.iloc`). In the following code, we are using `print()` to print out a subset of the `labels_vs_integers` DataFrame using each method to produce the same result. In the first case, we are using the `.iloc[]` method and passing two integer ranges, 1:4 for the rows (so we get 1, 2, and 3 since the end value is omitted), and 1 for the column. In the second case, we are using the `.loc[]` method and passing the `non_citrus` label for the row index values and `names` for the column we want:

```
print(labels_vs_integers.iloc[1:4, 1])
print()
print(labels_vs_integers.loc['non_citrus', 'names'])
```

This will produce the following output:

```
non_citrus     apples
non_citrus    bananas
non_citrus      pears
Name: names, dtype: object

non_citrus     apples
non_citrus    bananas
non_citrus      pears
Name: names, dtype: object
```

Figure 5.8 – Printing a subset of the labels_vs_integers DataFrame using integers (first) and labels (second)

Notice that although we don't see any row numbers (or column numbers) after labeling the rows, we can still use `.iloc on them`. Also, in the first case, we used the `1:4` notation, which has the same meaning as core Python list indexing, or the range of integers from 1 to 4, including 1 and excluding 4.

Now, let's see what happens if we remove a row. Here, in the first print statement, we will just use `.iloc` and print the value at row 1, column 1 of our current DataFrame (which is `apples`). In the second print statement, we will call `.loc` first to select just the `non_citrus` rows, and then call `.iloc` to print the value at row 1, column 1 of the subset of the DataFrame. Note the difference in the results:

```
print('using .iloc alone: ', labels_vs_integers.iloc[1, 1])
print('using .loc to subset first, then using .iloc: ',
      labels_vs_integers.loc['non_citrus', :].iloc[1, 1])
```

This will produce the following output:

```
using .iloc alone: apples
using .loc to subset first, then using .iloc: bananas
```

Notice that in the second case, the `bananas` value is now in location `[1, 1]`, whereas it was in location `[2, 1]` in the original DataFrame. pandas has automatically renumbered the integer values after we subset to just the `non_citrus` rows. To recap, the integer values are always there and can be referenced using the pandas `.iloc` method. They automatically adjust to reflect the actual number of rows and columns and always run from 0 to n-1, where *n* is the total number of rows or columns. The labels, on the other hand, are attached to *particular* rows or columns, and if a row (or column) is deleted, that label is removed; the rest remain *unchanged*.

Also, notice that in the second case, we used `:` for the columns, instead of a value. `.loc` and `.iloc` both accept this notation, in either the `rows` or `columns` argument, to mean `all`. So, in the second case, we asked for the `non_citrus` rows and `all` the columns in the first part of the call, and then subset that result to only get `[1, 1]` in the second part of the call. We've also seen that both `.loc` and `.iloc` have a similar syntax in that they take `[rows, columns]` as their arguments, and that in the case of `.loc`, since it is acting on labels, we can get results with multiple rows or columns if the labels are duplicated, as is the case for the `non_citrus` row label.

It is easy to confuse labels and integers if the labels are integer values. Since pandas will assign integers to the row numbers by default, the difference is not obvious at first. We should make the distinction that what is usually displayed for a DataFrame is the labels, even if those labels contain integer values. Another brief example will make this last point clear. Here, we are creating a DataFrame that contains some examples of feline and canine species of animals:

```
int_labels_vs_integers = \
pd.DataFrame({'species' : ['feline', 'canine',\
                           'canine', 'feline'],
              'name' : ['housecat', 'wolf',\
                        'dingo', 'tiger']})
int_labels_vs_integers
```

This will produce the following output:

Out[26]:

	species	name
0	feline	housecat
1	canine	wolf
2	canine	dingo
3	feline	tiger

Figure 5.9 – The int_labels_vs_integers DataFrame contains examples of animal species

Now, we will use `.iloc` to subset the DataFrame. To do this, we will demonstrate another method pandas provides for **subsetting**, which is to provide a list of values for the rows or columns. In the following code, we are passing a list to the rows:

```
int_labels_vs_integers = int_labels_vs_integers.iloc[[0, 2, 3],
:]
int_labels_vs_integers
```

This will produce the following subset of our DataFrame:

Out[27]:

	species	name
0	feline	housecat
2	canine	dingo
3	feline	tiger

Figure 5.10 – The subset comprising rows 0, 2, and 3 of the int_labels_vs_integers DataFrame

Here, we can see that the rows are labeled 0, 2, and 3 and that the value 1 is missing. That is because pandas is showing us the labels, which had values of 0, 1, 2, and 3, and the row with label 1 is being removed. The integer values have been *automatically* reset. We can see this by asking for row 1:

```
int_labels_vs_integers.iloc[1, :]
```

This will produce the following output:

```
Out[28]: species    canine
         name        dingo
         Name: 2, dtype: object
```

Figure 5.11 – Row 1 of the subsetted DataFrame

Here, we can see that integer row `1` is indeed the second row, but it is also labeled `2`. We can confirm the latter by using `.loc` instead of `.iloc`, as shown in the following code:

```
int_labels_vs_integers.loc[2, :]
```

This will produce the following output:

```
Out[29]: species    canine
         name        dingo
         Name: 2, dtype: object
```

Figure 5.12 – Row 2 of the subset of the subsetted DataFrame

Now that we have shown that you can use a range or a list to specify multiple rows, let's return to the GDP data and see how that works for columns. Note that if you haven't loaded the GDP data since restarting the notebook, you may need to return to the cell that contains the `pd.read_csv()` method and run it. Here, we have been asked to make a summary of the total GDP per quarter. The first step is to select just the columns with the quarters and the `GDP` values. We can do this by using `.iloc` on the columns and passing a list of `[0, 2]` to get the first and third columns. We can use the `.head()` method to check the result:

```
GDP_summary = GDP_data.iloc[:, [0, 2]]
GDP_summary.head()
```

This will produce the following output:

Out[32]:

	period	GDP
0	2015_Q1	31917.8
1	2015_Q2	32266.2
2	2015_Q3	32406.6
3	2015_Q4	32298.7
4	2016_Q1	32303.8

Figure 5.13 – The first and third columns of GDP_data stored in a new DataFrame, GDP_summary

As expected, this produced the new DataFrame since we knew the columns that we wanted were the first and third ones. A common pattern in data science is that we know what variables or column names we are working with, but they could be different column numbers if we have manipulated or modified the DataFrame. In such cases, a better coding practice would be to pass the list of names of the columns to `.loc`, which is the method that uses labels. This will always work so long as the columns exist. Considering the previous example again, we would pass the `['period', 'GDP']` list to `.loc`.

However, we used `.iloc` here to illustrate that it works the same for columns as it does for rows. Suppose we got distracted and came back to our work and were coding this first step to select the two columns and confused the labels and integer values. In the following code, we are repeating the selection but mistakenly using `.loc` instead of `.iloc`. This will typically generate errors since `.loc` expects labels, and the labels are not integer values:

```
GDP_summary = GDP_data.loc[:, [0, 2]]
```

This will produce the following output:

```
---------------------------------------------------------------------------
KeyError                                  Traceback (most recent call last)
<ipython-input-33-c6b03db243d3> in <module>
      3 # use the same column list of [0, 2]
      4 #
----> 5 GDP_summary = GDP_data.loc[:, [0, 2]]

~\Miniconda3\envs\keras-gpu-4\lib\site-packages\pandas\core\indexing.py in __getitem__(self, key)
   1760                 except (KeyError, IndexError, AttributeError):
   1761                     pass
-> 1762             return self._getitem_tuple(key)
   1763         else:
   1764             # we by definition only have the 0th axis

~\Miniconda3\envs\keras-gpu-4\lib\site-packages\pandas\core\indexing.py in _getitem_tuple(self, tup)
   1287                 continue
   1288 
-> 1289             retval = getattr(retval, self.name)._getitem_axis(key, axis=i)
   1290 
   1291         return retval

~\Miniconda3\envs\keras-gpu-4\lib\site-packages\pandas\core\indexing.py in _getitem_axis(self, key, axis)
   1952                     raise ValueError("Cannot index with multidimensional key")
   1953 
-> 1954                 return self._getitem_iterable(key, axis=axis)
   1955 
   1956             # nested tuple slicing

~\Miniconda3\envs\keras-gpu-4\lib\site-packages\pandas\core\indexing.py in _getitem_iterable(self, key, axi
s)
   1593         else:
   1594             # A collection of keys
-> 1595             keyarr, indexer = self._get_listlike_indexer(key, axis, raise_missing=False)
   1596             return self.obj._reindex_with_indexers(
   1597                 {axis: [keyarr, indexer]}, copy=True, allow_dups=True

~\Miniconda3\envs\keras-gpu-4\lib\site-packages\pandas\core\indexing.py in _get_listlike_indexer(self, key,
axis, raise_missing)
   1551 
   1552         self._validate_read_indexer(
-> 1553             keyarr, indexer, o._get_axis_number(axis), raise_missing=raise_missing
   1554         )
   1555         return keyarr, indexer

~\Miniconda3\envs\keras-gpu-4\lib\site-packages\pandas\core\indexing.py in _validate_read_indexer(self, key,
indexer, axis, raise_missing)
   1638             if missing == len(indexer):
   1639                 axis_name = self.obj._get_axis_name(axis)
-> 1640                 raise KeyError(f"None of [{key}] are in the [{axis_name}]")
   1641 
   1642             # We (temporarily) allow for some missing keys with .loc, except in

KeyError: "None of [Int64Index([0, 2], dtype='int64')] are in the [columns]"
```

Figure 5.14 – An error message passing a list of integers to the .loc method in pandas

Well, that's a lot of errors for such a simple mistake. Fortunately, most of the time when you make this mistake, you just need to look at the very last line:

```
KeyError: "None of [Int64Index([0, 2], dtype='int64')] are in the [columns]"
```

Figure 5.15 – The most important part of the previous error message

When you see this type of error, it's a flag that's indicating that you have mixed up the type of indexing (`DataFrame.loc` or `DataFrame.iloc`) and the **index** values you requested. The fix is simple in this case, as we have already suggested – use the actual column names in `.loc`:

```
GDP_summary = GDP_data.loc[:, ['period', 'GDP']]
GDP_summary
```

This gives us the desired result:

Out[34]:

	period	GDP
0	2015_Q1	31917.8
1	2015_Q2	32266.2
2	2015_Q3	32406.6
3	2015_Q4	32298.7
4	2016_Q1	32303.8
...	...	...
2129	2019_Q2	371.4
2130	2019_Q3	373.5
2131	2019_Q4	375.1
2132	2020_Q1	372.8
2133	2020_Q2	346.0

2134 rows × 2 columns

Figure 5.16 – The result of using DataFrame.iloc with the (correct) list of column names

You can generate similar but slightly different errors by passing labels to `.iloc`.

pandas is very flexible and powerful. In addition to allowing the Python-like ranges of integers and lists of values as the arguments for `.loc` and `.iloc`, in the case of `.loc`, you can pass ranges of labels. This can seem a bit strange at first, but it can come in handy. Here, we are reading the gas turbine data from the `gt_2015.csv` file, using `pd.read_csv()`, and listing the column names:

```
gas_turbine_data = pd.read_csv('Datasets/gt_2015.csv')
gas_turbine_data.columns
```

> **Note**
>
> The examples notebook, `Ch05Examples.ipynb`, that we are using for the examples can be found in the `Chapter05` folder of the workshop, while the exercises can be found in separate folders. Therefore, the path to the `gt_2015.csv` data is `'Datasets\\gt_2015.csv'` instead of `'..\\Datasets\\gt_2015.csv'`, as in the exercises. In the exercises, since we are one folder deeper in the workshop folder structure, we have to use `..\\` to tell the operating system to go up one level to find the `Datasets` folder. Here, we do not need to do that.

This will produce the following output:

```
Out[37]: Index(['AT', 'AP', 'AH', 'AFDP', 'GTEP', 'TIT', 'TAT', 'TEY', 'CDP', 'CO',
               'NOX'],
              dtype='object')
```

Figure 5.17 – The column names of the gas_turbine_data DataFrame

Now, let's say you are interested in analyzing all the columns except for the Carbon Monoxide and Nitrogen Oxide columns (`CO` and `NOX`). You decide to create another DataFrame that contains those columns and use `.loc` to accomplish that. You could create a list that contains all nine names, but pandas makes this simpler. In the following code, we are using `.loc` and asking for all the columns from `AT` to `CDP`:

```
non_emissions_turbine_data = gas_turbine_data.loc[:,
'AT':'CDP']
non_emissions_turbine_data.head()
```

This will produce the following output:

Out[38]:

	AT	AP	AH	AFDP	GTEP	TIT	TAT	TEY	CDP
0	1.95320	1020.1	84.985	2.5304	20.116	1048.7	544.92	116.27	10.799
1	1.21910	1020.1	87.523	2.3937	18.584	1045.5	548.50	109.18	10.347
2	0.94915	1022.2	78.335	2.7789	22.264	1068.8	549.95	125.88	11.256
3	1.00750	1021.7	76.942	2.8170	23.358	1075.2	549.63	132.21	11.702
4	1.28580	1021.6	76.732	2.8377	23.483	1076.2	549.68	133.58	11.737

Figure 5.18 – All the columns of gas_turbine_data except for CO and NOx

You may notice something interesting here – when we used the range with `.iloc` and passed an integer range such as `1:4`, we obtained rows 1, 2, and 3. In the preceding code, we passed the `'AT':'CDP'` range and obtained all the columns from `'AT'` to `'CDP'`, including the last one, `'CDP'`. In the case of integers, pandas follows the Python convention of including the first value and excluding the last value. However, in the case of labels, this could be confusing or even ambiguous, so pandas gives us *all* the requested labels. If you happen to work with labels that are integers, this can be confusing and appear contradictory, so it is important to keep track of whether you are working with the integer row and column numbers or the labels.

Boolean indexing

An extremely useful method that can be enabled by pandas using the methods we have seen so far is Boolean indexing. Boolean refers to Boolean logic, which, in turn, refers to using logical operations to determine an outcome. A logical operation involves using operators such as `OR` or `AND` and getting an outcome of `TRUE` or `FALSE`. In pandas, we can use a Boolean expression that evaluates as `TRUE` for the cases we want. We can reframe many of the cases we've seen so far in this way. For example, in *Figure 5.20*, we saw the result of the `gas_turbine_data.loc[:, 'AT':'CDP']` statement. Earlier, we learned that the syntax is `[rows, columns]` and that for `.loc[]`, we are specifying labels. So, this statement is the same as saying "give me the data from `gas_turbine_data` for all the rows and where the columns are in the values from `AT` to `CDP`." The second part of this statement is a logical expression – for any column, it is either in that range (`TRUE`) or not (`FALSE`). In pandas, this can be done explicitly with logical expressions. Note that there are specific logical operators that follow core Python – here, we are using `==`, which is the logical equals sign. (Others include `<`, `>`, `<=`, `>=`, `!=`, `and`, and `or`, as well as `&` and `|`, which are bitwise `AND` and `OR`, respectively.) Suppose we wanted to list only the `emissions` columns from `gas_turbine_data` (`CO` and `NOX`). We could do that in various ways, but to illustrate Boolean indexing, we will use `.loc[]` here and two Boolean expressions:

```
gas_turbine_data.loc[:,
                     (gas_turbine_data.columns == 'NOX') |
                     (gas_turbine_data.columns == 'CO')]
```

This will produce the following output:

Out[48]:

	CO	NOX
0	7.4491	113.250
1	6.4684	112.020
2	3.6335	88.147
3	3.1972	87.078
4	2.3833	82.515
...	...	...
7379	10.9930	89.172
7380	11.1440	88.849
7381	11.4140	96.147
7382	3.3134	64.738
7383	11.9810	109.240

7384 rows × 2 columns

Figure 5.19 – The result of using Boolean indexing to select two columns from gas_turbine_data

Although the result is straightforward here and requires more typing, there are times when Boolean indexing is extremely useful for selecting data. We will see more cases of its use later in this chapter.

So far, we've learned how to select groups of rows or columns using the pandas index and the `.loc` and `.iloc` methods. We also looked at the differences between using the integer row or column numbers and the labels, including the confusing case where the labels are integer values. In the next exercise, you will apply these concepts to the gas turbine data.

Exercise 5.02 – subsetting rows and columns

Following on from the previous exercise, you are the environmental compliance manager for a gas-fueled power plant and have received some data on the emissions and other measurements of a gas turbine in one of the generators for 2015. The data for this can be found in the `gt_2015.csv` file in the `Datasets` subdirectory in this book's GitHub repository. In this exercise, you will read data from a `.csv` file into a `DataFrame` object, subset some rows and columns, and compare the summary statistics of the sample to the whole dataset. You are aware that there are about 20 measurements per day in this file and you are interested in comparing the first 5 days to the overall year, so you need to select 100 rows of data.

> **Note**
>
> You can find the code for this exercise at `https://github.com/PacktWorkshops/The-Pandas-Workshop/tree/master/Chapter05/Exercise05_02`.

Follow these steps to complete this exercise:

1. In the `Chapter05` directory that you created earlier, create the `Exercise05_02` directory.
2. Open your Terminal (macOS or Linux) or Command Prompt (Windows), navigate to the `Chapter05` directory, and type `jupyter notebook`. The Jupyter notebook should open.
3. Select the `Exercise05_02` directory to change the working Jupyter directory to that folder. Then, click **New** > **Python 3** to create a new **Python 3** notebook.
4. For this exercise, all you will need is the `pandas` library. Load it into the first cell of the notebook:

```
import pandas as pd
```

5. Use the pandas `.read_csv()` method and prepend it to the `Datasets` path, as follows:

```
gas_turbine_data = pd.read_csv('../Datasets/gt_2015.csv')
```

> **Note**
>
> Please change the path of the dataset file (highlighted) based on where you have downloaded it on your system.

6. Now, print the shape of the DataFrame using the `.shape` attribute and print the column names using the `.columns` attribute:

```
print(gas_turbine_data.shape)
print(list(gas_turbine_data.columns))
```

This will produce the following output:

```
(7384, 11)
['AT', 'AP', 'AH', 'AFDP', 'GTEP', 'TIT', 'TAT', 'TEY', 'CDP', 'CO',
'NOX']
```

Figure 5.20 – The shape and columns of the gas_turbine_data DataFrame

7. Now, let's compare the summary statistics of the emissions data. Since there are over 7,000 rows, you will get the summary for the first 100 rows (recall that this is approximately 5 days' worth of data) and the entire dataset. Use the pandas `.describe()` method to output the statistics. You only want to see data for the `'CO'` and `'NOX'` columns, so use the `.loc` method to get those columns. For the sample of 100 rows, use the `.iloc` method. This can be done in one line of code, as follows:

```
print(gas_turbine_data.iloc[:100, :]\
.loc[:, ['CO', 'NOX']].describe())
print(gas_turbine_data.loc[:, ['CO', 'NOX']].describe())
```

This will produce the following output:

```
               CO         NOX
count  100.000000  100.000000
mean     3.774012   77.661970
std      1.774795   13.708632
min      0.475440   58.432000
25%      2.656625   64.672000
50%      3.501650   78.084000
75%      4.078250   85.121250
max     12.659000  118.270000
                CO          NOX
count  7384.000000  7384.000000
mean      3.129986    59.890509
std       2.234962    11.132464
min       0.212800    25.905000
25%       1.808175    52.399000
50%       2.533400    56.838500
75%       3.702550    65.093250
max      41.097000   119.680000
```

Figure 5.21 – The summary emissions statistics for the first 100 rows and all the data in the gas_turbine_data DataFrame

It is up to your professional judgment as to whether the sample of 100 rows is enough for further analysis. However, the statistics in both columns look a bit different from the value for the entire dataset.

In this exercise, you used both `.loc` and `.iloc` to select the rows and columns of a DataFrame at the same time, combined with another pandas operation. This type of pattern is very common in data analysis. Now, let's look at some of the other capabilities provided by pandas for using labels as the index.

Using labels as the index and the pandas multi-index

You might be wondering what the point of label indexing the rows is. Earlier, we saw a simple case with "citrus" and "non-citrus" fruits. The best reason to use these methods is for clarity – pandas makes it possible to use tabular data in a natural, intuitive form. So, instead of having just the default labels of "0" to "N-1" (where *N* is the number of rows), or only using the integer index, we can put labels that help us see the data in the same way as we are thinking about it or analyzing it.

However, another very powerful feature is provided by pandas for the index when you're using labels, called a **multi-index**. A multi-index can have multiple columns (for the row index) or multiple rows (for the column index). Effectively, a pandas multi-index allows you to represent higher dimensional data in a 2D DataFrame. It is also analogous to the result of doing a *pivot-table* operation in a spreadsheet. Suppose we have been given the following data, describing the weight of various animals by their species and where they live:

	species	location	weight	color	fur
0	dog	city	10	striped	long
1	chicken	town	11	solid	long
2	cat	city	12	striped	short
3	cat	farm	13	striped	short
4	chicken	farm	14	solid	long
...	...	...	...	...	...
95	pig	city	105	solid	short
96	chicken	city	106	solid	short
97	dog	town	107	striped	long
98	cat	town	108	solid	short
99	chicken	town	109	solid	short

100 rows × 5 columns

Figure 5.22 – Data on animals

The data in the preceding screenshot was generated using code that can be found in the `Examples.ipynb` file. For reference, here is that code:

```
species = pd.Series(['cat', 'dog', 'pig', 'chicken'])
species = species.sample(100,
                         replace = True,
                         random_state = 1).reset_index(drop =
```

```
True)
location = pd.Series(['city', 'town', 'farm'])
location = location.sample(100,
                                replace = True,
                                random_state = 2).reset_index(drop =
True)
weight = pd.Series(range(10, 110)).reset_index(drop = True)
fur = pd.Series(['long', 'short'])
fur = fur.sample(100,
                replace = True,
                random_state = 3).reset_index(drop = True)
color = pd.Series(['solid', 'spotted', 'striped'])
color = color.sample(100,
                    replace = True,
                    random_state = 42).reset_index(drop =
True)
animals = pd.DataFrame({'species' : species,
                        'location' : location,
                        'weight' : weight,
                        'color' : color,
                        'fur' : fur})
```

This data is five-dimensional – there is the species dimension, the location dimension, the color dimension, the fur dimension, and the weight dimension. pandas provides the `.pivot_table()` method, which allows us to summarize high-dimensional data into **hierarchical** columns and rows and represents the hierarchy on each axis using a multi-index. Anticipating a lot of resulting values, first, we use the `pandas .options()` method to set the format of floating-point values so that they only have two digits after the decimal point. Then, we apply the `.sort_values()` method to the `location` and `species` columns, which we want to be the rows, and we pass the `color` and `fur` columns as the columns to use as the column index. The remaining data – in this case, `weight` – is operated on by `aggfunc` (the aggregation function), which we pass the `mean` value to get averages in the cases where there is more than one value that has the same index values. Finally, we pass `fill_value = ''` to put blanks where a combination is not present in our original data:

```
pd.options.display.float_format = '{:,.2f}'.format
animals.sort_values(['location', 'species'])\
.pivot_table(index = ['location', 'species'],\
```

```
columns = ['color', 'fur'],\
aggfunc = 'mean',\
fill_value = '')
```

This will produce the following output:

Out[129]:

		weight					
	color	**solid**		**spotted**		**striped**	
	fur	**long**	**short**	**long**	**short**	**long**	**short**
location	**species**						
city	**cat**	74.00	23.50	33.00	84.00	20.00	12.00
	chicken	44.00	91.33	63.67		103.00	87.00
	dog	15.00	44.00	64.00	75.00	39.00	51.00
	pig		86.00	40.00		39.00	
farm	**cat**	49.00	85.00	85.00	82.00	69.00	24.33
	chicken	59.00		81.00	43.00	16.00	75.67
	dog	102.00	99.00	49.20	58.00	44.50	
	pig	30.00	36.00	65.00	54.50	47.00	49.00
town	**cat**		108.00	60.50	64.75	37.00	
	chicken	52.50	81.00				44.25
	dog	61.00	71.50		51.50	79.67	55.00
	pig	101.00		95.00	60.50		

Figure 5.23 – The animals data with a multi-index

As a result, `location` and `species` are a row hierarchy, and `color` and `fur` are a column hierarchy. In many cases, we did not have the given combination of `location`, `species`, `color`, and `fur`, so we have blanks. In cases where there were *multiple* values, we can see an average value. Compared to the original data, this format makes it easy to see where the data is concentrated. It also gives us the "typical" (average) values.

Consider the GDP by industry data we had earlier. Our focus is on the GDP figures, but we think about them in terms of time and particular industries. pandas allows us to label the data that way. Let's change the `pd.read_csv()` method's parameters to tell pandas to use the first two columns (`columns [0, 1]`) as the index. Notice that we get two index columns – `pandas` allows this multi-index structure, which then makes it simple to select by multiple levels. Let's also introduce another index method, `.sort_index()`. The `.sort_index()` method is especially useful in the case of a multi-index since it accepts a list of columns and sorts them in order, as you would expect:

```
GDP_by_industry = \
pd.read_csv('Datasets/US_GDP_Industry.csv',
            index_col = [0, 1]).sort_index(level = [0, 1])
GDP_by_industry
```

This will produce the following output:

Out[131]:

		GDP
period	**Industry**	
2015_Q1	**Accommodation**	256.20
	Accommodation and food services	973.60
	Administrative and support services	795.70
	Administrative and waste management services	883.00
	Agriculture, forestry, fishing, and hunting	466.30
...	**...**	...
2020_Q2	**Warehousing and storage**	135.10
	Waste management and remediation services	100.60
	Water transportation	32.00
	Wholesale trade	1,810.90
	Wood products	111.90

2134 rows × 1 columns

Figure 5.24 – The GDP by Industry data, read in by telling pandas to use the first two columns as the index and sorting by both index columns

Here, the first index, `period`, begins at the first period value, but now, within that group, we have the `Industry` index sorted as well. Now that we have this multiple-level index (called a multi-index in pandas), we can do things a bit more easily than we did previously. For example, let's say you wanted to get the GDP for the segments from `Farms` to `Finance and insurance` for `2017_Q2`. We can do that with one simple line as follows. A pandas multi-index allows you to pass *tuples* to specify the values you want from each level of the index. Recall that the `.loc` method (and `.iloc`) accepts a range of values – here, we are providing a range of tuples to get all the values for the chosen industries:

```
GDP_by_industry.loc[('2017_Q2', 'Farms'):
                    ('2017_Q2', 'Finance and insurance')]
```

This will produce the following output:

Out[98]:

		GDP
period	**Industry**	
2017_Q2	**Farms**	399.7
	Federal	1124.6
	Federal Reserve banks, credit intermediation, and related activities	926.5
	Finance and insurance	2827.1

Figure 5.25 – Using the pandas multi-index to select values based on both index columns simultaneously

This result is exactly as we would have envisioned it in our minds and shows the utility of more complex, label-based row indexing in pandas. Note that we could have achieved the same result without the labeled multi-index, by explicitly passing criteria for the period and time `Industry` columns, using Boolean indexing. Here, we are reading the same data again that we did the first time, using it (see *The index and its forms* section), and then using the default row index. Then, we are using the `.loc()` method and using Boolean indexing for the two columns we are using to select the data:

```
GDP_by_industry = pd.read_csv('Datasets\\US_GDP_Industry.csv')
GDP_by_industry.loc[((GDP_by_industry['Industry'] == 'Farms') |
                     (GDP_by_industry['Industry'] == 'Federal')
 |
                     (GDP_by_industry['Industry'] == ('Federal
Reserve banks, ' +
                                                      'credit
intermediation, ' +
                                                      'and
related activities')) |
                     (GDP_by_industry['Industry'] == 'Finance
and insurance')) &
                    (GDP_by_industry['period'] == '2017_Q2'),
:]
```

This will produce the following output:

Out[121]:

	period	Industry	GDP
75	2017_Q2	Farms	399.7
1197	2017_Q2	Finance and insurance	2827.1
1219	2017_Q2	Federal Reserve banks, credit intermediation, ...	926.5
1967	2017_Q2	Federal	1124.6

Figure 5.26 – Reading the GDP data without the multi-index and using a compound .loc[] method to select the same data as in Figure 5.29

In addition to requiring more code, notice that the formatting, as a hierarchy, is not present. This is because we are just listing the rows of the DataFrame structure that were selected. `period`, for example, is repeated for all four rows.

Additionally, this output provides another good example of the difference between the labels and the integer row numbers. On the left-hand side of the preceding screenshot, we can see the labels for the row numbers for the rows that were found by Boolean indexing. We can prove this by using those integers in an `.iloc[]` method, as follows:

```
GDP_by_industry.iloc[[75, 1197, 1219, 1967], :]
```

This will produce the same output it did previously:

Out[146]:

	period	Industry	GDP
75	2017_Q2	Farms	399.7
1197	2017_Q2	Finance and insurance	2827.1
1219	2017_Q2	Federal Reserve banks, credit intermediation, ...	926.5
1967	2017_Q2	Federal	1124.6

Figure 5.27 – The 2017_Q2 GDP for a few selected industries, which was obtained using .iloc[] with the row numbers from Figure 5.25

Now that we have learned more about how to use labels as row indexes, as well as the distinction between the integer row numbers and the row labels, we'll reinforce these concepts in *Exercise 5.03 – integer row numbers versus labels.*

Creating a multi-index from columns

In the previous section, you learned how to read in a DataFrame and create a multi-index by specifying multiple index columns when we read the data. However, a common pattern in data exploration is to read the data and then understand what the columns are. After that, you may decide to create a multi-index using some of the columns. Let's learn how to do that. First, we will read the `US_GDP_Industry.csv` file again, but this time, we won't specify any index columns. This means we get all the columns as data and the default row index:

```
GDP_by_industry = pd.read_csv('Datasets\\US_GDP_Industry.csv')
GDP_by_industry.head()
```

This will produce the following output:

Out[39]:

	period	Industry	GDP
0	2015_Q1	All industries	31917.8
1	2015_Q2	All industries	32266.2
2	2015_Q3	All industries	32406.6
3	2015_Q4	All industries	32298.7
4	2016_Q1	All industries	32303.8

Figure 5.28 – The US GDP data with the default row index

Here, we can see all the data in columns and the usual sequential integer row labels. Now, let's use a pandas method to create the multi-index. This is such a common need that pandas provides methods specifically for this. Here, we will use the `pd.MultiIndex.from_frame()` method to create an index object and then inspect it. Note that here, we are not modifying the DataFrame yet; instead, we are creating a standalone index object:

```
GDP_index = pd.MultiIndex.from_frame(GDP_by_industry[['peri-
od',\
                                                        'Indus-
try']])
GDP_index
```

This will produce the following output:

```
Out[40]: MultiIndex([('2015_Q1',         'All industries'),
                     ('2015_Q2',         'All industries'),
                     ('2015_Q3',         'All industries'),
                     ('2015_Q4',         'All industries'),
                     ('2016_Q1',         'All industries'),
                     ('2016_Q2',         'All industries'),
                     ('2016_Q3',         'All industries'),
                     ('2016_Q4',         'All industries'),
                     ('2017_Q1',         'All industries'),
                     ('2017_Q2',         'All industries'),
                     ...
                     ('2018_Q1', 'Government enterprises'),
                     ('2018_Q2', 'Government enterprises'),
                     ('2018_Q3', 'Government enterprises'),
                     ('2018_Q4', 'Government enterprises'),
                     ('2019_Q1', 'Government enterprises'),
                     ('2019_Q2', 'Government enterprises'),
                     ('2019_Q3', 'Government enterprises'),
                     ('2019_Q4', 'Government enterprises'),
                     ('2020_Q1', 'Government enterprises'),
                     ('2020_Q2', 'Government enterprises')],
                    names=['period', 'Industry'], length=2134)
```

Figure 5.29 – The multi-index created from the period and Industry columns

Note that in the preceding output, we can see that the index names are the column names that were used to create the index and that the length is the same as the DataFrame. Also, note that it isn't in the nice hierarchal format we saw earlier. We'll address this in a moment. For now, we will assign the index to the DataFrame. Before we do that, though, we will drop the two columns (`'period'` and `'Industry'`) as we don't need them anymore; this information is in the index. To assign the index, we can use the `.set_index()` method:

```
GDP_by_industry.drop(columns = ['period', 'Industry'],\
                     inplace = True)
GDP_by_industry.set_index(GDP_index, drop = True,\
                          inplace = True)
GDP_by_industry
```

This will produce the following output:

Out[41]:

		GDP
period	**Industry**	
2015_Q1	**All industries**	31917.8
2015_Q2	**All industries**	32266.2
2015_Q3	**All industries**	32406.6
2015_Q4	**All industries**	32298.7
2016_Q1	**All industries**	32303.8
...	...	...
2019_Q2	**Government enterprises**	371.4
2019_Q3	**Government enterprises**	373.5
2019_Q4	**Government enterprises**	375.1
2020_Q1	**Government enterprises**	372.8
2020_Q2	**Government enterprises**	346.0

2134 rows × 1 columns

Figure 5.30 – The new DataFrame with the multi-index

Here, we can see that the two index columns are similar to what we saw in the previous section. However, the hierarchy isn't displayed as it was previously. This is because when we created the index, it only took the information in the order it was in the columns. That said, the benefit of moving the data into the multi-index is that we can now perform a multi-level sort on the index and get the hierarchal display we saw previously. Let's use the `.sort()` method and pass the two index column names:

```
GDP_by_industry.sort_values(by = ['period', 'Industry'],\
                            inplace = True)
GDP_by_industry
```

This will produce the following output:

Out[42]:

period	Industry	GDP
2015_Q1	Accommodation	256.2
	Accommodation and food services	973.6
	Administrative and support services	795.7
	Administrative and waste management services	883.0
	Agriculture, forestry, fishing, and hunting	466.3
...	...	...
2020_Q2	Warehousing and storage	135.1
	Waste management and remediation services	100.6
	Water transportation	32.0
	Wholesale trade	1810.9
	Wood products	111.9

2134 rows × 1 columns

Figure 5.31 – The GDP_by_industry DataFrame with the sorted multi-index

Now, you see the same hierarchy you saw in the previous section. Of course, you could take any of the columns in the index and order them in any way you wanted. In many cases in data analysis, you will find that you are thinking of data hierarchically. That is the power of pandas – it allows you to structure and work with data in ways that match the analysis and your mental picture. Now, let's reinforce what you have learned by going through an activity.

Activity 5.01 – Creating a multi-index from columns

In this activity, you will read in a DataFrame from a file and then use some of the columns to create a sorted multi-index. Suppose you have been given a `.csv` file containing data about mushrooms, which, as you understand it, contains a classification of edible or poisonous mushrooms, as well as many visual features to allow them to be identified. Since you are a mushroom hunting enthusiast, you are very interested in analyzing and summarizing the data. You begin by reading the data in. Let's get started:

1. For this activity, all you will need is the `pandas` library. Load it into the first cell of the notebook.
2. Read in the `mushroom.csv` data from the `Datasets` directory and list the first five rows using `.head()`.

3. You will see the class column and many visible attributes. List all the columns to see what else there is to work with. The result should be as follows:

```
Out[12]: Index(['class', 'cap-shape', 'cap-surface', 'cap-color', 'bruises', 'odor',
               'gill-attachment', 'gill-spacing', 'gill-size', 'gill-color',
               'stalk-shape', 'stalk-root', 'stalk-surface-above-ring',
               'stalk-surface-below-ring', 'stalk-color-above-ring',
               'stalk-color-below-ring', 'veil-type', 'veil-color', 'ring-number',
               'ring-type', 'spore-print-color', 'population', 'habitat'],
              dtype='object')
```

Figure 5.32 – The columns of the mushroom dataset

4. In addition to `class`, you see `population` and `habitat`, which are not visible attributes. You decide to create a multi-index using `class`, `population`, and `habitat`.

5. Now, drop the columns that are in the index and set the `DataFrame` index to the multi-index. Make sure that you drop the existing default index.

6. Use the `.loc[]` notation you learned and list the data about the edible mushrooms.

   This will produce the following output:

Out[5]:

		cap-shape	cap-surface	cap-color	bruises	odor	gill-attachment	gill-spacing	gill-size	gill-color	stalk-shape	stalk-root	stalk-surface-above-ring	stalk-surface-below-ring	stalk-color-above-ring	stalk-color-below-ring	veil-type	veil-color	nu
population	**habitat**																		
n	g	x	s	y	t	a	f	c	b	k	e	c	s	s	w	w	p	w	
	m	b	s	w	t	l	f	c	b	n	e	c	s	s	w	w	p	w	
a	g	x	s	g	f	n	f	w	b	k	t	e	s	s	w	w	p	w	
n	g	x	y	y	t	a	f	c	b	n	e	c	s	s	w	w	p	w	
	m	b	s	w	t	a	f	c	b	g	e	c	s	s	w	w	p	w	
...	...	...	...	...	...	...	...	...	...	...	...	...	...	...	...	...	...	...	
v	l	x	s	n	f	n	a	c	b	y	e	?	s	s	o	o	p	o	
c	l	k	s	n	f	n	a	c	b	y	e	?	s	s	o	o	p	o	
v	l	x	s	n	f	n	a	c	b	y	e	?	s	s	o	o	p	n	
c	l	f	s	n	f	n	a	c	b	n	e	?	s	s	o	o	p	o	
	l	x	s	n	f	n	a	c	b	y	e	?	s	s	o	o	p	o	

4208 rows × 20 columns

Figure 5.33 – The mushroom data with a multi-index comprising population and habitat

> **Note**
>
> The solution to this activity can be found in the *Appendix*.

Now that you know how to utilize labels for the row index, as well as the **multi-index**, let's learn more about integer rows index versus labels.

Bracket and dot notation

In the previous section, we focused on the `DataFrame.loc` method. pandas offers two ways to select data – using just brackets, `[]`, and using what is called **dot notation** (pandas also refers to the latter as *attribute access* since `object.name` is Python syntax for accessing the `name` attribute in `object`).

Bracket notation

We have already introduced one form of bracket notation, which is using a column name inside brackets. There are several ways to apply bracket notation to a DataFrame, as follows:

- **Select entire columns**: `DataFrame['column_name']` or `DataFrame[[list of column names]]`. If a single column is selected, the result is a Series; otherwise, the result is a DataFrame. If an additional selection results in only one row, the result can be a Series. Also, if the DataFrame only contains one row, selecting one column returns a Series (even though the result is a single value).
- **Selecting a range of rows**: `DataFrame[start:end]`, where `start` and `end` are integers, but `end` is not included in the result. Note that this returns a DataFrame, even if the result is one row.
- **Extended indexing**: `DataFrame[start:end:by]`, where `start`, `end`, and `by` are integers, although `by` can be negative. When `by` is negative, this can be used to reverse the row order. Using a value of `by` other than plus or minus 1 results in every `by` row being returned. For example, `DataFrame[0:100:2]` will return every other element from 0 to 100.

Dot notation

Dot notation (attribute access) can do some of the things bracket notation can do but is more limited. Although dot notation is very compact and efficient for coding, it's important to be aware of its limitations and, if in doubt, consider using bracket notation or being explicit with the `.loc` or `.iloc` method. Here are the main ways in which you can use dot notation:

- **Select entire columns**: `DataFrame.column_name` returns a series unless only one row results from the selection (see the next bullet points).
- **Select one element of a column**: `DataFrame.iloc[row, :].column_name` returns an object that's the type of the cell – that is, `[row, 'column_name']`.

A very important limitation of using dot notation is that since we are using the same notation as Python attribute access, you cannot use this notation if the name of the column matches an existing method name or is an invalid identifier. Therefore, using `DataFrame.min` to attempt to select the `min` column will not work, and using `DataFrame.1` to attempt to select the `1` column will also not work. In both cases, bracket notation will work, with `min` or `1` quoted: `DataFrame['min']` or `DataFrame['1']`, respectively.

Let's review each method in detail.

Selecting complete columns

Let's return to the GDP by industry data again. Let's say you just want to select the `GDP` column for analysis. After reading in the original data again, let's combine checking the type of the result with printing the result in one print statement and use bracket notation to select the column:

```
GDP_by_industry = pd.read_csv('Datasets/US_GDP_Industry.csv')
print(type(GDP_by_industry['GDP']), '\n',
      GDP_by_industry['GDP'])
```

This will produce the following output:

```
<class 'pandas.core.series.Series'>
 0        31917.8
1        32266.2
2        32406.6
3        32298.7
4        32303.8
           ...
2129       371.4
2130       373.5
2131       375.1
2132       372.8
2133       346.0
Name: GDP, Length: 2134, dtype: float64
```

Figure 5.34 – Using bracket notation to select the GDP column

Here, we obtained the entire column as a pandas Series. We can obtain the same result using dot notation. Let's do this here by following the same pattern of checking the type and printing the result:

```
print(type(GDP_by_industry.GDP), '\n',
      GDP_by_industry.GDP)
```

This will produce the same output we saw previously:

```
<class 'pandas.core.series.Series'>
0        31917.8
1        32266.2
2        32406.6
3        32298.7
4        32303.8
          ...
2129       371.4
2130       373.5
2131       375.1
2132       372.8
2133       346.0
Name: GDP, Length: 2134, dtype: float64
```

Figure 5.35 – Using dot notation to select the GDP column

Now, suppose we realize that we want both periods and the GDP data. We can use bracket notation to select multiple columns. Note that in this case, we need to pass a list of columns, and since a Python list is enclosed in square brackets, this looks like "double brackets," but it's just pandas brackets with a list passed as the argument. Following the same pattern, let's print the type of the result, as well as the result, in one statement:

```
print(type(GDP_by_industry[['period', 'GDP']]), '\n',
      GDP_by_industry[['period', 'GDP']])
```

This will produce the following output:

```
<class 'pandas.core.frame.DataFrame'>
       period      GDP
0     2015_Q1  31917.8
1     2015_Q2  32266.2
2     2015_Q3  32406.6
3     2015_Q4  32298.7
4     2016_Q1  32303.8
...       ...      ...
2129  2019_Q2    371.4
2130  2019_Q3    373.5
2131  2019_Q4    375.1
2132  2020_Q1    372.8
2133  2020_Q2    346.0

[2134 rows x 2 columns]
```

Figure 5.36 – Selecting both the period and GDP columns using pandas bracket notation by supplying a list of column names

By using bracket notation, we can pass either a single column name or a list without having to use `DataFrame.loc`. Using dot notation, we can only select a single column but with bracket notation, we can select multiple columns. Also, notice that when you select multiple columns, the result is a DataFrame.

Selecting a range of rows

Using bracket notation with integer ranges selects rows. Here, we will select seven rows; this will return all the columns:

```
GDP_by_industry[3:10]
```

Out[231]:

	period	Industry	GDP
3	2015_Q4	All industries	32298.7
4	2016_Q1	All industries	32303.8
5	2016_Q2	All industries	32696.4
6	2016_Q3	All industries	33070.7
7	2016_Q4	All industries	33457.8
8	2017_Q1	All industries	33984.6
9	2017_Q2	All industries	34167.9

Figure 5.37 – Using pandas bracket notation to slice rows

Here, as expected, we get rows 3 through 9 by using `[3:10]` as the range. The result is still a DataFrame.

Exercise 5.03 – integer row numbers versus labels

In this exercise, you will create a simple DataFrame, create a new column representing the month of the year, and use that new column as the row index. The data for this exercise is some sales data over 3 years by quarter. You have been given data from a colleague in an email and must create the DataFrame yourself and enter the data. Ultimately, you'd like to summarize by quarter to see whether there is a trend.

> **Note**
>
> You can find the code for this exercise at `https://github.com/PacktWorkshops/The-Pandas-Workshop/tree/master/Chapter05/Exercise05_03`.

Follow these steps to complete this exercise:

1. In the `Chapter05` directory that you created earlier, create a directory called `Exercise05_03`.
2. Open your Terminal (macOS or Linux) or Command Prompt (Windows), navigate to the `Chapter05` directory, and type `jupyter notebook`. The Jupyter notebook should open.

3. Select the `Exercise05_03` directory to change the working Jupyter directory to that folder. Then, click **New** > **Python 3** to create a new **Python 3** notebook.

4. For this exercise, all you will need is the `pandas` library. Load it into the first cell of the notebook:

```
import pandas as pd
```

5. You have been given some sales data for your region and want to enter it into pandas. Rather than creating a file, you decide to create the DataFrame directly. Use the following code to create some data that matches what you have been given:

```
sales_data = \
pd.DataFrame({'date' : ['2017-03-31', '2017-06-30',
'2017-09-30',\
                         '2017-12-31', '2018-03-30',
 '2018-06-30',\
                         '2018-09-30', '2019-12-31',
 '2019-03-31',\
                         '2019-06-30', '2019-09-30',
 '2019-12-31'],\
               'sales' : [199190.4, 194356.6, 191611.7, \
                          198918.9, 200163.2, 201510.2, \
                          209749.8, 201897.8, 200098.8, \
                          219340.3, 211542.5, 211729.1]})
```

This will produce the following output:

Out[2]:

	date	sales
0	2017-03-31	199190.4
1	2017-06-30	194356.6
2	2017-09-30	191611.7
3	2017-12-31	198918.9
4	2018-03-30	200163.2
5	2018-06-30	201510.2
6	2018-09-30	209749.8
7	2019-12-31	201897.8
8	2019-03-31	200098.8
9	2019-06-30	219340.3
10	2019-09-30	211542.5
11	2019-12-31	211729.1

Figure 5.38 – The sales data in a pandas DataFrame

6. Now, add a `month` column to the DataFrame by using the `sales_data['month'] =` notation, as well as an expression to slice the 5th and 6th characters from each `date` value. This expression is using Python subsetting on a string variable since the dates are read as strings in this case. Note that this is an example of pandas **bracket notation**, which we saw in the previous section. Finally, note that we are using *list comprehension* to iterate over the index. In the next line, we must use `.set_index` to make the month column the index. By specifying `drop = True`, we can drop the original index labels:

```
sales_data['month'] = \
[sales_data.loc[i, 'date'][5:7] for i in sales_data.
index]
sales_data.set_index('month', drop = True, inplace =
True)
sales_data
```

Now, the output should look as follows:

month	date	sales
03	2017-03-31	199190.4
06	2017-06-30	194356.6
09	2017-09-30	191611.7
12	2017-12-31	198918.9
03	2018-03-30	200163.2
06	2018-06-30	201510.2
09	2018-09-30	209749.8
12	2019-12-31	201897.8
03	2019-03-31	200098.8
06	2019-06-30	219340.3
09	2019-09-30	211542.5
12	2019-12-31	211729.1

Figure 5.39 – The updated DataFrames with the month column

To review the transformations, we did the following:

- We used a *list comprehension* that is equivalent to a `for` loop, and we iterated over the index values of the original DataFrame. This has the effect of creating a value for every row.

- We extracted the month from each date string by specifying `sales_data.loc[i, 'month'][5:7]`, where `i` is the value we are iterating and `[5:7]` is a normal Python slice of a string. In typical Python indexing, this gives us the 5th and 6th characters.
- The results of the list comprehension are stored in the new `month` column using bracket notation. pandas allows us to specify a column name in brackets without `.loc`, which is equivalent to the notation we saw previously – that is, `sales_data.loc[:, 'month']`. This shorthand makes code more readable, as we'll learn in the next section.
- Lastly, we assigned the `month` column to the index and dropped the original (default) index. As a reminder, using the `inplace = True` parameter modifies the DataFrame as-is, instead of creating a copy. Many pandas methods support the `inplace` parameter, which can make code more readable, streamlined, and use less memory (since a copy is not created).

7. Now, compare selecting rows using `.iloc` with `index 3`, and using `.loc` with the `03` label:

```
print('using .iloc with index 3: ', sales_data.iloc[3,
:])
print('\nusing .loc with index 03: ', sales_data.
loc['03', :])
```

The will produce the following output:

```
using .iloc with index 3:
  date     2017-12-31
sales         198919
Name: 12, dtype: object

using .loc with index 03:
               date     sales
month
03      2017-03-31  199190.4
03      2018-03-30  200163.2
03      2019-03-31  200098.8
```

Figure 5.40 – Output of using .iloc[3,] versus .loc['03'.]

Here, we can see `.iloc`, when used with `index 3`, only returns one row, which is the fourth actual row, at integer row 3. On the other hand, in the `month` index, there are three rows with the same label of `03`, so we get all of them. We saw examples of duplicate index labels earlier, but this shows us that even though the `month` index looks like it contains integers, they are labels, and it is perfectly fine to have duplicates, which often makes sense.

8. Now, use the `.groupby()` method on the `month` index, along with the `.mean()` aggregate function, to see the average sales for each period of the year:

```
sales_data.groupby('month').mean()
```

This will produce the following output:

Out[18]:

month	sales
03	199817.466667
06	205069.033333
09	204301.333333
12	204181.933333

Figure 5.41 – The average sales by quarter

This shows the benefit of structuring the index labels in a logical, group-oriented way.

By completing this exercise, you have seen the key differences between integer row numbers (in particular, the value `3` in this exercise) and labels (`03` in this exercise), as well as the benefit that can come from creating a row index with labels that facilitate grouping the data.

Using extended indexing

The bracket notation also supports extended indexing by using three values separated by colons in pandas. The basic notation is `[start:end:by]`, where the first value is the starting row number (integer row), the second value is the ending row (not inclusive), and the selection goes by every `by` row. Suppose that, with our GDP data, we would like to select every third row from the first 50 rows. We can do this as follows:

```
GDP_by_industry[0:50:3]
```

This will produce the following output:

Out[234]:

	period	Industry	GDP
0	2015_Q1	All industries	31917.8
3	2015_Q4	All industries	32298.7
6	2016_Q3	All industries	33070.7
9	2017_Q2	All industries	34167.9
12	2018_Q1	All industries	35838.6
15	2018_Q4	All industries	37205.3
18	2019_Q3	All industries	37991.1
21	2020_Q2	All industries	34260.0
24	2015_Q3	Private industries	28826.0
27	2016_Q2	Private industries	29058.3
30	2017_Q1	Private industries	30263.5
33	2017_Q4	Private industries	31425.1
36	2018_Q3	Private industries	32940.9
39	2019_Q2	Private industries	33632.4
42	2020_Q1	Private industries	33685.4
45	2015_Q2	Agriculture, forestry, fishing, and hunting	457.9
48	2016_Q1	Agriculture, forestry, fishing, and hunting	443.7

Figure 5.42 – Using extended indexing and bracket notation to select every third row in the first 50 rows

Note that in this case, the last row is 48, since the next item in the series would be 51, which is outside the selection range.

A useful way to apply this method is to reverse a DataFrame by using `-1` as the `by` value. When the `by` value is negative, this means that every `by` row is going in reverse order:

```
GDP_by_industry[::-1]
```

This gives us the entire DataFrame in reverse order:

Out[190]:

	period	Industry	GDP
2133	2020_Q2	Government enterprises	346.0
2132	2020_Q1	Government enterprises	372.8
2131	2019_Q4	Government enterprises	375.1
2130	2019_Q3	Government enterprises	373.5
2129	2019_Q2	Government enterprises	371.4
...	...	...	...
4	2016_Q1	All industries	32303.8
3	2015_Q4	All industries	32298.7
2	2015_Q3	All industries	32406.6
1	2015_Q2	All industries	32266.2
0	2015_Q1	All industries	31917.8

2134 rows × 3 columns

Figure 5.43 – Reversing a DataFrame using extended indexing

Here, we can see that the start and end values are optional; if no value is given for the start and end, they take on the first and last values in the DataFrame, respectively.

We can select a range of rows while using a negative `by` value as well. For this, the start value must be greater than the end value, as follows:

```
GDP_by_industry[100:50:-3]
```

This will produce the following output:

Out[194]:

	period	Industry	GDP
100	2018_Q1	Forestry, fishing, and related activities	56.0
97	2017_Q2	Forestry, fishing, and related activities	55.7
94	2016_Q3	Forestry, fishing, and related activities	51.8
91	2015_Q4	Forestry, fishing, and related activities	52.8
88	2015_Q1	Forestry, fishing, and related activities	54.6
85	2019_Q4	Farms	405.9
82	2019_Q1	Farms	392.3
79	2018_Q2	Farms	405.0
76	2017_Q3	Farms	395.8
73	2016_Q4	Farms	375.1
70	2016_Q1	Farms	390.0
67	2015_Q2	Farms	405.4
64	2020_Q1	Agriculture, forestry, fishing, and hunting	467.7
61	2019_Q2	Agriculture, forestry, fishing, and hunting	448.4
58	2018_Q3	Agriculture, forestry, fishing, and hunting	449.9
55	2017_Q4	Agriculture, forestry, fishing, and hunting	455.1
52	2017_Q1	Agriculture, forestry, fishing, and hunting	455.1

Figure 5.44 – Using a negative by value while selecting a range of rows

Here, we have selected every third row, beginning with row 100, and going backward. The result ends at row 52 because the next in the sequence would be 49, which is outside the range we specified.

Type exceptions

There are certain cases where pandas may return a result of a different type than expected; in particular, returning a series instead of a DataFrame. This occurs when the result of the selection is one row, though it also depends on the nature of the selection. This can be confusing because it can generate an error if you pass the result to another operation that expects a DataFrame. Let's look at some examples to clarify the different outcomes.

Considering the GDP example again, here, we will use bracket notation to select multiple columns, which will produce a DataFrame, but use `.iloc[0, :]` in the same statement, which will return just row 0. This type of expression is called chaining. In this case, the result is a Series:

```
print(type(GDP_by_industry[['period','GDP']].iloc[0, :]), '\n',
      GDP_by_industry[['period','GDP']].iloc[0, :])
```

This will produce the following output:

```
<class 'pandas.core.series.Series'>
 period    2015_Q1
GDP        31917.8
Name: 0, dtype: object
```

Figure 5.45 – Getting a Series instead of a DataFrame by chaining a one-row selection onto the bracket notation selection using .iloc

We can get the same result if we use `.loc` instead of `.iloc`:

```
print(type(GDP_by_industry[['period', 'GDP']].loc[0, :]), '\n',
      GDP_by_industry[['period', 'GDP']].loc[0, :])
```

This will produce the same output we saw previously:

```
<class 'pandas.core.series.Series'>
 period    2015_Q1
GDP        31917.8
Name: 0, dtype: object
```

Figure 5.46 – Using .loc as a chained operation to select one row can also return a Series instead of a DataFrame

Recall that in this case, the index labels are the default, so using `0` is equivalent in `.loc` and `.iloc`.

Perhaps surprisingly, we can use bracket notation to select one row. Here, the result will be a DataFrame, as shown here:

```
new_df = GDP_by_industry[0:1]
print(type(new_df), '\n', new_df, '\n')
```

This will produce the following output:

```
<class 'pandas.core.frame.DataFrame'>
    period        Industry      GDP
0  2015_Q1  All industries  31917.8
```

Figure 5.47 – Using bracket notation to select a range of one row returns a DataFrame

Now, if we select some columns from our one-row DataFrame using bracket notation, we will get a DataFrame as the result:

```
print(type(new_df[['period', 'GDP']]), '\n',
      new_df[['period', 'GDP']], '\n')
```

This will produce the following output:

```
<class 'pandas.core.frame.DataFrame'>
    period      GDP
0  2015_Q1  31917.8
```

Figure 5.48 – Using bracket notation to select multiple columns from a one-row DataFrame returns a DataFrame

However, if we use either dot notation or bracket notation to return only one column from the one-row DataFrame, we will get a Series as the result:

```
print(type(new_df['GDP']), '\n', new_df['GDP'], '\n')
print(type(new_df.GDP), '\n', new_df.GDP)
```

These two statements produce the following output:

```
<class 'pandas.core.series.Series'>
 0    31917.8
Name: GDP, dtype: float64

<class 'pandas.core.series.Series'>
 0    31917.8
Name: GDP, dtype: float64
```

Figure 5.49 – Using either bracket or dot notation to return a single column from a one-row DataFrame returns a Series as the resulting type in both cases

To complete our discussion of indexing, let's look at two more examples. If we select one row and all the columns using `.iloc`, and also use dot notation to select one column, we do not get a Series, but rather an *object* that's the type of location in the DataFrame. Here, the result is a Numpy `float`:

```
print(type(GDP_by_industry.iloc[0, :].GDP), '\n',
      GDP_by_industry.iloc[0, :].GDP)
```

This will produce the following output:

```
<class 'numpy.float64'> 31917.8
```

Lastly, if we use extended indexing to select just one row of the DataFrame, and follow that by selecting one or more columns, the result is still a DataFrame:

```
print(type(GDP_by_industry[0:1:1][['GDP', 'period']]))
print(type(GDP_by_industry[0:1:1][['GDP']]))
```

These two lines produce the following output:

```
<class 'pandas.core.frame.DataFrame>
<class 'pandas.core.frame.DataFrame>
```

In this section, you learned how pandas provides more readable and streamlined access alternatives to using `.loc` and `.iloc` in many situations. We also covered several cases where you may get results of an unexpected type. If you are concerned about remembering all the possible combinations, don't worry. Just focus on the fact that in some cases, if you are using these selection methods and getting errors in other expressions, especially anything about the wrong type, try changing to `.loc` or `.iloc` and see whether that fixes the issue.

Changing DataFrame values using bracket or dot notation

Many of the methods we've discussed can be used to *change* the values in a DataFrame, as well as select slices or ranges. In the following screenshot, we can see the GDP data that we have been working with for 2015:

Out[43]:

	period	Industry	GDP
0	2015_Q1	All industries	31917.8
1	2015_Q2	All industries	32266.2
2	2015_Q3	All industries	32406.6
3	2015_Q4	All industries	32298.7
22	2015_Q1	Private industries	28392.6
...	...	...	...
2093	2015_Q4	General government	2164.7
2112	2015_Q1	Government enterprises	324.4
2113	2015_Q2	Government enterprises	326.4
2114	2015_Q3	Government enterprises	328.7
2115	2015_Q4	Government enterprises	330.2

388 rows × 3 columns

Figure 5.50 – The new GDP_2015 DataFrame

Now, suppose that as part of the economic analysis, we want to increase all the GDP values by 5,000. We can do this by selecting the GDP column using bracket notation on the left, and then doing the same and adding 5,000 on the right:

```
GDP_2015['GDP'] = GDP_2015['GDP'] + 5000
GDP_2015
```

This will produce the following output:

Out[44]:

	period	Industry	GDP
0	2015_Q1	All industries	36917.8
1	2015_Q2	All industries	37266.2
2	2015_Q3	All industries	37406.6
3	2015_Q4	All industries	37298.7
22	2015_Q1	Private industries	33392.6
...	...	...	...
2093	2015_Q4	General government	7164.7
2112	2015_Q1	Government enterprises	5324.4
2113	2015_Q2	Government enterprises	5326.4
2114	2015_Q3	Government enterprises	5328.7
2115	2015_Q4	Government enterprises	5330.2

388 rows × 3 columns

Figure 5.51 – The GDP_2015 DataFrame with every value in the GDP column increased by 5,000

Here, we can see the expected result – that is, all our GDP figures have been increased by 5,000. Thus, using bracket notation, we can choose where new data goes into an existing DataFrame. Now, continuing with our experiment, we need to set any updated values that are greater than 25,000 to 0. As we saw earlier, we can combine bracket notation with Boolean indexing with another bracket notation for column selection. So, let's use that to select values that are greater than 25,000 in the GDP column and assign them the value 0:

```
GDP_2015[GDP_2015['GDP'] > 25000]['GDP'] = 0
```

Unfortunately, this produces a rather obscure sounding error, as shown here:

```
C:\Users\bbate\Miniconda3\envs\keras-gpu-2\lib\site-packages\ipykernel_
launcher.py:1: SettingWithCopyWarning:
A value is trying to be set on a copy of a slice from a DataFrame.
Try using .loc[row_indexer,col_indexer] = value instead

See the caveats in the documentation: https://pandas.pydata.org/pandas-
docs/stable/user_guide/indexing.html#returning-a-view-versus-a-copy
  """Entry point for launching an IPython kernel.
```

Figure 5.52 – A pandas warning stating that "A value is trying to be set on a copy of a slice from a DataFrame"

The reasons behind this can be confusing. This problem arises when we chain operations together; pandas, working from left to right, evaluates `GDP_2015[GDP_2015[GDP] > 25000]` first and creates a copy in memory. Then, it selects the `GDP` column using `['GDP']`. When we try to assign it to the copy, it returns the previously shown warning. This warning is telling us that we may not get the expected result – in this case, what we want and *expect* is that the `GDP_2015` DataFrame has been modified. However, in this particular example, the data was *NOT* changed. We can confirm this by listing out the current contents of `GDP_2015`:

```
GDP_2015
```

You should see the following output:

Out[46]:

	period	Industry	GDP
0	2015_Q1	All industries	36917.8
1	2015_Q2	All industries	37266.2
2	2015_Q3	All industries	37406.6
3	2015_Q4	All industries	37298.7
22	2015_Q1	Private industries	33392.6
...	...	...	...
2093	2015_Q4	General government	7164.7
2112	2015_Q1	Government enterprises	5324.4
2113	2015_Q2	Government enterprises	5326.4
2114	2015_Q3	Government enterprises	5328.7
2115	2015_Q4	Government enterprises	5330.2

388 rows × 3 columns

Figure 5.53 – The GDP_2015 data remains unchanged after receiving the warning

Notice in the warning (*Figure 5.52*) that we have been given a hint on how to avoid this via the `Try using .loc[row_indexer,col_indexer] = value instead` statement. The reason this will make a difference is that by using `.loc`, we will combine both the Boolean expression and the column selection in one, and pandas will then make the assignment directly to our DataFrame, without complaint. We can do this as follows:

```
GDP_2015.loc[GDP_2015.GDP > 25000, 'GDP'] = 0
GDP_2015
```

This approach will produce the following (desired) output:

Out[47]:

	period	Industry	GDP
0	2015_Q1	All industries	0.0
1	2015_Q2	All industries	0.0
2	2015_Q3	All industries	0.0
3	2015_Q4	All industries	0.0
22	2015_Q1	Private industries	0.0
...	...	...	...
2093	2015_Q4	General government	7164.7
2112	2015_Q1	Government enterprises	5324.4
2113	2015_Q2	Government enterprises	5326.4
2114	2015_Q3	Government enterprises	5328.7
2115	2015_Q4	Government enterprises	5330.2

388 rows × 3 columns

Figure 5.54 – The desired result of setting values in the DataFrame when we use .loc to combine the Boolean operation and the column selection

Although there are ways to disable this warning, and you can *sometimes* safely ignore it, it's a good idea to find the offending code and try to resolve the warning if you want your code to be robust. Recall earlier when we learned that using bracket and dot notation can also sometimes return a Series when you may expect a DataFrame? We noted that `.loc` and `.iloc` can resolve this issue, and the same is true here. So, again, the thing to remember is that if you see this error, try reworking the code so that it uses `.loc`, and see whether that helps resolve the warning.

Exercise 5.04 – selecting data using bracket and dot notation

Consider that you are a data scientist working for a large agricultural firm that produces soybeans as a key product. You have been given a dataset containing information about various diseases or other problems with the soybean plants, as well as information on their growing conditions and characteristics. In particular, you are interested in a hypothesis that some conditions are related to hail damage. You must find all the cases where the reported condition has occurred and the plants have been damaged by hail.

> **Note**
> You can find the code for this exercise at `https://github.com/PacktWorkshops/The-Pandas-Workshop/tree/master/Chapter05/Exercise05_04`.

Follow these steps to complete this exercise:

1. In the `Chapter05` directory, create the `Exercise05_04` directory.
2. Open your Terminal (macOS or Linux) or Command Prompt (Windows), navigate to the `Chapter05` directory, and type `jupyter notebook`. The Jupyter notebook should open.
3. Select the `Exercise05_04` directory to change the working Jupyter directory to that folder. Then, click **New** > **Python 3** to create a new **Python 3** notebook.
4. For this exercise, all you will need is the `pandas` library. Load it into the first cell of the notebook:

```
import pandas as pd
```

5. You have been given a `.csv` file containing data on soybeans. Read the `soybean.csv` file into pandas using `read_csv()` and then list the file:

```
soybean_diseases = pd.read_csv('..\\Datasets\\soybean.
csv')
soybean_diseases
```

> **Note**
> Please change the path of the dataset file (highlighted) based on where you have downloaded it on your system.

This will produce the following output:

Out[3]:

	condition	date	plant-stand	precip	temp	hail	crop-hist	area-damaged	severity	seed-tmt	...	int-discolor	sclerotia	fruit-pods	fruitspots	seed	mold-growth	seed-discolor	seed-size	shr
0	diaporthe-stem-canker	6.0	0.0	2.0	1.0	0.0	1.0	1.0	1.0	0.0	...	0.0	0.0	0.0	4.0	0.0	0.0	0.0	0.0	
1	diaporthe-stem-canker	4.0	0.0	2.0	1.0	0.0	2.0	0.0	2.0	1.0	...	0.0	0.0	0.0	4.0	0.0	0.0	0.0	0.0	
2	diaporthe-stem-canker	3.0	0.0	2.0	1.0	0.0	1.0	0.0	2.0	1.0	...	0.0	0.0	0.0	4.0	0.0	0.0	0.0	0.0	
3	diaporthe-stem-canker	3.0	0.0	2.0	1.0	0.0	1.0	0.0	2.0	0.0	...	0.0	0.0	0.0	4.0	0.0	0.0	0.0	0.0	
4	diaporthe-stem-canker	6.0	0.0	2.0	1.0	0.0	2.0	0.0	1.0	0.0	...	0.0	0.0	0.0	4.0	0.0	0.0	0.0	0.0	
...	...	...	...	...	...	...	...	...	...	...	...	...	...	...	...	...	...	...	...	
302	2-4-d-injury	NaN	NaN	NaN	NaN	NaN	NaN	NaN	NaN	NaN	...	NaN	NaN	NaN	NaN	NaN	NaN	NaN	NaN	
303	herbicide-injury	1.0	1.0	NaN	0.0	NaN	1.0	0.0	NaN	NaN	...	NaN	NaN	3.0	NaN	NaN	NaN	NaN	NaN	
304	herbicide-injury	0.0	1.0	NaN	0.0	NaN	0.0	3.0	NaN	NaN	...	NaN	NaN	3.0	NaN	NaN	NaN	NaN	NaN	
305	herbicide-injury	1.0	1.0	NaN	0.0	NaN	0.0	0.0	NaN	NaN	...	NaN	NaN	3.0	NaN	NaN	NaN	NaN	NaN	
306	herbicide-injury	1.0	1.0	NaN	0.0	NaN	1.0	3.0	NaN	NaN	...	NaN	NaN	3.0	NaN	NaN	NaN	NaN	NaN	

307 rows × 36 columns

Figure 5.55 – The soybean_diseases DataFrame after reading in soybean.csv

Here, you can see the `condition` column. You want to see how many different conditions have been represented in the 307 cases in the data file.

6. Select the `condition` column and print out the number of unique values, as well as the values, in a nicely formatted output:

```
print('there are',\
      len(soybean_diseases['condition'].unique()),\
      'unique conditions',\
      [soybean_diseases['condition'].unique()[i]
      for i in range\
      (len(soybean_diseases['condition'].unique()))])
```

This will produce the following output:

```
there are 19 unique conditions ['diaporthe-stem-canker', 'charcoal-rot', 'rhizoctonia-root-rot', 'phytophthora-ro
t', 'brown-stem-rot', 'powdery-mildew', 'downy-mildew', 'brown-spot', 'bacterial-blight', 'bacterial-pustule', 'pu
rple-seed-stain', 'anthracnose', 'phyllosticta-leaf-spot', 'alternarialeaf-spot', 'frog-eye-leaf-spot', 'diaporthe
-pod-&-stem-blight', 'cyst-nematode', '2-4-d-injury', 'herbicide-injury']
```

Figure 5.56 – The unique values in the condition column of the soybean data

7. Before digging into the hail question, you are curious about cases of `brown-spot`. Find all the cases of `brown-spot` and print out the total number in a human-readable format:

```
brown_spots = \
    soybean_diseases\
.loc[soybean_diseases['condition'] == 'brown-spot', :]
print('there are',\
      brown_spots.shape[0],\
      'instances having brown-spot')
```

This will produce the following output:

```
there are 40 instances having brown-spot
```

8. Now, determine how many different conditions are associated with hail damage (represented as `hail == 1` in the data):

```
hail_related = soybean_diseases\
.loc[soybean_diseases['hail'] == 1, 'condition'].unique()
print('there are', len(hail_related),\
      'conditions associated with hail damage out of',\
      len(soybean_diseases['condition'].unique()),\
      'total conditions')
```

This will produce the following output:

```
there are 14 conditions associated with hail damage out
of 19 total conditions
```

9. Now, you decide to make a new DataFrame, `hail_cases`, that only contains the instances with hail damage. Compare the number of cases in the subset to the original total:

```
hail_cases = soybean_diseases\
.loc[soybean_diseases['hail'] == 1, :]
print('there are ',\
      hail_cases.shape[0],\
      ' hail-related cases out of ',
      soybean_diseases.shape[0],\
      ' total cases')
```

This will produce the following output:

```
there are 55 hail-related cases out of 307 total cases
```

10. Now, you need to loop through `hail_cases` and print out the conditions where the severity is high (`severity == 2`). Use a `for` loop to do this:

```
for i in range(hail_cases.shape[0]):
    if hail_cases.loc[i, 'severity'] == 2:
        print('case ', i, ' with condition ',
              hail_cases.loc[i, 'condition'], ' is
severe')
```

The result will be an error, as follows – note that we have shortened the error message here for brevity:

```
---------------------------------------------------------------------------
KeyError                                  Traceback (most recent call last)
~\Miniconda3\envs\keras-gpu-5\lib\site-packages\pandas\core\indexes\base.py in get_loc(self, key, method, tolerance)
   2897             try:
-> 2898                 return self._engine.get_loc(casted_key)
   2899             except KeyError as err:

~\Miniconda3\envs\keras-gpu-5\lib\site-packages\pandas\core\indexes\base.py in get_loc(self, key, method, tolerance)
   2898                 return self._engine.get_loc(casted_key)
   2899             except KeyError as err:
-> 2900                 raise KeyError(key) from err
   2901 
   2902         if tolerance is not None:

KeyError: 0
```

Figure 5.57 – The (shortened) error message that was obtained when we attempted to loop over the rows of the DataFrame

We can see that the basic error is a `KeyError` – we are passing 0 where the DataFrame does not have that value in the index.

11. Inspect the index to see whether you can debug this error by using the `.index` method:

```
hail_cases.index
```

This will produce the following output:

```
Out[14]: Int64Index([  7,  11,  13,  15,  16,  19,  22,  30,  43,  48,  55,  76,  77,
                      80,  90,  91,  95, 100, 101, 102, 104, 110, 114, 116, 127, 128,
                     133, 137, 142, 150, 151, 155, 157, 159, 161, 164, 166, 168, 170,
                     172, 173, 176, 177, 181, 183, 185, 190, 195, 197, 201, 203, 205,
                     206, 208, 214],
                    dtype='int64')
```

Figure 5.58 – The index of the hail_cases DataFrame

Here, you can see that the index does not start at 0 and that it is missing some values. This is the result of taking `hail_cases` as a subset but not resetting the index. Since we used `.loc[]` in our code, we are trying to access the labels shown here, but we used `range(hail_cases.shape[0])` as the loop value, which starts at 0 and goes to the number of rows minus 1.

12. You realize that the simplest way to complete this task is to iterate over the index, instead of a range. Change the code and rerun it:

```
for i in hail_cases.index:
    if hail_cases.loc[i, 'severity'] == 2:
        print('case ', i, ' with condition ',
              hail_cases.loc[i, 'condition'], ' is
severe')
```

This will produce the following output:

```
case  22  with condition  rhizoctonia-root-rot  is severe
case  43  with condition  phytophthora-rot  is severe
case  48  with condition  phytophthora-rot  is severe
case  55  with condition  phytophthora-rot  is severe
```

Figure 5.59 – The correct list of severe cases with hail damage

Noting that `phytophthora-rot` appears multiple times, it would be a good idea to talk to the plant biologist about whether this association is sensible.

By completing this exercise, you should be comfortable with the different uses of bracket and dot notation, both to subset data and to change values or modify a DataFrame.

Summary

In this chapter, you learned about the pandas methods for data indexing and selection by using the primary pandas data structure – the DataFrame. You compared the `DataFrame.loc()` and `DataFrame.iloc()` methods to access items in DataFrames by labels and integer locations, respectively. You also looked at some pandas shortcut methods, including bracket notation, dot notation, and extended indexing. Along the way, you saw how the pandas index is used behind the scenes to align data, and how that can be changed by changing or resetting the index. In addition, we showed you that in many cases, you can assign new values to a subset of data by using it on the left-hand side of an assignment statement (using the equals operator). This creates a very compact and easy-to-read coding style. We saw that an important pandas capability that involved using labels for the row or column index produced more robust code – instead of "hardcoding" the column numbers, they can be referred to by names, and the order does not matter. You also saw how pandas supports the multi-index, which creates a natural hierarchal structure for tabular data.

You should now be comfortable working with pandas data access in DataFrames as well as understand the common pitfalls and workarounds. In the next chapter (*Chapter 6*, *Data Selection – Series*), you will see how many of these concepts apply to Series with no or minimal changes.

6
Data Selection – Series

In this chapter, you'll use most of the methods you've learned about for DataFrames to select data from a pandas Series.

By the end of this chapter, you will have a complete understanding of the Series Index, know how to apply the dot, bracket, and extended indexing methods, and how to use `.loc[]` and `.iloc[]` to select data from a Series.

In this chapter, we will cover the following topics:

- Introduction to pandas Series
- The Series index
- Data selection in pandas Series
- Preparing Series from DataFrames and vice versa
- Activity 6.01 – Series data selection
- Understanding the differences between base Python and pandas data selection
- Activity 6.02 – DataFrame data selection

Introduction to pandas Series

In *Chapter 5, Data Selection – DataFrames*, we introduced several ways you can select data from pandas DataFrames. While a pandas Series can be thought of as a single column of a pandas DataFrame, it is a separate data structure. In this chapter, we are going to learn how to select data from a Series in detail. The key methods, such as `.loc[]` and `.iloc[]`, will still apply to a one-dimensional Series, as well as some of the more advanced methods such as Boolean indexing and extended indexing. Now that you have mastered the methods you can apply to DataFrames, learning about Series will be very similar and intuitive. Toward the end of this chapter, we will spend some time understanding the differences between pandas and base Python regarding selecting data. This will reinforce some of the ideas and methods you have learned about. Conceptually, the same ideas we used to select elements from a DataFrame can be used to select elements from a Series. The first thing to understand is that a pandas Series has an index.

The Series index

Let's say we have some monthly income data from a YouTube channel. We create a Series with some values (monthly earnings in USD) in a list, and an index of month abbreviations, also in a list, using a constructor similar to what we've used for DataFrames. Note that we can add a name for the Series using the `name` argument:

```
import pandas as pd
income = pd.Series([100, 125, 105, 111, 275, 137,
                     99, 10, 250, 100, 175, 200],
                   index = ['Jan', 'Feb', 'Mar', 'Apr', 'May',
'Jun',
                            'Jul', 'Aug', 'Sep', 'Oct', 'Nov',
'Dec'],
                   name = 'income')
income
```

The preceding code produces the following output:

```
Out[2]: Jan    100
        Feb    125
        Mar    105
        Apr    111
        May    275
        Jun    137
        Jul     99
        Aug     10
        Sep    250
        Oct    100
        Nov    175
        Dec    200
        Name: income, dtype: int64
```

Figure 6.1 – The YouTube data we have been asked to analyze

> **Note**
>
> All the examples for this chapter can be found in the `Examples.ipynb` notebook in the `Chapter06` folder, while the data files can be found in the `Datafiles` folder, in this book's GitHub repository. To ensure that the examples run correctly, you need to run the notebook from start to finish in order.

The preceding output looks similar to the output of a DataFrame. However, note that there are no column labels – the Series will always only be comprised of an index and the values, as well as the optional name. As with DataFrames, the default index for a Series takes integer values. Here, we are using the `.reset_index()` method to remove the months and reset the index to the default values:

```
income.reset_index(drop = True)
```

This produces the following output:

```
Out[3]: 0     100
        1     125
        2     105
        3     111
        4     275
        5     137
        6      99
        7      10
        8     250
        9     100
        10    175
        11    200
        Name: income, dtype: int64
```

Figure 6.2 – The result of resetting the index on the YouTube data

Here, we see that the index of the Series now contains integer values, which is the expected result of using `.reset_index()`.

Data selection in a pandas Series

We worked through data selection with pandas DataFrames in the previous chapter, *Chapter 5, Data Selection – DataFrames*. Most of these methods are very similar to those that can be used with a pandas Series, so we'll quickly demonstrate those and explain how some methods differ between DataFrames and a Series.

Brackets, dots, Series.loc, and Series.iloc

Previously, we saw that a column of a DataFrame is a Series, except when there is only one value. A Series can be created and manipulated on its own as well. A Series only has one dimension (that is, it's just a sequence) compared to DataFrames, which have two dimensions (rows and columns). Nonetheless, most of the concepts we've reviewed carry over very intuitively. Let's begin by reading some data into a Series using `.read_csv()`.

Here, we will read some energy cost data from the UK government (source: `https://www.gov.uk/government/statistical-data-sets/annual-domestic-energy-price-statistics`), stored in a `.csv` file. As we know, we can tell pandas not to read the index from the file. We will use two more options for the `.read_csv()` method. First, we will specify `usecols = [1]`, where `usecols` takes a list of columns to read from the file. So, here, we are skipping the first (0) column. Second, we will use the `squeeze = True` option, which tells pandas that if the result is just one column, then "squeeze" it into a Series, instead of defaulting to a DataFrame:

```
UK_energy = pd.read_csv('Datasets\\UK_energy.csv',
                        index_col = None,
                        usecols = [1],
                        squeeze = True)
print(type(UK_energy))
print(UK_energy.head())
```

> **Note**
>
> Please change the path of the dataset file (highlighted) based on where you have downloaded it on your system.

This should produce the following output:

```
<class 'pandas.core.series.Series'>
0     288.177459
1     316.485721
2     338.565899
3     336.866984
4     332.844765
Name: annual_cost, dtype: float64
```

Figure 6.3 – The first rows of the UK_energy Series

Note that the Series has a `Name` (`annual_cost`). This comes from the value in the first row of the `.csv` file, since we did not tell pandas to ignore it. Also, note that the `.head()` method works the same on a Series as it does for a DataFrame.

Now, let's look at some methods for selecting from a Series using `.loc`, `.iloc`, and extended indexing. Let's say that we are interested in looking at a few energy values for every other period in the data, based on a possible pattern we've been asked to investigate. Each of the five examples that follow will select and print the same data selection from the Series. We will start by using `.loc` and passing a list to print the third, fifth, and seventh items (index numbers `2`, `4`, and `6`, respectively) in `UK_energy`:

```
print('UK_energy.loc[[2, 4, 6]]\n\n', UK_energy.loc[[2, 4, 6]])
```

This will produce the following output:

```
UK_energy.loc[[2, 4, 6]]

 2     338.565899
4     332.844765
6     341.909881
Name: annual_cost, dtype: float64
```

Figure 6.4 – Using .loc with a list of items to select from a Series

Now, repeat the same selection, but this time using **extended indexing**. Recall from *Chapter 5, Data Selection – DataFrames*, in the *Dot and bracket notation* section, that extended indexing uses three possible values in the form of `[start:end:by]`, where `start` is the first index to select, `end` is the last (not inclusive), and `by` can be a number other than 1, including negative, meaning every "by" value (such as 2 for "every second value"):

```
print('UK_energy[2:7:2]\n\n', UK_energy[2:7:2])
```

This will produce the following output:

```
UK_energy[2:7:2]

 2     338.565899
4     332.844765
6     341.909881
Name: annual_cost, dtype: float64
```

Figure 6.5 – The same selection using extended indexing

Select the same items using bracket notation by passing a list of items to print:

```
print('UK_energy[[2, 4, 6]]\n\n', UK_energy[[2, 4, 6]])
```

This will produce the same output we saw previously:

```
UK_energy[[2, 4, 6]]

 2     338.565899
4     332.844765
6     341.909881
Name: annual_cost, dtype: float64
```

Figure 6.6 – Accomplishing the selection using bracket notation by passing a list of the desired items

Once again, let's select the same items using `.iloc` and pass a list of the desired items:

```
print('UK_energy.iloc[[2, 4, 6]]\n\n', UK_energy.iloc[[2, 4,
6]])
```

This will produce the same output again:

```
UK_energy.iloc[[2, 4, 6]]

 2     338.565899
4     332.844765
6     341.909881
Name: annual_cost, dtype: float64
```

Figure 6.7 – Making the same selection from the Series using .iloc

Finally, we can use extended indexing with `.iloc`:

```
print('UK_energy.iloc[2:7:2]\n\n', UK_energy.iloc[2:7:2])
```

Once again, we have the same three items in the output:

```
UK_energy.iloc[2:7:2]

 2     338.565899
4     332.844765
6     341.909881
Name: annual_cost, dtype: float64
```

Figure 6.8 – Selecting the desired items with extended indexing and .iloc

Here, we can see that `.loc`, `.iloc`, bracket notation (with Series items as the equivalent of DataFrame rows), and extended indexing work the same as they did with DataFrames in *Chapter 5, Data Selection – DataFrames*. The only method we haven't used is dot notation.

To demonstrate this, let's learn how to change the Series index, which can be done in the same way as we changed the DataFrame row index. The difference is that a Series does not have a `.set_index()` method. Instead, we must assign it directly to the index. Suppose we have been told that the UK energy data is from 1990 to 2019. In the following code, we are using list comprehension to generate a series of labels in the form of `year_XXXX` and assigning the result to the Series index:

```
UK_energy.index = ['year_' + str(i) for i in range(1990, 2020)]
UK_energy.index
```

This will produce the following output:

```
Out[27]: Index(['year_1990', 'year_1991', 'year_1992', 'year_1993', 'year_1994',
               'year_1995', 'year_1996', 'year_1997', 'year_1998', 'year_1999',
               'year_2000', 'year_2001', 'year_2002', 'year_2003', 'year_2004',
               'year_2005', 'year_2006', 'year_2007', 'year_2008', 'year_2009',
               'year_2010', 'year_2011', 'year_2012', 'year_2013', 'year_2014',
               'year_2015', 'year_2016', 'year_2017', 'year_2018', 'year_2019'],
              dtype='object')
```

Figure 6.9 – The updated index of the UK_energy Series

Now, we can select a year by using dot notation:

```
UK_energy.year_1997
```

This will produce the following output:

```
Out[28]:  326.4184542
```

We can also use a range of index labels in bracket notation. Here, we are listing the values for the index labels from `year_1997` to `year_2011`:

```
UK_energy['year_1997' : 'year_2011']
```

This will produce the following output:

```
Out[30]: year_1997    326.418454
         year_1998    306.393163
         year_1999    295.687501
         year_2000    290.333333
         year_2001    283.333333
         year_2002    281.666667
         year_2003    283.666667
         year_2004    291.666667
         year_2005    323.666667
         year_2006    382.000000
         year_2007    423.111111
         year_2008    487.333333
         year_2009    498.666667
         year_2010    484.000000
         year_2011    523.181818
         Name: annual_cost, dtype: float64
```

Figure 6.10 – Selecting a range using a range of labels

Here, we have gathered the data for all the labels from **year_1997** to **year_2011**. Note that labels don't have an implied order, so the result is whatever is between the two given labels.

Now, let's apply some of the methods we've just reviewed to an exercise.

Exercise 6.01 – basic Series data selection

In this exercise, you will read a simple Series from a `.csv` file using some basic selection methods. You will analyze some MRI data that measures the activity of the S1 temporal area in the brain; this data was collected during a sleep apnea study (see `https://plos.figshare.com/articles/dataset/_Global_Brain_Blood_Oxygen_Level_Responses_to_Autonomic_Challenges_in_Obstructive_Sleep_Apnea_/1154343/1`). The data is collected every 2 seconds. In the file that you will load, you only have the raw measurement values. You will investigate whether a cycle is occurring every 4 seconds (roughly every other measurement). In particular, you want to avoid any issues where values that are collected early in the measurement period may not be stable, so you will focus on the data toward the end of the values provided.

> **Note**
>
> You can find the code for this exercise at https://github.com/PacktWorkshops/The-Pandas-Workshop/tree/master/Chapter06/Exercise06_01.

Follow these steps to complete this exercise:

1. In the `Chapter06` directory, create the `Exercise06_01` directory.
2. Open your Terminal (macOS or Linux) or Command Prompt (Windows), navigate to the `Chapter06` directory, and type `jupyter notebook`. The Jupyter Notebook should open.
3. Select the `Exercise06_01` directory to change the working Jupyter directory to that folder. Then, click **New** > **Python 3** to create a new **Python 3** notebook.
4. For this exercise, all you will need is the `pandas` library. Load it into the first cell of the notebook:

```
import pandas as pd
```

5. You have been provided with a `.csv` file containing time series data of one patient in a sleep apnea study, where a metric called `BOLD` (blood oxygen level-dependent activity) was recorded, which is related to the oxygen level in the blood. Read the `PLOS_BOLD_S1_patient_1.csv` file into a Series:

```
BOLD = pd.read_csv('..//Datasets//PLOS_BOLD_S1_patient_1.
csv',
                  squeeze = True)
BOLD
```

> **Note**
>
> Please change the path of the dataset file (highlighted) based on where you have downloaded it on your system.

The output should look as follows:

```
Out[5]: 0        0.783670
        1        0.293040
        2        0.111169
        3       -0.169703
        4       -0.147029
                   ...
        139      0.723983
        140      0.687518
        141      0.515671
        142      0.432008
        143      0.146747
        Name: Y, Length: 144, dtype: float64
```

Figure 6.11 – The patient blood oxygen data in a pandas Series

6. As we mentioned earlier, the data is collected every 2 seconds. Assign a range from 0 while counting by 2 to the index. We can accomplish this by assigning directly to the index, since `BOLD` is a Series, and using the `range()` method to generate the values:

```
BOLD.index = range(0, 2*len(BOLD), 2)
BOLD
```

This will produce the following output:

```
Out[3]: 0        0.783670
        2        0.293040
        4        0.111169
        6       -0.169703
        8       -0.147029
                   ...
        278      0.723983
        280      0.687518
        282      0.515671
        284      0.432008
        286      0.146747
        Name: Y, Length: 144, dtype: float64
```

Figure 6.12 – The blood oxygen data with a new index equivalent to the time of the data in seconds

7. The initial hypothesis stated that there may be a recurring cycle of around 4 seconds. To investigate this, you decide to create a Series that contains every other element of `BOLD` (which is every 4 seconds) and use extended indexing to increment by 2:

```
B2 = BOLD[::2]
B2
```

This will produce the following output:

```
Out[7]: 0      0.783670
        4      0.111169
        8     -0.147029
        12    -0.032271
        16    -0.202202
                 ...
        268   -0.014538
        272    0.180167
        276    0.382172
        280    0.687518
        284    0.432008
        Name: Y, Length: 72, dtype: float64
```

Figure 6.13 – A new series created from BOLD that contains every other element (every 4 seconds)

8. As noted in the problem statement, you want to focus on data toward the end of the test. To look at that data, list the last 10 elements of the new Series in reverse order. Do this using extended indexing, starting with the end value and going backward:

```
B2[len(B2):(len(B2) - 10):-1]
```

This will produce the following output:

```
Out[10]: 284    0.432008
         280    0.687518
         276    0.382172
         272    0.180167
         268   -0.014538
         264   -0.080900
         260    0.069567
         256    0.153728
         252    0.220703
         Name: Y, dtype: float64
```

Figure 6.14 – The last 10 items of the new Series, in reverse order

Looking at this data, it is not evident that the values are repeating on a 4-second cycle. We could apply more sophisticated statistical methods, but for now, we can conclude that it is unlikely there is such a cycle.

In this exercise, we implemented some of the basic selection methods we have seen already for DataFrames, but on Series data. At this stage, you should be comfortable working with both Series and DataFrames and be able to use the index in a variety of ways to organize and select data for analysis.

From the preceding sections, you can see that once you have mastered working with DataFrames, working with a Series is very intuitive. You may have noticed that the Series has a name, and you might be wondering what the utility of that is. The main use case is when we define Series that are later combined into DataFrames; the names are used as the column names. However, we don't directly reference the name when working with the Series alone. We'll look at some examples of the name becoming a column name in the next section.

Preparing Series from DataFrames and vice versa

In *Chapter 5*, *Data Selection – DataFrames*, we saw examples of getting a Series by slicing the column of a DataFrame. Let's review this. You have been provided with a dataset (adapted from `https://archive.ics.uci.edu/ml/datasets/Water+Treatment+Plant`) regarding a water treatment facility and you've been asked to analyze its performance. The data contains various chemical measurements for the input, two settling stages, and the output, plus some performance indicators. We will begin by reading the `water-treatment.csv` file. After reading the data, we will use the `.fillna()` method, which replaces any missing values, which are converted into `NaN` values during the file read, into the value that's passed to `.fillna()`. We will use a value of `-9999` here:

```
water_data = pd.read_csv('Datasets\\water-treatment.csv')
water_data.fillna(-9999, inplace = True)
water_data
```

> **Note**
>
> Please change the path of the dataset file (highlighted) based on where you have downloaded it on your system.

This will produce the following output:

Out[6]:

	date	input_flow	input_Zinc	input_pH	input_BOD	input_COD	input_SS	input_VSS	input_SED	input_CON	...	output_COND	RD-DBO-P	RD-SS-P	RD-SED-P
0	1/1/1990	41230.0	0.35	7.6	120.0	344.0	136.0	54.4	4.5	993	...	903.0	-9999.0	62.8	93.3
1	1/2/1990	37386.0	1.40	7.9	165.0	470.0	170.0	76.5	4.0	1365	...	1481.0	-9999.0	50.0	94.4
2	1/3/1990	34535.0	1.00	7.8	232.0	518.0	220.0	65.5	5.5	1617	...	1492.0	32.6	62.4	95.0
3	1/4/1990	32527.0	3.00	7.8	187.0	460.0	180.0	67.8	5.2	1832	...	1590.0	13.2	57.6	95.5
4	1/7/1990	27760.0	1.20	7.6	199.0	466.0	186.0	74.2	4.5	1220	...	1411.0	38.2	46.6	95.0
...	...	...	...	...	...	...	...	...	...	...	...	...	...	...	...
522	10/25/1991	35400.0	0.70	7.6	156.0	364.0	194.0	63.9	5.5	1680	...	1840.0	47.3	61.3	94.0
523	10/26/1991	30964.0	3.30	7.7	220.0	540.0	184.0	62.0	3.5	1445	...	1337.0	-9999.0	38.6	93.3
524	10/27/1991	35573.0	7.30	7.6	176.0	333.0	178.0	64.0	3.5	1627	...	1799.0	-9999.0	40.4	95.0
525	10/29/1991	29801.0	1.60	7.7	172.0	400.0	136.0	70.1	1.5	1402	...	1468.0	32.4	40.4	88.0
526	10/30/1991	31524.0	1.60	7.9	-9999.0	478.0	204.0	64.7	6.0	1798	...	1568.0	-9999.0	43.9	65.3

527 rows × 39 columns

Figure 6.15 – The water-treatment dataset

As we saw in *Chapter 5*, *Data Selection – DataFrames*, we can use `.set_index()` on a DataFrame to replace an index with a column. Let's do this to make the index the date:

```
water_data.set_index('date', drop = True, inplace = True)
```

We've used bracket notation already to select individual columns from DataFrames, but now, we want to make the outcome of this action explicit. When we use a method that produces the data from a single column as a result, the result is a Series. Here, we will select the `'input_flow'` column and check its type:

```
type(water_data['input_flow'])
```

This will produce the following output:

```
Out[26]:   pandas.core.series.Series
```

With this method, we can begin to do some analysis regarding the pH data (a measure of the acidity of the water), beginning with creating two Series. First, we must create a new DataFrame called `acidity` that only contains the rows where `input_pH` is less than `7.5`. Then, we must create a Series called `pH` that's used as the `input_pH` column of the `acidity` DataFrame:

```
acidity = water_data.loc[water_data['input_pH'] < 7.5, :]
pH = acidity['input_pH']
```

There are a variety of ways to construct DataFrames from Series, but here, we'll be using the DataFrame **constructor** we've already seen. Create a new DataFrame called `pH_data` using the `pH` Series:

```
pH_data = pd.DataFrame({'pH' : pH})
pH_data.head()
```

This will produce the following output:

Out[41]:

	pH
date	
3/20/1990	7.4
4/13/1990	7.2
6/4/1990	7.3
6/8/1990	7.4
7/1/1990	7.3

Figure 6.16 – The pH DataFrame

Note how the `acidity` index, which came from `water_data`, was carried through with the Series and is now the index of the `pH_data` DataFrame. You plan to use the `pH_data` DataFrame further in some operations without dates, so you reset the index:

```
pH_data.reset_index(drop = True, inplace = True)
```

Let's consider a hypothesis where the input flow rate is correlated with the input pH. We think that the water becomes more acidic (which results in a lower pH) when certain households dispose of cleaning agents. However, when the amount of water to be treated is higher, we suspect that the impact will be less, and we want to investigate this idea. We need to perform a few operations on the pair of values, so we must make another DataFrame called `flow_hypothesis`, again using the DataFrame constructor, as follows:

```
flow_hypothesis = pd.DataFrame({'pH' : pH_data['pH'],
                                'flow' : acidity['input_
flow']})
print(flow_hypothesis.head())
print(flow_hypothesis.tail())
```

This will produce the following output:

```
     pH  flow
0  7.4   NaN
1  7.2   NaN
2  7.3   NaN
3  7.4   NaN
4  7.3   NaN
             pH     flow
8/21/1990 NaN  34352.0
8/24/1990 NaN  32802.0
8/28/1991 NaN  32922.0
8/29/1991 NaN  32190.0
8/4/1991  NaN  24978.0
```

Figure 6.17 – The flow_hypothesis DataFrame

This does not look quite as expected. We know we've replaced any missing values with `9999`, so why do we have `NaN` values now? Also, why does the index contain both numbers and dates? By default, pandas will align the indexes, which can be very useful in many instances. So, it looked at the indexes of `pH_data['pH']` and `acidity['input_flow']` and lined them up. The problem arose when we created the `acidity` DataFrame. The `acidity` DataFrame has index values that came from `water_data`, and `water_data` had the index set to the `date` column. Those row labels (dates) were passed to the Series we used to form the `flow` column in the `flow_hypothesis` DataFrame. Since we had reset the index of `pH_data`, when we used the `pH` column, it had *sequential* index labels. Therefore, while the `pH` Series and the `flow` Series we used in the DataFrame constructor have the same number of values, the index values no longer line up. This is not what we wanted.

Let's print out both indexes that were involved in creating the `flow_hypothesis` DataFrame.

```
print(list(pH_data.index))
print(acidity.index)
```

This will produce the following output:

```
pH index:  [0, 1, 2, 3, 4, 5, 6, 7, 8, 9, 10, 11, 12, 13, 14, 15, 16, 17, 18, 19, 20, 21, 22, 23, 24, 25, 26]
acidity index:  ['3/20/1990', '4/13/1990', '6/4/1990', '6/8/1990', '7/1/1990', '7/23/1990', '7/29/1990', '8/21/1990', '8/24/199
0', '10/7/1990', '3/26/1991', '4/12/1991', '5/9/1991', '5/23/1991', '6/14/1991', '6/24/1991', '7/1/1991', '7/5/1991', '7/19/199
1', '7/21/1991', '7/30/1991', '8/1/1991', '8/4/1991', '8/18/1991', '8/28/1991', '8/29/1991', '10/5/1991']
```

Figure 6.18 – The index of the pH_data DataFrame and the acidity DataFrame

Our problem is now very clear – the two indices have completely different values. When pandas puts columns together, if the index values of a given column are missing from another column, it fills them in with `NaN`. So the `flow_hypothesis` DataFrame values are not aligned, and there are more rows than we wanted, to make room for all the `NaN` values.

There are various ways to avoid this, depending on what information we want to retain. In our case, we can set the index on the `acidity` DataFrame to the `pH_data` DataFrame's index (note that we could also use `.reset_index()` since `pH_data` has had its index reset already):

```
acidity.set_index(pH_data.index, drop = True, inplace = True)
flow_hypothesis = \
    pd.DataFrame({'pH' : pH_data['pH'],
                  'flow' : acidity['input_flow']})
print(flow_hypothesis.head())
print(flow_hypothesis.tail())
```

You will see the following output:

```
    pH     flow
0  7.4  39165.0
1  7.2  34667.0
2  7.3  51520.0
3  7.4  35789.0
4  7.3  30201.0
     pH     flow
22  7.3  24978.0
23  7.3  27527.0
24  7.4  32922.0
25  7.3  32190.0
26  7.3  33695.0
```

Figure 6.19 – The new version of flow_hypothesis

The main downside of the path we took is that the resulting data contains the reset index. In some cases, there will be an **index** that we specifically want to retain. For example, we stored the dates in the index and realized we would like to analyze them by date.

In other words, we want to ensure we retain the index values from the original `water_data` DataFrame, which are now the indexes from the `acidity` DataFrame. We can easily accomplish this if we take both Series from the `acidity` DataFrame. Here, we are recreating `acidity` because we changed that index earlier:

```
acidity = \
    water_data.loc[water_data['input_pH'] < 7.5, :]
pH = acidity['input_pH']
flow = acidity['input_flow']
flow_hypothesis = pd.DataFrame({'pH' : pH,
```

```
                                    'flow' : flow})
print(flow_hypothesis.head())
print(flow_hypothesis.tail())
```

This will produce the following output:

```
            pH     flow
date
3/20/1990  7.4  39165.0
4/13/1990  7.2  34667.0
6/4/1990   7.3  51520.0
6/8/1990   7.4  35789.0
7/1/1990   7.3  30201.0
            pH     flow
date
8/4/1991   7.3  24978.0
8/18/1991  7.3  27527.0
8/28/1991  7.4  32922.0
8/29/1991  7.3  32190.0
10/5/1991  7.3  33695.0
```

Figure 6.20 – The new flow_hypothesis DataFrame

We now have a clean dataset, with all the data we need for further investigation, including dates if we need to look at time trends.

We will conclude this topic by stressing the importance of **indexes** in a wide range of operations. The tools you have learned about in this section will become key parts of your pandas work.

Exercise 6.02 – using a Series index to select values

In this exercise, you will read a Series with a text-based **index** from a file, and apply methods to select data using the **index** values. You have been given a file called `fruit_orders.csv` that contains produce orders for a wholesale supplier over 1 week. Here, the first column is the name of a fruit, while the second column is the quantity ordered. You have been given a few simple tasks: calculate the total quantity of apples, peaches, and oranges that have been ordered, figure out which of these has the highest total orders, and create code that will extract all the pear and peach orders.

While doing these calculations, you also plan to experiment with different ways of leveraging a text-based index to find desired values, by both directly stating the values (`pear`, `peach`, and so on) and using a pandas method, `.startswith()`, that can apply a text comparison to the index and return matching values. By comparing different methods, you can choose the best method for a given case.

> **Note**
>
> You can find the code for this exercise at `https://github.com/PacktWorkshops/The-Pandas-Workshop/tree/master/Chapter06/Exercise06_02`.

Follow these steps to complete this exercise:

1. In the `Chapter06` directory, create a new directory called `Exercise06_02`.
2. Open your Terminal (macOS or Linux) or Command Prompt (Windows), navigate to the `Chapter02` directory, and type `jupyter notebook`. The Jupyter Notebook should open.

   Select the `Exercise06_02` directory to change the working Jupyter directory to that folder. Then, click **New** > **Python 3** to create a new **Python 3** notebook.
3. For this exercise, all you will need is the `pandas` library. Load it into the first cell of the notebook:

```
import pandas as pd
```

4. Read the data from the `.csv` file. Keeping in mind that the first column is to become our index, specify `index_col = 0`. To get the result as a Series instead of a DataFrame, use `squeeze = True`:

```
fruit_orders = pd.read_csv('../Datasets/fruit_orders.
csv',
                           index_col = 0,
                           squeeze = True)
fruit_orders.head(10)
```

This will produce the following output:

```
Out[2]: fruit
        orange    149
        apple      98
        orange     69
        peach     103
        peach     124
        orange     81
        pear      144
        orange     67
        peach     113
        peach     127
        Name: qty_ordered, dtype: int64
```

Figure 6.21 – The labeled fruit orders Series

5. Calculate and print out the total number of apples that have been ordered this week. This can be done using bracket notation, to select apples, and the `sum()` method:

```
apples = sum(fruit_orders['apple'])
print('the total number apples orders this week is:',
apples)
```

The output should be as follows:

```
the total number apples orders this week is:  1175
```

Note that because this is a labeled Series, you could have used Series dot notation to select apples instead of bracket notation (`fruit_orders.apple` versus `fruit_orders['apple']`).

6. Compute the total number of oranges and peaches that have been ordered and print out a statement regarding which is greater:

```
oranges = sum(fruit_orders['orange'])
peaches = sum(fruit_orders['peach'])
if oranges > peaches:
    print('there are more oranges (' + str(oranges) +
          ') ordered than peaches (' + str(peaches) +
')')
elif peaches > oranges:
    print('there are more peaches (' + str(peaches) +
          ') ordered than oranges (' + str(oranges) +
```

```
')')else:
    print('there are the same number of orders for
peaches and oranges')
```

This will produce the following output:

```
there are more (1125) ordered than peaches (1011)
```

7. Next, select all the rows that contain pears or peaches, and store them in a new Series. You can accomplish this in two different ways. Here, we will use `pd.concat()` to concatenate two sub-Series – one using bracket notation to select `'pear'` and the other to select `'peach'`.

   Alternatively, you can iterate through `fruit_orders` using list comprehension and the `.startswith()` method, which can operate on a string, and return `True` if it starts with the string passed to the method. In list comprehension, we select each item in `fruits_orders` in succession, so `.startswith()` is getting a *single* string on each iteration. It would not work if we were to pass all of `fruits_orders` to `.startswith()` in one go:

```
p_fruits_1 = pd.concat([fruit_orders['pear'], fruit_
orders['peach']])
p_fruits_2 = \
pd.Series(fruit_orders[[i
                        for i in range(len(fruit_orders))
                        if fruit_orders.index[i].
startswith('p')]])
```

   Consider the complexity of the two methods. If there were a large number of choices that began with a common letter or letters, the `.startswith()` method might be preferred. In this case, the direct method is simpler.

8. Print out and compare the Series that results from the preceding code:

```
print(p_fruits_1)
print(p_fruits_2)
```

You will see the following output:

```
pear      51
pear      92
pear      14
pear      74
pear      99
pear       2
pear      52
pear      37
pear      63
pear      59
pear      75
peach     60
peach     20
peach     82
peach     86
peach     21
peach      1
peach     87
peach     21
peach     48
dtype: int32
pear      51
pear      92
pear      14
peach     60
peach     20
peach     82
peach     86
pear      74
pear      99
pear       2
peach     21
pear      52
peach      1
peach     87
pear      37
pear      63
pear      59
pear      75
peach     21
peach     48
dtype: int32
```

Figure 6.22 – Comparison of the two methods to get all peaches and pears

Note that the order of the two Series is different because in the first case, we separately selected pears and then peaches, and concatenated them in that order, while in the second case, we got everything at once by using logic in the **index** strings. In this example, you have used the different methods you have learned about to select items using a label-based Series index. In this case, with only two choices, the direct method of referencing the index values is simpler, but if our data were large and contained many values, it would be much more efficient.

Activity 6.01 – Series data selection

In this activity, you will read some US population data for large cities for the years 2010 and 2019 and analyze it. The goal is to determine the population growth for the top three cities compared to all the top 20 from 2010 to 2019. To do this, you must compute the population of the three largest cities for 2010 and 2019, as well as the population of the 20 largest cities for both years. Using these values, you can compute the growth rates and compare them.

Follow these steps to complete this activity:

1. For this activity, all you will need is the `pandas` library. Load it into the first cell of the notebook.
2. Read in a pandas Series from the `US_Census_SUB-IP-EST2019-ANNRNK_top_20_2010.csv` file. This data is from the US Census Bureau (source: `https://www2.census.gov/programs-surveys/popest/datasets/2010/2010-eval-estimates/`). The city names are in the first column, so read them so that they are used as the indexes. List the resulting Series.
3. Calculate the total population of the three largest cities in the 2010 Series (New York, Los Angeles, and Chicago) and save the result in a variable.
4. Read in the corresponding data for 2019 from the `US_Census_SUB-IP-EST2019-ANNRNK_top_20_2019.csv` file, again using the first column as the index and reading the data into a Series. This data is from the US Census Bureau (source: `https://www2.census.gov/programs-surveys/popest/tables/2010-2019/cities/totals/`).
5. Calculate the total population for the same three cities in the 2019 Series and save the result in a variable.
6. Using the saved values, calculate the percent change from 2010 to 2019 for the three cities. Also, calculate the percent change for all the cities. Print out a comparison of the changes for the three cities versus all cities. The result should be similar to the following:

```
top 3 changed 2.2 %
 vs. all changed 8.0 %
```

Figure 6.23 – Changing the population from 2010 to 2019 for the largest 3 and the largest 20 US cities

> **Note**
>
> You can find the code for this activity in this book's GitHub repository and the solution in the *Appendix*.

Now that we know how to work with Series, let's explore some of the key differences between Series and base Python and DataFrames.

Understanding the differences between base Python and pandas data selection

For the most part, once you have learned a bit of pandas notation for slicing and indexing, pandas objects work nearly transparently with core Python. Since the indexing of some different object types looks similar, here, we'll touch on some of the differences so that you can avoid surprises in the future.

Lists versus Series access

Python lists look superficially like Series. When you're using bracket notation to **index** a Series, it works much the same way as indexing a list. Here, we make a simple list using the `range()` function, then print out 11 values within the list:

```
my_list = list(range(100))
print(my_list[12:33])
```

This will produce the following output:

```
[12 13, 14, 15, 16, 17, 18, 19, 20, 21, 22]
```

Now, let's attempt the same thing, but using `.iloc[]`:

```
print(my_list.iloc[12:33])
```

This will produce the following error:

```
---------------------------------------------------------------------------
AttributeError                            Traceback (most recent call last)
<ipython-input-30-1b2c688411ee> in <module>
      2 # try to print using .iloc
      3 #
----> 4 print(my_list.iloc[12:23])

AttributeError: 'list' object has no attribute 'iloc'
```

Figure 6.24 – List objects do not have the pandas Series.iloc() method

Here, we can see that using bracket notation on the list works as expected, but the `Series.iloc[]` method does not apply. Similarly, `Series.loc[]` and dot notation do not apply to lists.

DataFrames versus dictionary access

Core Python does not have a 2D tabular structure, similar to a pandas DataFrame. However, many of the functionalities of DataFrames can be accomplished with dictionaries, but pandas makes this much simpler. Dictionaries and DataFrames are similar enough that pandas provides methods to convert dictionaries into DataFrames and vice versa. Here, we will read in the GDP by Industry data again (the source data can be gathered from the US BEA: `https://www.bea.gov/data/gdp/gdp-industry`), then use a similar list comprehension to what we used previously to extract just `2015_Q1`. After that, we will take the first five items from that list. This will give us a small DataFrame to help us see what is going on:

```
GDP_data = pd.read_csv('Datasets\\US_GDP_Industry.csv')
GDP_2015_Q1_rows = [i for i in GDP_data.index
                    if GDP_data.loc[i, 'period'] == '2015_Q1']
[:5]
GDP_2015_Q1_1st_5 = GDP_data.copy()
GDP_2015_Q1_1st_5 = GDP_2015_Q1_1st_5.iloc[GDP_2015_Q1_rows, :]
GDP_2015_Q1_1st_5
```

This will produce the following output:

Out[18]:

	period	Industry	GDP
0	2015_Q1	All industries	31917.8
22	2015_Q1	Private industries	28392.6
44	2015_Q1	Agriculture, forestry, fishing, and hunting	466.3
66	2015_Q1	Farms	411.7
88	2015_Q1	Forestry, fishing, and related activities	54.6

Figure 6.25 – The first five items in the Q1_2015 GDP by Industry data

Now, let's convert this into a dictionary using the pandas `.to_dict()` method:

```
GDP_dict = GDP_2015_Q1_1st_5.to_dict()
GDP_dict
```

This will produce the following output:

```
Out[16]: {'period': {0: '2015_Q1',
  22: '2015_Q1',
  44: '2015_Q1',
  66: '2015_Q1',
  88: '2015_Q1'},
 'Industry': {0: 'All industries',
  22: 'Private industries',
  44: 'Agriculture, forestry, fishing, and hunting',
  66: 'Farms',
  88: 'Forestry, fishing, and related activities'},
 'GDP': {0: 31917.8, 22: 28392.6, 44: 466.3, 66: 411.7, 88: 54.6}}
```

Figure 6.26 – The result of converting a pandas DataFrame to a Python dictionary

Here, we can see that the column names have become dictionary **keys**, each column is another dictionary in the dictionary, and the previous row numbers are keys of the nested dictionaries. This leads to the similarities and differences in indexing and access. Let's learn how to access all the values associated with one key, as opposed to accessing one column of the original DataFrame. First, we must print the `'period'` information from the dictionary:

```
print(GDP_dict['period'])
```

This will produce the following output:

```
{0:  '2015_Q1',  22:  '2015_Q1',  44:  '2015_Q1',  66:  '2015_
Q1',,88:  '2015_Q1'}
```

Now, let's print the same information using the DataFrame:

```
print(GDP_2015_Q1_1st_5['period'])
```

This will produce the following output:

```
0     2015_Q1
22    2015_Q1
44    2015_Q1
66    2015_Q1
88    2015_Q1
Name: period, dtype: object
```

Figure 6.27 – The period values from the DataFrame of the GDP data

Note that the first result is a Python dictionary, as denoted by the curly braces, while the second result is a Series. We can verify this using `type()`:

```
print(type(GDP_dict['period']))
```

This will produce the following output:

```
<class 'dict'>
```

Now, let's use `type()` again on the original `'period'` DataFrame column:

```
print(type(GDP_2015_Q1_1st_5['period']))
```

This will produce the following output:

```
<class 'pandas.core.series.Series'>
```

The similarities in data access between using a dictionary and a DataFrame more or less end here, as shown in this example. This simple example shows how much pandas simplifies working with tabular data compared to if we were working in core Python and managing the structure in a dictionary.

Activity 6.02 – DataFrame data selection

In this activity, you need to analyze data from this year's survey of Abalone oysters for the National Marine Fisheries Service (the source data can be found in the UCI repository: `https://archive.ics.uci.edu/ml/datasets/abalone`). In particular, you want to get some summary values for the dimensions of male and female samples in the data, depending on the number of rings in the oysters' shells. The ring count is a measure of age, and reviewing this data provides comparisons to previous years to help you understand the health of the population. The data contains several observations, including sex, length, diameter, weight, shell weight, and the number of rings.

To complete this activity, follow these steps:

1. For this activity, all you will need is the `pandas` library. Load it into the first cell of the notebook.
2. Read the `abalone.csv` file into a DataFrame called `abalone` and view the first five rows.
3. Create a `MultiIndex` from the `Sex` and `Rings` columns since these are the variables that you want to summarize the data for. Be sure to drop the `Sex` and `Rings` columns once the index has been created.

   This will produce the following output:

Out[5]:

Sex	Rings	Length	Diameter	Height	Whole weight	.Shucked weight	Viscera weight	Shell weight
M	15	0.455	0.365	0.095	0.5140	0.2245	0.1010	0.150
	7	0.350	0.265	0.090	0.2255	0.0995	0.0485	0.070
F	9	0.530	0.420	0.135	0.6770	0.2565	0.1415	0.210
M	10	0.440	0.365	0.125	0.5160	0.2155	0.1140	0.155
I	7	0.330	0.255	0.080	0.2050	0.0895	0.0395	0.055
	8	0.425	0.300	0.095	0.3515	0.1410	0.0775	0.120
F	20	0.530	0.415	0.150	0.7775	0.2370	0.1415	0.330
	16	0.545	0.425	0.125	0.7680	0.2940	0.1495	0.260
M	9	0.475	0.370	0.125	0.5095	0.2165	0.1125	0.165
F	19	0.550	0.440	0.150	0.8945	0.3145	0.1510	0.320

Figure 6.28 – The result of creating a multi-index with Sex and Rings on the abalone data

4. You plan to focus on oysters that have more than 15 rings. Since you want statistics for each sex, you need to know the values of `Rings` in the data for each sex. Use `abalone.loc['sex'].index` to get a list of all the values for each sex (replace `Sex` with `M` and then with `F`). This works well because you have a two-level index, so by passing a value to filter on `Sex`, you get the relevant items in the next level index, which is `Rings`.

   To filter the data, you will need a list of the unique values of the rings. Python provides the `set()` method, which conveniently produces a set of unique values, so you can apply it as `set(abalone.loc['sex'].index)` to store the unique values of rings for each sex in a variable.

5. You also need the maximum number of rings for each sex. You can get that with `max(abalone.loc['sex'].index])`, which works in the same way as getting all the values. Store this value for each sex in a variable.

6. Now, you need to find the values for each sex that are greater than 15 and are in the unique values for that sex. You can use list comprehension to iterate over the possible values and keep only those that belong to a given sex. This looks like `[i for i in range(min_rings, max_rings + 1)  if i in all_rings]`, where `all_rings` is the list of unique values for the sex, `min_rings` is 16 (one more than 15), and `max_rings` is the maximum value for the sex. Do this and save the result for each sex.

7. Now, you need to select the data for each sex for the `.Shucked weight`, `Length`, `Diameter`, and `Height` columns. For each column, you want the mean value. Thanks to your multi-index, you can do this like so:

   I. `abalone.loc['sex']` can be used to select one sex (`M` or `F`).

   II. `.loc[rings]` can be used to select just the values of the rings you obtained in the list comprehension.

   III. Then, you can use bracket notation with a list of the columns to select the columns, so `[['Length', 'Diameter', 'Height', '.Shucked weight']]`.

   IV. Finally, add the `.mean(axis = 0)` method to tell pandas to take the column's means values.

   The entire operation for each sex looks like this:

   ```
   abalone.loc['sex'][rings][[ 'Length', 'Diameter',
   'Height', '.Shucked weight']].mean(axis = 0)
   ```

   Perform this operation for each sex, using the correct list of rings, and save each result in a separate variable.

8. Print out a comparison of the values between the two sexes. The result should look something like this:

   ```
   for oysters with 16 or more rings

   males weigh 0.458 vs. females weigh 0.449
   males are 0.603 long  vs. females are 0.603 long
   males are 0.478 in diameter  vs. females are 0.479 in diameter
   males are 0.176 in height  vs. females are 0.174 in height
   ```

   Figure 6.29 – The size summaries for the larger oysters

> **Note**
>
> Callout: You can find the solution for this activity in the *Appendix*.

Summary

In this chapter, we have learned about the pandas methods of data indexing and selection using a Series. We compared the `Series.loc()` and `Series.iloc()` methods for accessing items in a Series by labels and integer locations, respectively. We also used pandas shortcut methods, including bracket notation and extended indexing. We reviewed that most methods for DataFrames work similarly and intuitively for a pandas Series, and we highlighted a few key differences. After understanding **indexes** and how to access them, we illustrated differences between core pandas data structures such as lists and dictionaries, as well as some things to keep in mind regarding pandas and core Python.

At this point, you should be comfortable working with pandas data access as well as understand the common pitfalls and workarounds. With these tools in hand, you are ready to tackle data projects of any complexity. In the next chapter, *Chapter 7, Data Transformation*, you will apply some of these methods and learn about other tools you can use to organize data into a clean form for analysis, including dealing with missing data and creating summaries via pivot tables.

7
Data Exploration and Transformation

In the previous chapter, you were introduced to data selection methods in pandas. In this chapter, you will learn more about data transformation in pandas and how to get comfortable with data manipulation to achieve a dataset that is ready for data analysis. By the end of this chapter, you will understand how to deal with messy or missing data and how to summarize it for the purpose of your analysis.

In this chapter, we will cover the following topics:

- Dealing with messy data
- Dealing with missing data
- Summarizing data
- Activity 7.01 – data analysis using pivot tables

Introduction to data transformation

When working with data science, it is important to ensure that your dataset has been cleaned of all the messy data, that is, all of the missing data has been handled correctly. Otherwise, you could end up getting unexpected results when summarizing your dataset and deriving insights. For example, if you want to calculate an average but haven't cleaned up missing data that might be arbitrarily represented as a specific number, such as -999, you could calculate an incorrect aggregation (such as an average) that will include that specific number, -999. Having a good understanding of that arbitrary convention (with -999 representing the missing data) will allow you to exclude that number from any calculation to avoid reporting incorrect aggregations. A good understanding of how to handle messy and missing data in pandas will increase the confidence and accuracy of your analysis.

Dealing with messy data

Messy data can occur for a wide variety of reasons. For example, there are various forms of missing data, such as N/A, NA, None, Null, or any arbitrary number (in other words, -1, 999, 10,000, and more). It is important for analysts to understand the business meaning of the dataset they are handling during the data preparation process. By knowing the nature of missing values, the way that missing values are shown, and the data collection procedures that have triggered the occurrence of missing values, they can choose the best way to interpret this type of data.

Working on data without column headers

Often, the column headers in your data hold the preliminary information and business meaning. However, there is a chance that the column headers will be absent. This results in no specific information that can be derived to help understand the relationship between the headers and the content of the data.

Let's start with the example that we previously saw in *Chapter 4, pandas Data Types*:

```
# Importing pandas
import pandas as pd

# Defining lists
row1 = list([1001.0, 'Pandas Banking', 235000, 248000, 5.5,
2013,3,10, 0])
row2 = list([1002.0, 'Pandas Grocery', 196000, 205000, 4.5,
2016,4,30, 0])
```

```
row3 = list([1003.0, 'Pandas Telecom', 167000, 193000, 15.5,
2010,11,24, 0])
row4 = list([1004.0, 'Pandas Transport', 79000, 90000, 13.9,
2018,1,15, 1])
row5 = list([1005.0, 'Pandas Insurance', 241000, 264000, 9.5,
2009,6,1, 0])

# Defining a DataFrame
data_frame = pd.DataFrame(data=[row1, row2, row3, row4, row5])

# Display DataFrame values
data_frame
```

The output will be as follows:

	0	1	2	3	4	5	6	7	8
0	1001.0	Pandas Banking	235000	248000	5.5	2013	3	10	0
1	1002.0	Pandas Grocery	196000	205000	4.5	2016	4	30	0
2	1003.0	Pandas Telecom	167000	193000	15.5	2010	11	24	0
3	1004.0	Pandas Transport	79000	90000	13.9	2018	1	15	1
4	1005.0	Pandas Insurance	241000	264000	9.5	2009	6	1	0

Figure 7.1 – A DataFrame without column headers

As you can see, the DataFrame doesn't show any column headers when it is generated. Luckily, pandas will generate unique numerical column headers in order for you to use the DataFrame.

Use the `info()` method to get more details about the DataFrame:

```
# Display DataFrame info
data_frame.info()
```

The output will be as follows:

```
<class 'pandas.core.frame.DataFrame'>
RangeIndex: 5 entries, 0 to 4
Data columns (total 9 columns):
 #   Column  Non-Null Count  Dtype
---  ------  --------------  -----
 0   0       5 non-null      float64
 1   1       5 non-null      object
 2   2       5 non-null      int64
 3   3       5 non-null      int64
 4   4       5 non-null      float64
 5   5       5 non-null      int64
 6   6       5 non-null      int64
 7   7       5 non-null      int64
 8   8       5 non-null      int64
dtypes: float64(2), int64(6), object(1)
memory usage: 488.0+ bytes
```

Figure 7.2 – DataFrame information

Notice that pandas is using the column index number as the column header.

As demonstrated in *Chapter 4, pandas Data Types*, where you learned about dtypes, column 0 is currently a float. It needs to be converted into an int.

Convert column 0 into an int, as follows:

```
# Convert column 0 into a int
data_frame[0] = data_frame[0].astype('int')
data_frame[0]
```

The output will be as follows:

```
0    1001
1    1002
2    1003
3    1004
4    1005
Name: Customer ID, dtype: int64
```

Figure 7.3 – The conversion of the column into an int

Now, try summing columns 2 and 3 together (the result will not be stored inside the DataFrame):

```
# Summing column 2 and column 3 together
data_frame[2] + data_frame[3]
```

The output will be as follows:

```
0    483000
1    401000
2    360000
3    169000
4    505000
dtype: int64
```

Figure 7.4 – A summation of columns 2 and 3

Here, you can see that every operation can still be performed on the columns by using their generated headers instead of their real names. Of course, working with data without column headers is not recommended as it leads to confusion and error.

Next, you will perform data manipulation to fix the missing headers:

1. First, define a list of column headers that will replace the missing column headers:

```
# Create list of column headers
column_names = ["Customer ID", "Customer Name", "2018 Revenue", "2019 Revenue", "Growth", "Start Year", "Start Month", "Start Day", "New Customer"]
column_names
```

The output will be as follows:

```
['Customer ID',
 'Customer Name',
 '2018 Revenue',
 '2019 Revenue',
 'Growth',
 'Start Year',
 'Start Month',
 'Start Day',
 'New Customer']
```

Figure 7.5 – Defining the column headers

2. Now, use the `.columns` attribute of the DataFrame to replace the missing column headers:

```
# Replace missing column headers
data_frame.columns = column_names
data_frame
```

The output will be as follows:

	Customer ID	Customer Name	2018 Revenue	2019 Revenue	Growth	Start Year	Start Month	Start Day	New Customer
0	1001	Pandas Banking	235000	248000	5.5	2013	3	10	0
1	1002	Pandas Grocery	196000	205000	4.5	2016	4	30	0
2	1003	Pandas Telecom	167000	193000	15.5	2010	11	24	0
3	1004	Pandas Transport	79000	90000	13.9	2018	1	15	1
4	1005	Pandas Insurance	241000	264000	9.5	2009	6	1	0

Figure 7.6 – The DataFrame after replacing the column headers

The following snippet is another example of how you can import a CSV file without column headers:

```
# Importing pandas
import pandas as pd

#Define the csv URL
file_url = ' https://raw.githubusercontent.com/PacktWorkshops/
The-Pandas-Workshop/master/Chapter07/Data/retail_purchase_
missing_headers.csv '

# Import the csv as a DataFrame
data_frame = pd.read_csv(file_url)

#Display the DataFrame
data_frame
```

The output will be as follows:

	10001	24/05/20	Wheat	4.8lb	€17	Fline Store
0	10002	05/05/20	Fruit Juice	3.1lb	€19	Dello Superstore
1	10003	27/04/20	Vegetables	1.2lb	€15	Javies Retail
2	10004	05/05/20	Oil	3.1lb	€17	Javies Retail
3	10005	27/04/20	Wheat	4.8lb	€13	Javies Retail
4	10006	14/01/20	Butter	3.6lb	€27	Oldi Superstore
5	10007	20/04/20	Oil	4.8lb	€21	Dello Superstore
6	10008	05/05/20	Wheat	3.6lb	€25	Oldi Superstore
7	10009	17/04/20	Fruits	1.2lb	€24	Oldi Superstore
8	10010	15/06/20	Oil	4.4lb	€25	Kanes Store
9	10011	17/06/20	Oil	4.4lb	€16	Fline Store
10	10012	11/06/20	Cheese	2.3lb	€20	Fline Store
11	10013	19/03/20	Rice	4.4lb	€27	Kanes Store
12	10014	01/01/20	Cheese	1.2lb	€10	Fline Store
13	10015	07/07/20	Fruit Juice	3.6lb	€27	Oldi Superstore

Figure 7.7 – An alternative DataFrame example

As you can see, you have an issue with the column headers, as pandas is interpreting the first row of our CSV file as the column header. To avoid this situation, you can add the `header=None` parameter to specify to pandas that you don't have any column headers when you import the CSV file:

```
# Import the csv as a DataFrame
data_frame = pd.read_csv(file_url, header=None)

#Display the DataFrame
data_frame
```

The output will be as follows:

	0	1	2	3	4	5
0	10001	24/05/20	Wheat	4.8lb	€17	Fline Store
1	10002	05/05/20	Fruit Juice	3.1lb	€19	Dello Superstore
2	10003	27/04/20	Vegetables	1.2lb	€15	Javies Retail
3	10004	05/05/20	Oil	3.1lb	€17	Javies Retail
4	10005	27/04/20	Wheat	4.8lb	€13	Javies Retail
5	10006	14/01/20	Butter	3.6lb	€27	Oldi Superstore
6	10007	20/04/20	Oil	4.8lb	€21	Dello Superstore
7	10008	05/05/20	Wheat	3.6lb	€25	Oldi Superstore
8	10009	17/04/20	Fruits	1.2lb	€24	Oldi Superstore
9	10010	15/06/20	Oil	4.4lb	€25	Kanes Store
10	10011	17/06/20	Oil	4.4lb	€16	Fline Store
11	10012	11/06/20	Cheese	2.3lb	€20	Fline Store
12	10013	19/03/20	Rice	4.4lb	€27	Kanes Store
13	10014	01/01/20	Cheese	1.2lb	€10	Fline Store
14	10015	07/07/20	Fruit Juice	3.6lb	€27	Oldi Superstore

Figure 7.8 – Replacing the column headers with numbers

Here, pandas has imported the CSV file and generated the column headers. Now, you can replace the column headers:

```
# Create list of column headers
column_names = ["Receipt Id", "Date of Purchase", "Product
Name", "Product Weight", "Total Price", "Retail shop name"]

# Replace missing column headers
data_frame.columns = column_names
data_frame
```

The output will be as follows:

	Receipt Id	Date of Purchase	Product Name	Product Weight	Total Price	Retail shop name
0	10001	24/05/20	Wheat	4.8lb	€17	Fline Store
1	10002	05/05/20	Fruit Juice	3.1lb	€19	Dello Superstore
2	10003	27/04/20	Vegetables	1.2lb	€15	Javies Retail
3	10004	05/05/20	Oil	3.1lb	€17	Javies Retail
4	10005	27/04/20	Wheat	4.8lb	€13	Javies Retail
5	10006	14/01/20	Butter	3.6lb	€27	Oldi Superstore
6	10007	20/04/20	Oil	4.8lb	€21	Dello Superstore
7	10008	05/05/20	Wheat	3.6lb	€25	Oldi Superstore
8	10009	17/04/20	Fruits	1.2lb	€24	Oldi Superstore
9	10010	15/06/20	Oil	4.4lb	€25	Kanes Store
10	10011	17/06/20	Oil	4.4lb	€16	Fline Store
11	10012	11/06/20	Cheese	2.3lb	€20	Fline Store
12	10013	19/03/20	Rice	4.4lb	€27	Kanes Store
13	10014	01/01/20	Cheese	1.2lb	€10	Fline Store
14	10015	07/07/20	Fruit Juice	3.6lb	€27	Oldi Superstore

Figure 7.9 – Adding the correct headers

Now that you have learned how to create and assign column headers in your data, it is strongly recommended that you always have meaningful column headers, as this will help you to better understand your dataset.

Multiple values in one column

Sometimes, you will run into a column containing data that can be split into further columns. For instance, an address column can be split into a "Street Number" column, a "Street Name" column, a "City" column, and a "State" column. There is no rule regarding how to identify these columns as it is up to your own understanding of the dataset to make the decision to split it or not.

Let's start with an example in which you are given a dataset comprising three columns: `full_name`, `address`, and `creation_date_time`. Consider the following:

	full_name	address	creation_date_time
0	Pasquale Cooper	1268 Burgoyne Promenade, San Leandro, Florida	2004-05-29 02:07:28
1	Giuseppe Wood	738 Opalo Circle, Brooklyn Center, Kansas	2008-04-24 19:42:11
2	Lindsey Garza	747 Desmond Nene, Olive Branch, Wisconsin	2013-08-23 09:41:48
3	Randy Mcpherson	171 Byron Street, Pleasanton, Vermont	2010-06-21 22:52:23
4	Cristobal Walsh	55 Crestwell Square, Oxford, Alaska	2014-12-13 09:47:34

Figure 7.10 – An example DataFrame

You have been asked to transform it as follows:

	first_name	last_name	street	city	state	creation_date	creation_time
0	Pasquale	Cooper	1268 Burgoyne Promenade	San Leandro	Florida	2004-05-29	02:07:28
1	Giuseppe	Wood	738 Opalo Circle	Brooklyn Center	Kansas	2008-04-24	19:42:11
2	Lindsey	Garza	747 Desmond Nene	Olive Branch	Wisconsin	2013-08-23	09:41:48
3	Randy	Mcpherson	171 Byron Street	Pleasanton	Vermont	2010-06-21	22:52:23
4	Cristobal	Walsh	55 Crestwell Square	Oxford	Alaska	2014-12-13	09:47:34

Figure 7.11 – DataFrame output after splitting

Perform the following steps:

1. Start by importing pandas and the dataset (CSV). Then, display the DataFrame:

```
# Importing pandas
import pandas as pd

#Define the csv URL
file_url = ' https://raw.githubusercontent.com/
PacktWorkshops/The-Pandas-Workshop/master/Chapter07/Data/
multiple_values_in_column.csv '

# Import the csv as a DataFrame
data_frame = pd.read_csv(file_url)

#Display the DataFrame
data_frame
```

The output will be as follows:

	full_name	address	creation_date_time
0	Pasquale Cooper	1268 Burgoyne Promenade, San Leandro, Florida	2004-05-29 02:07:28
1	Giuseppe Wood	738 Opalo Circle, Brooklyn Center, Kansas	2008-04-24 19:42:11
2	Lindsey Garza	747 Desmond Nene, Olive Branch, Wisconsin	2013-08-23 09:41:48
3	Randy Mcpherson	171 Byron Street, Pleasanton, Vermont	2010-06-21 22:52:23
4	Cristobal Walsh	55 Crestwell Square, Oxford, Alaska	2014-12-13 09:47:34

Figure 7.12 – Importing the original DataFrame

As you can see, each column in the DataFrame is made up of multiple values inside the same columns. The `full_name` column contains the first and last names, and the address column contains the street, the city, and the state. Finally, the `creation_date_time` column contains the date (year, month, and day) and the time (hours, minutes, and seconds).

2. Now, you want to split all of these values into their own columns. To do that, you will use the `str.split()` method with the `expand=True` parameter (this will return a DataFrame), which you learned about in *Chapter 4, pandas Data Types*, when you looked at string methods.

   Start by splitting the `full_name` column into two new columns, called `first_name` and `last_name`:

```
# Split the column into 2 new columns
data_frame[['first_name','last_name']]=data_frame.full_
name.str.split(expand=True)
data_frame
```

   The output will be as follows:

	full_name	address	creation_date_time	first_name	last_name
0	Pasquale Cooper	1268 Burgoyne Promenade, San Leandro, Florida	2004-05-29 02:07:28	Pasquale	Cooper
1	Giuseppe Wood	738 Opalo Circle, Brooklyn Center, Kansas	2008-04-24 19:42:11	Giuseppe	Wood
2	Lindsey Garza	747 Desmond Nene, Olive Branch, Wisconsin	2013-08-23 09:41:48	Lindsey	Garza
3	Randy Mcpherson	171 Byron Street, Pleasanton, Vermont	2010-06-21 22:52:23	Randy	Mcpherson
4	Cristobal Walsh	55 Crestwell Square, Oxford, Alaska	2014-12-13 09:47:34	Cristobal	Walsh

Figure 7.13 – Splitting the full_name column

   By default, the `str.split()` method will split the values separated by a whitespace.

3. You can delete the `full_name` column, as you no longer need it, with the `drop()` function. You can add the `inplace=True` parameter to store the result in the existing DataFrame instead of creating a new DataFrame:

```
# Delete the column
data_frame.drop('full_name', axis=1, inplace=True)
data_frame
```

The output will be as follows:

	address	creation_date_time	first_name	last_name
0	1268 Burgoyne Promenade, San Leandro, Florida	2004-05-29 02:07:28	Pasquale	Cooper
1	738 Opalo Circle, Brooklyn Center, Kansas	2008-04-24 19:42:11	Giuseppe	Wood
2	747 Desmond Nene, Olive Branch, Wisconsin	2013-08-23 09:41:48	Lindsey	Garza
3	171 Byron Street, Pleasanton, Vermont	2010-06-21 22:52:23	Randy	Mcpherson
4	55 Crestwell Square, Oxford, Alaska	2014-12-13 09:47:34	Cristobal	Walsh

Figure 7.14 – Deleting the full_name column

4. Next, you need to split the address column into three new columns, called `street`, `city`, and `state`:

```
# Split the column into 3 new columns
data_frame[['street', 'city','state']] = data_frame.
address.str.split(pat = ", ", expand=True)
data_frame
```

The output will be as follows:

	address	creation_date_time	first_name	last_name	street	city	state
0	1268 Burgoyne Promenade, San Leandro, Florida	2004-05-29 02:07:28	Pasquale	Cooper	1268 Burgoyne Promenade	San Leandro	Florida
1	738 Opalo Circle, Brooklyn Center, Kansas	2008-04-24 19:42:11	Giuseppe	Wood	738 Opalo Circle	Brooklyn Center	Kansas
2	747 Desmond Nene, Olive Branch, Wisconsin	2013-08-23 09:41:48	Lindsey	Garza	747 Desmond Nene	Olive Branch	Wisconsin
3	171 Byron Street, Pleasanton, Vermont	2010-06-21 22:52:23	Randy	Mcpherson	171 Byron Street	Pleasanton	Vermont
4	55 Crestwell Square, Oxford, Alaska	2014-12-13 09:47:34	Cristobal	Walsh	55 Crestwell Square	Oxford	Alaska

Figure 7.15 – Splitting the address column

This time, you specified the `pat` parameter to split the values with `,` instead of the whitespace.

5. Delete the address column, as you no longer need it with the `drop()` function:

```
# Delete the column
data_frame.drop('address', axis=1, inplace=True)
data_frame
```

The output will be as follows:

	creation_date_time	first_name	last_name	street	city	state
0	2004-05-29 02:07:28	Pasquale	Cooper	1268 Burgoyne Promenade	San Leandro	Florida
1	2008-04-24 19:42:11	Giuseppe	Wood	738 Opalo Circle	Brooklyn Center	Kansas
2	2013-08-23 09:41:48	Lindsey	Garza	747 Desmond Nene	Olive Branch	Wisconsin
3	2010-06-21 22:52:23	Randy	Mcpherson	171 Byron Street	Pleasanton	Vermont
4	2014-12-13 09:47:34	Cristobal	Walsh	55 Crestwell Square	Oxford	Alaska

Figure 7.16 – Deleting the address column

6. Now you need to split the `creation_date_time` column, which is a string, into two columns called `creation_date` and `creation_time`. To do that, start by converting the column into `datetime64` so that it is usable by pandas:

```
# Converting the column into datetime
data_frame['creation_date_time'] = pd.to_datetime(data_
frame['creation_date_time'], format='%Y-%m-%d %H:%M:%S')
data_frame['creation_date_time']
```

The output will be as follows:

```
0    2004-05-29 02:07:28
1    2008-04-24 19:42:11
2    2013-08-23 09:41:48
3    2010-06-21 22:52:23
4    2014-12-13 09:47:34
Name: creation_date_time, dtype: datetime64[ns]
```

Figure 7.17 – Converting creation_date_time into datetime format

7. Create two new columns using the `dt.date` and `dt.time` methods:

```
# Create date and time columns from datetime
data_frame['creation_date'] = data_frame.creation_date_
time.dt.date
data_frame['creation_time'] = data_frame.creation_date_
time.dt.time
data_frame
```

The output will be as follows:

	creation_date_time	first_name	last_name	street	city	state	creation_date	creation_time
0	2004-05-29 02:07:28	Pasquale	Cooper	1268 Burgoyne Promenade	San Leandro	Florida	2004-05-29	02:07:28
1	2008-04-24 19:42:11	Giuseppe	Wood	738 Opalo Circle	Brooklyn Center	Kansas	2008-04-24	19:42:11
2	2013-08-23 09:41:48	Lindsey	Garza	747 Desmond Nene	Olive Branch	Wisconsin	2013-08-23	09:41:48
3	2010-06-21 22:52:23	Randy	Mcpherson	171 Byron Street	Pleasanton	Vermont	2010-06-21	22:52:23
4	2014-12-13 09:47:34	Cristobal	Walsh	55 Crestwell Square	Oxford	Alaska	2014-12-13	09:47:34

Figure 7.18 – Splitting the creation_date_time column

8. Finally, delete the `creation_date_time` column:

```
# Delete the column
data_frame.drop('creation_date_time', axis=1,
inplace=True)
data_frame
```

The output will be as follows:

	first_name	last_name	street	city	state	creation_date	creation_time
0	Pasquale	Cooper	1268 Burgoyne Promenade	San Leandro	Florida	2004-05-29	02:07:28
1	Giuseppe	Wood	738 Opalo Circle	Brooklyn Center	Kansas	2008-04-24	19:42:11
2	Lindsey	Garza	747 Desmond Nene	Olive Branch	Wisconsin	2013-08-23	09:41:48
3	Randy	Mcpherson	171 Byron Street	Pleasanton	Vermont	2010-06-21	22:52:23
4	Cristobal	Walsh	55 Crestwell Square	Oxford	Alaska	2014-12-13	09:47:34

Figure 7.19 – Deleting the creation_date_time column

Now that you have learned how to handle missing headers, in the next section, we will cover how to handle duplicate observations in both rows and columns.

Duplicate observations in both rows and columns

Duplications in data could cause errors in the statistics conclusion. This section shows how duplicated rows or columns can be removed to ensure that only unique values are retained.

Let's say you are given the following dataset:

	id	city	state	city	state
0	1	Hutchinson	Texas	Hutchinson	Texas
1	2	Yorkville	South Dakota	Yorkville	South Dakota
2	1	Hutchinson	Texas	Hutchinson	Texas
3	3	Round Lake	Kansas	Round Lake	Kansas
4	4	Orinda	Montana	Orinda	Montana
5	3	Round Lake	Kansas	Round Lake	Kansas

Figure 7.20 – A DataFrame with duplicates

You have been asked to transform it as follows:

	id	city	state
0	1	Hutchinson	Texas
1	2	Yorkville	South Dakota
3	3	Round Lake	Kansas
4	4	Orinda	Montana

Figure 7.21 – A DataFrame without duplicates

Perform the following steps:

1. Start by importing pandas and the dataset (CSV). Then, display the DataFrame:

```
# Importing pandas
import pandas as pd

#Define the csv URL
file_url = ' https://raw.githubusercontent.com/
PacktWorkshops/The-Pandas-Workshop/master/Chapter07/Data/
duplicate_observations.csv'

# Import the csv as a DataFrame
data_frame = pd.read_csv(file_url)

# Forcing duplicates columns names
data_frame.rename(columns={'city.1': 'city', 'state.1':
'state'}, inplace=True)

#Display the DataFrame
data_frame
```

The output will be as follows:

	id	city	state	city	state
0	1	Hutchinson	Texas	Hutchinson	Texas
1	2	Yorkville	South Dakota	Yorkville	South Dakota
2	1	Hutchinson	Texas	Hutchinson	Texas
3	3	Round Lake	Kansas	Round Lake	Kansas
4	4	Orinda	Montana	Orinda	Montana
5	3	Round Lake	Kansas	Round Lake	Kansas

Figure 7.22 – Importing the DataFrame

As you can see, there are two issues of duplication in this DataFrame. The first issue is the duplicate columns of city and state. The second is that there are duplicates in the rows, as you have additional rows for IDs 1 and 3.

2. To remove these points of duplication, start by identifying the duplicate columns using the `.duplicated()` method:

```
# Check if columns are duplicates
data_frame.columns.duplicated()
```

The output will be as follows:

```
array([False, False, False,  True,  True])
```

Figure 7.23 – Checking for duplicate columns

Note that the `.duplicated()` method only works on similar column names. If the duplicate columns had their own unique name, the `.duplicated()` method would not be able to detect any duplicates.

3. Use the Boolean results to select only the non-duplicate columns:

```
# Select only the non duplicate columns
data_frame = data_frame.loc[:,~data_frame.columns.
duplicated()]
data_frame
```

The output will be as follows:

	id	city	state
0	1	Hutchinson	Texas
1	2	Yorkville	South Dakota
2	1	Hutchinson	Texas
3	3	Round Lake	Kansas
4	4	Orinda	Montana
5	3	Round Lake	Kansas

Figure 7.24 – Viewing the non-duplicate columns

4. Now that you have removed the duplicate columns, you can move on to removing the duplicate rows using the `drop_duplicates()` function:

```
# Remove row duplicates
data_frame = data_frame.drop_duplicates()
data_frame
```

The output will be as follows:

	id	city	state
0	1	Hutchinson	Texas
1	2	Yorkville	South Dakota
3	3	Round Lake	Kansas
4	4	Orinda	Montana

Figure 7.25 – Removing the duplicate rows

Now that you have learned how to handle duplicates, in the following exercise, you can practice doing so on your own.

Exercise 7.01 – working with messy addresses

In this exercise, you will load a CSV file that does not have headers and fix the different data issues in order to make a DataFrame usable for data analysis.

> **Note**
>
> The data is contained inside the `messy_addresses.csv` file, which you can find in our GitHub repository at `https://raw.githubusercontent.com/PacktWorkshops/The-Pandas-Workshop/master/Chapter07/Data/messy_addresses.csv`.

The following steps will help you to complete the exercise:

1. Open a new Jupyter notebook file and import `pandas`:

```
import pandas as pd
```

2. Next, load the CSV file as a DataFrame with the header parameter set to `None`:

```
file_url = ' https://raw.githubusercontent.com/
PacktWorkshops/The-Pandas-Workshop/master/Chapter07/Data/
messy_addresses.csv'
data_frame = pd.read_csv(file_url, header=None)
```

3. Use the `head()` function to display the first five rows of the DataFrame:

```
data_frame.head()
```

The output will be as follows:

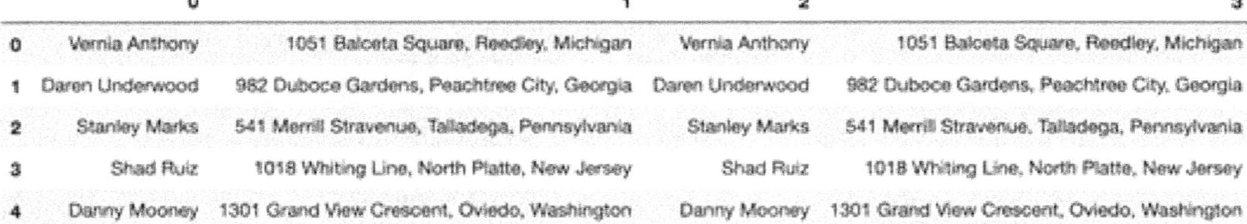

	0	1	2	3
0	Vernia Anthony	1051 Balceta Square, Reedley, Michigan	Vernia Anthony	1051 Balceta Square, Reedley, Michigan
1	Daren Underwood	982 Duboce Gardens, Peachtree City, Georgia	Daren Underwood	982 Duboce Gardens, Peachtree City, Georgia
2	Stanley Marks	541 Merrill Stravenue, Talladega, Pennsylvania	Stanley Marks	541 Merrill Stravenue, Talladega, Pennsylvania
3	Shad Ruiz	1018 Whiting Line, North Platte, New Jersey	Shad Ruiz	1018 Whiting Line, North Platte, New Jersey
4	Danny Mooney	1301 Grand View Crescent, Oviedo, Washington	Danny Mooney	1301 Grand View Crescent, Oviedo, Washington

Figure 7.26 – Viewing the first five rows of the DataFrame

Next, you will need to start removing the duplicate columns.

4. Use the `tail()` function to display the last five rows of the DataFrame:

```
data_frame.tail(5)
```

The output will be as follows:

	0	1	2	3
45	Augustus Conley	421 Powhattan Sideline, Caldwell, Mississippi	Augustus Conley	421 Powhattan Sideline, Caldwell, Mississippi
46	Lyndia Humphrey	323 Lori Plantation, Vernon Hills, Oregon	Lyndia Humphrey	323 Lori Plantation, Vernon Hills, Oregon
47	Vito Cochran	136 Dr Tom Waddell Bypass, Pembroke Pines, New...	Vito Cochran	136 Dr Tom Waddell Bypass, Pembroke Pines, New...
48	Preston Randall	1178 Burke Boulevard, Sugar Land, Oklahoma	Preston Randall	1178 Burke Boulevard, Sugar Land, Oklahoma
49	Derrick Holman	401 Cayuga Viaduct, Pittsburg, Indiana	Derrick Holman	401 Cayuga Viaduct, Pittsburg, Indiana

Figure 7.27 – Viewing the last five rows of the DataFrame

As you can see, there are three issues with this DataFrame:

- Missing headers
- Duplicate columns
- Multiple values in the same column

5. Use the `drop()` function to remove the duplicate columns (2 and 3) and then display the first five rows of the DataFrame:

```
data_frame.drop([2,3], axis=1, inplace=True)
data_frame.head()
```

The output will be as follows:

	0	1
0	Vernia Anthony	1051 Balceta Square, Reedley, Michigan
1	Daren Underwood	982 Duboce Gardens, Peachtree City, Georgia
2	Stanley Marks	541 Merrill Stravenue, Talladega, Pennsylvania
3	Shad Ruiz	1018 Whiting Line, North Platte, New Jersey
4	Danny Mooney	1301 Grand View Crescent, Oviedo, Washington

Figure 7.28 – Removing the duplicate columns

Now you can give names to the remaining columns.

6. Create the list of column headers (`full_name` and `address`):

```
column_names = ["full_name", "address"]
print(column_names)
```

The output of `column_names` will be as follows:

```
['full_name', 'address']
```

Figure 7.29 – Creating the column headers

7. Replace the missing column headers. Then, display the DataFrame:

```
data_frame.columns = column_names
data_frame.head()
```

The output will be as follows:

	full_name	address
0	Vernia Anthony	1051 Balceta Square, Reedley, Michigan
1	Daren Underwood	982 Duboce Gardens, Peachtree City, Georgia
2	Stanley Marks	541 Merrill Stravenue, Talladega, Pennsylvania
3	Shad Ruiz	1018 Whiting Line, North Platte, New Jersey
4	Danny Mooney	1301 Grand View Crescent, Oviedo, Washington

Figure 7.30 – Adding the column headers to the DataFrame

Now you can turn your attention to the issue of multiple values in the same column.

8. Split the values of `full_name` into two columns called `first_name` and `last_name`. Then, display the DataFrame:

```
data_frame[['first_name','last_name']]=data_frame.full_
name.str.split(expand=True)
data_frame.head()
```

The output will be as follows:

	full_name	address	first_name	last_name
0	Vernia Anthony	1051 Balceta Square, Reedley, Michigan	Vernia	Anthony
1	Daren Underwood	982 Duboce Gardens, Peachtree City, Georgia	Daren	Underwood
2	Stanley Marks	541 Merrill Stravenue, Talladega, Pennsylvania	Stanley	Marks
3	Shad Ruiz	1018 Whiting Line, North Platte, New Jersey	Shad	Ruiz
4	Danny Mooney	1301 Grand View Crescent, Oviedo, Washington	Danny	Mooney

Figure 7.31 – Splitting the full_name column

9. Delete the `full_name` column and then display the DataFrame:

```
data_frame.drop('full_name', axis=1, inplace=True)
data_frame.head()
```

The output will be as follows:

	address	first_name	last_name
0	1051 Balceta Square, Reedley, Michigan	Vernia	Anthony
1	982 Duboce Gardens, Peachtree City, Georgia	Daren	Underwood
2	541 Merrill Stravenue, Talladega, Pennsylvania	Stanley	Marks
3	1018 Whiting Line, North Platte, New Jersey	Shad	Ruiz
4	1301 Grand View Crescent, Oviedo, Washington	Danny	Mooney

Figure 7.32 – Removing the full_name column

10. Split the values of `address` into three new columns, called `street`, `city`, and `state`. Then, display the DataFrame:

```
data_frame[['street', 'city','state']] = data_frame.
address.str.split(pat = ", ", expand=True)
data_frame.head()
```

The output will be as follows:

	address	first_name	last_name	street	city	state
0	1051 Balceta Square, Reedley, Michigan	Vernia	Anthony	1051 Balceta Square	Reedley	Michigan
1	982 Duboce Gardens, Peachtree City, Georgia	Daren	Underwood	982 Duboce Gardens	Peachtree City	Georgia
2	541 Merrill Stravenue, Talladega, Pennsylvania	Stanley	Marks	541 Merrill Stravenue	Talladega	Pennsylvania
3	1018 Whiting Line, North Platte, New Jersey	Shad	Ruiz	1018 Whiting Line	North Platte	New Jersey
4	1301 Grand View Crescent, Oviedo, Washington	Danny	Mooney	1301 Grand View Crescent	Oviedo	Washington

Figure 7.33 – Splitting the address column

11. Delete the `address` column and then display the DataFrame:

```
data_frame.drop('address', axis=1, inplace=True)
data_frame.head()
```

The output will be as follows:

	first_name	last_name	street	city	state
0	Vernia	Anthony	1051 Balceta Square	Reedley	Michigan
1	Daren	Underwood	982 Duboce Gardens	Peachtree City	Georgia
2	Stanley	Marks	541 Merrill Stravenue	Talladega	Pennsylvania
3	Shad	Ruiz	1018 Whiting Line	North Platte	New Jersey
4	Danny	Mooney	1301 Grand View Crescent	Oviedo	Washington

Figure 7.34 – Deleting the address column

Now, all the messy data issues have been fixed, and the DataFrame is ready to be used.

In this exercise, you dealt with messy data regarding missing headers, multiple values in one column, and duplicate columns.

Now that you have learned how to handle duplicate observations in both rows and columns, you can move on to look at how to handle multiple variables stored in one column.

Multiple variables stored in one column

Another situation that you might encounter is when multiple variables are stored inside a single column. The solution will be to split that column's variables into multiple columns and rows. This situation is the reverse operation of a pivot transformation, which we will discuss later.

Consider the following example. You are given a dataset containing the yearly sales of a few demographics:

	YEAR	M0-24	M25-54	M55	F0-24	F25-54	F55
0	2018	282	812	993	712	466	373
1	2019	243	196	365	340	969	659

Figure 7.35 – A sample DataFrame showing sales

And you have been asked to transform it, as follows:

	YEAR	sales	gender	age_group
0	2018	282	M	0-24
1	2019	243	M	0-24
2	2018	812	M	25-54
3	2019	196	M	25-54
4	2018	993	M	55
5	2019	365	M	55
6	2018	712	F	0-24
7	2019	340	F	0-24
8	2018	466	F	25-54
9	2019	969	F	25-54
10	2018	373	F	55
11	2019	659	F	55

Figure 7.36 – The DataFrame following its transformation

1. Start by importing pandas and the dataset (CSV). Then, display the DataFrame:

```
# Importing pandas
import pandas as pd

#Define the csv URL
file_url = ' https://raw.githubusercontent.com/
PacktWorkshops/The-Pandas-Workshop/master/Chapter07/Data/
multiple_variables_in_column.csv'

# Import the csv as a DataFrame
data_frame = pd.read_csv(file_url)

#Display the DataFrame
data_frame
```

The output will be as follows:

	YEAR	M0-24	M25-54	M55	F0-24	F25-54	F55
0	2018	282	812	993	712	466	373
1	2019	243	196	365	340	969	659

Figure 7.37 – Importing the data

As you can see, the column headers (with the exception of `YEAR`) contain the gender and the age group. You need to manipulate the DataFrame if you want to be able to perform any data analysis on the demographics. The data format shown in the preceding table is known as a wide format. For the majority of cases where statistical analysis needs to be carried out, data must be in a format that we call long format, which we see next. To convert wide-format data into long-format data, the `melt()` function of pandas is useful.

The `id_vars` parameter indicates which columns should be unaltered following the `melt` operation, while `var_name` indicates what the name of the column should be after columns other than `id_vars` have been combined together to form a single column. The `value_name` parameter indicates the name of the column that contains the corresponding data.

2. First, you will convert the demographics column headers into rows with the `melt()` function. Use the `YEAR` column as an identifier alongside the `var_name=["demographic"]` (the column to unpivot) and `value_name="sales"` (the name of the new column) parameters:

```
# Convert the demographics columns into rows
data_frame = data_frame.melt(id_vars=["YEAR"],var_
name=["demographic"],value_name="sales")
data_frame
```

The output will be as follows:

	YEAR	demographic	sales
0	2018	M0-24	282
1	2019	M0-24	243
2	2018	M25-54	812
3	2019	M25-54	196
4	2018	M55	993
5	2019	M55	365
6	2018	F0-24	712
7	2019	F0-24	340
8	2018	F25-54	466
9	2019	F25-54	969
10	2018	F55	373
11	2019	F55	659

Figure 7.38 – Converting the demographic values from the column headers into an actual column

3. Split the demographic column into two columns, named `gender` and `age_group`:

```
# Split the column into new columns
data_frame['gender'] = data_frame.demographic.str[0].
astype(str)
data_frame['age_group'] = data_frame.demographic.str[1:].
astype(str)
data_frame
```

The output will be as follows:

	YEAR	demographic	sales	gender	age_group
0	2018	M0-24	282	M	0-24
1	2019	M0-24	243	M	0-24
2	2018	M25-54	812	M	25-54
3	2019	M25-54	196	M	25-54
4	2018	M55	993	M	55
5	2019	M55	365	M	55
6	2018	F0-24	712	F	0-24
7	2019	F0-24	340	F	0-24
8	2018	F25-54	466	F	25-54
9	2019	F25-54	969	F	25-54
10	2018	F55	373	F	55
11	2019	F55	659	F	55

Figure 7.39 – Splitting the demographic column

4. The final step will be to delete the demographic column:

```
# Delete the column
data_frame.drop('demographic', axis=1, inplace=True)
data_frame
```

The output will be as follows:

	YEAR	sales	gender	age_group
0	2018	282	M	0-24
1	2019	243	M	0-24
2	2018	812	M	25-54
3	2019	196	M	25-54
4	2018	993	M	55
5	2019	365	M	55
6	2018	712	F	0-24
7	2019	340	F	0-24
8	2018	466	F	25-54
9	2019	969	F	25-54
10	2018	373	F	55
11	2019	659	F	55

Figure 7.40 – Removing the demographic column

Now that you have learned how to handle multiple variables stored in a single column, in the next section, we will show you how to handle the same observation in multiple tables.

Multiple DataFrames with identical structures

Data is not always structured according to our needs. When there are multiple tables showing the same meaning and structure, it is recommended that you concatenate them into one master DataFrame.

For example, let's say you are given two files containing the yearly sales (the filename contains the year) of four stores, as follows:

	store_id	sales
0	1	282
1	2	243
2	3	391
3	4	973

Figure 7.41 – A sample DataFrame for 2018 sales

For 2019, it is as follows:

	store_id	sales
0	1	272
1	2	370
2	3	178
3	4	622

Figure 7.42 – A sample DataFrame for 2019 sales

And you have been asked to transform it as follows:

	store_id	sales	year
0	1	282	2018
1	1	272	2019
2	2	243	2018
3	2	370	2019
4	3	391	2018
5	3	178	2019
6	4	973	2018
7	4	622	2019

Figure 7.43 – The merged DataFrame following its transformation

1. Start by importing pandas and the first dataset (CSV). Then, display the DataFrame:

```
# Importing pandas
import pandas as pd

#Define the csv URL
file_url_2018 = ' https://raw.githubusercontent.com/
PacktWorkshops/The-Pandas-Workshop/master/Chapter07/Data/
data_frame_2018.csv'
data_frame_2018 = pd.read_csv(file_url_2018)

#Display the DataFrame
data_frame_2018
```

The output will be as follows:

	store_id	sales
0	1	282
1	2	243
2	3	391
3	4	973

Figure 7.44 – The first DataFrame

2. Import the second dataset (CSV) and then display the DataFrame:

```
#Define the csv URL
file_url_2019 = 'https://raw.githubusercontent.com/
PacktWorkshops/The-Pandas-Workshop/master/Chapter07/Data/
data_frame_2019.csv'
data_frame_2019 = pd.read_csv(file_url_2019)
#Display the DataFrame
data_frame_2019
```

The output will be as follows:

	store_id	sales
0	1	272
1	2	370
2	3	178
3	4	622

Figure 7.45 – The second DataFrame

3. Create a year column in each table according to the name of the DataFrame:

```
# Creating a column for the year
data_frame_2018["year"]="2018"
data_frame_2018
```

The output will be as follows:

	store_id	sales	year
0	1	282	2018
1	2	243	2018
2	3	391	2018
3	4	973	2018

Figure 7.46 – Creating a year column in the first dataset

We will do the same for the 2019 table, as follows:

```
# Creating a column for the year
data_frame_2019["year"]="2019"
data_frame_2019
```

The output will be as follows:

	store_id	sales	year
0	1	272	2019
1	2	370	2019
2	3	178	2019
3	4	622	2019

Figure 7.47 – Creating a year column in the second dataset

4. Now, combine the two DataFrames together using the `concat()` function:

```
# Concatenating the DataFrame
data_frame = pd.concat([data_frame_2018, data_
frame_2019])
data_frame
```

The output will be as follows:

	store_id	sales	year
0	1	282	2018
1	2	243	2018
2	3	391	2018
3	4	973	2018
0	1	272	2019
1	2	370	2019
2	3	178	2019
3	4	622	2019

Figure 7.48 – Combining both DataFrames

5. To make it easier to read, you can sort the values by the `store_id` and `year` columns. This will result in a DataFrame that is, initially, sorted by `store_id`. Then, for each `store_id` value, their sales will be sorted by year:

```
# Sorting the DataFrame
data_frame = data_frame.sort_values(by=['store_id',
'year'])
data_frame
```

The output will be as follows:

	store_id	sales	year
0	1	282	2018
0	1	272	2019
1	2	243	2018
1	2	370	2019
2	3	391	2018
2	3	178	2019
3	4	973	2018
3	4	622	2019

Figure 7.49 – Sorting the values

6. You can see that after sorting the DataFrame, its index is messed up, as there are duplicates (0,0,1,1,2,2,3,3) instead of (0,1,2,3,4,5,6,7,8). So, the final step will be to reset the DataFrame index:

```
# Resetting the DataFrame's index
data_frame = data_frame.reset_index(drop = True)
data_frame
```

The output will be as follows:

	store_id	sales	year
0	1	282	2018
1	1	272	2019
2	2	243	2018
3	2	370	2019
4	3	391	2018
5	3	178	2019
6	4	973	2018
7	4	622	2019

Figure 7.50 – Resetting the DataFrame index

So far, we have learned how to work with multiple DataFrames that have similar structures. Now, we will implement that knowledge in the following exercise.

Exercise 7.02 – storing sales by demographics

In this exercise, you will load two CSV files and correct any data issues in order to make a single DataFrame usable for data analysis.

> **Note**
>
> The data is contained inside the `store_sales_demographics_2019.csv` and `store_sales_demographics_2018.csv` files, which you can find inside our GitHub repository at `https://raw.githubusercontent.com/PacktWorkshops/The-Pandas-Workshop/master/Chapter07/Data/store_sales_demographics_2019.csv`.

The following steps will help you to complete the exercise:

1. Open a new Jupyter notebook file and import pandas:

```
import pandas as pd
```

2. Next, load both CSV files as DataFrames:

```
file_url_2018 = 'https://raw.githubusercontent.com/
PacktWorkshops/The-Pandas-Workshop/master/Chapter07/Data/
store_sales_demographics_2018.csv'
file_url_2019 = 'https://raw.githubusercontent.com/
PacktWorkshops/The-Pandas-Workshop/master/Chapter07/Data/
store_sales_demographics_2019.csv'

data_frame_2018 = pd.read_csv(file_url_2018)
data_frame_2019 = pd.read_csv(file_url_2019)
```

3. Use the `head()` function to display the first five rows of the first DataFrame:

```
data_frame_2018.head()
```

The output will be as follows:

	store_id	M0-24	M25-54	M55	F0-24	F25-54	F55
0	1	34	27	60	54	17	98
1	2	54	73	89	25	12	78
2	3	86	66	68	81	32	75
3	4	19	58	55	37	70	12
4	5	91	17	46	67	19	14

Figure 7.51 – The first five rows of the first DataFrame

Now, you can start removing the duplicate columns.

4. Use the `head()` function to display the first five rows of the second DataFrame:

```
data_frame_2019.head()
```

The output will be as follows:

	store_id	M0-24	M25-54	M55	F0-24	F25-54	F55
0	1	46	16	28	62	98	76
1	2	44	92	60	26	86	50
2	3	53	85	50	84	34	44
3	4	88	71	45	48	19	34
4	5	37	18	45	45	10	11

Figure 7.52 – The first five rows of the second DataFrame

As you can see, there are two issues with these DataFrames:

- We have the same observations inside multiple tables.
- Multiple variables have been stored in one column.

5. To fix the first issue, create a new column called `year` for the first DataFrame and then display it:

```
data_frame_2018["year"] = 2018
data_frame_2018
```

The output will be as follows:

	store_id	M0-24	M25-54	M55	F0-24	F25-54	F55	year
0	1	34	27	60	54	17	98	2018
1	2	54	73	89	25	12	78	2018
2	3	86	66	68	81	32	75	2018
3	4	19	58	55	37	70	12	2018
4	5	91	17	46	67	19	14	2018

Figure 7.53 – Creating a year column in the first dataset

6. Create a new column called `year` for the second DataFrame and then display it:

```
data_frame_2019["year"] = 2019
data_frame_2019
```

The output of `column_names` will be as follows:

	store_id	M0-24	M25-54	M55	F0-24	F25-54	F55	year
0	1	46	16	28	62	98	76	2019
1	2	44	92	60	26	86	50	2019
2	3	53	85	50	84	34	44	2019
3	4	88	71	45	48	19	34	2019
4	5	37	18	45	45	10	11	2019

Figure 7.54 – Creating a year column in the second dataset

7. Concatenate the two DataFrames and display the result:

```
data_frame = pd.concat([data_frame_2018, data_
frame_2019])
data_frame
```

The output will be as follows:

	store_id	M0-24	M25-54	M55	F0-24	F25-54	F55	year
0	1	34	27	60	54	17	98	2018
1	2	54	73	89	25	12	78	2018
2	3	86	66	68	81	32	75	2018
3	4	19	58	55	37	70	12	2018
4	5	91	17	46	67	19	14	2018
0	1	46	16	28	62	98	76	2019
1	2	44	92	60	26	86	50	2019
2	3	53	85	50	84	34	44	2019
3	4	88	71	45	48	19	34	2019
4	5	37	18	45	45	10	11	2019

Figure 7.55 – Combining the two DataFrames

8. To handle the issue of multiple variables, first, convert the demographic columns into rows. Then, display the DataFrame:

```
data_frame = data_frame.melt(id_vars=["year", "store_
id"],var_name=["demographic"],value_name="sales")
data_frame.head(6)
```

The output will be as follows:

	year	store_id	demographic	sales
0	2018	1	M0-24	34
1	2018	2	M0-24	54
2	2018	3	M0-24	86
3	2018	4	M0-24	19
4	2018	5	M0-24	91
5	2019	1	M0-24	46

Figure 7.56 – Converting demographic into a column

9. Split the demographic column into two columns, called gender and age_group. Then, display the DataFrame:

```
data_frame['gender'] = data_frame.demographic.str[0].
astype(str)
data_frame['age_group'] = data_frame.demographic.str[1:].
astype(str)
data_frame.head(6)
```

The output will be as follows:

	year	store_id	demographic	sales	gender	age_group
0	2018	1	M0-24	34	M	0-24
1	2018	2	M0-24	54	M	0-24
2	2018	3	M0-24	86	M	0-24
3	2018	4	M0-24	19	M	0-24
4	2018	5	M0-24	91	M	0-24
5	2019	1	M0-24	46	M	0-24

Figure 7.57 – Splitting the demographic column

10. Delete the demographic column and then display the DataFrame:

```
data_frame.drop('demographic', axis=1, inplace=True)
data_frame.head()
```

The output will be as follows:

	year	store_id	sales	gender	age_group
0	2018	1	34	M	0-24
1	2018	2	54	M	0-24
2	2018	3	86	M	0-24
3	2018	4	19	M	0-24
4	2018	5	91	M	0-24

Figure 7.58 – Deleting the demographic column

11. Sort the DataFrame by `year`, `store_id`, and `gender`. Then, display the DataFrame:

```
data_frame = data_frame.sort_values(by=['year', 'store_
id', 'gender'])
data_frame.head()
```

The output will be as follows:

	year	store_id	sales	gender	age_group
30	2018	1	54	F	0-24
40	2018	1	17	F	25-54
50	2018	1	98	F	55
0	2018	1	34	M	0-24
10	2018	1	27	M	25-54

Figure 7.59 – Sorting the DataFrame

12. Reorder the columns in this order: `("store_id", "age_group", "gender", "year", "sales")`. Then, display the DataFrame:

```
data_frame = data_frame[["store_id", "age_group",
"gender", "year", "sales"]]
data_frame.head()
```

The output will be as follows:

	store_id	age_group	gender	year	sales
30	1	0-24	F	2018	54
40	1	25-54	F	2018	17
50	1	55	F	2018	98
0	1	0-24	M	2018	34
10	1	25-54	M	2018	27

Figure 7.60 – Reordering the DataFrame

13. Reset the DataFrame's index and then display the DataFrame:

```
data_frame = data_frame.reset_index(drop = True)
data_frame.head()
```

The output will be as follows:

	store_id	age_group	gender	year	sales
0	1	0-24	F	2018	54
1	1	25-54	F	2018	17
2	1	55	F	2018	98
3	1	0-24	M	2018	34
4	1	25-54	M	2018	27

Figure 7.61 – Resetting the DataFrame index

Now all the messy data issues have been fixed and the DataFrame is ready to be used.

In this exercise, you dealt with messy data in terms of multiple variables being stored in the same column and identical observations found within multiple tables.

Now that you have learned how to handle messy data, it's time to move on to missing data.

Dealing with missing data

You might have already encountered missing data in your studies or career while performing any kind of data analysis. Missing data is a very common issue that you will encounter in most datasets. It's extremely rare to find a "perfect" dataset. Missing data is not just a nuisance. It is a serious problem that you need to account for as it can affect your results.

What is missing data?

Before you can learn how to deal with missing data, first, you need to understand each of its three types:

- **Missing at Random** (**MAR**): This refers to data that is missing due to other variables you have information about. For example, in a survey, if you found that some specific demographics show a tendency to not reply to a question, then the missing data is considered to be MAR. An easy way to remember this is that if you can explain why the data is missing by using other variables, but not to the value of the variable with missing values itself, then it is MAR.

- **Missing Completely at Random** (**MCAR**): This refers to data that is missing due to randomness, which means that the missing data is independent of any variables. An easy way to remember this is that if we cannot explain why the data is missing by using other variables, then it is MCAR. This type has the least impact on any results, as the missing data is completely random and cannot be used to derive any correlation. This might be due to an issue with the system that is loading or the data being generated—for example, if we are collecting data from sensors and one sensor is out of service because of a failure.
- **Missing Not at Random** (**MNAR**): This means that the data is missing due to another reason that cannot be explained by your other variables, as with MAR, but might have a reason why it is not currently represented in a variable in your dataset. For example, generally, high-income earners do not want to reveal their income in a survey. This is the worst type as it is very hard to find, but luckily, it is quite rare.

In general, it is considered safe to delete missing data in the MAR situation or the MCAR situation, as it won't impact the results of your analysis (unless we don't have enough data points after removing them), whereas deleting the missing data in the MNAR situation is not recommended as this can lead to bias.

Now that you understand the different types of missing data, you can learn how to deal with them.

Strategies for missing data

There are two main strategies for dealing with missing data: deletion and imputation.

If you choose to use the deletion strategy, then you need to decide whether you want to delete the data by one of the following methods:

- Listwise deletion (complete case analysis) removes the entire row if there is at least one missing value in any column. This method works well with MCAR, but it introduces bias for MAR and MNAR.
- Pairwise deletion (available case analysis) removes the entire row only if we are using the variable where the missing value is. This method can cause incorrect estimates in statistical inferences.
- The dropping variable deletion strategy removes the variable (column) from the entire dataset if there are a lot of missing values in this specific column.

For example, let's say you are given a file containing the population of a few states, as follows:

	id	city	state	population
0	1.0	Hutchinson	Texas	20938.0
1	NaN	Yorkville	Illinois	20119.0
2	3.0	Round Lake	Illinois	NaN
3	4.0	Orinda	California	19926.0

Figure 7.62 – A sample DataFrame of the population

You have been asked to use the deletion strategy to handle missing data, which should result in a DataFrame, as follows:

	city	state
0	Hutchinson	Texas
1	Yorkville	Illinois
2	Round Lake	Illinois
3	Orinda	California

Figure 7.63 – The DataFrame following the removal of missing data

Start by importing the dataset (CSV) and then display the DataFrame:

```
# Importing pandas
import pandas as pd

#Define the csv URL
file_url = 'https://raw.githubusercontent.com/PacktWorkshops/
The-Pandas-Workshop/master/Chapter07/Data/deletion.csv'
data_frame = pd.read_csv(file_url)

#Display the DataFrame
data_frame
```

The output will be as follows:

	id	city	state	population
0	1.0	Hutchinson	Texas	20938.0
1	NaN	Yorkville	Illinois	20119.0
2	3.0	Round Lake	Illinois	NaN
3	4.0	Orinda	California	19926.0

Figure 7.64 – Importing the data

If you were to choose the listwise deletion strategy, you would need to remove any rows with missing values:

```
# Drop any rows with missing data
data_frame.dropna()
```

The output will be as follows:

	id	city	state	population
0	1.0	Hutchinson	Texas	20938.0
3	4.0	Orinda	California	19926.0

Figure 7.65 – The DataFrame after listwise deletion

If you were to choose the pairwise deletion strategy and your goal is to analyze the total population, then you would need to remove rows with missing values in the population variable:

```
# Drop any rows with missing data in the column population
data_frame[~data_frame['population'].isnull()]
```

The output will be as follows:

	id	city	state	population
0	1.0	Hutchinson	Texas	20938.0
1	NaN	Yorkville	Illinois	20119.0
3	4.0	Orinda	California	19926.0

Figure 7.66 – The DataFrame after pairwise deletion

If you were to choose the dropping variable deletion strategy, then you would need to remove rows with missing values in the population variable:

```
# Drop any columns with missing data
data_frame.dropna(axis = 1)
```

The output will be as follows:

	city	state
0	Hutchinson	Texas
1	Yorkville	Illinois
2	Round Lake	Illinois
3	Orinda	California

Figure 7.67 – The DataFrame following dropping variable deletion

Often, deletion strategies result in the loss of data; therefore, they become problematic when the dataset is relatively small or the number of missing cases is higher. Now that you have seen how the deletion strategy works, you can move on to the imputation strategy, which is the process of replacing missing data with substituted values. If you choose to use the imputation strategy, then you need to decide by which of the following methods you want to impute the data:

- Fixed imputation replaces any missing values with a fixed value. Usually, this method is used for categorical data.
- Statistical imputation replaces missing values with statistical inferences (that is, the average, the median, and the mode) based on the non-missing data. Usually, this method is used for numerical data.
- Regression imputation replaces missing values by performing a regression based on the non-missing data. Usually, this method is used for numerical data where the data can be modeled by a regression.
- Model-based imputation: In this method, we train the machine learning models to predict the value of the missing data. This is applicable to both numerical and categorical data.

For example, let's say you are given a file containing the population of a few states, as follows:

	id	city	state	population
0	1.0	Hutchinson	Texas	20938.0
1	NaN	Yorkville	Illinois	20119.0
2	3.0	Round Lake	Illinois	NaN
3	4.0	Orinda	NaN	19926.0

Figure 7.68 – A sample DataFrame

You have been asked to use the imputation strategy to handle the missing data, which should result in a DataFrame, as follows:

	id	city	state	population
0	1.0	Hutchinson	Texas	20938.0
1	-999.0	Yorkville	Illinois	20119.0
2	3.0	Round Lake	Illinois	-999.0
3	4.0	Orinda	Missing Value	19926.0

Figure 7.69 – The DataFrame after imputation

14. Start by importing the dataset (CSV) and then display the DataFrame:

```
# Importing pandas
import pandas as pd

#Define the csv URL
file_url = 'https://raw.githubusercontent.com/
PacktWorkshops/The-Pandas-Workshop/master/Chapter07/Data/
imputation.csv'
data_frame = pd.read_csv(file_url)

#Display the DataFrame
data_frame
```

The output will be as follows:

	id	city	state	population
0	1.0	Hutchinson	Texas	20938.0
1	NaN	Yorkville	Illinois	20119.0
2	3.0	Round Lake	Illinois	NaN
3	4.0	Orinda	NaN	19926.0

Figure 7.70 – Importing the data

15. As you can see, there is missing data in three columns: `("id", "state", "population")`. The first step is to select the rows with missing data using `isnull()` with `any(axis = 1)` (to return each row that has missing values):

```
data_frame[data_frame.isnull().any(axis = 1)]
```

	id	city	state	population
1	NaN	Yorkville	Illinois	20119.0
2	3.0	Round Lake	Illinois	NaN
3	4.0	Orinda	NaN	19926.0

Figure 7.71 – Selecting all the rows with missing values

16. Start by filling any missing values in the `state` column using the fixed imputation method with `'Missing Value'`:

```
# Filling missing values
data_frame['state'] = data_frame.state.fillna('Missing
Value')
data_frame
```

The output will be as follows:

	id	city	state	population
0	1.0	Hutchinson	Texas	20938.0
1	NaN	Yorkville	Illinois	20119.0
2	3.0	Round Lake	Illinois	NaN
3	4.0	Orinda	Missing Value	19926.0

Figure 7.72 – Replacing NaN with a label

The value doesn't matter as long it is clearly identifiable as the value for missing data.

17. Fill the missing values in the `id` column using the fixed imputation method with `-999`:

```
# Filling missing values
data_frame['id'] = data_frame.id.fillna(-999)
data_frame
```

The output will be as follows:

	id	city	state	population
0	1.0	Hutchinson	Texas	20938.0
1	-999.0	Yorkville	Illinois	20119.0
2	3.0	Round Lake	Illinois	NaN
3	4.0	Orinda	Missing Value	19926.0

Figure 7.73 – Replacing NaN with a label

As `id` is only composed of positive integers, it is recommended that you use any negative values to flag missing data.

18. For the `population` column, you can choose the fixed imputation method with `-999`:

```
# Creating a copy of the dataframe
data_frame_999 = data_frame.copy()
# Filling missing values
data_frame_999['population'] = data_frame.population.
fillna(-999)
data_frame_999
```

The output will be as follows:

	id	city	state	population
0	1.0	Hutchinson	Texas	20938.0
1	-999.0	Yorkville	Illinois	20119.0
2	3.0	Round Lake	Illinois	-999.0
3	4.0	Orinda	Missing Value	19926.0

Figure 7.74 – Replacing NaN with a label (fixed imputation)

Alternatively, you can choose the statistical imputation method with the average (you need to round off the calculated average in order to retain the integer format):

```
# Creating a copy of the dataframe
data_frame_mean = data_frame.copy()
# Filling missing values
data_frame_mean['population'] = round(data_frame.
population.fillna(data_frame.population.mean()),0)
data_frame_mean
```

The output will be as follows:

	id	city	state	population
0	1.0	Hutchinson	Texas	20938.0
1	-999.0	Yorkville	Illinois	20119.0
2	3.0	Round Lake	Illinois	20328.0
3	4.0	Orinda	Missing Value	19926.0

Figure 7.75 – Replacing NaN with a label (statistical imputation)

Note that, in this situation, you are introducing a bias in the population of Round Lake, as you used an average calculated based on the populations of other cities to impute the missing value. The fixed imputation method is the preferred method for these situations.

Now that you have learned how to handle missing data, we can move on to the next topic: summarizing data with pandas.

Summarizing data

Summarizing data is one of the most important tasks in data analysis, as this is the step where a data analyst will convert a large amount of data into a few main aggregates that represent a summary of the data. First, you will learn about the basics of data aggregation with pandas. Then, we will move on to a more advanced topic with pivot tables.

Grouping and aggregation

In general, datasets are made of a single observation per row, which means that you can end up with datasets comprising millions of rows. Of course, deriving any data analysis on dozens of rows is not the same as millions of rows. In these situations, grouping/summarizing rows together based on common variables is a good solution.

Consider the following example. You are given a file containing the yearly sales of a number of stores, as follows:

	store_id	sales	year
0	1	282	2018
1	1	272	2019
2	2	243	2018
3	2	370	2019
4	3	391	2018
5	3	178	2019
6	4	973	2018
7	4	622	2019

Figure 7.76 – A sample DataFrame of sales

And you have been asked to summarize the sales for each store, which should result in a DataFrame as follows:

	sales							
	count	mean	std	min	25%	50%	75%	max
store_id								
1	2.0	277.0	7.071068	272.0	274.50	277.0	279.50	282.0
2	2.0	306.5	89.802561	243.0	274.75	306.5	338.25	370.0
3	2.0	284.5	150.613744	178.0	231.25	284.5	337.75	391.0
4	2.0	797.5	248.194480	622.0	709.75	797.5	885.25	973.0

Figure 7.77 – A DataFrame with summarized values

1. Start by importing the dataset (CSV) and then display the DataFrame:

```
# Importing pandas
import pandas as pd

#Define the csv URL
file_url = 'https://raw.githubusercontent.com/
PacktWorkshops/The-Pandas-Workshop/master/Chapter07/Data/
grouping.csv'
data_frame = pd.read_csv(file_url)
```

```
#Display the DataFrame
data_frame
```

The output will be as follows:

	store_id	sales	year
0	1	282	2018
1	1	272	2019
2	2	243	2018
3	2	370	2019
4	3	391	2018
5	3	178	2019
6	4	973	2018
7	4	622	2019

Figure 7.78 – Importing the data

Let's say that you want to group your data and calculate a few aggregations in order to have a better understanding of your dataset.

2. Start by defining a group based on `store_id`:

```
# Grouping the data
grouped = data_frame.groupby('store_id')[['sales']]
grouped
```

The output will be as follows:

```
<pandas.core.groupby.generic.DataFrameGroupBy object at 0x7fddf8d71dd0>
```

Figure 7.79 – Defining a group

As you can see, this doesn't return a DataFrame but a `DataFrameGroupBy` object (please note that, in your case, the number at the end of the preceding output could be different). This object is like a special view of the DataFrame: it contains the group but does not hold any computations until we specify an aggregation:

- You can specify an aggregation such as `sum()` to get the total number of sales for each store:

```
# Sum aggregation
grouped.sum()
```

The output will be as follows:

store_id	sales
1	554
2	613
3	569
4	1595

Figure 7.80 – Finding the total number of sales of all stores

- You can specify an aggregation such as `mean()` to get the average number of sales for each store:

```
# Average aggregation
grouped.mean()
```

The output will be as follows:

store_id	sales
1	277.0
2	306.5
3	284.5
4	797.5

Figure 7.81 – Finding the average number of sales for all stores

- You can specify an aggregation such as `min()` to get the minimum number of sales for each store:

```
# Min aggregation
grouped.min()
```

The output will be as follows:

	sales
store_id	
1	272
2	243
3	178
4	622

Figure 7.82 – Finding the minimum number of sales for each store

- You can specify an aggregation such as `max()` to get the maximum number of sales for each store:

```
# Min aggregation
grouped.max()
```

The output will be as follows:

	sales
store_id	
1	282
2	370
3	391
4	973

Figure 7.83 – Finding the maximum number of sales for all stores

- You can specify an aggregation such as `std()` to get the standard deviation of sales for each store:

```
# Standard deviation aggregation
grouped.std()
```

The output will be as follows:

store_id	sales
1	7.071068
2	89.802561
3	150.613744
4	248.194480

Figure 7.84 – Finding the standard deviation of sales for all stores

- You can specify an aggregation such as `var()` to get the variance in sales for each store:

```
# Variance aggregation
grouped.var()
```

The output will be as follows:

store_id	sales
1	50.0
2	8064.5
3	22684.5
4	61600.5

Figure 7.85 – Finding the variance in sales for all stores

- You can specify the `describe()` method to get a list of aggregations:

```
# Describe aggregation
grouped.describe()
```

The output will be as follows:

	sales							
	count	mean	std	min	25%	50%	75%	max
store_id								
1	2.0	277.0	7.071068	272.0	274.50	277.0	279.50	282.0
2	2.0	306.5	89.802561	243.0	274.75	306.5	338.25	370.0
3	2.0	284.5	150.613744	178.0	231.25	284.5	337.75	391.0
4	2.0	797.5	248.194480	622.0	709.75	797.5	885.25	973.0

Figure 7.86 – Viewing all of the aggregations at once

- You can also use the `agg()` method to specify a list of aggregations to get multiple aggregations:

```
# List of aggregations
grouped.agg(['sum', 'mean','min', 'max', 'std'])
```

The output will be as follows:

	sales				
	sum	mean	min	max	std
store_id					
1	554	277.0	272	282	7.071068
2	613	306.5	243	370	89.802561
3	569	284.5	178	391	150.613744
4	1595	797.5	622	973	248.194480

Figure 7.87 – Viewing specific aggregations

Now that you have learned about grouping and aggregations, you can move on to pivot tables in pandas. These are critical for any data analyses or data insights that you will derive.

Exploring pivot tables

A pivot table is a table of statistics that summarizes the data of a more extensive table. This summary might include sums, averages, or other statistics that the pivot table groups together in a meaningful way. You might be familiar with it if you have already performed any data analysis with Excel. Likewise, pandas has a function that allows it to easily generate pivot tables through which you can derive insights from your data. The following example will demonstrate how to create pivot tables using the pivot table function in pandas. The `pivot_table` function is quite similar to the `group_by` function, which you saw earlier, but it offers significantly more customization.

Let's say you are given a file containing the yearly sales of a few products for two brands, as follows:

	brand	type	sales	units	year
0	Pandas	Product A	476	46	2010
1	Pandas	Product B	794	39	2010
2	Pandas	Product C	199	62	2010
3	Pandas	Product A	686	26	2011
4	Pandas	Product B	207	93	2011
5	Pandas	Product C	199	62	2011
6	Python	Product A	300	33	2010
7	Python	Product B	949	51	2010
8	Python	Product C	168	30	2010
9	Python	Product A	921	51	2011
10	Python	Product B	266	24	2011
11	Python	Product C	674	39	2011

Figure 7.88 – A sample DataFrame of two brand sales

And you have been asked to create a pivot table to summarize the sales for each brand, which should result in a DataFrame as follows:

		sum		min		max	
		sales	units	sales	units	sales	units
brand	type						
Pandas	Product A	1162	72	476	26	686	46
	Product B	1001	132	207	39	794	93
	Product C	398	124	199	62	199	62
Python	Product A	1221	84	300	33	921	51
	Product B	1215	75	266	24	949	51
	Product C	842	69	168	30	674	39
Total		5839	556	168	24	949	93

Figure 7.89 – A DataFrame summarizing the sales for both brands

Start by importing the dataset (CSV) and then display the DataFrame:

```
# Importing pandas
import pandas as pd

#Define the csv URL
file_url = 'https://raw.githubusercontent.com/PacktWorkshops/
The-Pandas-Workshop/master/Chapter07/Data/pivot.csv'
data_frame = pd.read_csv(file_url)

#Display the DataFrame
data_frame
```

The output will be as follows:

	brand	type	sales	units	year
0	Pandas	Product A	476	46	2010
1	Pandas	Product B	794	39	2010
2	Pandas	Product C	199	62	2010
3	Pandas	Product A	686	26	2011
4	Pandas	Product B	207	93	2011
5	Pandas	Product C	199	62	2011
6	Python	Product A	300	33	2010
7	Python	Product B	949	51	2010
8	Python	Product C	168	30	2010
9	Python	Product A	921	51	2011
10	Python	Product B	266	24	2011
11	Python	Product C	674	39	2011

Figure 7.90 – Importing the data

Begin by aggregating the sales values by brand:

```
# Pivot table average sales by brand
pd.pivot_table(data_frame, index = 'brand', values = 'sales')
```

The output will be as follows:

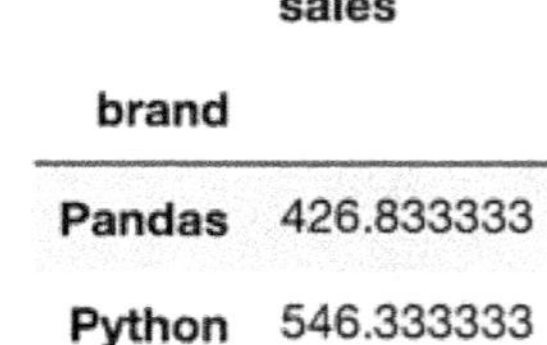

	sales
brand	
Pandas	426.833333
Python	546.333333

Figure 7.91 – Aggregating the sales values by brand

By default, the aggregation performed by `pivot_table()` is the average.

You can specify another aggregation with the `aggfunc` parameter. Try a `sum` aggregation to get the total number of sales by brand:

```
# Pivot table total sales by brand
pd.pivot_table(data_frame, index = 'brand', values = 'sales',
aggfunc='sum')
```

The output will be as follows:

	sales
brand	
Pandas	2561
Python	3278

Figure 7.92 – An alternative aggregation

You can also specify multiple aggregations in the same pivot table:

```
# Pivot table total, min and max sales by brand
pd.pivot_table(data_frame, index = 'brand', values = 'sales',
aggfunc = ['sum', 'min', 'max'])
```

The output will be as follows:

	sum	min	max
	sales	sales	sales
brand			
Pandas	2561	199	794
Python	3278	168	949

Figure 7.93 – Viewing multiple aggregations

As you can see, you end up with a DataFrame with a multi-index, especially on the columns where the first level comprises the aggregations and the second level is the column you want to have the aggregation on `sales`.

In addition, you can also specify multiple values on which you want to calculate aggregations:

```
# Pivot table total, min and max sales and unites by brand
pd.pivot_table(data_frame, index = ['brand'], values =
['sales','units'], aggfunc = ['sum', 'min', 'max'])
```

The output will be as follows:

	sum		min		max	
	sales	**units**	**sales**	**units**	**sales**	**units**
brand						
Pandas	2561	328	199	26	794	93
Python	3278	228	168	24	949	51

Figure 7.94 – Specific aggregations

You can also specify more than one index. Try `'type'` and `'brand'`:

```
# Pivot table total, min and max sales and unites by type and
brand
pd.pivot_table(data_frame, index = ['type','brand'], values =
['sales','units'], aggfunc = ['sum', 'min', 'max'])
```

The output will be as follows:

		sum		min		max	
		sales	**units**	**sales**	**units**	**sales**	**units**
type	**brand**						
Product A	**Pandas**	1162	72	476	26	686	46
	Python	1221	84	300	33	921	51
Product B	**Pandas**	1001	132	207	39	794	93
	Python	1215	75	266	24	949	51
Product C	**Pandas**	398	124	199	62	199	62
	Python	842	69	168	30	674	39

Figure 7.95 – Adding index column headers

Finally, you can also add a `Total` row to the pivot table:

```
# Pivot table total, min and max sales and unites by type and
brand
pd.pivot_table(data_frame, index = ['brand','type'], values =
['sales','units'], aggfunc = ['sum', 'min', 'max'], margins =
True, margins_name='Total')
```

The output will be as follows:

		sum		min		max	
		sales	units	sales	units	sales	units
brand	**type**						
Pandas	**Product A**	1162	72	476	26	686	46
	Product B	1001	132	207	39	794	93
	Product C	398	124	199	62	199	62
Python	**Product A**	1221	84	300	33	921	51
	Product B	1215	75	266	24	949	51
	Product C	842	69	168	30	674	39
Total		5839	556	168	24	949	93

Figure 7.96 – Adding a column of total values

Now that you have learned how to build a pivot table, you can practice your skills by working on the following activity.

Activity 7.01 – data analysis using pivot tables

In this activity, you will build pivot tables in order to perform data analysis. We will work on the *Student Performance* dataset from the GitHub repository.

> **Note**
> More details about the *Student Performance* dataset can be found at `https://archive.ics.uci.edu/ml/datasets/Student+Performance`.

Your tasks will be to do the following:

1. Open a Jupyter notebook.
2. Import the `pandas` package.
3. Load the CSV file (using the `;` delimiter to separate the columns) as a DataFrame.
4. Modify the DataFrame to contain only these columns: `school`, `sex`, `age`, `address`, `heath`, `absences`, `G1`, `G2`, and `G3`.
5. Display the first 10 rows of the DataFrame.
6. Build a pivot table that is indexed on `school`.
7. Build a pivot table that is indexed on `school` and `age`.
8. Build a pivot table that is indexed on `school`, `sex`, and `age`, with the mean and sum aggregation on the `absences` column.

The expected output is as follows:

school	sex	age	mean	sum
GP	F	15	3.894737	148.0
		16	5.888889	318.0
		17	7.120000	356.0
		18	8.137931	236.0
		19	13.083333	157.0
	M	15	2.863636	126.0
		16	4.980000	249.0
		17	6.138889	221.0
		18	6.500000	182.0
		19	12.166667	73.0
		20	0.000000	0.0
		22	16.000000	16.0
MS	F	17	5.625000	45.0
		18	1.785714	25.0
		19	2.000000	4.0
		20	4.000000	4.0
	M	17	2.750000	11.0
		18	4.818182	53.0
		19	4.250000	17.0
		20	11.000000	11.0
		21	3.000000	3.0

Figure 7.97 – The final outcome of the DataFrame

> **Note**
>
> The solution to this activity can be found in the *Appendix*.

Now that you have seen a simple data analysis build using pivot tables, you can also try to build different pivot tables and find more insights.

In this activity, you handled missing data and summarized data using pivot tables to derive insights.

Summary

This chapter covered the fundamental techniques of data transformation and how to apply them. Through a series of examples and exercises dealing with messy and missing data, you explored several possible types of issues, along with various strategies by which to correct them. You learned about grouping, aggregation, and pivot table methods, which are essential in order to summarize data, and you applied this knowledge to an activity with the goal of deriving insights from raw data.

In the next chapter, you will learn how to visualize data using pandas and matplotlib.

8
Understanding Data Visualization

In the previous chapter, you were introduced to data transformation methods in pandas. In this chapter, you will learn more about data visualization in pandas and use different types of charts such as line, bar, pie, scatter, and box to perform exploratory data analysis. In this chapter, we shall also touch upon different ways you can plot these charts using the `plot()` function by pandas and matplotlib. We will learn the differences between these two methods and learn which one to use, depending on the desired outcome. The plots that we are going to learn about in this chapter will help us analyze our data to find out useful insights, such as the distribution of certain features over the population using histograms and finding outliers using boxplots. By the end of this chapter, you will know how to select the best chart type for your data, build it, and customize it for the purpose of your analysis.

This chapter consists of the following topics:

- Introduction to data visualization
- Understanding the basics of pandas visualization
- Exploring matplotlib
- Visualizing data of different types
- Activity 8.01 – Using data visualization for exploratory data analysis

Introduction to data visualization

Humans can process a large amount of information using their sense of vision. Data visualization utilizes humans' innate skills to enhance the efficiency of data processing and organization. A classic visualization process starts by filtering data, transforming it into visual forms, and eventually displaying the data interactively to end users. With data visualization, users find it easier to understand and interpret the meaning of the underlying data. Good data visualization helps identify patterns, trends, and extreme values in a concise presentation. This is important in every aspect, especially when the data is big in volume or highly complex. Making sense of a large amount of data in a small amount of time is a huge business value.

pandas offers various options for visualizing data. To ensure your visualizations are accurate and that they correctly convey the insights gained from the underlying data, it is critical to identify and clean messy and missing data first. Performing visualization using the pandas library makes it really simple to generate plots out of a pandas DataFrame and Series. This will help you discover trends and correlations between different variables in your dataset.

Understanding the basics of pandas visualization

pandas has built-in plot generation capabilities that can be used to visualize both DataFrames and series alike. pandas comes with a built-in `plot` function that acts as a wrapper on top of the matplotlib `plot` function. This means that pandas is actually using the matplotlib library but with a simplified syntax. This presents the advantage of being much easier to use (less code and simpler syntax) compared to matplotlib. It provides a wide range of functionality and flexibility to plot data analytics charts with given data.

To start off using pandas in-built visualizations, you will need to know several key parameters for the `.plot()` function, which can be called from a DataFrame. Some of these are listed as follows:

- `kind`: This is the type of plot (`bar`, `barh`, `pie`, `scatter`, `kde`, and so on).
- `color`: This is the color of the plot.
- `linestyle`: This is the style of the line used in the plot (`solid`, `dotted`, and `dashed`).
- `legend`: This is a Boolean parameter to specify whether the legend should be shown or not.
- `title`: This is to give a title name to the chart.

Let's start with an implementation of the `plot` function:

1. Here, we are creating a DataFrame that shows the yearly sales for a company, from 2000 to 2019:

```
# Importing libraries
import pandas as pd
import numpy as np
import matplotlib.pyplot as plt

# Defining a DataFrame
data_frame = pd.DataFrame({
    'Year':['2000','2001','2002','2003','2004','2005','20
06','2007','2008','2009',
            '2010','2011','2012','2013','2014','2015','20
16','2017','2018','2019'],
    'Sales':[4107,6492,1476,8508,7416,2747,1606,7947,9506
,5441,7617,847,4389,3139,7546,3150,4426,4969,8457,5491]})

# Display DataFrame values
data_frame
```

The output will be as follows:

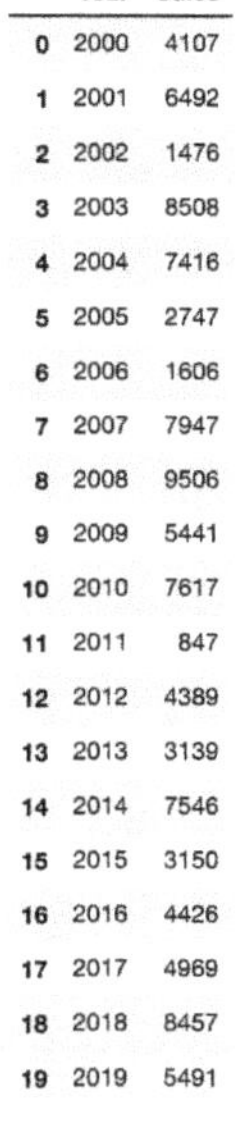

	Year	Sales
0	2000	4107
1	2001	6492
2	2002	1476
3	2003	8508
4	2004	7416
5	2005	2747
6	2006	1606
7	2007	7947
8	2008	9506
9	2009	5441
10	2010	7617
11	2011	847
12	2012	4389
13	2013	3139
14	2014	7546
15	2015	3150
16	2016	4426
17	2017	4969
18	2018	8457
19	2019	5491

Figure 8.1 – A DataFrame containing yearly sales

2. The `plot()` function can help us visualize this DataFrame as a chart, as follows:

```
data_frame.plot()
```

The plot should look as follows:

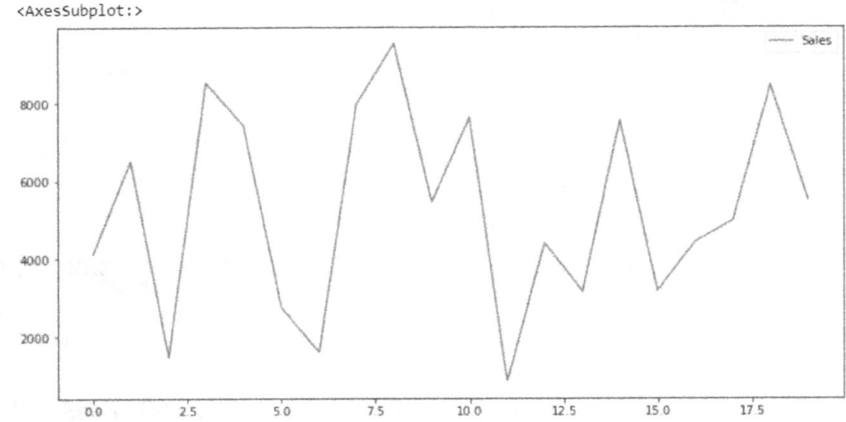

Figure 8.2 – Plotting the yearly sales

Another thing you might have noticed is the `<AxesSubplot:>` text on top of the chart. What you see is the matplotlib object returned by pandas.

3. If you want a cleaner-looking plot, you can easily remove it by simply adding a semicolon after the `plot()` function, as follows:

```
data_frame.plot();
```

The output will be as follows:

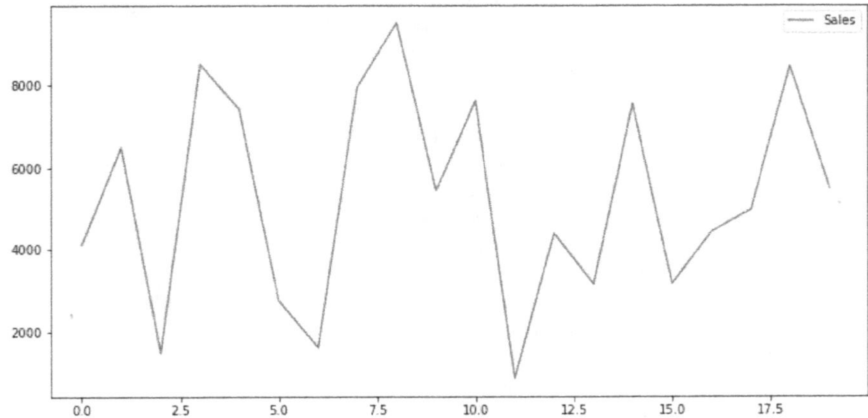

Figure 8.3 – Plotting the yearly sales without <AxesSubplot:>

As you can see, pandas plots a line chart by default and uses the `Sales` column as the *y* axis, which is also automatically used as a legend. Furthermore, the DataFrame index is plotted along the *x* axis.

4. Let's use the `Year` column as the *x* axis by specifying the `x` parameter, as follows:

```
data_frame.plot(x = 'Year');
```

The output will be as follows:

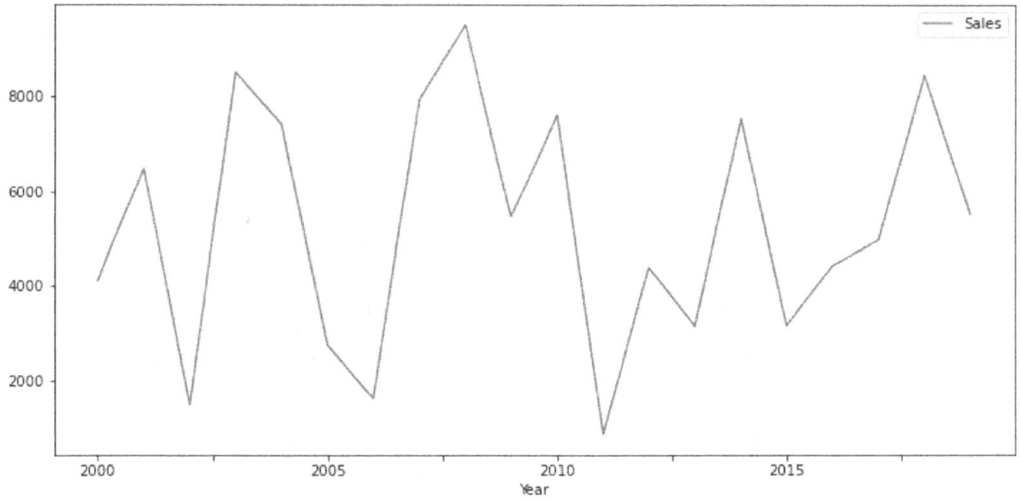

Figure 8.4 – Plotting the yearly sales with Year in the x axis

5. To make sure your chart makes even more sense to the readers, you can add a title to it using the `title` parameter, as follows:

```
data_frame.plot(x = 'Year', title = 'Yearly Sales');
```

The output will be as follows:

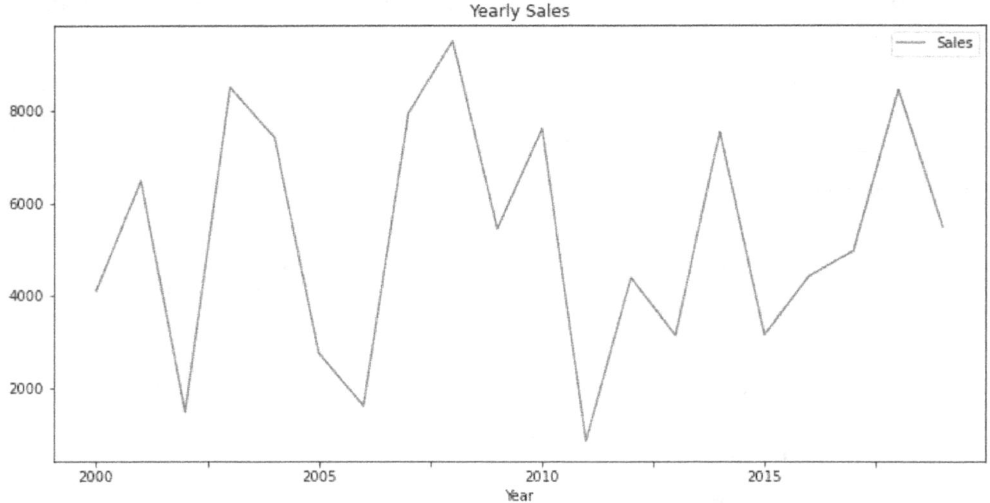

Figure 8.5 – Plotting the yearly sales with a title

6. If you want to remove the legend, you can simply use the `legend` parameter and assign a `False` value to it:

```
data_frame.plot(x = 'Year', title = 'Yearly Sales',
legend = False);
```

The output will be as follows:

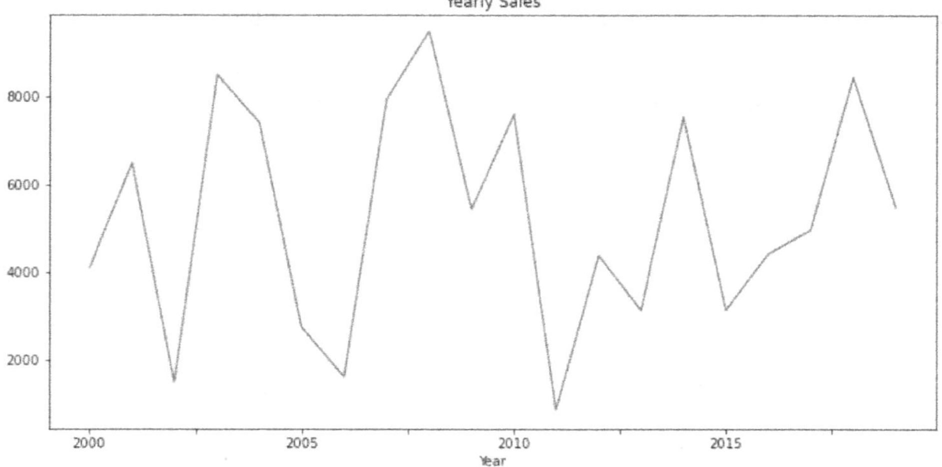

Figure 8.6 – Plotting the yearly sales without a legend

If it is a bit hard to visualize the values on the chart, you can use the `grid` parameter to add a grid to it, as follows:

```
data_frame.plot(x = 'Year', title = 'Yearly Sales',
legend = False, grid = True);
```

The resulting plot will be as follows:

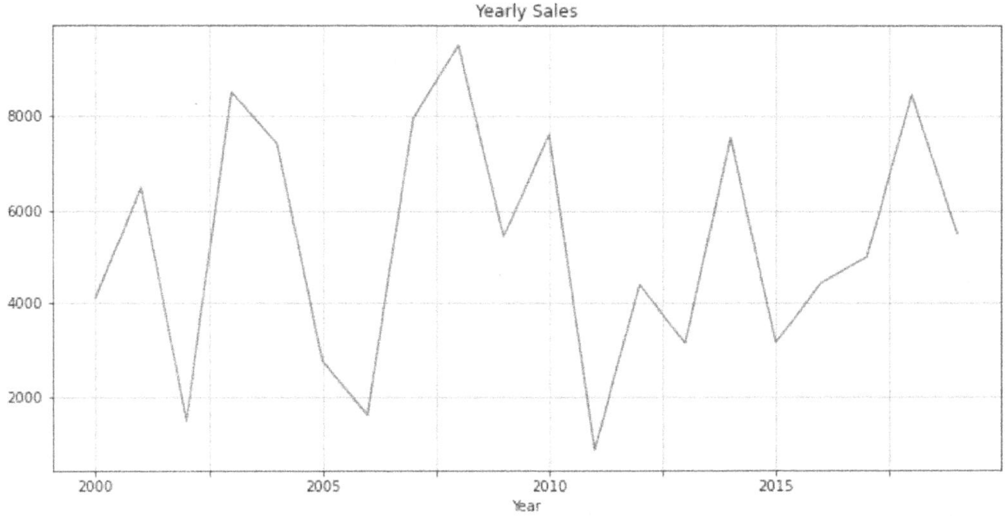

Figure 8.7 – Plotting the yearly sales with a grid

7. You can change the color of the line in the chart by using the `color` parameter; here, we will choose the gray color using the Tableau notation for `gray`:

```
data_frame.plot(x = 'Year', title = 'Yearly Sales',
legend = False, grid = True, color = 'tab:gray');
```

The output will be as follows:

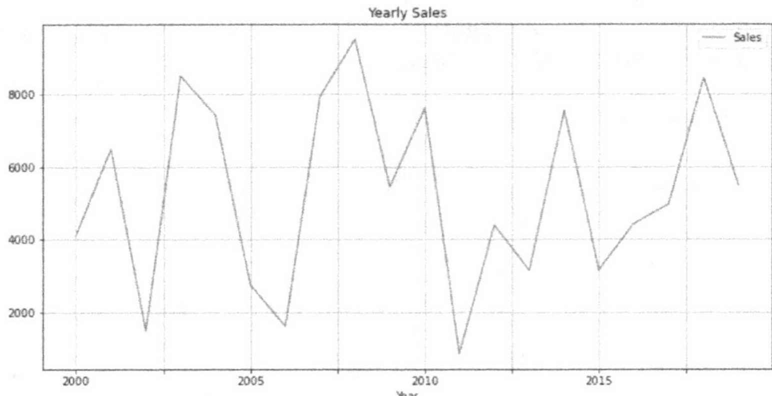

Figure 8.8 – Plotting the yearly sales in gray

> **Note**
>
> You may use any other color, such as `'tab:orange'`, `'tab:blue'`, and `'tab:cyan'`, simply by choosing one of the colors from the Tableau color palette. The full list of colors from the Tableau palette is as follows: `'tab:blue'`, `'tab:orange'`, `'tab:green'`, `'tab:red'`, `'tab:purple'`, `'tab:brown'`, `'tab:pink'`, `'tab:gray'`, `'tab:olive'`, and `'tab:cyan'`.

8. You can also change the style of the line by using the `linestyle` parameter:

```
data_frame.plot(x = 'Year', title = 'Yearly Sales', grid
= True, color = 'tab:gray', linestyle = 'dotted');
```

The output will be as follows:

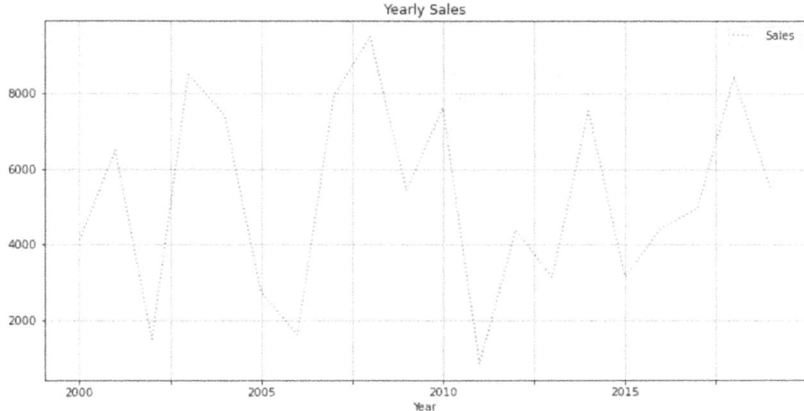

Figure 8.9 – Plotting the yearly sales as a dotted line

Similarly, you can change the line to a dashed line:

```
data_frame.plot(x = 'Year', title = 'Yearly Sales', grid
= True, color = 'tab:gray', linestyle = 'dashed');
```

The output will be as follows:

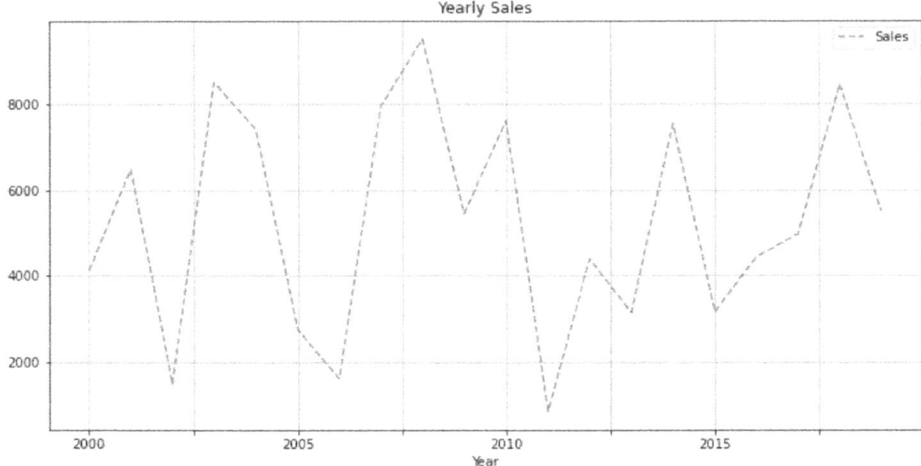

Figure 8.10 – Plotting the yearly sales as a dashed line

9. The line chart is the pandas default chart type, but you can change it to another type of chart by changing the `kind` attribute. For example, to plot a bar chart, add the following:

```
data_frame.plot(kind = 'bar', x = 'Year', y ='Sales',
title = 'Yearly Sales', color = 'tab:gray');
```

You should see output as follows:

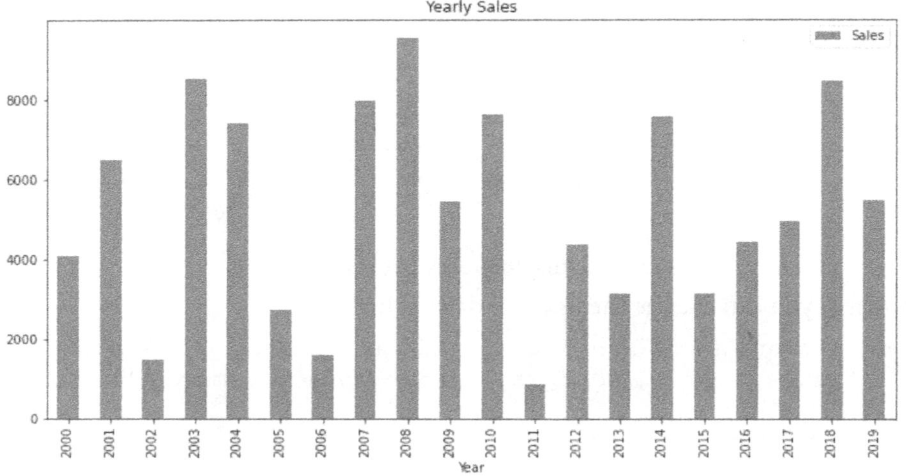

Figure 8.11 – Plotting the yearly sales as a bar chart

10. Now, let's try to plot a horizontal bar chart (using the `kind` attribute again):

```
data_frame.plot(kind = 'barh', x = 'Year', title =
'Yearly Sales', color = 'tab:gray');
```

The output will be as follows:

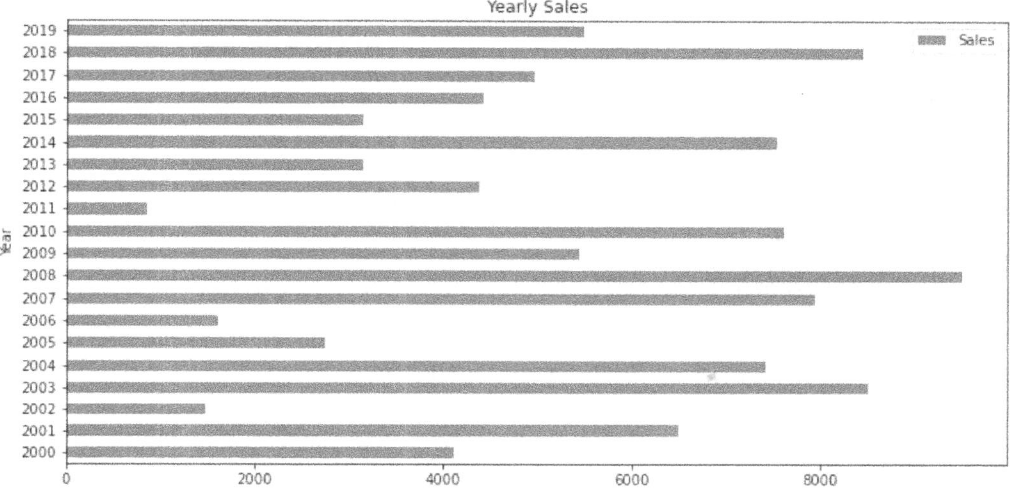

Figure 8.12 – Plotting the yearly sales as an horizontal bar chart

11. Let's try to plot an `area` chart:

```
data_frame.plot(kind = 'area', x = 'Year', title =
'Yearly Sales', color = 'tab:gray');
```

The output will be as follows:

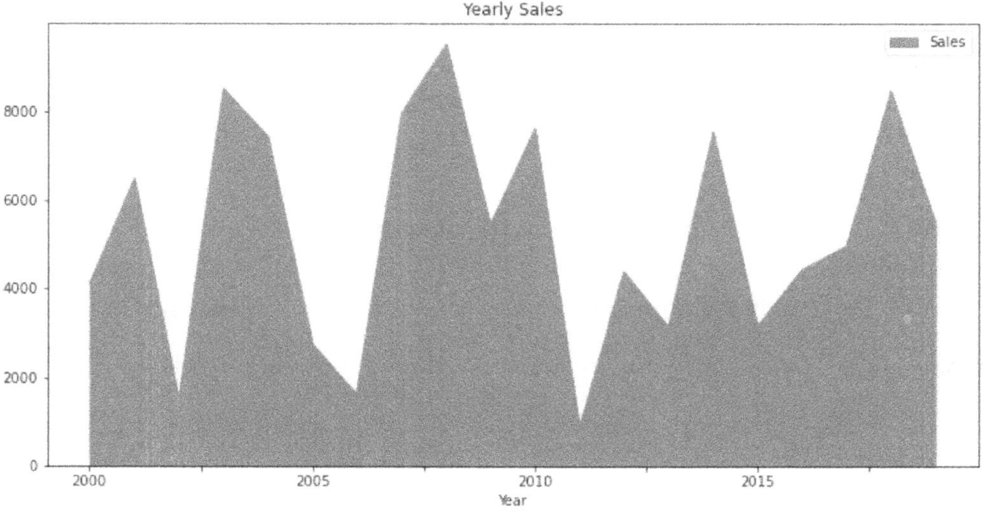

Figure 8.13 – Plotting the yearly sales as an area chart

The following list summarizes the different kinds of plots you can make using the pandas `plot()` function:

- `'line'`: Line plot (default)
- `'bar'`: Vertical bar plot
- `'barh'`: Horizontal bar plot
- `'hist'`: Histogram
- `'box'`: Boxplot
- `'kde'`: Kernel density estimation plot
- `'density'`: The same as `'kde'`
- `'area'`: Area plot
- `'pie'`: Pie plot
- `'scatter'`: Scatter plot
- `'hexbin'`: Hexbin plot

In this section, you learned how to use basic parameters such as `legend`, `kind`, and `linestyle` to customize the appearance of your plot. In the next exercise, we will practice using these on a real dataset.

Exercise 8.01 – Building histograms for the Titanic dataset

The goal of this exercise is to derive insights into the ages of the passengers who survived when the Titanic sank. Using the Titanic dataset, you'll need to handle the missing data and build a histogram that will help you achieve this goal. Use those histograms to answer the following questions:

1. What is the predominant age group of the passengers who were on board the Titanic?
2. What is the predominant age group of the passengers who survived the accident?

Specifically, we will create histograms to derive some insights, such as the number of passengers who survived and the ages of these passengers.

> **Note**
>
> The data is contained inside `titanic.csv`, which you can find on GitHub at `https://raw.githubusercontent.com/PacktWorkshops/The-pandas-Workshop/master/Chapter08/Data/titanic.csv`.

The following steps will help you complete the exercise:

1. Open a new Jupyter notebook file.
2. Import the `pandas`, `numpy`, and `matplotlib` packages:

```
import pandas as pd
import numpy as np
import matplotlib.pyplot as plt
```

3. Next, load the CSV file as a DataFrame with the header parameter set to `None`:

```
file_url = 'titanic.csv'
data_frame = pd.read_csv(file_url)
```

> **Note**
>
> Please change the path of the CSV file here, mentioning the appropriate location if it's not in your current directory.

4. Use the `head()` function to display the first five rows of the DataFrame, and check whether the data has been properly loaded:

```
data_frame.head()
```

The output will be as follows:

	survived	ticket_class	gender	age	number_sibling_spouse	number_parent_children	passenger_fare	port_of_embarkation	age_group
0	0	3	male	22.0	1	0	7.2500	S	18-59
1	1	1	female	38.0	1	0	71.2833	C	18-59
2	1	3	female	26.0	0	0	7.9250	S	18-59
3	1	1	female	35.0	1	0	53.1000	S	18-59
4	0	3	male	35.0	0	0	8.0500	S	18-59

Figure 8.14 – Displaying the first five rows of the DataFrame

5. Start by removing any rows with missing data by using the `.dropna()` function, which you learned about in the previous chapter, and display the DataFrame by entering the following code:

```
data_frame = data_frame.dropna()
data_frame
```

You should see the following output:

	survived	ticket_class	gender	age	number_sibling_spouse	number_parent_children	passenger_fare	port_of_embarkation	age_group
0	0	3	male	22.0	1	0	7.2500	S	18-59
1	1	1	female	38.0	1	0	71.2833	C	18-59
2	1	3	female	26.0	0	0	7.9250	S	18-59
3	1	1	female	35.0	1	0	53.1000	S	18-59
4	0	3	male	35.0	0	0	8.0500	S	18-59
...	...	...	...	...	...	...	...	...	...
885	0	3	female	39.0	0	5	29.1250	Q	18-59
886	0	2	male	27.0	0	0	13.0000	S	18-59
887	1	1	female	19.0	0	0	30.0000	S	18-59
889	1	1	male	26.0	0	0	30.0000	C	18-59
890	0	3	male	32.0	0	0	7.7500	Q	18-59

712 rows × 9 columns

Figure 8.15 – Displaying the DataFrame without duplicates

6. Plot a histogram chart for the ages of the passengers on board the Titanic:

```
data_frame.age.plot(kind = 'hist', title = 'Histogram
plot for ages of the passengers onboard Titanic');
```

The output will be as follows:

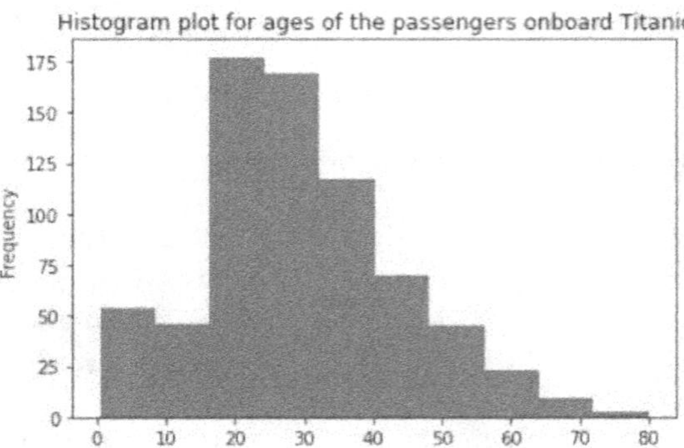

Figure 8.16 – Plotting a histogram for the age groups

We can see that most passengers are aged between 20 to 40 years, and there is also a very considerable fraction that are under 5 years of age.

7. Plot a histogram chart for the ages of the passengers who survived:

```
data_frame.loc[data_frame['survived'] == 1].age.plot(kind
= 'hist', title = 'Histogram for age of passengers who
survived');
```

This should result in the following output:

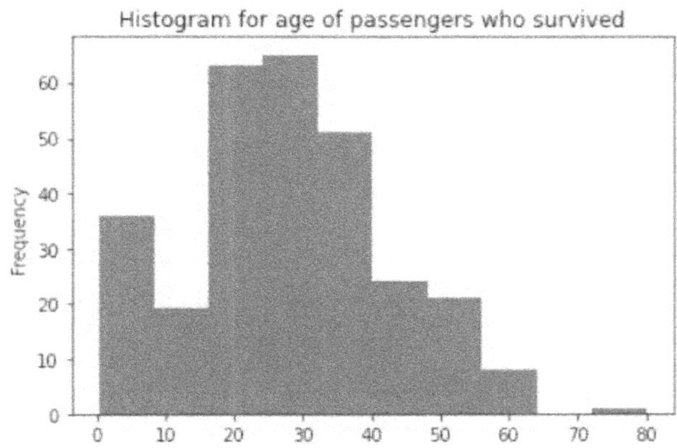

Figure 8.17 – Plotting a histogram for surviving passengers

If we focused on passengers who survived, we can see that most children under the age of 10 did survive, contrasting with the group aged over 65, where only 1 managed to survive.

Now that we have seen the basics of pandas visualization, we can move on to our next topic, matplotlib.

Exploring matplotlib

Matplotlib is one of the most frequently used Python libraries. It can generate plotting diagrams with great flexibility. The pandas `plot()` function is a wrapper on top of matplotlib with some bare minimum functionality. While it does simplify the syntax, it also restrains the numerous possibilities of matplotlib. If you want to build complex visualizations, then matplotlib will be your best choice, as it allows controls over all kinds of properties, such as the size, the type of figures and markers, the line width, the colors, and the styles. We will see some of the customizations that can be easily done with matplotlib compared to pandas:

1. Let's start with an example. Consider the following snippet:

```
# Importing libraries
import pandas as pd
```

```
import numpy as np
import matplotlib.pyplot as plt

# Defining a DataFrame
data_frame = pd.DataFrame({
    'Year':['2010','2011','2012','2013','2014','2015','20
16','2017','2018','2019'],
    'Sales':[4107,1606,7947,9506,5441,7617,8437,4389,3139
,7546]})

# Display DataFrame values
data_frame
```

In the preceding snippet, we created a DataFrame to show the yearly sales values, from 2010 to 2019. The output will be as follows:

	Year	Sales
0	2010	4107
1	2011	1606
2	2012	7947
3	2013	9506
4	2014	5441
5	2015	7617
6	2016	8437
7	2017	4389
8	2018	3139
9	2019	7546

Figure 8.18 – A DataFrame containing yearly sales

2. You can use the `plot()` function of matplotlib followed by a semicolon to plot a line plot, as follows:

```
x = data_frame['Year']
y = data_frame['Sales']
plt.plot(x, y);
```

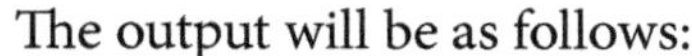

The output will be as follows:

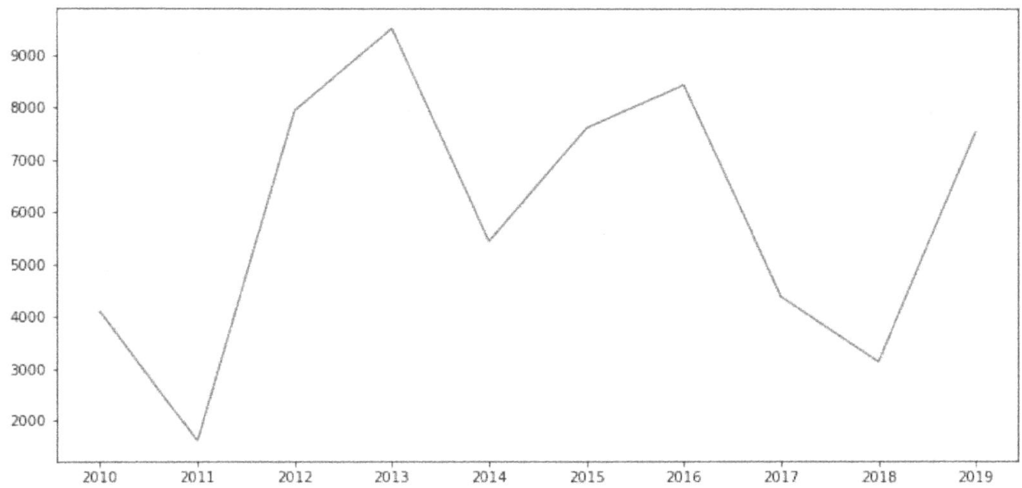

Figure 8.19 – Plotting the yearly sales

As you can see, we need to define what our *x* and *y* axes are in order to show the values of `Sales` and `Year` instead of the DataFrame index.

3. We can add more customization to our chart by giving it a title and also a name for each axis:

```
x = data_frame['Year']
y = data_frame['Sales']
plt.title('Yearly Sales')
plt.xlabel('Years')
plt.ylabel('Sales in Units')
plt.plot(x, y);
```

The output for this will be as follows:

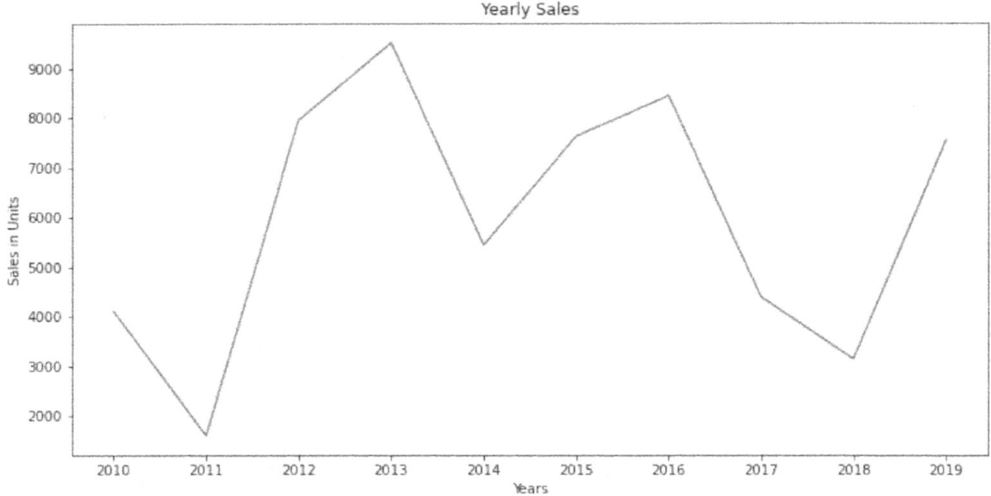

Figure 8.20 – Plotting the yearly sales with a title and axis labels

4. Now, let's say that we want to zoom in to the area where **Sales** fall between 6,000 and 10,000; we can achieve that by defining `plt.ylim()`.

   This should result in the following output:

Figure 8.21 – Plotting the yearly sales with a zoom on sales between 6,000 and 10,000

Note that `ylim()` and `xlim()` are used to limit the value range of the vertical and horizontal axes resepectively.

5. We can change the color of the line to gray (or any other color of our choice):

```
x = data_frame['Year']
y = data_frame['Sales']
plt.title('Yearly Sales')
plt.xlabel('Years')
plt.ylabel('Sales in Units')
plt.plot(x, y, color = 'tab:gray');
```

You should see output as follows:

Figure 8.22 – Plotting the yearly sales in gray

6. We can also change the width of the line to make it thicker; we can achieve that by using `linewidth`:

```
x = data_frame['Year']
y = data_frame['Sales']
plt.title('Yearly Sales')
plt.xlabel('Years')
plt.ylabel('Sales in Units')
plt.plot(x, y, color = 'tab:gray', linewidth = 5);
```

The output will be as follows:

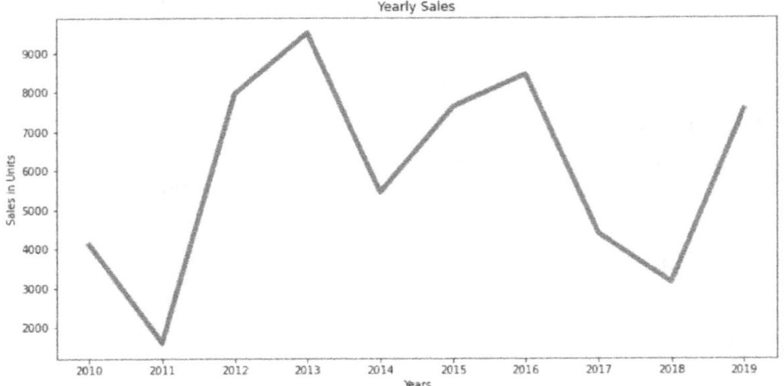

Figure 8.23 – Plotting the yearly sales as a gray bold line

7. Now, let's change the style of the line into a dash line; we can achieve that by using `linestyle`:

```
x = data_frame['Year']
y = data_frame['Sales']
plt.title('Yearly Sales')
plt.xlabel('Years')
plt.ylabel('Sales in Units')
plt.plot(x, y, color = 'tab:gray', linewidth = 2,
linestyle = '--');
```

The output will be as follows:

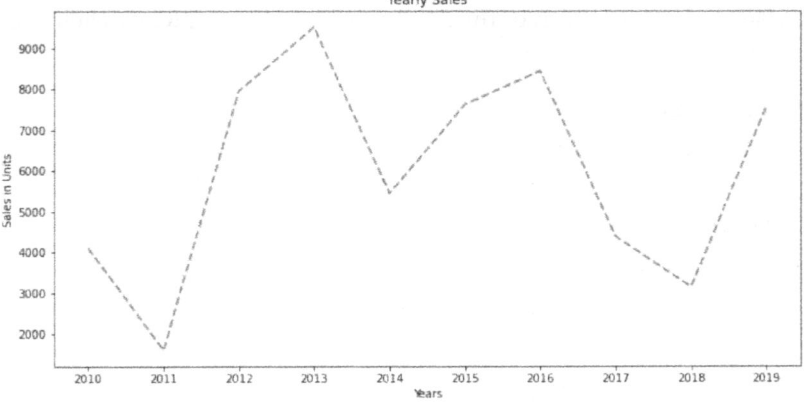

Figure 8.24 – Plotting the yearly sales as a gray dashed line

8. Now, let's add markers for each data point in the chart; we can achieve that by using the `marker` parameter:

```
x = data_frame['Year']
y = data_frame['Sales']
plt.title('Yearly Sales')
plt.xlabel('Years')
plt.ylabel('Sales in Units')
plt.plot(x, y, color = 'tab:gray', linewidth = 2,
linestyle = '--', marker='o');
```

The output will be as follows:

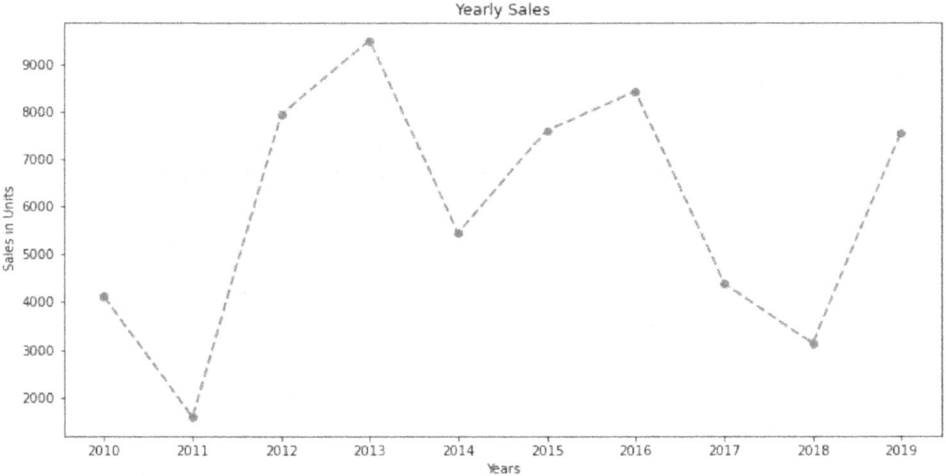

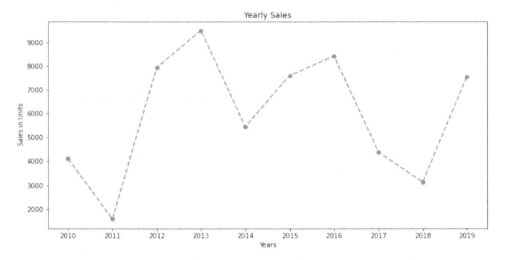

Figure 8.25 – Plotting the yearly sales as a gray dashed line with markers

The functions you have learned should help you customize your graphs with greater flexibility now. You will be able to zoom into the graphs, change the line and marker styles, and much more. Based on the complexity of the plot you want to build, you can either use the pandas `plot()` function or MATPLOTLIB. The more complex your chart is, the more likely you will need to use matplotlib, as it will provide you with many more customization options compared to the pandas `plot()` function.

So far, we have learned how to plot numeric data. But as an analyst, you will be required to visualize more than that. Next, let's see how we can visualize data of different types.

Visualizing data of different types

In the previous section, we saw how to use pandas and matplotlib to create charts for data visualization. In a data analytics project, data visualization can be used either for data analysis or to communicate insights. Presenting results in a visual way that stakeholders can easily understand and interpret is definitely a must-have skill for any good data analyst. However, you cannot choose any random chart or plot to visualize all of the different types of data that an analyst may encounter. Different chart or plot types are suitable for communicating the insight for different types of data – that is, when communicating the reach of social media on different age groups, it is preferable to use a pie chart instead of a bar or a box. On the other hand, line plots are more suitable for visualizing gradual change. The trick of data visualization is to know exactly which type of plot is appropriate for each data type you will encounter. This is exactly what we are going to learn in the following sections, starting with numerical data types.

Visualizing numerical data

As the name suggests, numeric data refers to information that is measurable in number form. One of the ways to identify numeric data is to test whether the values can be added together or simply by checking the data type using the `dtype` property of a pandas DataFrame. The types of numeric data include floats, integers, and negative numbers.

For numerical data, a histogram is usually used to show the distribution of a given variable. However, if the data is continuous (age, weight, or height – basically, any value till infinity), a grouping will be required in order to have a distinct group of values. In the following example, we will see how pandas automatically creates these groups called bins. We will also learn how to specify custom bins to overwrite the default bins specified by pandas. The *x* axis of a histogram will show the possible values of a numerical column, and the *y* axis will plot the number of observations that fall under each value:

1. Let's start with an example where we use a DataFrame to display the height of 20 individuals:

```
# Importing libraries
import pandas as pd
import numpy as np
import matplotlib.pyplot as plt

# Defining a DataFrame
data_frame =
pd.DataFrame({'Height':[175,208,159,159,178,179,168,198,
155,165,195,203,190,157,153,194,177,184,170,158]})
```

```
# Display DataFrame values
data_frame
```

This should result in output as follows:

	Height
0	175
1	208
2	159
3	159
4	178
5	179
6	168
7	198
8	155
9	165
10	195
11	203
12	190
13	157
14	153
15	194
16	177
17	184
18	170
19	158

Figure 8.26 – A DataFrame containing the height of 20 individuals

2. Now, we can plot a histogram showing the distribution of heights:

```
data_frame.plot(kind='hist');
```

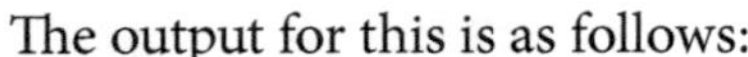

The output for this is as follows:

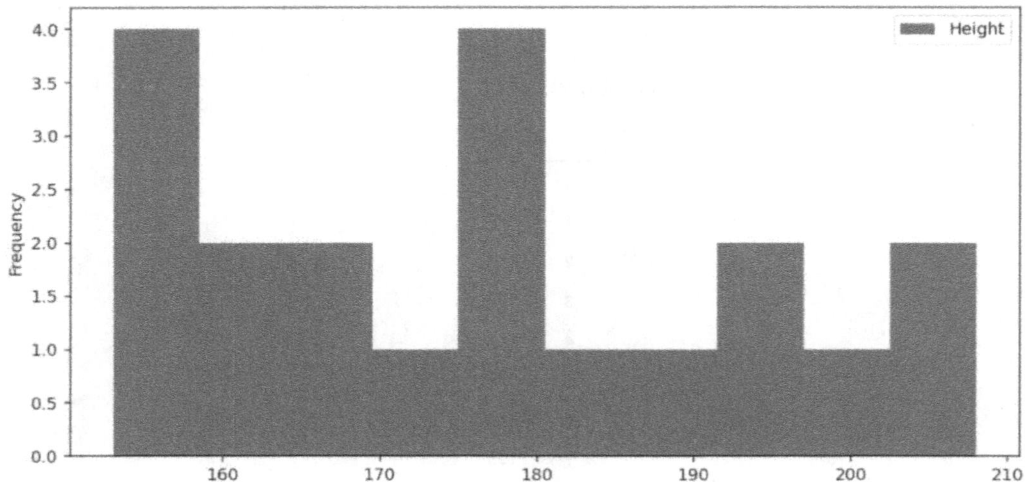

Figure 8.27 – Plotting the distribution of heights as a histogram

3. By default, pandas will set up 10 bins for the histogram, but we can change it by using the `bins` parameter:

```
data_frame.plot(kind='hist', bins=5);
```

The output will be as follows:

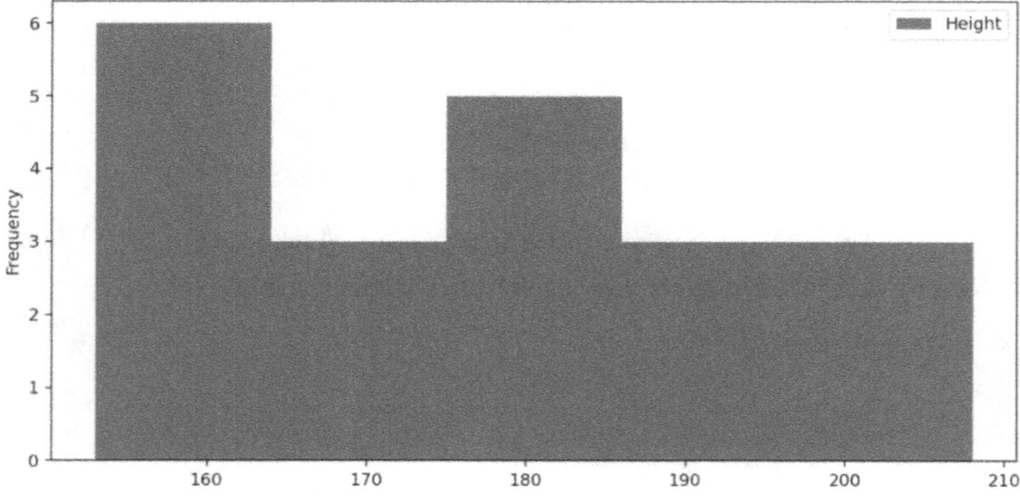

Figure 8.28 – Plotting the distribution of grouped heights as a histogram

When you encounter a dataset with a numerical variable and a time component such as dates, year, or months, you usually want to show what the trends are and how they are changing over time. Analyzing the trends will help you understand correlations between the numerical and the time component. Take the example of daily stock prices or the rate of change in the gross domestic product of a nation. An upward trend means a positive correlation and a downward trend means a negative correlation between your numerical variable and the time component. As we shall see in the next example with yearly sale numbers for a corporation, for this type of data visualization, line graphs are the most preferable choice:

1. Let's start with an example where we have the yearly sales over 20 years (2000–2019):

```
# Importing libraries
import pandas as pd
import numpy as np
import matplotlib.pyplot as plt

# Defining a DataFrame
data_frame = pd.DataFrame({
'Year':['2000','2001','2002','2003','2004','2005','2006',
'2007','2008','2009',
            '2010','2011','2012','2013','2014','2015','20
16','2017','2018','2019'],
'Sales':[175,208,159,159,178,179,168,198,155,165,195,203,
190,157,153,194,177,184,170,158]})

# Display DataFrame values
data_frame
```

The output will be as follows:

	Year	Sales
0	2000	175
1	2001	208
2	2002	159
3	2003	159
4	2004	178
5	2005	179
6	2006	168
7	2007	198
8	2008	155
9	2009	165
10	2010	195
11	2011	203
12	2012	190
13	2013	157
14	2014	153
15	2015	194
16	2016	177
17	2017	184
18	2018	170
19	2019	158

Figure 8.29 – A DataFrame containing yearly sales

2. Now, we can plot a line chart:

```
data_frame.plot(kind='line', x = 'Year');
```

The output will be as follows:

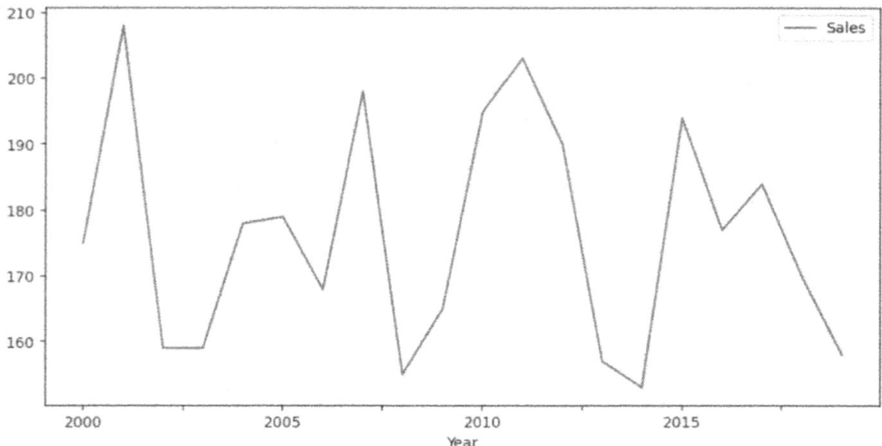

Figure 8.30 – Plotting the yearly sales

For numerical data with two variables, a scatter plot is usually used to show the relationship (correlation or trend patterns) between these two variables, as it is more convenient to plot a large amount of data points using a scatter plot while not bothering to have lines connecting the dots and abstracting the view. Scatter plots are also useful for seeing the relative distribution of data over a time frame:

1. Let's start with an example where we have the height and weight of 20 individuals and see whether there is any correlation between the two parameters:

```
# Defining a DataFrame
data_frame = pd.DataFrame({
'Weight':[67,75,119,69,106,111,120,80,108,100,79,53,75,89
,120,67,70,77,65,71],
'Height':[175,208,159,159,178,179,168,198,155,165,195,203
,190,157,153,194,177,184,170,158]})

# Display DataFrame values
data_frame
```

We will see output as shown in the following figure:

	Weight	Height
0	67	175
1	75	208
2	119	159
3	69	159
4	106	178
5	111	179
6	120	168
7	80	198
8	108	155
9	100	165
10	79	195
11	53	203
12	75	190
13	89	157
14	120	153
15	67	194
16	70	177
17	77	184
18	65	170
19	71	158

Figure 8.31 – A DataFrame containing the weight and height of 20 individuals

2. Now, we can plot a scatter plot chart:

```
data_frame.plot(kind='scatter', x = 'Height', y =
'Weight');
```

The output will be as follows:

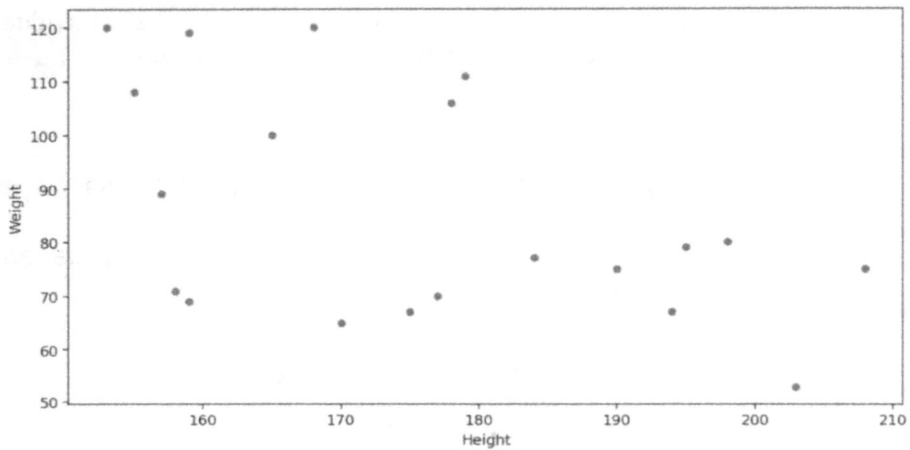

Figure 8.32 – Plotting a scatter plot of the weight versus the height

As you can see, there is a negative correlation between the weight and the height of the individuals. The taller an individual, the more likely their weight will be low.

You have now learned to visualize numerical data in order to build histograms, line graphs, and scatter plots.

So far, we have learned how to deal with data represented by numbers, but often, data can be represented in terms of labels – for example, data that represents education levels as primary, secondary, and postgraduate. In the next section, we will learn how to deal with such types of data.

Visualizing categorical data

Categorical data refers to the type of data that can be separated into groups. Only a fixed number of possible values and limited grouping are regarded as categorical data. The examples of categorical data include race, gender, marital status, and occupation. To clarify further, strings are not necessarily regarded as categorical data, as *categorical* implies a sense of grouping. Names of persons are strings but not categorical because names are most likely to be unique, whereas age groups such 0–10, 11–20, and 21–30 are categorical, as these imply groups. Sometimes, numeric data is grouped into small groups to form categorical data, in order to generate a more informative summary in data analysis.

For categorical data, a bar chart is usually used to show the distribution of each category and allows easy comparison between them.

Let's start with an example where we have the sales figures of 20 brands:

```
# Defining a DataFrame
data_frame = pd.DataFrame({
'Brand':['A','B','C','D','E','F','G','H','I','J','K','L','M',
'N','O','P','Q','R','S','T'],
'Sal
es':[1725,2108,1459,1859,1778,1279,1968,1198,1055,1865,1395,
2803,1590,2157,978,1894,1177,1084,1790,1578]})

# Display DataFrame values
data_frame
```

The output will be as follows:

	Brand	Sales
0	A	1725
1	B	2108
2	C	1459
3	D	1859
4	E	1778
5	F	1279
6	G	1968
7	H	1198
8	I	1055
9	J	1865
10	K	1395
11	L	2803
12	M	1590
13	N	2157
14	O	978
15	P	1894
16	Q	1177
17	R	1084
18	S	1790
19	T	1578

Figure 8.33 – DataFrame containing the sales of different brands

Now, we can plot a bar chart:

```
data_frame.plot(kind='bar', x = 'Brand');
```

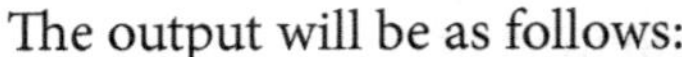
The output will be as follows:

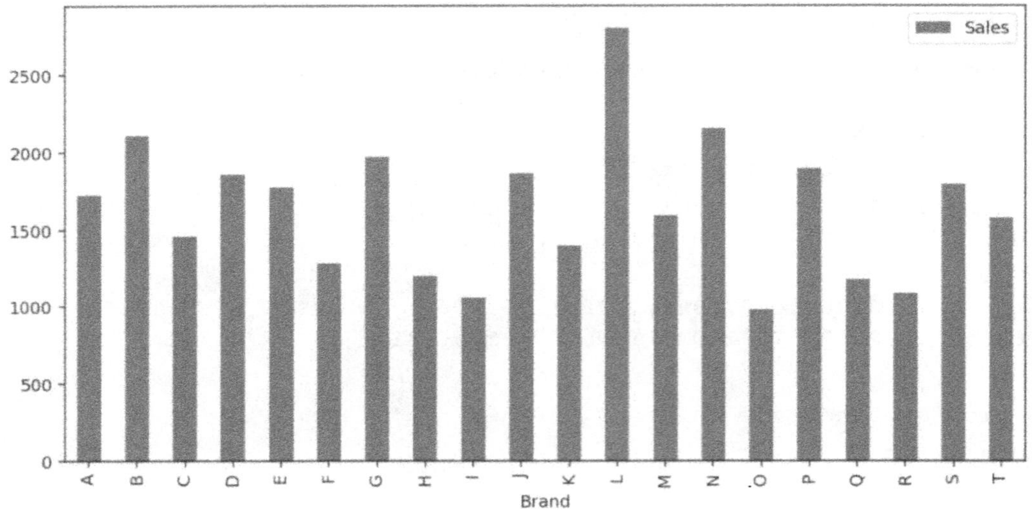

Figure 8.34 – Plotting the sales for each brand

If we have more than one column, the pandas `plot()` function will add every column in the chart as bars and add them to the legend.

Let's try adding an extra column to our previous example:

```
# Adding new column
data_frame['Quantity'] = [110,330,100,570,940,970,790,370,130,2
00,840,330,220,940,480,670,900,640,680,180]

# Display DataFrame values
data_frame
```

The output will be as follows:

	Brand	Sales	Quantity
0	A	1725	110
1	B	2108	330
2	C	1459	100
3	D	1859	570
4	E	1778	940
5	F	1279	970
6	G	1968	790
7	H	1198	370
8	I	1055	130
9	J	1865	200
10	K	1395	840
11	L	2803	330
12	M	1590	220
13	N	2157	940
14	O	978	480
15	P	1894	670
16	Q	1177	900
17	R	1084	640
18	S	1790	680
19	T	1578	180

Figure 8.35 – A DataFrame containing the sales and quantities of different brands

Now, we can replot the bar chart:

```
data_frame.plot(kind='bar', x = 'Brand');
```

The output will be as follows:

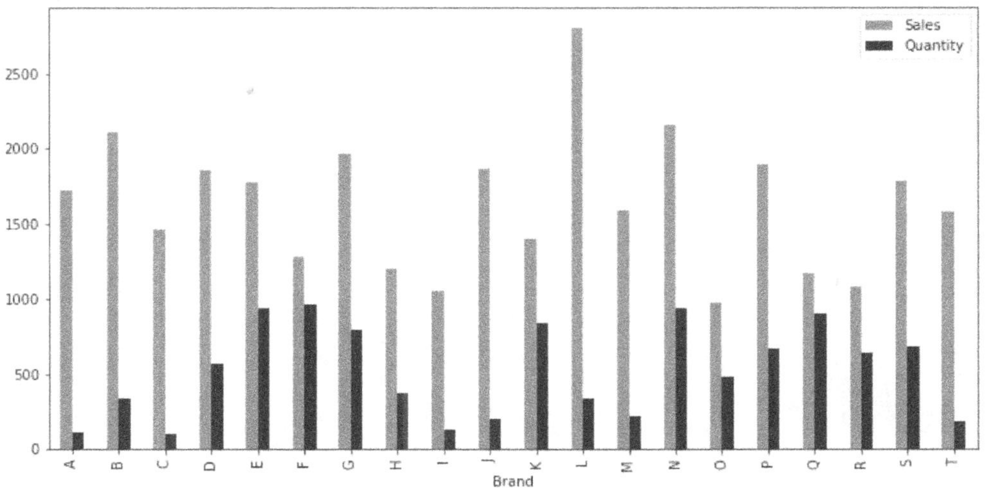

Figure 8.36 – Plotting the sales and quantities for each brand

We can also plot a horizontal bar chart:

```
data_frame.plot(kind='barh', x = 'Brand');
```

The output will be as follows:

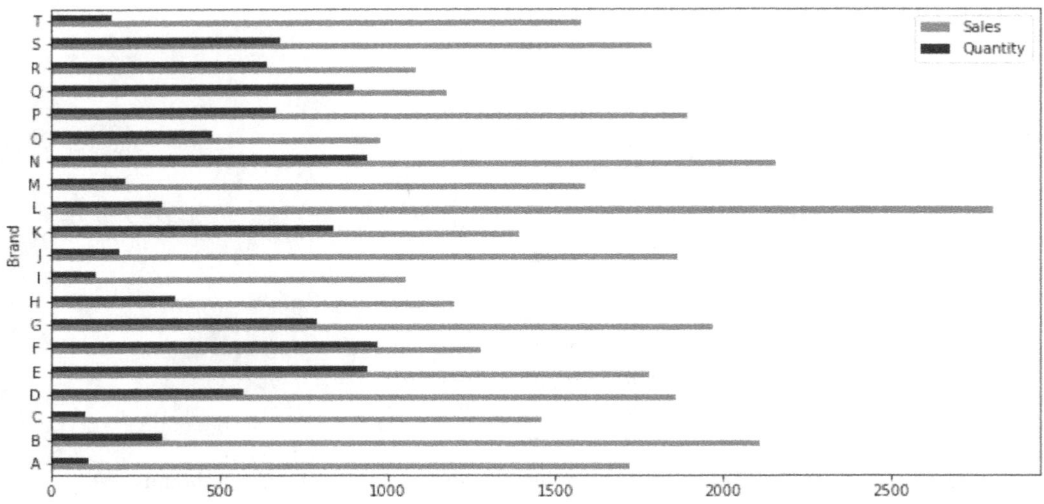

Figure 8.37 – Plotting the sales and quantities for each brand vertically

For categorical data where we want to know the proportion of each category, a pie chart is usually used to show the contribution of each category to the total.

Let's start with an example where we have the sales figures of 10 brands:

```
# Defining a DataFrame
data_frame = data_frame = pd.DataFrame({
'Brand':['1','2','3','4','5','6','7','8','9','10'],
'Sales':[1725,218,1459,185,1778,179,1968,198,155,165]})

# Display DataFrame values
data_frame
```

The output will be as follows:

	Brand	Sales
0	1	1725
1	2	218
2	3	1459
3	4	185
4	5	1778
5	6	179
6	7	1968
7	8	198
8	9	155
9	10	165

Figure 8.38 – A DataFrame containing the sales of different brands

Now, we can plot a pie chart (we have opted to not show the legend by setting the `False` option):

```
data_frame.plot(kind="pie", y = 'Sales', legend = False);
```

The output will be as follows:

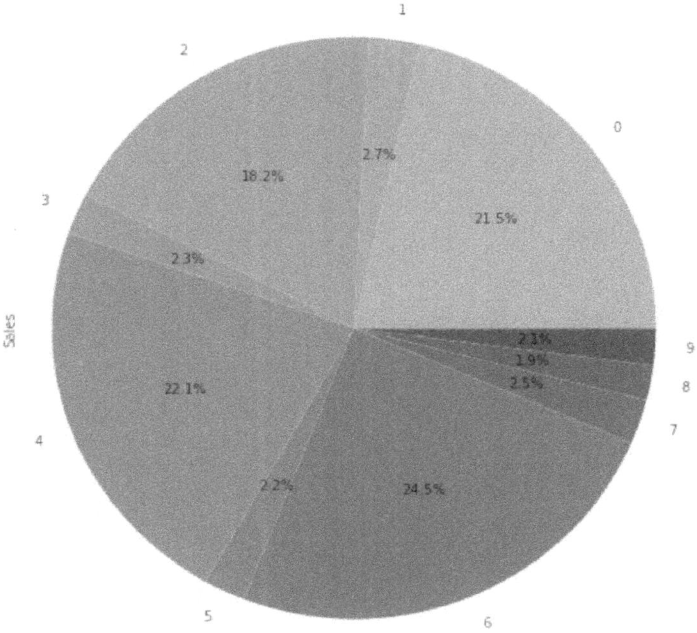

Figure 8.39 – Plotting the proportion of sales for each brand

As you can see, most sales come from the **2**, **0**, **4**, and **6** brands.

Now that we have learned how to visualize categorical data, let's move on to learn about how to visualize statistical data.

Visualizing statistical data

Statistics is a field of study that makes use of mathematical equations for data analysis. Presenting statistical data is common across all industries, such as finance and government and research institutions. Organizations often use statistical data for testing, predictive analysis, and more.

For statistical data, a boxplot chart is frequently used to show overall statistical information on the distribution of the data. It is also used to detect outliers in the data.

Let's start with an example where we have the height and gender of 20 individuals, and we will see, using boxplots, the distribution of `Height` for males and females:

```
# Defining a DataFrame
data_frame = data_frame = pd.DataFrame({
'Height':[175,208,159,130,178,179,168,100,155,165,195,250,190,
157,153,194,177,184,170,210],
'Gender':['F','M','F','F','M','M','F','M','F','F','M','M','F',
'M','M','M','F','M','M','F']})

# Display DataFrame values
data_frame
```

The output will be as follows:

	Height	Gender
0	175	F
1	208	M
2	159	F
3	130	F
4	178	M
5	179	M
6	168	F
7	100	M
8	155	F
9	165	F
10	195	M
11	250	M
12	190	F
13	157	M
14	153	M
15	194	M
16	177	F
17	184	M
18	170	M
19	210	F

Figure 8.40 – A DataFrame containing the height and gender of 20 individuals

Now, we can plot a boxplot chart (sometimes referred to as Cat Whiskers plot) grouped by `Gender`:

```
data_frame.boxplot(by="Gender", column="Height");
```

The output should be as follows:

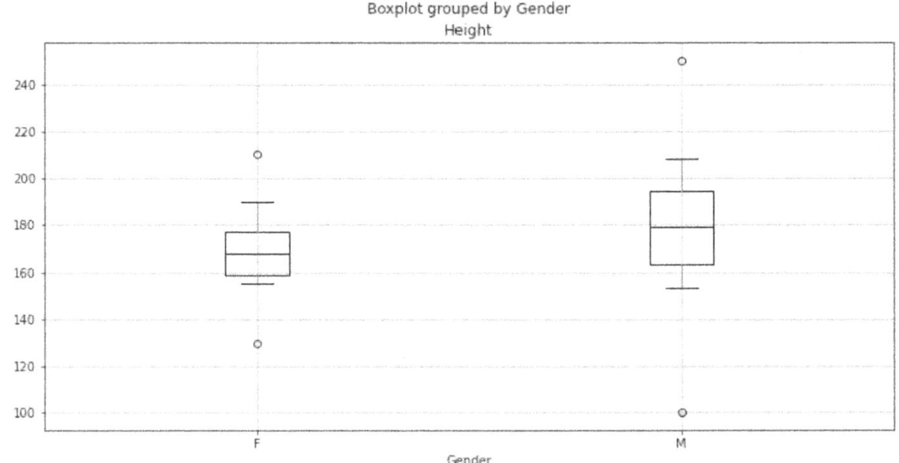

Figure 8.41 – Plotting the distribution of height for each gender

Let's have a deeper look at this boxplot to understand each component of it:

- The bottom horizontal line represents the minimum value excluding any outliers. If we consider this for the female population, the lowest height (excluding the outliers) was 157 cm.
- The bottom edge of the rectangle represents the first quartile or 25%. If we consider this for the female population, 25% of the population was under a height of 160 cm.
- The middle line inside the rectangle represents the second quartile, also called the median. If we consider this for the female population, half the population was under a height of 170 cm.
- The top edge of the rectangle represents the third quartile or 75%. If we consider this for the female population, 75% of the population was under a height of 178 cm.
- The top horizontal line represents the maximum value, excluding any outliers. If we consider this for the female population, the highest height (excluding the outliers) was 190 cm.
- The circles at the top and bottom represent the outliers.

In this section, you learned about handling statistical data. The preceding examples only show a small part of statistical concepts; it is worthwhile to learn other statistical evaluation methods, such as variances, standard deviations, and correlations. Data visualization with statistical data is a vast topic and beyond the scope of this chapter.

Exercise 8.02 – Boxplots for the Titanic dataset

In this exercise, we will load the Titanic dataset, handle the missing data, and build a few boxplots in order to find the correlation of different factors contributing to the chances of survival.

> **Note**
> The data is contained inside `titanic.csv`, which you can find on our GitHub: `https://raw.githubusercontent.com/PacktWorkshops/The-pandas-Workshop/master/Chapter08/Data/titanic.csv`.

The following steps will help you complete the exercise:

1. Open a new Jupyter notebook file.

2. Import the `pandas`, `numpy`, and `matplotlib` packages:

```
import pandas as pd
import numpy as np
import matplotlib.pyplot as plt
```

3. Next, load the CSV file as a DataFrame:

```
file_url = 'titanic.csv'
data_frame = pd.read_csv(file_url)
```

4. Use the `head()` function to display the first five rows of the DataFrame, and check that the data was properly loaded:

```
data_frame.head()
```

The output will be as follows:

	survived	ticket_class	gender	age	number_sibling_spouse	number_parent_children	passenger_fare	port_of_embarkation	age_group
0	0	3	male	22.0	1	0	7.2500	S	18-59
1	1	1	female	38.0	1	0	71.2833	C	18-59
2	1	3	female	26.0	0	0	7.9250	S	18-59
3	1	1	female	35.0	1	0	53.1000	S	18-59
4	0	3	male	35.0	0	0	8.0500	S	18-59

Figure 8.42 – A DataFrame displaying the first five rows of the dataset

5. Remove the rows with missing data, and then display the DataFrame:

```
data_frame = data_frame.dropna()
data_frame
```

This will result in the following output:

	survived	ticket_class	gender	age	number_sibling_spouse	number_parent_children	passenger_fare	port_of_embarkation	age_group
0	0	3	male	22.0	1	0	7.2500	S	18-59
1	1	1	female	38.0	1	0	71.2833	C	18-59
2	1	3	female	26.0	0	0	7.9250	S	18-59
3	1	1	female	35.0	1	0	53.1000	S	18-59
4	0	3	male	35.0	0	0	8.0500	S	18-59
...	...	...	...	...	...	...	...	...	...
885	0	3	female	39.0	0	5	29.1250	Q	18-59
886	0	2	male	27.0	0	0	13.0000	S	18-59
887	1	1	female	19.0	0	0	30.0000	S	18-59
889	1	1	male	26.0	0	0	30.0000	C	18-59
890	0	3	male	32.0	0	0	7.7500	Q	18-59

712 rows × 9 columns

Figure 8.43 – A DataFrame displaying the dataset without missing data

6. Plot a boxplot group by `survived` on the `'age'` column:

```
data_frame.boxplot(by='survived', column='age');
```

This should result in the following output:

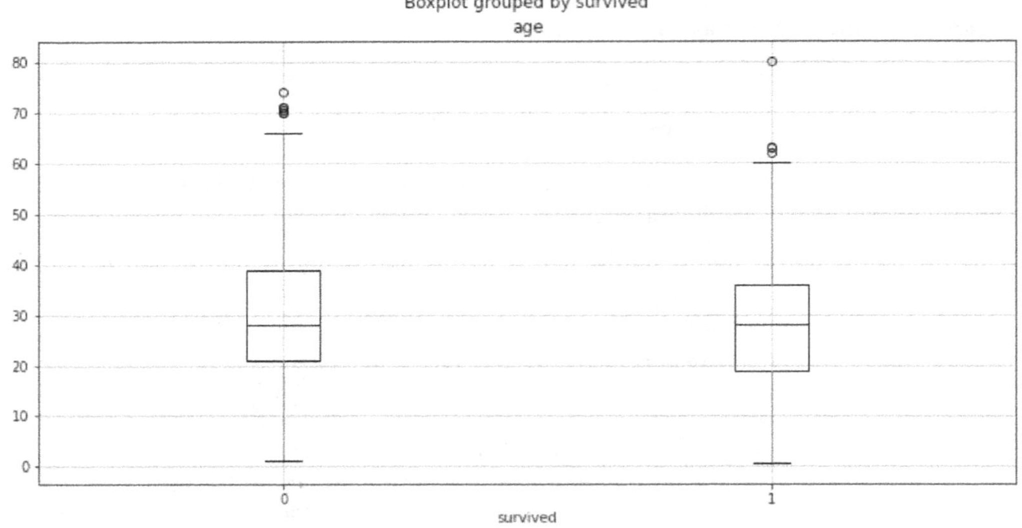

Figure 8.44 – Plotting the distribution of age for each outcome

We can see that, in general, younger passengers had a higher chance of surviving with 75% of the survivors being under 36 compared to 39 in the other group. Moreover, older passengers, regardless of the outcome, are classed as outliers. This might be due to the very small population of elderly.

7. Plot a boxplot by `'survived'` and on the `'passenger_fare'` column:

```
data_frame.boxplot(by='survived', column='passenger_
fare');
```

This should result in the following output:

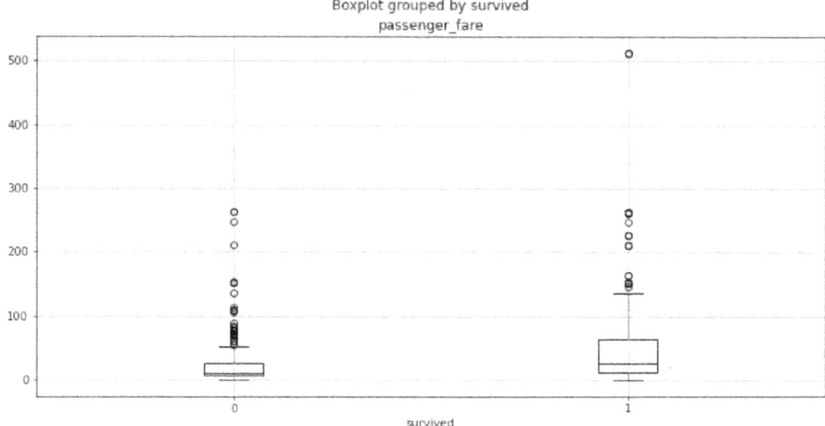

Figure 8.45 – Plotting the distribution of passenger fares for each outcome

It seems that the higher the passenger fare was, the higher the chance of survival was for the passenger. This can be easily seen from the position of the box on the survivor group, which is higher than the other group's box.

8. Plot a boxplot by `survived` and on the `ticket_class` column, as follows:

```
data_frame.boxplot(by='survived', column='ticket_class');
```

You should see output as follows:

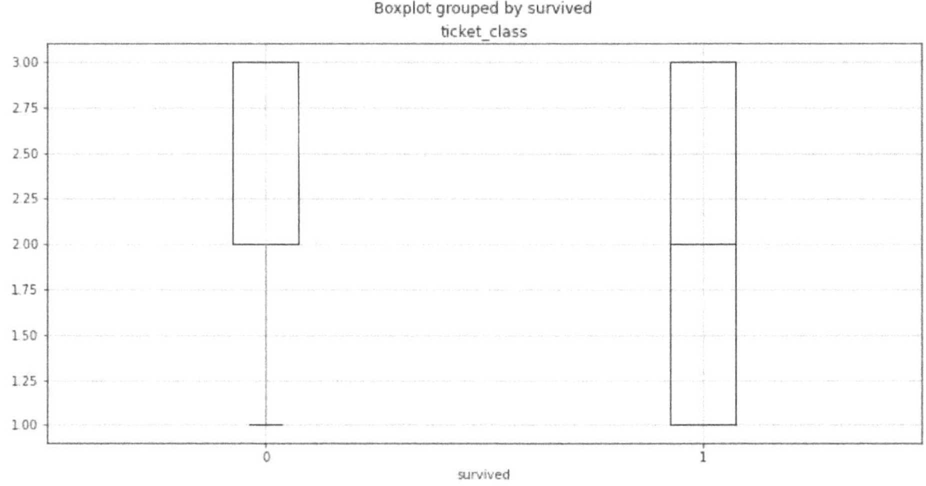

Figure 8.46 – Plotting the distribution of the ticket class for each outcome

We can see that, in general, passengers from first class and second class had a higher chance of surviving, with half of them being at least from the first or second class.

Now that we have seen how we can use a specific chart type (boxplot) to derive quick insights, we can move on to our next topic about visualizing multiple data plots.

Visualizing multiple data plots

Up till now, we have been working with a fairly simple, linear approach to data visualization. However, most data analytics insights are gleaned by comparing two or more series of data against each other. We will see shortly how this can be achieved with pandas visualization – for example, a brand might want to compare its stores' performances against each other.

If we want to have multiple data plots on the same chart, then we need to use the pandas `pivot()` function to tell pandas that we are using multiple sets of data instead of a single one.

Let's start with an example where we have the yearly sales of two stores:

```
# Defining a DataFrame
data_frame = data_frame = pd.DataFrame({
'Store':['A','A','A','A','A','A','A','A','A','A','B','B','B',
'B','B','B','B','B','B','B'],
'Year':[2010,2011,2012,2013,2014,2015,2016,2017,2018,2019,2010,
2011,2012,2013,2014,2015,2016,2017,2018,2019],
'Sal
es':[175,208,159,159,178,179,168,198,155,165,195,203,190,157,
153,194,177,184,170,158]})

# Display DataFrame values
data_frame
```

The output will be as follows:

	Store	Year	Sales
0	A	2010	175
1	A	2011	208
2	A	2012	159
3	A	2013	159
4	A	2014	178
5	A	2015	179
6	A	2016	168
7	A	2017	198
8	A	2018	155
9	A	2019	165
10	B	2010	195
11	B	2011	203
12	B	2012	190
13	B	2013	157
14	B	2014	153
15	B	2015	194
16	B	2016	177
17	B	2017	184
18	B	2018	170
19	B	2019	158

Figure 8.47 – A DataFrame containing the yearly sales of two stores

Let's plot a line chart with two lines by using the `pivot()` function:

```
df = data_frame.pivot(index='Year', columns='Store',
values='Sales')
df.plot();
```

The output will be as follows:

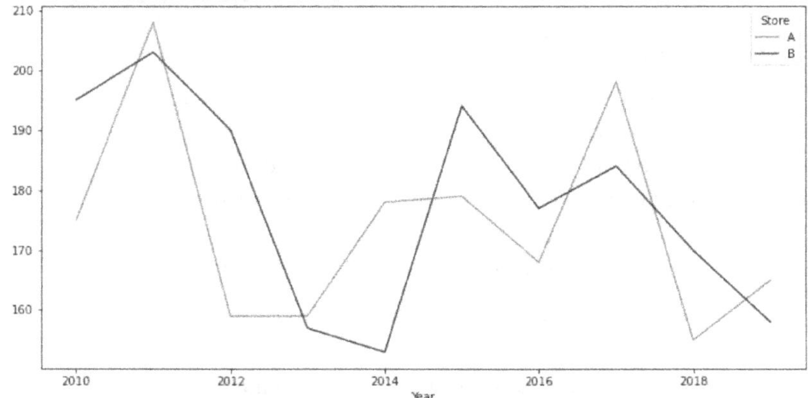

Figure 8.48 – Plotting the yearly sales of each store

We can also plot a bar chart with multiple bars:

```
df = data_frame.pivot(index='Year', columns='Store',
values='Sales')
df.plot(kind = 'bar');
```

The output will be as follows:

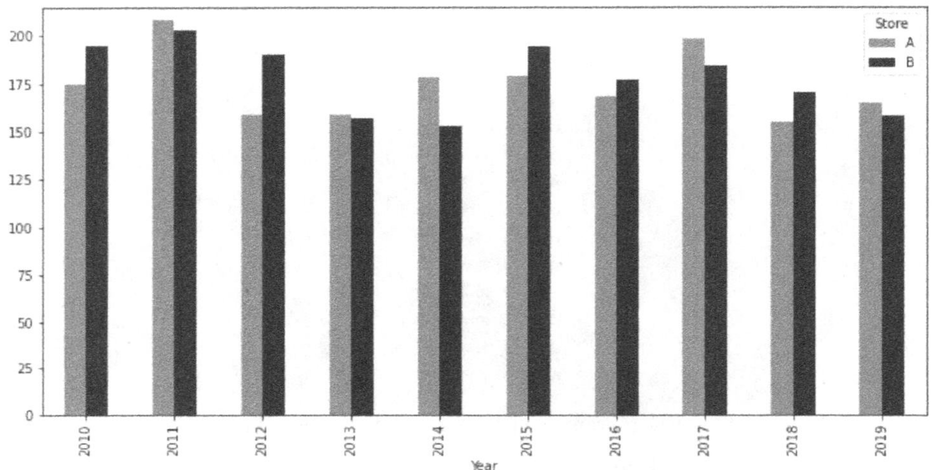

Figure 8.49 – Plotting the yearly sales of each store as a bar chart

In this section, you learned about handling multiple data plots. In the next activity, you will apply all of the skills learned up till now for analyzing data in a real-world dataset.

Activity 8.01 – Using data visualization for exploratory data analysis

In this activity, we will apply what we have learned in this chapter to building different types of plots in order to perform an exploratory data analysis on a sale price. We will work on the Manufactured Housing Survey dataset, published by the United States Census Bureau, that can be found in the GitHub repository at `https://raw.githubusercontent.com/PacktWorkshops/The-pandas-Workshop/master/Chapter08/Data/PUF2020final_v1coll.csv`.

> **Note**
>
> More details about the Ames Housing dataset can be found at `https://www.census.gov/data/datasets/2020/econ/mhs/puf.html`.

The goal of this activity is to analyze the different factors contributing to a sale price in the housing market. We will use different types of plots in order to achieve it.

Your tasks will be as follows:

1. Open a Jupyter notebook.
2. Import the pandas, numpy, and matplotlib packages.
3. Load the CSV file as a DataFrame.
4. For the sake of simplicity, keep only the following columns:
    - `REGION`
    - `SQFT`
    - `BEDROOMS`
    - `PRICE`
5. Display the first 10 rows of the DataFrame.
6. Plot a histogram chart for `PRICE` to see the distribution of house prices.
7. Plot a histogram chart for `SQFT` to see the distribution of house areas.
8. Plot a scatter plot chart for `PRICE` and `SQFT` to understand whether there is any correlation between the two.
9. Plot a boxplot chart for `BEDROOMS` on `PRICE` to understand the distribution of prices across bedroom sizes.
10. Plot a scatter chart for `BEDROOMS` and `PRICE` to understand whether there is any correlation between the two.
11. Plot a boxplot chart for `REGION` on `PRICE` to understand the distribution of prices across regions.
12. Plot a scatter chart for `REGION` on `PRICE` to understand whether there is any correlation between the two.
13. Plot a horizontal bar chart for `REGION` and `PRICE` by using the `pivot` function to see how the average price varies across regions.
14. Plot a line chart for `REGION` and `PRICE` by using the `pivot` function to validate the previous.

> **Note**
>
> The solution for this activity can be found in the *Appendix*.

In this activity, we used `pandas` to plot different types of charts on a list of factors contributing to the sale price of a house in Iowa. We also determined these factors' relationship with the sale price.

Summary

In this chapter, we have learned the fundamentals of pandas visualization and how to create charts. After going through the basics of creating charts in pandas, we looked at how we can further customize charts by using the matplotlib package. Then, we learned what the main charts are for each type of data, such as numerical data, categorical data, and statistical data, before learning how to handle multiple data plots.

Finally, we applied our learnings to an activity with the purpose of applying what we learned in this chapter to a business case, where the goal was to determine how different factors affect a price. In the next chapter, you will learn how to model data to derive insights.

Part 3 – Data Modeling

This section looks at how we can use various techniques to work with and model data so that we can gain essential insights from it. We will cover a lot of advanced features and utilities to get the most out of our data.

This section contains the following chapters:

- *Chapter 9, Data Modeling – Preprocessing*
- *Chapter 10, Data Modeling – Modeling Basics*
- *Chapter 11, Data Modeling – Regression Modeling*

9

Data Modeling - Preprocessing

In this chapter, you will learn two important processes used to prepare data for modeling – splitting and scaling. You will learn how to use the `sklearn` methods – `.StandardScaler` and `.MinMaxScaler` for scaling, and `.train_test_split` for splitting. You will also be introduced to the reasons behind scaling and exactly what these methods do. As part of exploring splitting and scaling, you will use `sklearn LinearRegression` and `statsmodels` to create simple linear regression models.

By the end of this chapter, you will be comfortable preparing datasets to begin modeling. The main ideas you will learn in this chapter are as follows:

- Exploring independent and dependent variables
- Understanding data scaling and normalization
- Activity 9.01 – Data splitting, scaling, and modeling

An introduction to data modeling

Consider a statement such as *the weather depends on the season*. If we wanted to confirm this statement with data, we would collect information about the weather during different times of the year. The statement is asserting a model – a model of the weather that says we can say something about the weather if we know the season. Proposing and evaluating models is data modeling.

Often, we would like to understand the relationships within our data (numbers and other types of information), and in the previous chapter, we used visualization methods for that. Here, we can go a little deeper and ask questions such as *are the independent variables correlated with each other?* or *is the output a linear function of the input?* In some cases, we can answer these questions with charts; in other cases, we may construct a mathematical model. A mathematical model is simply a function that transfers some input data into an output.

Data modeling, the topic of this chapter and the next, includes both descriptive and model-based approaches. Some of the descriptive approaches fully overlap into visualization (*Chapter 7, Data Visualization*), but here we'll talk more about the insights we can gain from describing data, such as "the target variable appears to be normally distributed." The latter statement can be taken to mean we can formulate a hypothesis that the data we want to analyze is being generated by an underlying stochastic (random or randomly generated) process that generates data conforming to a normal distribution. This is equivalent to forming a model of the data.

Model-based approaches use statistical and other types of models to attempt to build a transfer function that transforms the independent data (often called *X*) into the observed dependent data (often called *Y*). If we say that *Y* depends on *X*, and *X* is tabular data, then representing *X* and *Y* in pandas is a natural way to work with our data. This makes it easy to manipulate and prepare data, and then fit models to *X* using *Y* as the ground truth.

Exploring dependent and independent variables

In this chapter, you will learn about dependent and independent variables. You will learn about the need for scaling and normalization of data, in addition to performing those operations. You will also use some basic modeling methods to analyze your data.

At a high level, we can say a **dependent variable** is related to one or more independent variables in a linear or non-linear way. Linear models are easy to understand. A linear model relating one Y to one X is just a line. With multiple X variables, each one has a coefficient that gives its effect on Y, and since all those effects are independent, we just add all the effects together in a multivariate linear model. In a non-linear model, Y depends on X in a more complex way, such as Y being a function of X^2. We can create non-linear models nearly as easily as linear models in pandas using some simple additional modules. We'll explore how to do that in the following chapter.

Much of the time while working with data in pandas, our goal is to build a model to predict or explain something. Other times, we are exploring data to find insights directly, which was covered in the previous chapter. Typically, we call the thing that we want to predict the **dependent variable** because the assumption is there is a relationship between the other data and this variable. Conversely, the things we will use as input to a model are the **independent variables**. Here, independent refers to cases where we have more than one X variable, and the statement means each variable is independent of all the other X variables. We can look at that in more detail, using an example.

Suppose you are given some data from a metal production facility, containing the percentages of two metal components and the resulting hardness of the alloy. You are interested in the relationship of the final hardness to the proportions of the two component metals. You read the data into a pandas DataFrame, as follows:

```
import pandas as pd
metal_data = pd.read_csv('Datasets\\metal_alloy.csv')
metal_data.head(10)
```

> **Note**
>
> All the examples for this chapter can be found in the `Examples.ipynb` notebook in the `Chapter09` folder, and the data files can be found in the `Datasets` folder. To have the examples run correctly, you need to run the notebook from start to finish in order. In the code presented in this chapter, change the highlighted path as needed to match where you stored the data.

The data is shown in *Figure 9.1*:

Out[2]:

	metal_1	metal_2	alloy_hardness
0	0.958000	0.140659	1.254157
1	0.920147	0.107089	0.956846
2	0.590646	0.483316	1.952517
3	0.787427	0.239446	1.636522
4	0.223974	0.817454	2.367797
5	0.339729	0.694622	2.115060
6	0.242666	0.837370	2.899579
7	0.721072	0.365196	1.758518
8	0.666492	0.430698	1.591216
9	0.650387	0.414661	1.780010

Figure 9.1 – Data for production of a metal alloy

In *Figure 9.1*, we see the fractions of the two component metals in the first two columns and the hardness values in the last column. Let's begin by just looking at the data graphically. Here, we use `matplotlib.pyplot`, create a figure with three subplots on one line (the `.subplots(1, 3` part of the code), and then plot each data pair in one plot:

```
import matplotlib.pyplot as plt
fig, ax = plt.subplots(1, 3, figsize = (15, 5))
ax[0].scatter(metal_data['metal_1'],
              metal_data['metal_2'],
              label = 'metal 2 pct vs. metal 1')
ax[0].legend()
ax[1].scatter(metal_data['metal_1'],
              metal_data['alloy_hardness'],
              label = 'hardness vs. metal 1 pct')
ax[1].legend()
ax[2].scatter(metal_data['metal_2'],
              metal_data['alloy_hardness'],
              label = 'hardness vs. metal 2 pct')
ax[2].legend()
plt.show()
```

The preceding code snippet produces the following output:

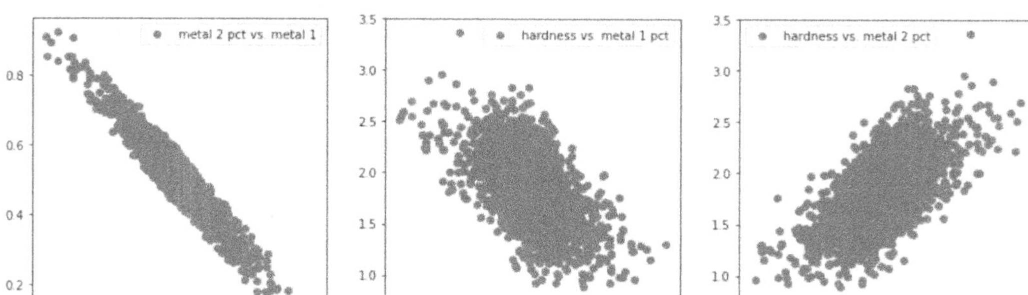

Figure 9.2 – Pairwise relationships of the metals data

From the first panel, we see that as the **metal 2** percentage decreases, the **metal 1** increases. That makes sense if there are only two components and we are working in fractions or percentages. However, there appears to be noise in the data, possibly from measurement errors when the alloys are composed. In the two rightmost panels, we see that the hardness decreases as **metal 1** increases, and conversely increases as **metal 2** increases. The two rightmost panels appear to be near-mirrored images of each other. The latter observation makes sense if, as before, there are only two components and we are working in fractions or percentages.

Another way to look at the relationship between variables is to use a **correlation** analysis. Correlation analysis is a method to quantify pairwise relationships of data. A correlation coefficient of 1 means two sets of data are perfectly aligned – when one goes up, so does the other, and vice versa. Intuitively, a correlation coefficient of -1 means two sets of data are aligned opposite one another – when one goes up, the other goes down, and vice versa. From this description, you can also see that the correlation coefficient ranges from -1 to 1. Pandas offers us the `.corr()` method to use with `DataFrames`, as follows:

```
correlation = metal_data['metal_1'].corr(metal_data['metal_2'])
print('correlation between x1 and x2: ', correlation)
```

This gives us the following output:

```
correlation between x1 and x2:  -0.9335045017430936
```

We see that while both `x1` and `x2` appear highly correlated with `y` (which is good from a modeling point of view), they are also highly correlated (a correlation coefficient of -0.93) with each other, which might be a concern. What does *might be a concern* mean? In some modeling methods, highly correlated variables can lead to unstable results or incorrect conclusions. We will go deeper into linear regression in the next chapter, but here, we use the `statsmodels` package to build a simple linear regression model of `y` versus the two `x` variables. Don't worry if the syntax is unfamiliar to you now:

```
import statsmodels.api as sm
X = sm.add_constant(metal_data.loc[:, ['metal_1', 'metal_2']])
lin_model = sm.OLS(metal_data['alloy_hardness'], X)
my_model = lin_model.fit()
print(my_model.summary())
print(my_model.params)
```

The preceding code imports the `statsmodels` module that has a linear regression method built in. The syntax is nearly self-explanatory; the only unfamiliar part might be the first call to the `add_constant()` method. `statsmodels`, unlike some other linear regression methods, does not automatically fit a constant, which would be the y-intercept if we were fitting only one `x` variable. The way to add the constant term is by calling `.add_constant()`. The resulting model and its statistical features look as follows:

```
                            OLS Regression Results
==============================================================================
Dep. Variable:         alloy_hardness   R-squared:                       0.394
Model:                            OLS   Adj. R-squared:                  0.394
Method:                 Least Squares   F-statistic:                     929.6
Date:                Sun, 01 Aug 2021   Prob (F-statistic):          1.23e-311
Time:                        10:02:39   Log-Likelihood:                -44.409
No. Observations:                2858   AIC:                             94.82
Df Residuals:                    2855   BIC:                             112.7
Df Model:                           2
Covariance Type:            nonrobust
==============================================================================
                 coef    std err          t      P>|t|      [0.025      0.975]
------------------------------------------------------------------------------
const         -0.3434      0.147     -2.339      0.019      -0.631      -0.055
metal_1        1.1086      0.139      7.951      0.000       0.835       1.382
metal_2        3.0783      0.136     22.618      0.000       2.811       3.345
==============================================================================
Omnibus:                        1.075   Durbin-Watson:                   2.016
Prob(Omnibus):                  0.584   Jarque-Bera (JB):                1.023
Skew:                           0.044   Prob(JB):                        0.600
Kurtosis:                       3.031   Cond. No.                         66.1
==============================================================================

Notes:
[1] Standard Errors assume that the covariance matrix of the errors is correctly specified.
const     -0.343381
metal_1    1.108639
metal_2    3.078313
dtype: float64
```

Figure 9.3. Result of fitting a simple linear model to our data.

There are a few things to note in this output. First, there are a dizzying number of statistical terms and numbers. Don't get overly concerned about those at this point; there are cases where you need deeper statistical knowledge, but *in many cases, you do not*. The main point to note is that the **R-squared** (also **R2**, or $\mathbf{R^2}$) value is fairly low, at 0.39. In other words, our simple model accounts for only 39% of the variation seen in `y`. In most cases, you can interpret an R2 value very simply – it expresses the fraction of the variation in `y`, as explained by the model.

It's also important to understand from this output what the actual model is, which is the following:

$$y = -0.343381 + 1.108639 * metal\_1 + 3.078313 * metal\_2$$

We see those values in the `coef` column in the output, as well as at the end of the output where we printed the parameters.

Finally, we see that the **p-values** for both `metal_1` and `metal_2` are 0.000. One way to interpret the p-values for linear regression is, *if all model assumptions are met, the p-value is the probability that the coefficients are 0 versus the fitted values*. The issue here is that we already know `metal_1` and `metal_2` are perfectly correlated in theory (as `metal_2` is 1 – `metal_1`, and vice versa) and are nearly perfectly correlated in our real data (a correlation of -0.93), and a key assumption in linear regression is that the variables (also referred to as **covariates**) are not highly correlated. When variables are perfectly correlated or nearly so, we say they are **collinear**, and we call the case **multicollinearity**. So, what should we do?

The answer to this question is, *it depends*. If your only goal is a model that will correctly predict the data on which it was constructed, then you do not have to do anything. Although you might not know it from reading all the posts on data sites or social media, multicollinearity does not mean the model is bad at predicting the values. Instead, the existence of multicollinearity means issues arise regarding the coefficients and how to interpret them. In the case of multicollinearity, the values of the coefficients can be sensitive to small changes in data (such as repeating an experiment). Also, the interpretation of the coefficients can be misleading when there are collinear variables. If you are testing the effectiveness of a new medicine, this might be important, but in many data science applications, it is less so.

It's worth noting that in this simple case, the original data was synthetically generated using the 1.0 and 3.0 coefficients, as compared to the model results of 1.11 and 3.08, so the results in this case were not affected by the **collinearity** of the two `x` variables.

If we had a larger number of variables, creating plots as shown previously can become cumbersome. Fortunately, there are some tools available. Before using the tools, we need some data. A great way to reinforce the idea of correlated data and multicollinearity is to create some synthetic data that exhibits correlations. Let's see how to do that. The following code works as follows. We first load the `numpy` module and then set the **random seed**. The value used (here, `42`) does not matter, but repeating the code execution with the same seed will produce the same results; otherwise, the values would vary due to the use of `.random`.

Next, we create a `DataFrame` with one column (`x1`) with numbers from 0 to 999. The first for loop adds columns 2 through to 10, filling them with the same values as x1 but adding random noise to all the values, where the noise ranges from -50 to 50 and, therefore, averaging to 0.

The second loop modifies the even x columns by adding more noise – in that loop, we add additional noise, but it is non-zero because we specify the mean to be -100 in the `.random.normal()` method, and then we multiply that distribution by a factor that will range between 0 and 10 (from the `10 * np.random.uniform(0, 1)` code). The expectation then is that the odd-numbered x variables (in our scheme of `x1` to `x10`) will be highly correlated, but the even columns will not be very correlated, because we added the same level of noise to all the odd columns but varying amounts of noise and offset to the even columns:

```
import numpy as np
np.random.seed(42)
multi_coll_data = pd.DataFrame({'x1' : range(1000)})
for i in range(9):
    multi_coll_data['x' + str(i + 2)] =
np.add(list(range(1000)),
                                                    np.random.
uniform(-50, 50, 1000))
for i in range(0, 9, 2):
    multi_coll_data['x' + str(i + 2)] = np.add(multi_coll_
data['x' + str(i + 2)],
                                                    10 * np.random.
uniform(0, 1) *
                                                    np.random.
normal(-100, 100, 1000))
multi_coll_data['y'] = range(1000)
print(multi_coll_data.head())
print(multi_coll_data.tail())
```

Running this snippet will result in the following output:

```
   x1           x2         x3          x4         x5           x6         x7  \
0   0   690.303674 -31.486707 -731.643758  17.270299  1436.411756 -10.636448
1   1  -685.241074   5.190095 -458.895861  30.668140  -716.580334  -1.656434
2   2   936.292932  39.294584 -712.144359 -22.953210  -122.183985  37.454739
3   3 -1798.095409  26.222489 -269.751619  15.487410  -464.936948 -12.999561
4   4 -2114.215496  34.656115 -480.576137  11.174598  -768.245414  40.964968

           x8         x9         x10  y
0  -492.026404 -46.120055   22.754113  0
1 -3610.645334 -30.322747 -472.866262  1
2  -762.459068  35.124581 -170.442837  2
3 -2052.517125  29.676836 -758.140719  3
4  -679.874801 -10.935731  -69.331760  4
      x1           x2           x3           x4           x5           x6  \
995  995  -298.409060  1010.695516   814.386703   989.210703   782.825485
996  996  -286.163537  1041.661462  1279.540113   979.440118 -1432.060863
997  997  1018.026789   953.895802   805.609186   986.457232   752.152415
998  998 -1630.960898   953.705472  1000.300624  1000.994059   236.396803
999  999 -1273.687387   977.218707   540.179294   965.136736   -51.844077

              x7           x8           x9         x10    y
995  1013.443536    94.047877   972.315962  463.710712  995
996   996.322041   683.241456   966.951922  962.712279  996
997  1023.514885   378.872916   992.532875  797.471943  997
998   996.529063  1041.017110  1038.843755  988.338761  998
999   963.938164   914.585894   959.448032  430.768433  999
```

Figure 9.4 – Synthetic data with varying correlations among the x variables

> **Note**
>
> Note that we started with `np.random.seed(42)`. Because we are using some random number generation functions, if we want to repeat running the code and get the same result, we need to add this step to initialize the random number generator to a fixed value. The reason this works is that random number generators are not truly random but have a starting point that is, by default, also chosen at random. By fixing the starting point, often called the **seed**, the random number generator becomes deterministic. Many data modeling algorithms and other methods use some form of random initialization. If you are working on a problem and the results change each time to run your code, try using the Numpy `random.seed()` method. Note that it's not generally important what seed value you choose, just that a value is fixed. This also makes your code reproducible by someone else on a different computer at another time, a common requirement in data science.

It's not evident what is correlated to what in *Figure 9.4*. That's why visualization is so important. The pandas plotting wrappers of `matplotlib` don't offer many options to plot a lot of variables. We can try using the `Pandas DataFrame.plot()` method:

```
multi_coll_data.plot(x = 'x1')
```

Upon running this code, you should see output as follows:

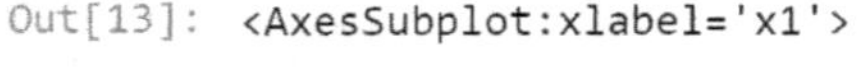

Out[13]: <AxesSubplot:xlabel='x1'>

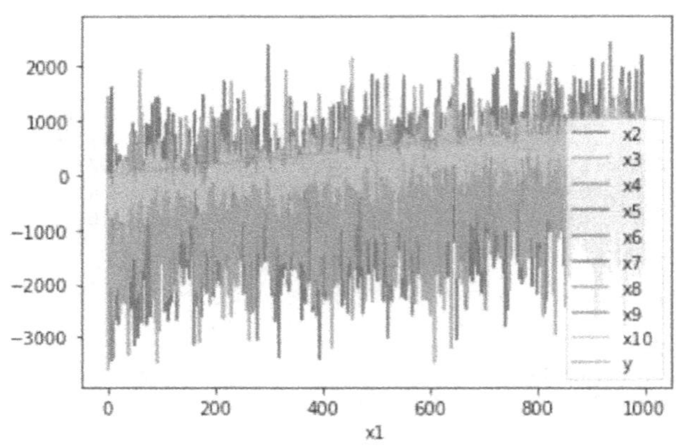

Figure 9.5 – Using the default pandas DataFrame.plot() method to look at correlations

If this plot was all we had to work with, we might conclude everything is correlated to `x1` because all the variables seem to trend similarly upward, and with the noise and overlap, it's hard to discern more detail. Because there are limited plot types and controls available in the `Pandas .plot()` methods, it's useful to access some additional libraries for more complex plotting. The `seaborn` library offers some nice methods for multiple variables; we'll use a couple here. In the following code, we use the `.corr()` function to get the correlation coefficients between all the `x` variables; note that the `.drop(columns = ['y'])` tells `Pandas` to remove the `y` column before sending the data to the `.corr()` function. We then load the `seaborn` library and use the `.heatmap()` method, passing the `corr DataFrame` to it:

```
corr = multi_coll_data.drop(columns = ['y']).corr()
import seaborn as sns
plt.figure(figsize = (11, 11))
sns.heatmap(corr, square = True)
```

This generates a grid, as follows:

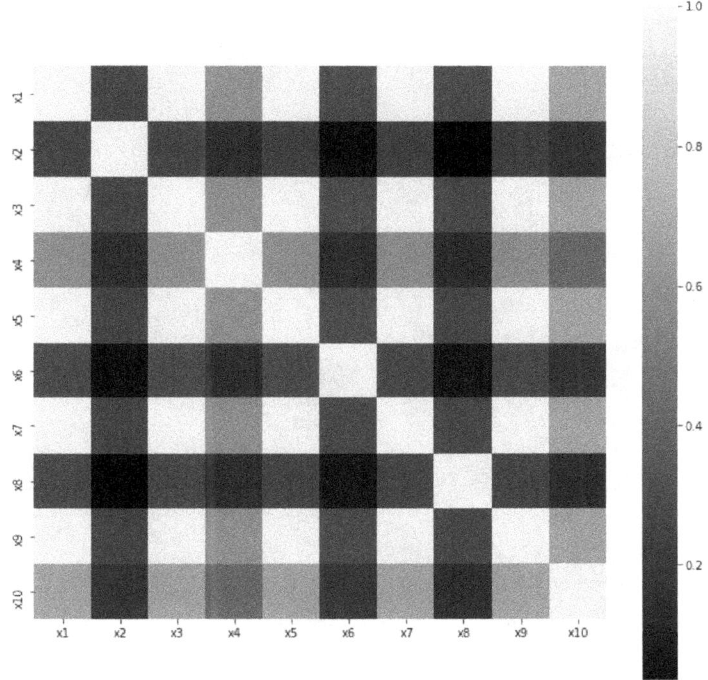

Figure 9.6 – The seaborn heatmap of the correlations among the x variables

Since we added varying levels of non-zero noise to the even-numbered columns (`X2`, `X4`, `X6`, `X8`, and `X10`), it makes sense that they are less correlated to the other variables (the darkest squares are near 0 correlation). Note that `x4` and `x10` are more correlated than the others; that's why we used `np.random.uniform(0, 1)`, which (randomly) in some cases reduced the noise. Since all the columns were highly correlated initially, the higher the noise, the less the final correlation.

We can approach this another way with `seaborn` and look at the data versus the correlation values directly. Here, we use the `.pairplot()` method, again using only the x columns. Pair plots are very useful in data visualization, as you can look at the pairwise relationships among many variables at one time and see those that stand out. By default, the `seaborn.pairplot()` method plots each pair in the associated grid intersection and plots the distribution of a variable on the corresponding diagonal cell:

```
plt.figure(figsize = (11, 11))
sns.pairplot(multi_coll_data.drop(columns = ['y']))
```

This produces the following visualization:

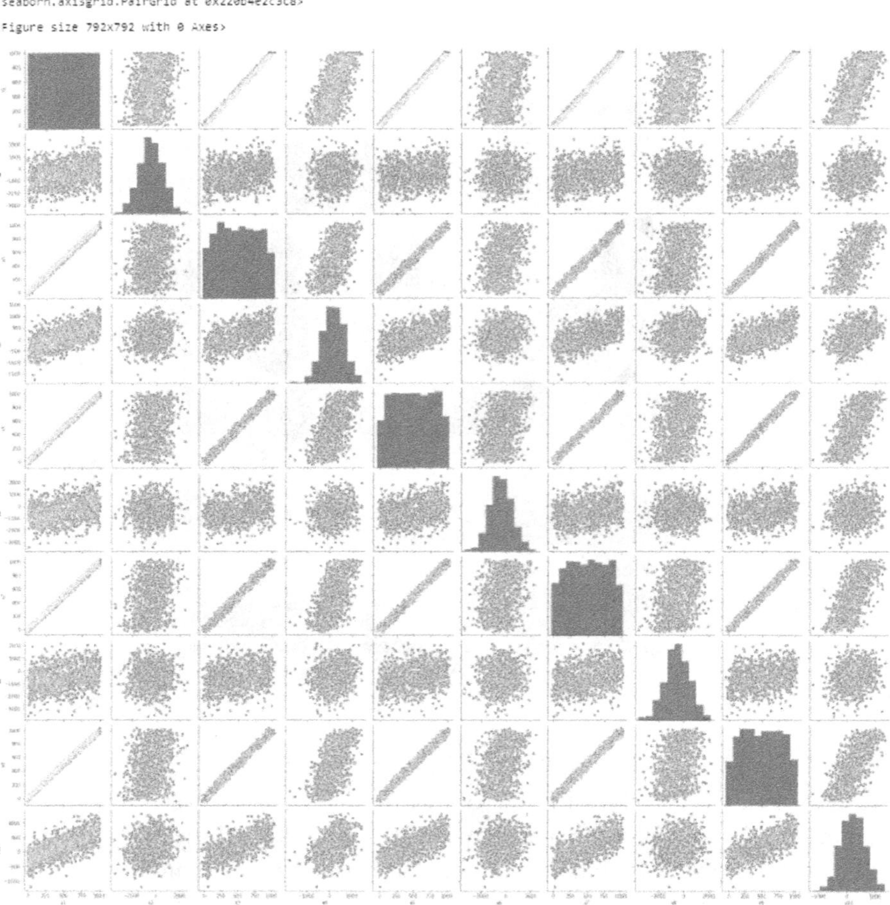
Out[74]: <seaborn.axisgrid.PairGrid at 0x220b4e2c3c8>
<Figure size 792x792 with 0 Axes>

Figure 9.7 – The seaborn pair plot of our x data

In the `pairplot()` method, each variable is a row and a column. On the diagonal are histograms of the individual variables, and in the upper and lower triangle are the plots of each variable against the others. Note that in *Figure 9.7*, the upper-right charts have the same information as those in the lower left because they just have the variables on the opposite axes. Just by looking, we see some are nearly straight lines, indicating high correlation; some are fat lines with lots of noise, which indicates some but lower correlation than before; and some seem to be blobs, indicating little or no correlation. This makes pairwise correlations very easy to see and provides more information than the heatmap, as it depicts the histogram of the plot as well.

In this section, we've seen how linear regression may be impacted by correlation (collinearity) among the independent variables, as well as some easy but powerful ways to inspect even larger datasets for these properties. This sort of review is typically one of the early stages of so-called **exploratory data analysis**, where an analyst "looks" at the data and tries to get a sense of the behaviors and relationships, using methods such as correlation analysis and visualization. The next step is to prepare for analyzing data by modeling. In the next section, we will discuss methods to organize data so that we can validate models.

Training, validation, and test splits of data

So far, we have used very simple and artificial examples to highlight the concepts of independent and dependent variables, and correlation and multicollinearity. However, real data is often more complicated, with truly random noise and often unknown or hidden factors that affect the behavior of the dependent variable but may not be fully represented in the independent variables. This means we may be asked to build models of the data that are based on incomplete or noisy information. We would like to not only build good models but also be able to understand the expected performance (such as the **accuracy**) of predictions we will make in the future.

Noise and missing information notwithstanding, in many cases, it is possible to fit a model that essentially memorizes the data and provides very accurate values of the dependent variable for each instance in our data. Suppose we are asked to predict sales for 10 stores next month and are given sales and pricing data from those stores for the last 3 months, on a daily basis. We know that store managers change pricing based on competitor pricing and recent sales trends, as well as planned changes (such as promotions or clearances). However, in our data, we do not have the competitor pricing or the management changes - we only have whatever the prices were and the sales each day. Any model we make, no matter how well it matches the data we are given, does not "know" the competitor pricing, the timing, or the amount of price changes that were in response to competitors.

With this sales forecasting scenario, over the limited time period, we may still be able to build a model that fits the past data nearly perfectly. While initially that may sound great, it can present a serious problem – when we want to make predictions of the dependent variable using newly collected data for the independent variables, the model may not perform as well. The basic concept of how to address this is to fit the model on only some of the available data and hold some data back as a test of model performance. The following figure illustrates the main ideas of splitting data this way and evaluating models:

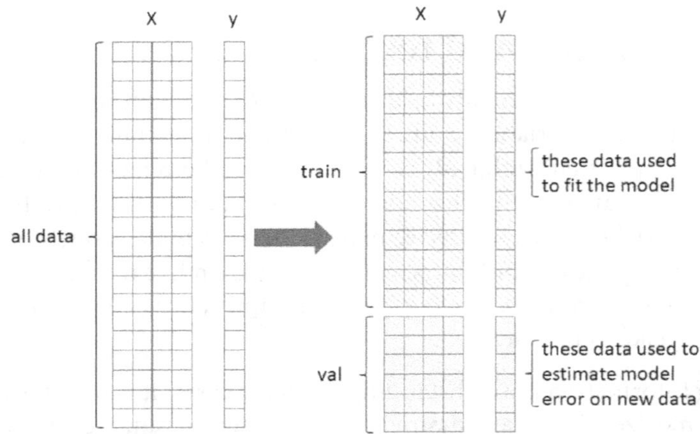

Figure 9.8 – Splitting data into train and validation sets to evaluate model performance

In *Figure 9.8*, you can see that the original data is split into two sets; one is used to fit the model, and the other is used to evaluate the model performance. Although shown as a split that preserves the order of the data, in practice the split is usually done randomly. You might wonder why the first split is labeled **train**. So far, we have used simple models that don't have adjustable parameters (hyperparameters) and are deterministic – if we provide the same data, we will get the same model coefficients. More complex models have adjustable hyperparameters and also require multiple iterations to find a solution for the coefficients. The process of iteratively finding the best coefficients is called *training* the model, and it's common to read about models *learning* from the data. Hence, the data which we use for this iterative step is called training data, or the train split.

Let's begin exploring data splitting using an example. Here, we use `Pandas` to read a data file containing information on miles-per-gallon performance of various cars, along with some specifications of the cars:

```
import pandas as pd
my_data = pd.read_csv('Datasets/auto-mpg.data.csv')
my_data.head()
```

The preceding snippet produces the following output:

Out[3]:

	mpg	cyl	disp	hp	weight	accel	my	name
0	18.0	8	307.0	130	3504	12.0	70	chevrolet chevelle malibu
1	15.0	8	350.0	165	3693	11.5	70	buick skylark 320
2	18.0	8	318.0	150	3436	11.0	70	plymouth satellite
3	16.0	8	304.0	150	3433	12.0	70	amc rebel sst
4	17.0	8	302.0	140	3449	10.5	70	ford torino

Figure 9.9 – The car mileage dataset

> **Note**
> In your code, change the highlighted path as needed to match where you stored the data.

This data is adapted from the UCI Data Repository car mileage data (`https://archive.ics.uci.edu/ml/machine-learning-databases/auto-mpg/`). The data consist of numeric variables and vehicle names; the numeric variables provide the miles per gallon (`mpg`) for the particular car, along with several variables that might be useful to predict mpg (the number of cylinders (`cyl`), engine displacement in cubic inches (`disp`), engine power in horsepower (`hp`), the vehicle weight in pounds (`weight`), the acceleration in seconds to reach 60 miles per hour (`accel`), and the model year of the car with an assumed 19 prefix (`my`)).

Earlier, we used `seaborn` to look at data; here, we show another approach – defining a function to plot all the histograms. The following code loops over the variables we pass in, checks to see whether the bins (the number of slices in the histogram) are too many and adjusts accordingly, then uses a `Pandas .hist()` method (which uses `matplotlib`) to plot the **histogram** in its grid location, and adds a per-chart title showing the variable. We call the function passing in a DataFrame, the variables we wish to plot, rows and columns for the grid, and the number of bins. We use the `Pandas` slice notation (`[:-1]`) to pass all but the last column as our data (It doesn't make sense to "plot" the vehicle names). Note that for some variables, there may be only a few unique values, which is why the function modifies the bins in those cases.

You can attempt this yourself and compare it with the notebook on GitHub: `https://github.com/PacktWorkshops/The-Pandas-Workshop/blob/master/Chapter09/Examples.ipynb`.

This will result in the following output:

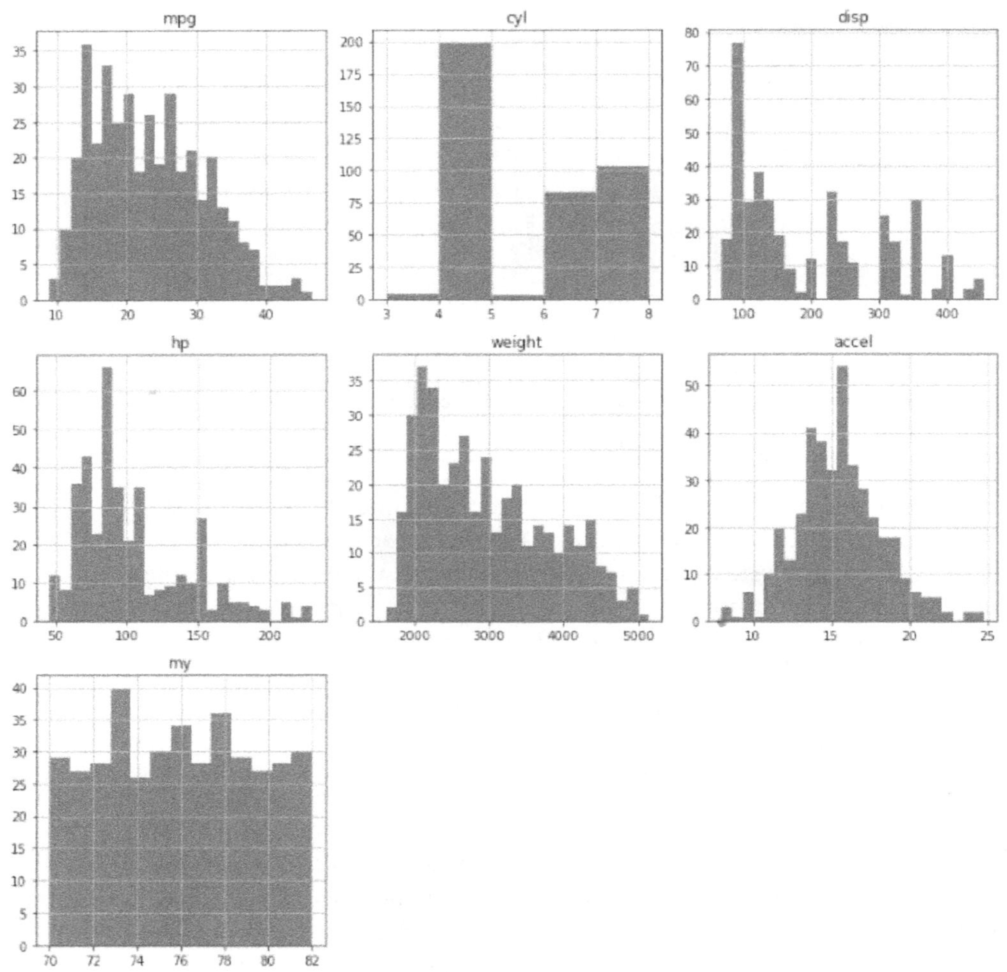

Figure 9.10 – Histograms of the car mileage data

In each histogram in *Figure 9.10*, the x axis shows the values over which the particular variable ranges, and the y-values are the number of data points that fall in a given "bin." For example, in the last histogram, for the model year, the bins are 1 year wide (the axis is labeled at every other bin), so there are, for example, 29 cars in the dataset for the model year 1970, the first bin on the histogram. We see, for example, that the model years are uniformly represented in the data and that acceleration is somewhat like a normal distribution of values. The most common number of cylinders is **4**, followed by **8**. The other variables are skewed to the right.

Similar to how we used `seaborn` before, we can make another function to make pair plots. In this code, we again loop over our variables, but we don't make the upper-right charts, which are relatively redundant. The `.plot()` `Pandas` method is used to make the scatter plots, and we create a title showing the variables in the chart.

You can check out the code here: `https://github.com/PacktWorkshops/The-Pandas-Workshop/blob/master/Chapter09/Examples.ipynb`.

Running the preceding snippet will lead to the following output:

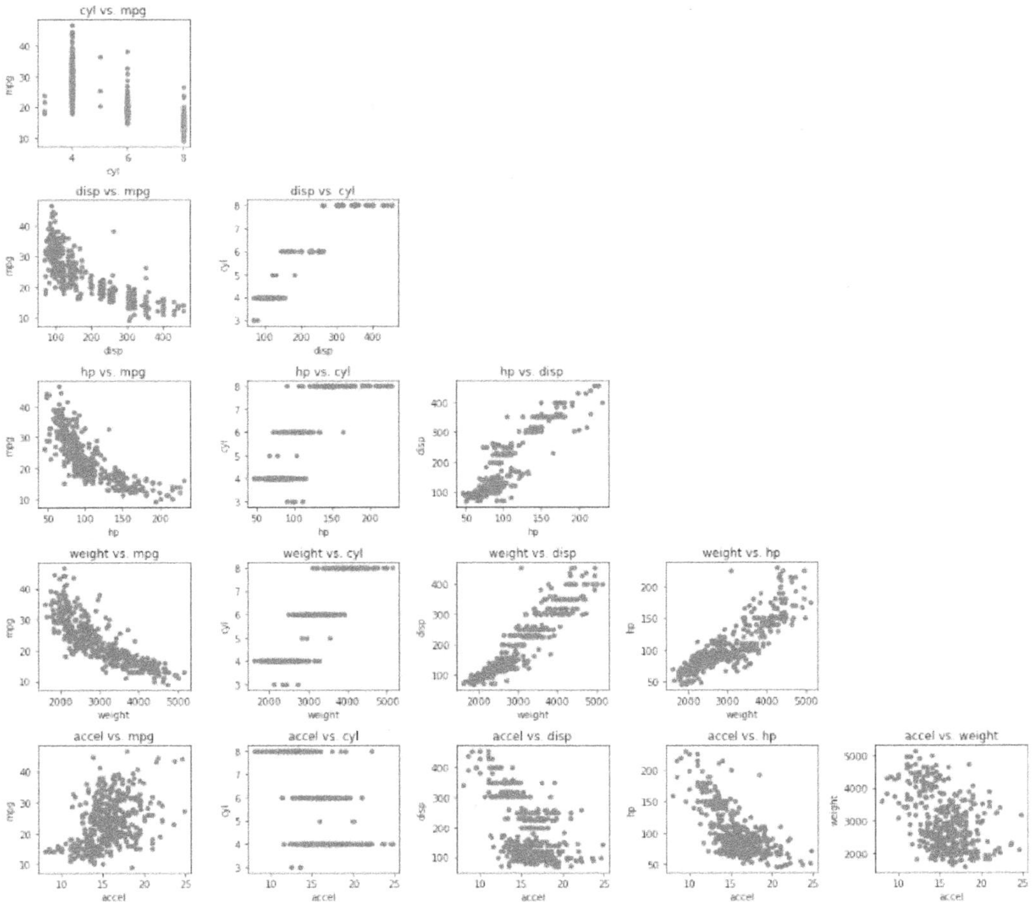

Figure 9.11 – The correlation scatterplots of the car mileage data

We can see that several variables appear highly correlated, which isn't too surprising. For example, in the third row on the right of *Figure 9.11*, we see the chart of horsepower versus displacement, and generally, horsepower increases with displacement. Another interesting feature is that heavier cars tend to have more horsepower.

Just as we did before, we can build a simple linear regression model for this data. However, now that we have seen the statistical output of `statsmodels`, let's use the predictive modeling interface provided by `sklearn`. We first set a random seed as before, then we create two arrays – the first, called `train`, randomly samples 70 percent of the index values for `my_data`, and the second, called `validation`, contains the other values. In the code here, `X_train` and `y_train` are then created using the `train` array. We use the `.random.choice()` `numpy` method to get the 70% split and then simply pick the index values that are not in `train` for the validation split. For the `X` data, we drop the target (`mpg`) and the `name` column, and for the `y` data, we take just the `mpg` column:

```
import numpy as np
np.random.seed(42)
train = np.random.choice(my_data.shape[0], int(0.7 * my_data.
shape[0]))
validation = [i for i in range(my_data.shape[0]) if i not in
train]
X_train = my_data.iloc[train, :].drop(columns = ['name',
'mpg'])
y_train = np.reshape(np.array(my_data.loc[train, 'mpg']), (-1,
1))
```

You might wonder where the 70/30 split comes from. In fact, there is no fixed rule for the split proportions; if the data is very large, it may be acceptable to use even more than 70% to fit the model. On the other hand, if the dataset has limited rows, the split might be chosen to ensure a reasonable number of rows in the validation split. With limited data, it can also arise that by random chance, the train split has a different distribution of the variables than the validation set. For now, we'll assume the 70/30 split is sufficient.

Here, we import the `LinearRegression` class from `sklearn.linear_model`, and we name it `OLS` for convenience. We create an instance of `OLS` in `lin_model` and then call the `.fit()` method, passing our `X` and `y` data. Finally, we print out the key results:

```
from sklearn.linear_model import LinearRegression as OLS
lin_model = OLS()
my_model = lin_model.fit(X, y)
print('R2 score is ', my_model.score(X, y))
print('model coefficients:\n', my_model.coef_, '\nintercept: ',
my_model.intercept_)
```

This generates the following:

```
R2 score is  0.831869958782409
model coefficients:
 [[-3.53519873e-01 -4.91464180e-04 -1.15484755e-02 -6.08231188e-03
   2.60263994e-02  6.81342318e-01]]
intercept:  [-7.066461]
```

Figure 9.12 – The linear regression model to predict mileage

Using what we already learned about the `statsmodels` `OLS` method, we see that for the sample of the data we used, we are accounting for about 83% of the variation in mileage with this simple model.

Commonly, we are interested in the errors a model makes. Error is usually the difference between the predicted value and the target value and is often called **residual** error, and the set of errors for all the data points are the **residuals**. Intuitively, we want models that make the errors as small as possible and given some set of constraints. An error measure that is frequently useful in data models is the **Root Mean Squared Error** (**RMSE**). This is computed by squaring the error at each point, averaging those squares, and then taking the square root. We can get RMSE easily from `sklearn`. Here, we import the `mean_squared_error` `sklearn` function, then we create `RMSE` by calling `mean_squared_error`, and pass it the target (`y`) and the predictions, which we get using the `.predict()` method on our fitted model. `squared = False` tells the method to take the square root of the result:

```
from sklearn.metrics import mean_squared_error
RMSE = mean_squared_error(y, my_model.predict(X), squared =
False)
print('the root mean square error is ', RMSE)
```

This should result in output as follows:

```
the root mean square error is  3.2361376539382127
```

A nice property of RMSE is that it is in units of `y` – in this case, mpg. So, we see that our model predicted the mileage with an error of 3.2 mpg, expressed as the RMSE. So, what about data we didn't use to fit our model? In data analysis, it's best practice to test a model on data not used to fit it. In this case, we split it into two sets – 70% was used to fit the model. Generally, that set is called `train`, or `X_train`. When we have two sets, the other retained data is commonly called **validation** data, although it can be called test data. Here, we create the validation set, using the `oos` row vector we created earlier, then we use `val_X` to make predictions, and compute `val_RMSE`, comparing the validation predictions to `y_val`:

```
X_val = my_data.iloc[oos, :].drop(columns = ['name', 'mpg'])
y_val = my_data.loc[oos, 'mpg']
val_pred = my_model.predict(X_val)
val_RMSE = mean_squared_error(val_pred, y_val, squared = False)
print('the validation RMSE is ', val_RMSE)
```

Output for this snippet is as follows:

```
the validation RMSE is  3.530822072558969
```

We see that the error is 3.5 mpg, or 10% worse than the predictions in the `train` data. This demonstrates an important reason to hold out the validation set – we would report our expected error on future predictions to be 3.5 mpg as the RMSE, instead of the 3.2 we obtained on the `train` data. To summarize:

- Training data: 70% of original data, chosen at random, with the RMSE = 3.2 mpg
- Validation data: 30% of data after selecting the 70% (hence, also random), with the RMSE = 3.5 mpg

Generally, we would say that the 3.5 mpg result is a better estimate of the future performance of the model. Let's say we received the necessary features of a new model year of cars and wanted to estimate their mileage. We would use our model but state the results with an estimate of the RMSE error of 3.5 mpg, not 3.2 mpg. This is the value in splitting the validation set, as it gives us a more realistic error estimate.

So far, we have been using linear regression (ordinary least squares), and there have been no arbitrary parameters other than the 70/30 split. In more complex models, there are many adjustable parameters, usually called **hyperparameters**, as they are chosen before a given model is fit. In such cases, it's common to try a range of all the hyperparameters in some sort of search to find the best model. The test of what is the best model is done using the validation set. However, this introduces another issue, which is that by fitting many models and choosing the one that gives the best validation performance, the model "learns" about the nature of the validation data. Therefore, this data is no longer a good test of future model performance. This leads to the best practice of making three splits – train, validation, and test. The following figure illustrates that; compare it to *Figure 9.8*:

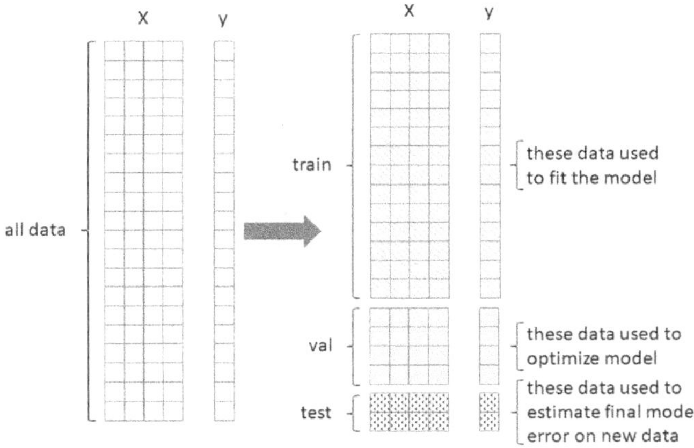

Figure 9.13 – Using three splits instead of two

It's important to note that the test split in *Figure 9.13* is only used to evaluate and report the expected model performance. If it is used to choose the model, then it is no longer a valid estimate of future performance. The following exercise will reinforce using three splits. Another consideration occurs when splitting **time series** data. Usually, with time-series data, you want to predict the future. If you randomly sample data into train and validation, for example, your training set will likely have points that overlap the times of points in the validation data. This is a case of **information leakage**, which will be discussed in more detail later. Generally, for time series, you split by time, not randomly, and use the most recent times to optimize models and test their performance. The second part of the following exercise addresses that case.

Exercise 9.01 – Creating training, validation, and test data

This exercise is in two parts; the first part is tabular data where time isn't important, and the second part uses time series data – data that is ordered in time. In the first part, you are given weather data for Austin, Texas, and want to organize train, validation, and test splits of the data, with the target being "`Event`", which includes values such as Rain, Fog, and Thunderstorm:

1. For this exercise, all you will need is the `pandas` library, a module from `sklearn`, and `matplotlib`. Load them in the first cell of the notebook:

```
import pandas as pd
from sklearn.model_selection import train_test_split
import matplotlib.pyplot as plt
```

   We are going to use the `sklearn` `train_test_split()` method instead of manually splitting the data.

2. Read the `austin_weather.csv` file into a DataFrame called `weather_data`:

```
weather_data = pd.read_csv('Datasets\\austin_weather.
csv')
weather_data.head()
```

   This should produce the following output:

Out[2]:

	Date	TempHighF	TempAvgF	TempLowF	DewPointHighF	DewPointAvgF	DewPointLowF	HumidityHighPercent	HumidityAvgPercent	HumidityLowPercent	...
0	2013-12-21	74	60	45	67	49	43	93	75	57	...
1	2013-12-22	56	48	39	43	36	28	93	68	43	...
2	2013-12-23	58	45	32	31	27	23	76	52	27	...
3	2013-12-24	61	46	31	36	28	21	89	56	22	...
4	2013-12-25	58	50	41	44	40	36	86	71	56	...

5 rows × 21 columns

Figure 9.14 – The weather dataset

3. As the target is the `Event` type, you decide to ignore the date and use just the numeric data in the model. So, you drop the `Date` column. The target variable is the last column, `Events`. Display all the unique values in `Events`:

```
weather_data.drop(columns = ['Date'], inplace = True)
weather_data['Events'].unique()
```

This produces the following output:

```
Out[12]: array(['Rain , Thunderstorm', ' ', 'Rain', 'Fog', 'Rain , Snow',
       'Fog , Rain', 'Thunderstorm', 'Fog , Rain , Thunderstorm',
       'Fog , Thunderstorm'], dtype=object)
```

Figure 9.15 – The values of Events

We want to replace `' '` with `'None'`. Make that change to the original dataset using the `.replace()` pandas method:

```
weather_data['Events'].replace(' ', 'None', inplace = True)
weather_data.head()
```

We should see the change in the `Events` column:

Out[14]:

ressureLowInches	VisibilityHighMiles	VisibilityAvgMiles	VisibilityLowMiles	WindHighMPH	WindAvgMPH	WindGustMPH	PrecipitationSumInches	Events
29.59	10	7	2	20	4	31	0.46	Rain , Thunderstorm
29.87	10	10	5	16	6	25	0	None
30.41	10	10	10	8	3	12	0	None
30.3	10	10	7	12	4	20	0	None
30.27	10	10	7	10	2	16	T	None

Figure 9.16 – The updated Events values

4. Now, split the data randomly into train/validation/test sets with splits of 0.7/0.2/0.1 respectively. Use the `sklearn` `train_test_split()` method. This method can take multiple inputs and return multiple outputs. You want to pass in `X` and `y` – `X` is `weather_data` without the `Events` column, and `y` is the `Events` column. The method can make one split on one or two datasets, so by passing `X` and `y` and specifying the split, we get back four datasets – `X` becomes `train_X` and `val_X`, and `y` becomes `train_y` and `val_y`. The `train_size` variable is the fraction of the original data to sample into the train set, and the `test_size` is the fraction to put in the other split. These can be specified separately, so you specify `train` as 0.7, and `test` as 0.2, which becomes the validation split. Since 0.7 plus 0.2 is 0.9, there is still a fraction of 0.1 remaining, which you use as the test split. This is done by using the `Pandas` index of the resulting `train_X` and `val_X`, and then dropping all those rows to leave the test sets. Once you have made the splits, verify the results. You can check out the code for this here: `https://github.com/PacktWorkshops/The-Pandas-Workshop/blob/master/Chapter09/Exercise9.01.ipynb`.

5. To verify, calculate the percentages and use the `.intersection()` pandas method to compare the index of each pair of values. The result should be as follows:

```
train set is 69.98%
val set is 20.02%
```

```
test set is 10.01%
train rows in val set:   []
train rows in test set:   []
val rows in test set:   []
```

Note that the three row comparisons all return empty lists and the actual percentages are very close to what we asked for (often, the percentages will not be exact due to a finite number of points in the data).

6. Display the first five rows of the `val` set:

```
val_X.head()
```

Out[7]:

	TempHighF	TempAvgF	TempLowF	DewPointHighF	DewPointAvgF	DewPointLowF	HumidityHighPercent	HumidityAvgPercent	HumidityLowPercent	SeaL
677	81	66	51	64	54	49	96	66	35	
1046	91	81	71	73	71	64	100	72	44	
610	101	89	76	76	72	65	94	64	33	
49	65	51	37	42	36	29	85	63	40	
1284	91	81	71	74	72	67	100	75	50	

Figure 9.17 – The resulting validation set

At this point, you have the train, validation, and test splits of the weather data. You can move on to modeling. Here, we will instead perform a similar process on another dataset, which is time series data, meaning the data is ordered by time and we want to retain that information.

7. In this second part of the exercise, you want to analyze some stock closing price data for the S&P 500. Load `spx.csv` to a DataFrame called `stock_data`:

```
stock_data = pd.read_csv('Datasets\\spx.csv')
stock_data.date = pd.to_datetime(stock_data.date)
stock_data.head()
```

You should see the following output:

Out[17]:

	date	close
0	1986-01-02	209.59
1	1986-01-03	210.88
2	1986-01-06	210.65
3	1986-01-07	213.80
4	1986-01-08	207.97

Figure 9.18 – SPX stock data

8. Split the stock data into a training and validation set. Because it is a time series, we want to preserve the order and use the most recent data for validation. In this case, use the last 9 months of data as the validation set, and use training data after December 31, 2009. First, inspect the dates so that we can determine the cutoff date:

```
stock_data['date'].describe()
```

You should see the following output:

```
count                  8192
unique                 8192
top        1989-12-27 00:00:00
freq                      1
first      1986-01-02 00:00:00
last       2018-06-29 00:00:00
Name: date, dtype: object
```

9. Now, perform the split. Since the data runs to the end of June 2018, the validation split should be from October 1, 2017 to the end:

```
train_data = stock_data[(stock_data['date'] < '2017-10-
01') & (stock_data['date'] > '2009-12-31')]
val_data = stock_data[stock_data['date'] >= '2017-10-01']
```

10. Now, visualize the resulting `train` and `val` sets. Create a line plot, label each series, and make the validation series a different color to highlight it, resulting in the following output (you can check out the code at `https://github.com/PacktWorkshops/The-Pandas-Workshop/blob/master/Chapter09/Exercise9.01.ipynb`):

Figure 9.19 – The resulting data splits

In this exercise, you've learned a couple of ways to split data into train, validation, and test sets. The `train_test_split sklearn` method makes this easy for tabular data, while for time series, you split the data manually to ensure no overlap of dates in the splits. The resulting splits will then be ready to begin exploring modeling options. You will learn more about modeling later in this chapter and the next. In the next section, we will look more closely at information leakage and how to avoid it.

> **Note**
>
> You can find the code for this exercise here: `https://github.com/PacktWorkshops/The-Pandas-Workshop/blob/master/Chapter09/Exercise9.01.ipynb`.

Avoiding information leakage

The process of finding an optimized model often requires searching over many possible values of adjustable model parameters. In our simple linear regression examples so far, we have not worked with adjustable parameters, but in more general cases, most models (even linear regression – more on that later) do have parameters that can be adjusted before or during model training. Let's suppose we have two adjustable parameters that could each take on, independently, any of three possible values. Therefore, to test all possible models, in this simplified case, we would have to train nine models using training data. Since we will choose the model that performs best on the validation set, we actually "leak" information about the validation data to the model, due to us optimizing the model for the validation set.

In data science, the general way to avoid such leakage is to split data into three parts – training, validation, and test sets, as you did in the first exercise. The training set is as we have stated already, the validation set is what is used to optimize the model, and the test is used only once – to estimate the performance of the model on unseen data. You might be tempted in such a scenario to look at test set performance along the way, but that can lead to consciously or unconsciously optimizing the model for the test set, which defeats the purpose.

A very common split scheme is 70% training, 20% validation, and 10% test, which we used in the earlier exercise. You can see this may lead to problems with small datasets. For example, if you revisit the car mpg data from the beginning of this section and make a histogram of the model year for the train and validation sets, you will see there are some significant differences. Here, we use the `Pandas .plot()` method on each dataset:

```
X_train.my.plot(kind = 'hist', alpha = 0.5)
plt.show()
X_val.my.plot(kind = 'hist', alpha = 0.5)
plt.show()
```

This produces the following:

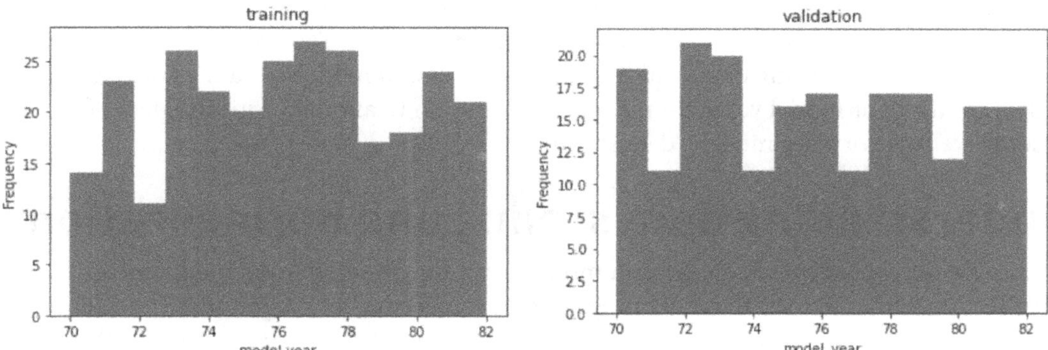

Figure 9.20 – A histogram of the model year from the train set (left) and the validation set (right)

You can see in *Figure 9.20* that there is a smaller proportion of the model year 1970 but a higher proportion of the model year 1971 in the train set versus the validation set, along with other smaller differences. If the model year is very important to predict mpg, then the train data and validation data may not be representative of the entire population, giving distorted model results.

Therefore, using three splits is not universal, and the percentages are not a fixed rule either. The split that we used in the previous exercise might mainly be applied in a production setting, where understanding the performance of the model on unseen or new data is very important. According to articles and popular media, it is common that you will see only a training/validation split (sometimes referred to as training/test splits).

Complete model validation

The processes we have described so far are only part of a fully robust and rigorous model validation approach. It may have occurred to you that a random selection of data splits can affect results. We've already touched on the repeatability issues around the seed for the random number generator; however, since we might expect variation in our results with different random splits, a methodology called cross-validation has become fairly standard in data science. We won't go into the details and nuances of model validation here, but briefly, cross-validation involves repeating the splitting process multiple times and averaging results to better describe how a model performs.

The general problem of optimizing models to perform well on unseen data is referred to as **generalization**, and it is considered that models that do a good job making predictions on unseen data *generalize* well. Understanding how and why some models generalize better than others is an active area of work within data science. You have now learned the basic elements of exploring data, building simple regression models, and the core elements of data splitting for model validation. Now, let's move on to another critical aspect of creating well-performing models, data scaling.

Understanding data scaling and normalization

If we inspect the coefficients of our `mpg` model, from the *Training, validation, and test splits* section earlier, we see that they range over several orders of magnitude. The code here iterates over the variable names and coefficient values, taking advantage of Python's `.enumerate()` method that iterates over the column names but also returns a counter, which we capture in `coef` and use to index the model coefficients. For reference, the code prints the range of the variable in data used to fit the model:

```
print('var\t  coef\t\t\t    range')
for coef, var in enumerate(my_data.columns[1:-1]):
    print(var, '\t', round(my_model.coef_[0][coef], 5),
          '\twith range ', round(float(my_data[var].max() -
                                  my_data[var].min()), 2)
```

You should see the following output for this code:

```
var          coef             range
cyl           -0.35352     with range  5.0
disp           -0.00049     with range  387.0
hp           -0.01155     with range  184.0
weight       -0.00608     with range  3527.0
accel       0.02603     with range  16.8
my           0.68134     with range  12.0
```

Although many models can handle raw data over many orders of magnitude, and even non-numerical data, it can help interpret coefficients to scale data before modeling. Also, in some models, scaling data in various ways may be required, and even when not, it may improve model performance. Scaling refers to adjusting the range of values of data, such as by subtracting the mean value, then dividing by the range (max minus min) of the data. Such a scale transformation makes the data have a mean of 0 and a range of 1. However, the relative distribution of the values is unchanged. Scaling data can avoid extremely large or small model coefficients, which in some cases can degrade model performance.

Different ways to Scale Data

`Pandas` doesn't offer direct methods to scale data but works well with `sklearn`, which has a number of methods for this purpose. The most typical way to scale data is to pass a DataFrame to a `sklearn` method. We'll see how to do this shortly. In addition, you can always scale data yourself if desired, and we'll demonstrate one way to do that before covering the `sklearn` methods.

Scaling data yourself

Let's take a look at what scaling manually using code might look like. In the following, you will apply what is called min/max scaling to the car mpg data from the previous section. To store the information used for scaling, we create a dictionary; we then step through each column in the DataFrame, collect the min, max, and range of the column in the dictionary using the `.update()` method, apply scaling to each column, and then print out the results. In simple min/max scaling, we subtract the minimum value and then divide by the range, so the data will range from 0 to 1 after scaling:

```
scales = dict()
X = my_data.iloc[train, 1:-1]
for col in my_data.columns[1:-1]:
    min = my_data[col].min()
    max = my_data[col].max()
    range = max - min
    scales.update({col : dict({'Xmin' : my_min,
                               'Xmax' : my_max,
                               'Xrange' : my_range})})
    X[col] = (my_data[col] - min) / range
scales = pd.DataFrame.from_dict(scales).T
print(scales)
X.describe().T
```

Running this code will lead to the following output:

```
           Xmin     Xmax  Xrange
cyl         3.0      8.0     5.0
disp       71.0    455.0   384.0
hp         48.0    230.0   182.0
weight   1613.0   5140.0  3527.0
accel       9.5     23.7    14.2
my         70.0     82.0    12.0
```

Out[50]:

	count	mean	std	min	25%	50%	75%	max
cyl	274.0	0.494161	0.329783	0.0	0.200000	0.300000	0.600000	1.0
disp	274.0	0.318041	0.260051	0.0	0.088542	0.208333	0.486979	1.0
hp	274.0	0.298187	0.196083	0.0	0.148352	0.258242	0.340659	1.0
weight	274.0	0.384055	0.237395	0.0	0.182733	0.339665	0.575631	1.0
accel	274.0	0.440552	0.183957	0.0	0.316901	0.443662	0.563380	1.0
my	274.0	0.519161	0.298829	0.0	0.250000	0.500000	0.750000	1.0

Figure 9.21 – The results of manually scaling data

In *Figure 9.21*, the first table shows all the values used to scale each column. The second table shows a summary of the data after scaling. So, the first row shows cylinders that range from 3 to 8 with a range of 5, and in the second table, we see that after subtracting 3 and dividing by 5, **cyl** now ranges from 0 to 1. Recall from the previous section that some of the variable distributions were skewed; there is evidence of that, where the mean values are considerably different from 0.5 – for example, **hp** has a mean of 0.3, even though it ranges from 0 to 1, a direct result of the distribution of the data.

If we wanted to scale to a different range, we'd have to add a bit more code to handle that. The general equation for min/max scaling is, where `X` is the individual data values, `Xmin` is the column minimum, `Xmax` is the column maximum, and `max` and `min` are the values we want the data to range across after scaling:

$$X_{scaled} = \frac{(X - X_{min}) * (\max - \min)}{X_{max} - X_{min}}$$

Note that when the desired value of `max` is 1 and `min` is 0, the term on the right side of the numerator is 1 and can be ignored.

We skipped the column with names, as we aren't using it and scaling would not apply anyway. We've also not scaled the target (mpg), as in general, there is no reason to do so for a continuous target variable. We created a Python dictionary to store the scaling parameters. We would want that to use later to scale back to real units if needed and to store for use in a data processing pipeline.

> **Note**
>
> After using scaled or transformed data to train a model, to make predictions with the model, any new data must be scaled exactly the same way as the training data. Therefore, we need to save scaling parameters or some other way to scale new data. If we scaled new data from scratch using its min and max values, we would get incorrect predictions from the model, since it would likely have different properties compared to the training data.

Min/max scaling

Let's look at fitting our mpg data again, this time scaling the data using the MinMaxScaler() sklearn method. Here, X is generated again from the original my_data DataFrame and then passed as a DataFrame to MinMaxScaler(). By passing no values for min and max, the MinMaxScaler defaults to scaling to (0, 1):

```
from sklearn.preprocessing import MinMaxScaler
scaler = MinMaxScaler()
X = my_data.iloc[train, 1:-1]
scaler.fit(X)
X_scaled = scaler.transform(X)
X_scaled = pd.DataFrame(X_scaled)
X_scaled.columns = my_data.columns[1:-1]
X_scaled.describe().T
```

Running the preceding snippet will result in the following output:

Out[52]:

	count	mean	std	min	25%	50%	75%	max
cyl	274.0	0.494161	0.329783	0.0	0.200000	0.300000	0.600000	1.0
disp	274.0	0.318041	0.260051	0.0	0.086542	0.208333	0.486979	1.0
hp	274.0	0.298187	0.196083	0.0	0.148352	0.258242	0.340659	1.0
weight	274.0	0.384055	0.237395	0.0	0.182733	0.339665	0.575631	1.0
accel	274.0	0.440552	0.183957	0.0	0.316901	0.443662	0.563380	1.0
my	274.0	0.519161	0.298829	0.0	0.250000	0.500000	0.750000	1.0

Figure 9.22 – Scaled X data

The data are all now scaled between 0 and 1, and so *Figure 9.22* is identical to *Figure 9.21*. Note how much easier this was in terms of code. Using this method, we created an object, `scaler`, which stores the scaling model after scaling the data. To get the scaling parameters, we call the `.fit()` method using `X`, which is analogous to using `.fit()` with other models, as we did before for `LinearRegression()`. The scaled data is then generated using the `.transform()` method with the fitted `scaler`. To get the parameters, we access attributes of the fitted `scaler`:

```
print(scaler.data_range_)
print(scaler.data_min_)
```

You should see the following output:

```
[   5.    384.    182.   3527.      14.2    12. ]
[   3.     71.     48.   1613.       9.5    70. ]
```

The parameters are returned as `Numpy` arrays, so if we wanted to associate them with names, we'd have to do that by getting the variable (column) names from the original data. Also, note that the result of the `.transform()` method is returned as a `Numpy` array as well, so we converted it back to a `Pandas DataFrame` and put the column names back for convenience.

A min/max Scaling use case – neural networks

Although there can be various reasons to use this scaling, and various benefits, one particular case deserves mention. In artificial neural networks, the model is constructed of an array of so-called neurons, which are actually math functions that, in general, perform summation and then a non-linear transformation. The non-linear transformation function for each neuron is called the **activation function**. A common activation function is the Sigmoid function, shown here:

$$f(x) = \frac{e^x}{e^x + 1}$$

This function transforms the input (x) into a smooth function that is bounded between 0 and 1, as shown here:

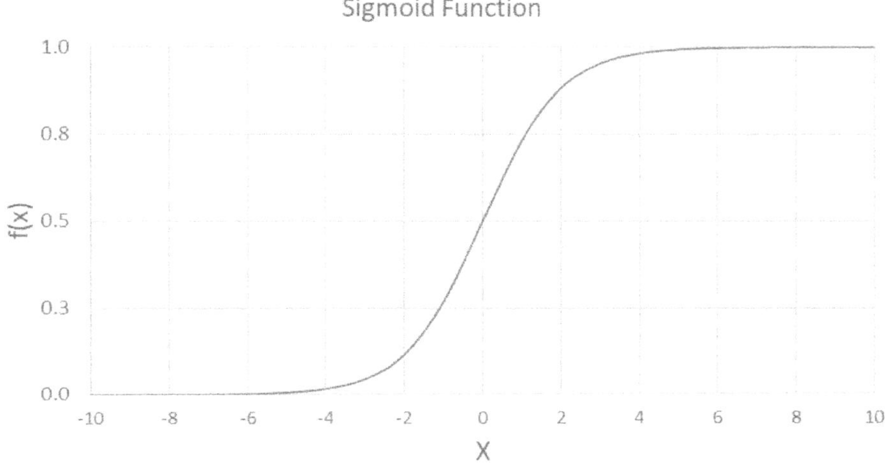

Figure 9.23 – The Sigmoid function

It is easy to see that large negative values are all transformed to 0, and large positive values are all transformed to 1. This means if we use unscaled `X` values to begin with, the values of the neurons can be "saturated," meaning they get stuck at 0 or 1, and we lose the ability to fine-tune the parameters of the network. In fact, many other models sometimes benefit from various scaling or other transformations, so it's important to be familiar with them. Now, let's revisit the linear regression model we used before, now using the scaled data:

```
lin_model = OLS()
y = np.reshape(np.array(my_data.loc[train, 'mpg']), (-1, 1))
my_model = lin_model.fit(X_scaled, y)
print('var\t   coef\t\t\t     range\t\t       impact')
for coef, var in enumerate(my_data.columns[1:-1]):
    print(var, '\t', round(my_model.coef_[0][coef], 5),
          '\twith range ', round(float(X_scaled[var].max() -
                                        X_scaled[var].min()),
2),
          '\ttotal impact', round(float(my_model.coef_[0][coef]
*
                                         (X_scaled[var].max() -
                                          X_scaled[var].min())),
2))
```

This code generates the following output:

```
Var            coef                    range
cyl             -1.7676          with range  1.0
disp              -0.18872          with range  1.0
hp             -2.10182          with range  1.0
weight         -21.45231      with range  1.0
accel        0.36957          with range  1.0
my             8.17611           with range  1.0
```

You can see that the coefficient values are different from before, but you can verify the performance is the same as before, using the `score` method you used earlier:

```
my_model.score(X_scaled, y)
```

This produces the same R2 value as before:

```
0.8318699587824089
```

Standardization – addressing variance

There are a number of methods to scale data. In the previous section, we used `MinMaxScaler()` with defaults, which scales from 0 to 1. We could also have added the parameters for the `min` and `max` values we wanted to change the scaling. Another common scaling approach for preparing data for modeling is **standardization**. Standardization is implemented in the `sklearn StandardScaler()` method and, by default, transforms numeric data to have a mean of 0 and a **standard deviation** of 1. This has the benefit of preserving information about the degree of scatter in the data, while transforming the values to be more similar across different underlying scales.

As an example, you can visualize three distributions that have the same mean but different standard deviations, where the standard deviation of a variable is a measure of the scatter in the data. Here, we load the data from `distributions.csv` and then visualize them. The code makes a grid of subplots in one row, with the number of columns in the data, and then iterates over the columns to plot the histograms:

```
distributions = pd.read_csv('Datasets\\distributions.csv')
fig, ax = plt.subplots(1, distributions.shape[1],
                       figsize = (15, 3),
                       sharey = True)
for i in range(distributions.shape[1]):
```

```
    _ = ax[i].hist(distributions.iloc[:, i], bins = 50)
    _ = ax[i].set_title('variable ' + str(i))
_ = ax[0].set_ylabel('count')
plt.show()
```

The code produces the following:

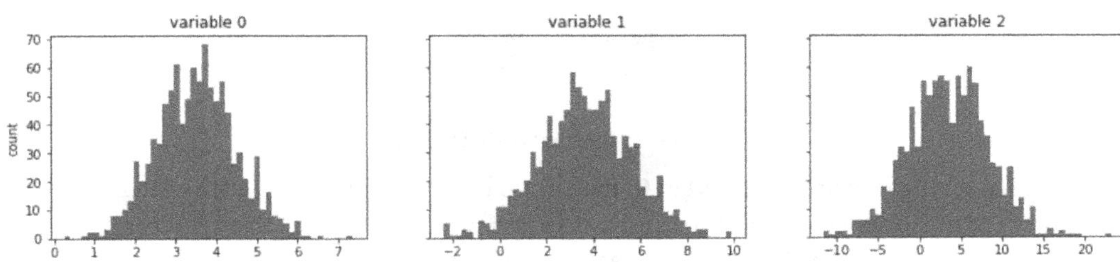

Figure 9.24 – Three data distributions with the same mean

We can now scale these distributions with `.StandardScaler()` to get a clear picture of the transformation. The code here instantiates `scaler`, fits it and transforms the `DataFrame` in one step (using `.fit_transform()`), and then we repeat the same visualization. We have to save and reapply the column names because `.StandardScaler()` returns a `numpy` array, not a `DataFrame`:

```
from sklearn.preprocessing import StandardScaler
scaler = StandardScaler()
colnames = distributions.columns
distributions = pd.DataFrame(scaler.fit_
transform(distributions))
distributions.columns = colnames
fig, ax = plt.subplots(1, distributions.shape[1],
                       figsize = (15, 3),
                       sharey = True)
for i in range(distributions.shape[1]):
    _ = ax[i].hist(distributions.iloc[:, i], bins = 50)
    _ = ax[i].set_title('variable ' + str(i))
_ = ax[0].set_ylabel('count')
plt.show()
```

Note that the usage of `.StandardScaler()` with default parameters is identical to the usage of `MinMaxScaler()`. The code produces the following:

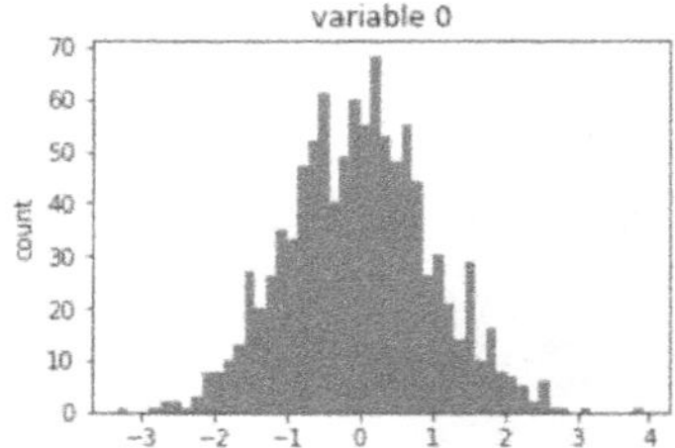

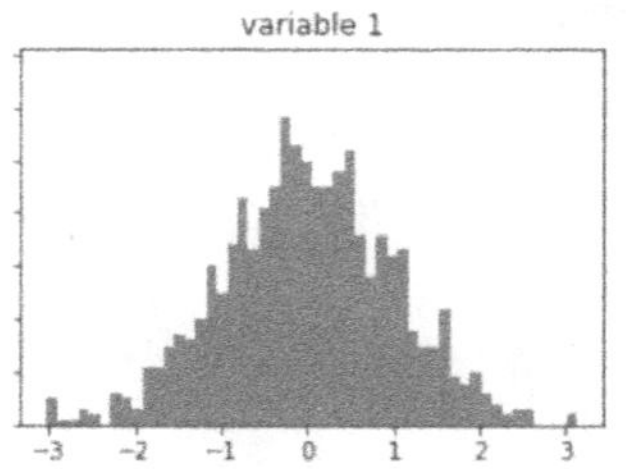

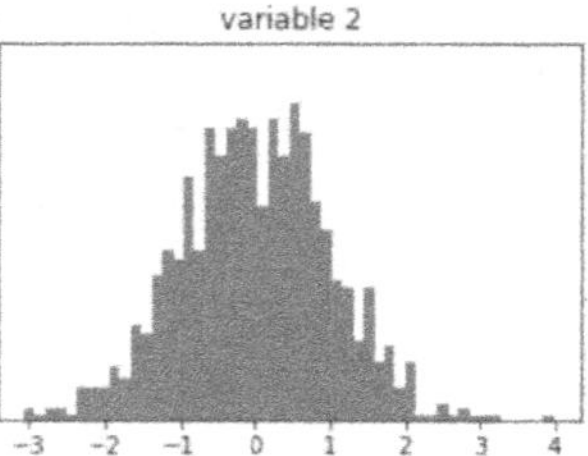

Figure 9.25 – The scaled distributions

You can see that all three are centered at **0**; it's a little less obvious that the standard deviation of each is one. Here, the `Pandas .describe()` method is used to compare the scaled data. First, set the `Pandas` display option for floats to make the output easier to read (fewer digits):

```
pd.set_option('display.float_format', lambda x: '%.2f' % x)
distributions.describe().T
```

This produces the following:

```
     count      mean      std      min      25%      50%      75%      max
values_1     1000.00      -0.00      1.00      -3.33      -0.68      0.01
0.64     3.92
values_2     1000.00      -0.00      1.00      -3.02      -0.68      -0.01
0.66     3.13
values_3     1000.00     0.00     1.00     -3.08     -0.67     -0.01
0.67     3.99
```

You can see that the means are 0 and the standard deviations are 1. However, note that the `min` and `max` values are not all the same – `.StandardScaler()` encodes the amount of scatter and retains the point-to-point relationships, by scaling to a fixed standard deviation instead of `min` and `max`. Thus, `.StandardScaler()` typically retains more information about the original data than `.MinMaxScaler()` but still aligns data to more similar scales.

Now, reload the `mpg` data and then scale as before, simply replacing `MinMaxScaler()` with `StandardScaler()`. As before, convert the result to a `DataFrame`, then restore the column names, and inspect the result using `.describe().T`:

```
my_data = pd.read_csv('Datasets\\auto-mpg.data.csv')
X = my_data.iloc[sample, :].drop(columns = ['name', 'mpg'])
scaler = StandardScaler()
scaler.fit(X)
X_scaled = scaler.transform(X)
X_scaled = pd.DataFrame(X_scaled)
X_scaled.columns = my_data.columns[1:-1]
X_scaled.describe().T
```

Out[87]:

	count	mean	std	min	25%	50%	75%	max
cyl	274.00	0.00	1.00	-1.50	-0.89	-0.59	0.32	1.54
disp	274.00	0.00	1.00	-1.23	-0.88	-0.42	0.65	2.63
hp	274.00	-0.00	1.00	-1.52	-0.77	-0.20	0.22	3.59
weight	274.00	0.00	1.00	-1.62	-0.85	-0.19	0.81	2.60
accel	274.00	-0.00	1.00	-2.40	-0.67	0.02	0.67	3.05
my	274.00	0.00	1.00	-1.74	-0.90	-0.06	0.77	1.61

Figure 9.26 – The results of using StandardScaler on the car mpg dataset

As expected, the variable means are 0 and the standard deviations are 1. Now, repeat the same simple linear model as before to compare the model coefficients and model performance. Note that compared to the earlier code, this is exactly the same, except that `X_train` is replaced by `X_scaled`:

```
y = np.reshape(np.array(my_data.loc[train, 'mpg']), (-1, 1))
lin_model = OLS()
my_model = lin_model.fit(X_scaled, y)
print('R2 score is ', my_model.score(X_scaled, y))
print('model coefficients:\n', my_model.coef_, '\nintercept: ',
my_model.intercept_)
RMSE = mean_squared_error(y, my_model.predict(X_scaled),
squared = False)
print('the root mean square error is ', RMSE)
```

Running this results in the following output:

```
R2 score is  0.831869958782409
model coefficients:
 [[-0.58185994 -0.0489877  -0.41137864 -5.08336838  0.06786155  2.438796  ]]
intercept:  [24.02262774]
the root mean square error is  3.2361376539382127
```

Figure 9.27 – The results of a simple linear model using standardized data

We can see that these results, as far as R2 and RMSE, are the same as those we obtained earlier. At this point, you may wonder why you are bothering with scaling, since you keep getting the same results. At the beginning of *Understanding data scaling and normalization*, it was noted that some models, including linear regression, are not sensitive to data scaling. Your goal so far has been to understand the different scaling methods, and using linear regression demonstrates that we are not changing the fundamental nature of the answer we are seeking. However, most more-complex models, such as RandomForest, Extreme Gradient Boosting, and neural networks will benefit from scaling data. (You will work with a RandomForest regression model in *Chapter 10, Data Modeling – Model Basics*.) You should now be confident that you can use different scaling methods and compare model results with and without scaling. The last topic in data scaling is how to transform data back to the original units, which is covered in the next section.

Transforming back to real units

What if we wanted to get our `X` data back into the original units? To do this, we simply need to invert the equation for the transform and apply it using the scaling parameters. We've already seen the general equation for the `min/max` transform. Here is the equation for the standardization transform:

$$X_{scaled} = \frac{(X - \mu)}{s}$$

Here, μ is the mean of the data, and `s` is the standard deviation. However, if we use the `sklearn` methods, we don't have to do this ourselves, as illustrated by the following snippet. Here, we use the method from the `scaler`, `.inverse_transform()`, to restore the car data we transformed in the previous section. As with `.transform()` or `.fit_transform()`, the result is a `numpy` array, so we have to convert back to a `DataFrame` and restore the column names:

```
X = scaler.inverse_transform(X_scaled)
X = pd.DataFrame(X)
```

```
X.columns = my_data.columns[1:-1]
X.head()
```

You'll see the following output upon running this code:

Out[32]:

	cyl	disp	hp	weight	accel	my
0	8.00	400.00	150.00	4997.00	14.00	73.00
1	4.00	98.00	65.00	2380.00	20.70	81.00
2	4.00	151.00	85.00	2855.00	17.60	78.00
3	6.00	232.00	100.00	2789.00	15.00	73.00
4	8.00	304.00	150.00	3892.00	12.50	72.00

Figure 9.28 – The X data transformed back to original units

This concludes our introduction to scaling data for modeling. We've seen along the way how to construct simple linear regression models. Now, let's look into tools in `Pandas` as well as some additional `sklearn` methods that are useful for data modeling.

Exercise 9.02 – Scaling and normalizing data

The `Pandas DataFrame` structure makes it easy to apply functions to subsets of columns of data. In this exercise, you will use such functionality to scale data. We choose to scale the data because we want to have a common dataset, regardless of the model we choose. Here, you will work again with the weather data from the Austin weather dataset. You need to prepare the data prior to considering models to predict the events. You will load the data, address some issues with the data types, and then apply a scaler to transform the data:

1. For this exercise, all you will need is the `pandas` library, `numpy`, two modules from `sklearn`, and `matplotlib`. Load them in the first cell of the notebook:

   ```
   import pandas as pd
   from sklearn.model_selection import train_test_split
   from sklearn.preprocessing import StandardScaler
   import matplotlib.pyplot as plt
   import numpy as np
   ```

   You are going to use the `sklearn StandardScaler()` method to scale data as preparation for modeling:

2. It's a good practice to look at data before scaling, so you want to implement the `utility` function seen earlier in the chapter to plot a grid of histograms. The following code loops over the variables you pass in, checks to see whether there are too many bins (the number of slices in the histogram) and adjusts accordingly, uses a `Pandas .hist()` method (which uses `matplotlib`) to plot the histogram in its grid location, and adds a per-chart title that shows the variable. You call the function by passing in a `DataFrame`, the variables you wish to plot, rows and columns for the grid, and the number of bins. The `Pandas` slice notation (`[:-1]`) is used to pass all but the last column as your data (it doesn't make sense to "plot" the vehicle names). Note that for some variables, there may be only a few unique values, which is why the function modifies the bins in those cases:

```
def plot_histogram_grid(df, variables, n_rows, n_cols,
bins):
    fig = plt.figure(figsize = (11, 11))
    for i, var_name in enumerate(variables):
        ax = fig.add_subplot(n_rows, n_cols, i + 1)
        if len(np.unique(df[var_name])) <= bins:
          use_bins = len(np.unique(df[var_name]))
        else:
          use_bins = bins
        df[var_name].hist(bins = use_bins, ax = ax)
        ax.set_title(var_name)
    fig.tight_layout()
    plt.show()
```

3. Now, load the `austin_weather.csv` file into a `DataFrame` called `weather_data`, change `Events` as we did before, and inspect the result:

```
weather_data = pd.read_csv('Datasets\\austin_weather.
csv')
weather_data.drop(columns = ['Date'], inplace = True)
weather_data['Events'] = ['None'
                          if weather_data['Events'][i] is
' '
                         else weather_data['Events'][i]
                         for i in range(weather_data.
shape[0])]
weather_data.describe().T
```

The result should be as follows:

Out[3]:

	count	mean	std	min	25%	50%	75%	max
TempHighF	1319.0	80.862775	14.766523	32.0	72.0	83.0	92.0	107.0
TempAvgF	1319.0	70.642911	14.045904	29.0	62.0	73.0	83.0	93.0
TempLowF	1319.0	59.902957	14.190648	19.0	49.0	63.0	73.0	81.0

Figure 9.29 – Using the .describe() method on the data

4. From the preceding output, you can see that most of the columns were not read in numerically, as only the `TempHighF`, `TempAvgF` and `TempLowF` columns are present in the `describe` result. If there were a new, unknown dataset. you'd have to do more EDA to investigate what is in the data and how to address it. In this case, the issue is caused by the use of `'-'` to represent missing data and the `T` value in precipitation columns to represent 'trace'. Use the `Pandas .replace( )` method to replace `'-'` with `np.nan` and `T` with `0`. After the replacement, print a list of rows with missing data.

```
weather_data.iloc[:, :-1] = \
    weather_data.iloc[:, :-1].replace(['-', 'T'],
                                      [np.nan, 0]).
astype(float)
print(weather_data.loc[weather_data.isna().any(axis = 1),
:].index)
```

Running this code will result in the following output:

```
Int64Index([174, 175, 176, 177, 596, 597, 598, 638, 639,
741, 742, 953,
            1001, 1107],
           dtype='int64')
```

Here, the data columns are sliced using `:-1` for the columns in `.iloc[]`, which skips the `Events` column, and then `.replace()` is used to change the values. The `Pandas .replace()` method can take lists for the things to replace and the replacement values, which means both `'-'` and `T` at the same time. The `na_rows` code uses the `Pandas .isnna()` method, which creates a `DataFrame` the same shape as what is passed, with `True` or `False` in it, and then the `.any(axis = 1)` method chooses any element where the value is `True`, and by passing `axis = 1`, that gives us the rows (look across the rows for any `True` values). Finally, we extract the index values with `.index` and just print the result. You can see that there aren't very many with missing values now, so dropping those rows is a good approach.

5. Drop the rows with missing values, verify the result using `.describe().T`, and then plot histograms of all the variables using the `utility` function. Use the `Pandas .dropna()` method with `axis = 0`, telling the method to drop rows with missing values. Before printing, change the `Pandas` float format to 2 digits to make the output easier to read:

```
weather_data.dropna(axis = 0, inplace = True)
pd.set_option('display.float_format', lambda x: '%.2f' %
x)
print(weather_data.describe().T)
```

The output should be similar to the following:

	count	mean	std	min	25%	50%	75%	max
TempHighF	1305.00	80.79	14.71	32.00	72.00	83.00	92.00	107.00
TempAvgF	1305.00	70.56	14.01	29.00	62.00	73.00	83.00	93.00
TempLowF	1305.00	59.82	14.19	19.00	49.00	62.00	73.00	81.00
DewPointHighF	1305.00	61.52	13.58	13.00	53.00	66.00	73.00	80.00
DewPointAvgF	1305.00	56.64	14.86	8.00	46.00	61.00	69.00	76.00
DewPointLowF	1305.00	50.94	16.19	2.00	38.00	56.00	65.00	75.00
HumidityHighPercent	1305.00	87.83	11.05	37.00	85.00	90.00	94.00	100.00
HumidityAvgPercent	1305.00	66.66	12.50	27.00	59.00	67.00	74.00	97.00
HumidityLowPercent	1305.00	44.98	17.01	10.00	33.00	44.00	55.00	93.00
SeaLevelPressureHighInches	1305.00	30.11	0.18	29.63	29.99	30.08	30.21	30.83
SeaLevelPressureAvgInches	1305.00	30.02	0.17	29.55	29.91	30.00	30.10	30.74
SeaLevelPressureLowInches	1305.00	29.93	0.17	29.41	29.82	29.91	30.02	30.61
VisibilityHighMiles	1305.00	9.99	0.16	5.00	10.00	10.00	10.00	10.00
VisibilityAvgMiles	1305.00	9.16	1.46	2.00	9.00	10.00	10.00	10.00
VisibilityLowMiles	1305.00	6.84	3.68	0.00	3.00	9.00	10.00	10.00
WindHighMPH	1305.00	13.25	3.43	6.00	10.00	13.00	15.00	29.00
WindAvgMPH	1305.00	5.01	2.08	1.00	3.00	5.00	6.00	12.00
WindGustMPH	1305.00	21.38	5.89	9.00	17.00	21.00	25.00	57.00
PrecipitationSumInches	1305.00	0.12	0.43	0.00	0.00	0.00	0.00	5.20

Figure 9.30 – The cleaned data before scaling

6. Visualize the variable distributions using the `utility` function to generate a grid of histograms:

```
plot_histogram_grid(df = weather_data.iloc[:, :-1],
                    varaibles = weather_data.iloc[:,
:-1].columns,
                    n_rows = 5,
                    n_cols = 5,
                    bins = 25)
```

This produces the following:

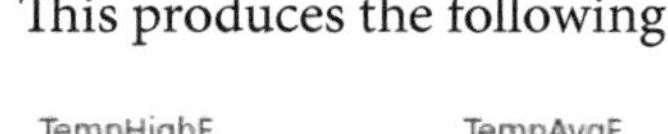

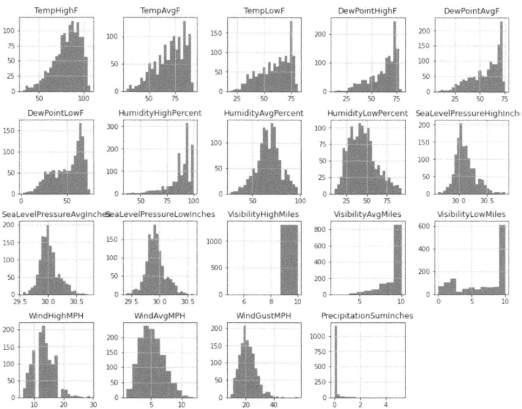

Figure 9.31 – The weather data variables before scaling

You can see some interesting features of the data that might affect modeling; for instance, `PrecipitationSumInches` is mostly 0, and `WindHigh` has an odd gap near 10 MPH. Several of the variables are skewed. As a first step, we'll choose to proceed with scaling the data.

7. Recall from the information leakage discussion that when splitting data into, say, train and validation, it's important to split first and then scale; otherwise, the scaler has information about the train data leaking into the validation data. Split the data 70/30. Use `.train_test_split()`. Remember to split the `y` values (`Events`) as well:

```
train_X, val_X, train_y, val_y = \
    train_test_split(weather_data.drop(columns =
'Events'),
                                 weather_data['Events'],
```

```
                                     train_size = 0.7,
                                     test_size = 0.2,
                                     random_state = 42)
```

8. Now, scale all the numeric data using the `StandardScaler` method, and display the first five rows of the result:

```
scaler = StandardScaler()
scaler = scaler.fit(train_X)
scaled_train = pd.DataFrame(scaler.transform(train_X))
scaled_train.columns = weather_data.columns[:-1]
scaled_val = pd.DataFrame(scaler.transform(val_X))
scaled_val.columns = weather_data.columns[:-1]
scaled_train.head()
```

The result should be as follows:

Out[20]:

	TempHighF	TempAvgF	TempLowF	DewPointHighF	DewPointAvgF	DewPointLowF	HumidityHighPercent	HumidityAvgPercent	HumidityLowPercent	SeaLevel
0	0.83	0.81	0.78	0.99	0.97	0.81	0.57	0.11	-0.23	
1	0.76	0.89	0.92	0.99	1.04	1.18	0.29	0.43	0.42	
2	-1.76	-1.76	-1.69	-2.24	-1.93	-1.73	-1.08	-1.01	-0.76	
3	0.35	0.17	0.01	-0.18	-0.11	-0.25	-0.35	-0.69	-0.76	
4	-0.80	-0.62	-0.35	-0.26	-0.38	-0.18	0.20	0.43	0.48	

Figure 9.32 – The scaled train split of the weather data

At this point, the data is in a form you can use for initial modeling – you have train and validation splits, and the data is scaled. You should be comfortable with the key concepts of addressing missing or incorrectly formatted or typed data, making two or three splits of the data, fitting a scaler to the train data, and then applying the fitted scaler to the validation (and test) split.

> **Note**
>
> You can find the code for this exercise here: `https://github.com/PacktWorkshops/The-Pandas-Workshop/blob/master/Chapter09/Exercise9.02.ipynb`.

Activity 9.01 – Data splitting, scaling, and modeling

You are charged with analyzing the performance of a combined cycle power plant and are given data on the full-load electrical power production along with environmental variables (such as temperature or humidity). In the first part of the activity, you will split the data manually and with `sklearn`, then you will scale the data, construct a simple linear model, and output the results:

1. For this activity, all you will need is the `Pandas` library, the modules from `sklearn`, and `numpy`. Load them in the first cell of the notebook.
2. Use the `power_plant.csv` dataset – `'Datasets\\power_plant.csv'`. Read the data into a `Pandas DataFrame`, print out the shape, and list the first five rows.

   The independent variables are as follows:

   - AT – ambient temperature
   - V – exhaust vacuum level
   - AP – ambient pressure
   - RH – relative humidity

   The dependent variable is EP – electrical power produced.
3. Split the data into a `train`, `val`, and `test` set with fractions of 0.8, 0.1, and 0.1 respectively, using `Python` and `Pandas` but not `sklearn` methods. You will use 0.8 for the train split because there is a large number of rows, so the validation and test splits will still have enough rows.
4. Repeat the split in step 3 but use `train_test_split`. Call it once to split the `train` data, and then call it again to split what remains into `val` and `test`.
5. Ensure that the row counts are correct in all cases.
6. Fit `.StandardScaler()` to the train data from step 3, and then transform `train`, `validation`, and `test X`. Do not transform the `EP` column, as it is the target.
7. Fit a `.LinearRegression()` model to the scaled train data, using the `X` variables to predict `y` (the `EP` column).
8. Print the `R2` score and the `RMSE` of the model on the `train`, `validation`, and `test` datasets.

> **Note**
>
> You can find the solution for this activity in the *Appendix*.

Summary

In this chapter, you learned how to split and scale data for downstream modeling tasks. You now can split data manually if that is appropriate but are also familiar with the `sklearn` methods to simplify the splitting tasks. You also saw how different scaling methods work and learned why `min/max` scaling might be used in some models and standardization in other models. You've seen how to make simple linear regression models, a topic to which we will return in the next chapter. Along the way, you learned why it is important to split data and hold some back from the modeling step in order to measure performance for new data. You now have the basic toolkit for preparing data for modeling, which is where we will begin the next chapter.

10
Data Modeling – Modeling Basics

In this chapter, you will learn how to discover patterns in data using **resampling** and **smoothing**. The `.resample()`, `.rolling()`, and `.ewm()` pandas methods will be introduced and you will learn how to use them to filter out the noise and perform other useful explorations of data series. You will learn how sampling can sometimes include data from future times, which is a problem for predictive modeling, and how to address that. At the end of the chapter, you will see how a combination of scaling (introduced in *Chapter 9*, *Data Modeling – Preprocessing*), and smoothing can show interesting similarities between different data series, which might otherwise be overlooked.

By the end of this chapter, you will be skilled at applying scaling, sampling, and smoothing in a variety of ways to your data analyses.

This chapter covers the following topics:

- Learning the modeling basics
- Predicting future values of time series
- Activity 10.01 – Normalizing and smoothing data

Introduction to data modeling

Data is often provided to you in a form that isn't completely suitable for analysis and modeling. As an example, suppose you are trying to summarize and analyze the sales of students selling cookies in an effort to raise money for a school trip. You would like to get an idea of the expected sales per student per week, in order to recognize students putting in effort and achieving higher sales. Unfortunately, the data for any given student comes in at somewhat random times, making comparisons more difficult. You decide to take each student's sales and fill in the missing days by interpolating between the days for which you have data. The process is quite tedious, and part-way through, you realize you will also have to go back and divide each day by the weekly total, otherwise you are inflating the total sales. Pandas provides the `.resample()` method you saw in *Chapter 9, Data Modeling – Preprocessing*, and by combing that with a `.rolling()` smoothing function and computing a rolling average (using `.mean()`), you can get the desired daily data in one line of code.

This approach can be used in many situations. Sometimes, you think the most recent data is the most informative, and you would like a rolling average that puts more weight on more recent data points. Pandas allows the addition of windowing functions with `.rolling()` to accomplish these goals, as well as the `.ewm()` method (**exponentially** weighted windows) to simplify the variable weighting in some cases. In this chapter, you will see examples of using these methods, as well as the general power of rolling aggregation to smooth data and reveal hidden patterns.

Learning the modeling basics

So far, we've talked about data modeling in a somewhat abstract sense. In this and the next chapter, we will focus on the tools that help us gain insights from data and construct some basic predictive models using that data. We will begin by defining the modeling landscape in more depth, then look at some of the tools provided directly in pandas.

Modeling tools

In *Chapter 9, Data Modeling – Preprocessing*, we introduced the **scikit-learn** (**sklearn**) `LinearRegression` method and showed how to fit a simple multiple linear regression model. While there is a vast range of modeling tools available for Python, sklearn is perhaps one of the most used for everything from regression to classification and even basic neural networks. The sklearn ecosystem is described (see `https://scikit-learn.org/stable/`) as follows:

- Simple and efficient tools for predictive data analysis
- Accessible to everybody, and reusable in various contexts

- Built on NumPy, SciPy, and matplotlib
- Open source, commercially usable – BSD license

The tools are organized into broad categories: classification, regression, and clustering.

You will learn more preprocessing and regression methods in this and the next chapter. Although pandas can be used to prepare data for **classification** and **clustering** methods, we won't go deeper into those here. There is nothing from the data viewpoint in classification or clustering that we have not already covered or will cover in this chapter and the next. We have already seen some elements in regression and preprocessing. In *Chapter 11*, *Regression Models*, we will revisit those methods and more. Here, we'll go deeper into pandas window functions in an important data modeling application, smoothing.

Pandas modeling tools

There are two major groups of functionalities in pandas useful for data modeling: time series functionality and sampling methods. We'll look at each in turn.

There is a range of methods for handling time series and especially manipulating dates / `datetimes` in pandas. What distinguishes time series methods in pandas is that they are *date/time-aware*, meaning they can be used to conduct operations over specific intervals of time or dates and using the times/dates in the data to determine data used in sampling or smoothing intervals. Here, we'll focus only on the `.resample()` method, which allows easy conversion of time-based dates to different periods.

As an example, suppose you are investigating the interesting case of the rapid rise in the global population of persons aged 100+ years (data adapted from `https://population.un.org/wpp/Download/Standard/Interpolated/`). You are given a dataset with dates in yearly increments. Here, we read the data from a file and look at the first rows and the data types:

```
import pandas as pd
pop_data = pd.read_csv('Datasets/world_pop_100_plus.csv')
print(pop_data.head())// print(pop_data.dtypes)
```

This produces the following:

```
        date  population aged 100+ (000)
0  7/1/1950                           34
1  7/1/1951                           31
2  7/1/1952                           29
3  7/1/1953                           27
4  7/1/1954                           25
date                          object
population aged 100+ (000)     int64
dtype: object
```

Figure 10.1 – World population aged 100+ years

Note the data types at the bottom of the output. The `date` column shows the `object` type. That's because, by default, pandas reads the dates as strings and shows it as the `object` type. To use them as a date, we first need to convert them to `datetime`. In the following, we use the `.to_datetime()` pandas method and we provide a format as a string with information for pandas as to how to interpret the string data. We then use `DataFrame` with the updated date values to make a simple plot:

```
pop_data['date'] = pd.to_datetime(pop_data['date'], format =
"%m/%d/%Y")
fig, ax = plt.subplots(figsize = (11, 11))
ax.scatter(pop_data.date, pop_data['population aged 100+
(000)'])
plt.show()
```

This will result in the following output:

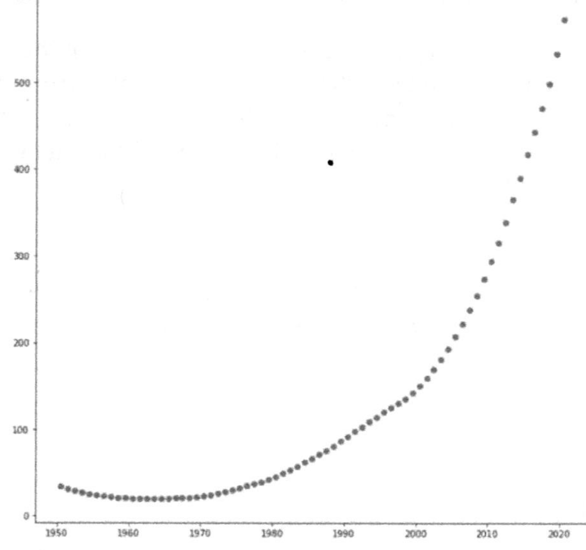

Figure 10.2 – Plot of simple time series data

Notice the use of `format = "%m/%d/%Y"` in the `datetime` conversion. This is a very common pattern and, as you can see, it is fairly intuitive. Pandas provides a range of these % format definitions, and they are combined with a literal string describing how the string looks (such as the slashes that are in the date strings). Also, once the column is converted to `datetime`, `pandas` and `matplotlib` recognize it as a date and apply some automatic formatting in the plot, which is convenient.

Now, suppose you wanted to present 5-year averages instead of the yearly values. Pandas makes such date frequency conversion easy with the `.resample()` method. Here, `.resample()` is used specifying a period of `5Y` for 5 years. The column to use for the dates is specified in `on = 'date'`, and `closed = 'right'` tells pandas to include data that falls on the last date of each period. The `.resample()` method is a form of **aggregation** method, and usually, you add an aggregation function along with such methods. Here, we use `.mean()` at the end, which tells pandas to apply the `.mean()` method to the data in each window after sampling according to the period:

```
five_yr_avg = pop_data.resample('5Y',
                                on = 'date',
                                closed = 'right').mean()
five_yr_avg.head()
```

Running this will show the following output:

Out[16]:

date	population aged 100+ (000)
1950-12-31	34.0
1955-12-31	27.2
1960-12-31	21.4
1965-12-31	20.0
1970-12-31	21.6

Figure 10.3 – Five-year averages of the original data

The `.resample()` method takes a string that defines a sampling period, here `'1W'` to mean one week. Other options include `D` (day), `H` (Hour), `M` (Month), and so on. When using `.resample()`, you can think of the period as the width (in time) of a window in which pandas collects the data, then applies the aggregation function (here, `.mean()`) to the data in each window. The `closed = 'right'` option tells pandas which way the windows should be bound – that is, whether they should include the left value or the right value or `None`. If you consider some data may fall on the boundary of a window, you need to be explicit as to whether those points are in the current window calculation or the next. (Note that a point can only be one or the other window.) Importantly, the default for the closed option can *depend on the data* so, if unsure, either specify what you want or look up the documentation.

The `on = 'date'` option is only required because pandas, by default, uses the index to do the sampling, and if a `datetime` index does not exist already, you must provide a `datetime` column, or you can specify a column to override the index. Not immediately apparent in *Figure 10.3* is that `date` is now `index`, no longer a column. You could get it back as a column, but in most cases, there is no reason to do so, and additional methods will use it, so it's convenient that pandas handles that for you.

Using a similar method, we can get the maximum value in every 5-year period and compare to the averages for each period. Here, we use `.resample()` again, but use `.max()` as the aggregation, then plot both results:

```
five_yr_max = pop_data.resample('5Y', on = 'date', closed =
'right').max()
fig, ax = plt.subplots(figsize = (11, 11))
ax.scatter(five_yr_max.index, five_yr_max['population aged 100+
(000)'],
          label = '5 year maximum')
ax.plot(five_yr_avg.index, five_yr_avg['population aged 100+
(000)'],
        lw = 0.5, color = "red",
       label = '5 year averages')
ax.legend()
plt.show()
```

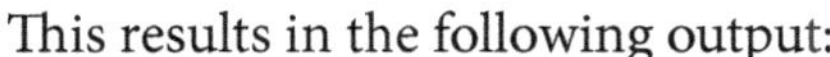
This results in the following output:

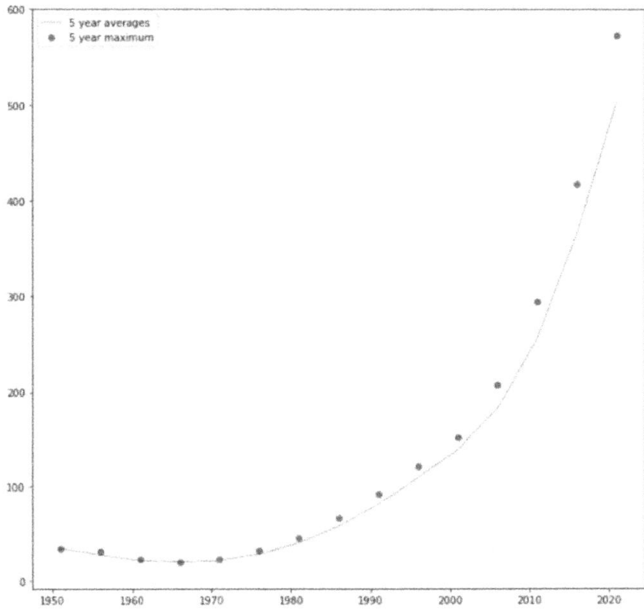

Figure 10.4 – Comparing 5-year maximums (dots) to the averages (line) for the data

You note that it seems as if the rate of increase is also increasing in the recent decades. Let's see how you can use pandas to get the incremental change from period to period. Here, we use the `.pct_change()` pandas method to get the values we want, and again make a simple plot:

```
fig, ax = plt.subplots(figsize = (11, 11))
ax.plot(pop_data.date, 100 * pop_data['population aged 100+
(000)'].pct_change())
ax.set_title('Percent change year to year of\nworld population
aged 100+',
            fontsize = 16)
ax.set_ylabel('Percent change from previous year',
             fontsize = 14)
ax.tick_params(labelsize = 12)
plt.show()
```

The output for this will be as follows:

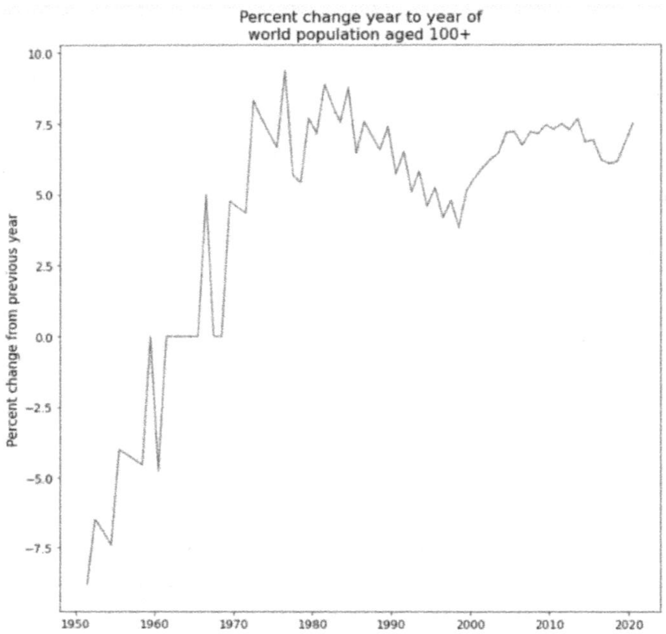

Figure 10.5 – Percent change year to year for the data

You see that the rate of growth peaked around 1980, then declined, and has increased again since 2000. As it was used to generate *Figure 10.5*, `.pct_change()` uses the interval that's already present in the data, not the dates. If the dates were unevenly spaced, it would be preferable to use `.resample()` first to create a uniform date index, then `.pct_change()` would use that uniform index. Note that the data in *Figure 10.5* is noisy and contains some spikes. Perhaps this is a reporting issue with some years getting spikes.

Rather than dig in to find that, let's just repeat this analysis on a basis of 5 years, as follows:

```
five_yr_mean = pop_data.resample('5Y', on = 'date', closed =
'right').mean()
fig, ax = plt.subplots(figsize = (11, 11))
ax.plot(five_yr_mean.index,
        100 * five_yr_mean['population aged 100+ (000)'].pct_
change())
ax.set_title('Percent change for five year periods\nof world
population aged 100+',
             fontsize = 16)
ax.set_ylabel('Percent change from previous 5 years',
```

```
                fontsize = 14)
ax.tick_params(labelsize = 12)
plt.show()
```

This should result in the following output:

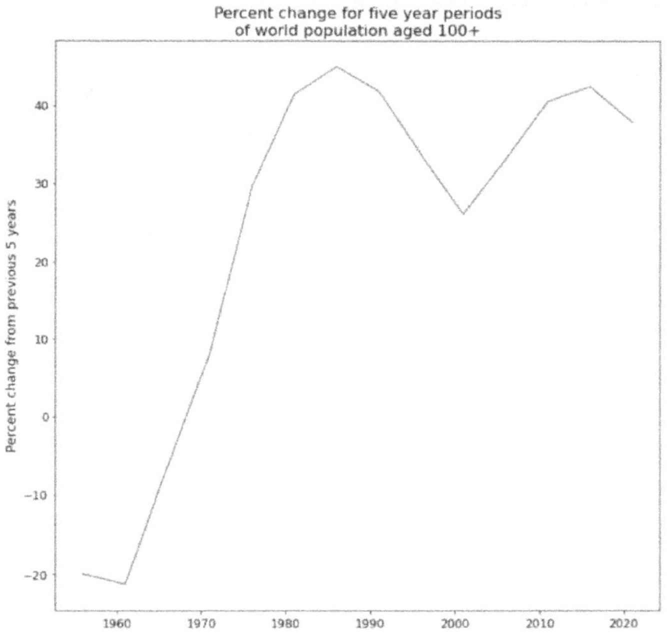

Figure 10.6 – Percent change between 5-year periods

Figure 10.6 is probably a better summary of the changes over time versus the noisier representation in *Figure 10.5*. To recap so far, we are able to read in a population data file with dates as strings and some values, convert the dates to `datetimes` with one pandas method, then use `.resample()` with `.mean()` followed by `.pct_change()` to get a plot that shows a peak around 1985 followed by a dip, and a similar peak recently. Note that you needed roughly eight lines of code to get to a presentation-ready analysis.

It is also possible to **upsample** data using pandas, that is, to increase the frequency compared to the original time intervals. A possible reason to upsample occurs when using multiple data series that are stored with different frequencies, and you don't want to downsample the higher-frequency data. You have to specify a method such as `.ffill()` (to fill forward) to tell pandas how to fill from the last known value, or `.interpolate()` to do some form of interpolation. As these methods know nothing about the underlying data generation process or periodic behavior of the data, you need to exercise caution and judgment as to the best approach.

You can guess that using `.resample()` to **downsample** to a lower frequency (longer period) and then trying to **upsample** loses some information. As an example of using `.resample()` to **upsample** data, here, you resample the 5-year means back to yearly values and plot the year-on-year changes again. Here, the code also plots the original data and adds a legend to make comparison easy. You can check the code here: `https://github.com/PacktWorkshops/The-Pandas-Workshop/blob/master/Chapter10/Examples.ipynb`.

You will see the following output:

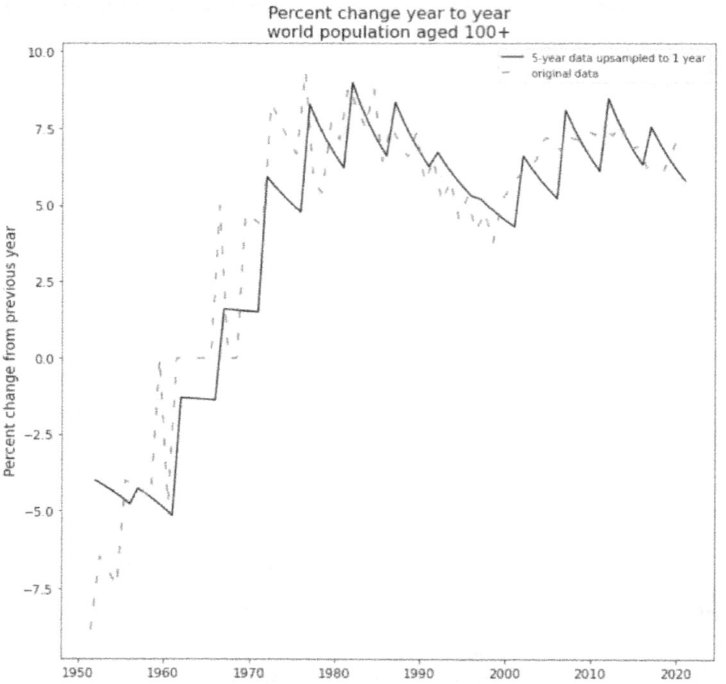

Figure 10.7 – Result of upsampling data that was already downsampled

We can see the new curve is similar to the original but has lost some of the fine structure. Note that you used the `.interpolate()` method with default parameters to fill in the missing values. Most pandas methods have many options you can explore for the method of interpolation, in this case, the default being linear interpolation, which simply connects the previous and next point with a line.

A pandas method related to `.resample()` is `.groupby()`. Like `.resample()`, `.groupby()` gathers the data according to what is passed in the method, and typically an aggregation function (such as `.mean()`) is applied to get a result. Here, the `.dt.year` pandas method is used to extract the year to a new column, then by dividing by 10, rounding, and multiplying by 10 a `'decade'` value is produced. Finally, `.groupby()` is applied to the `'decade'` column, and the means are calculated:

```
pop_data['decade'] = pop_data['date'].dt.year
pop_data['decade'] = round((pop_data['decade'] / 10), 0) * 10
pop_data.groupby('decade').mean()
```

This produces the following:

Out[51]:

decade	population aged 100+ (000)
1950.0	29.200000
1960.0	21.000000
1970.0	24.000000
1980.0	47.000000
1990.0	92.333333
2000.0	156.272727
2010.0	299.222222
2020.0	489.166667

Figure 10.8 – Mean 100+ population by decade

A useful capability of `.groupby()` versus `.resample()` is that `.groupby()` can operate on multiple columns at one time. Suppose you are given a more detailed set of 100+ population data containing the population in each year and the gender. Here, the data is read into a DataFrame, and the `date` column is converted to a datetime as before:

```
pop_data_by_gender = pd.read_csv('Datasets/world_pop_100_plus_
by_gender.csv')
pop_data_by_gender['date'] = pd.to_datetime(pop_data_by_
gender['date'], format = "%m/%d/%Y")
pop_data_by_gender.head(6)
```

This shows the first six lines of the data:

Out[56]:

	date	population aged 100+ (000)	gender
0	1950-07-01	9	male
1	1950-07-01	25	female
2	1951-07-01	8	male
3	1951-07-01	23	female
4	1952-07-01	8	male
5	1952-07-01	21	female

Figure 10.9 – 100+ population with gender

You want to make a summary similar to *Figure 10.8*, but keep men and women separate. Here, the `decade` column is created as before, and `.groupby()` is then called on both '`decade`' and '`gender`' at the same time:

```
pop_data_by_gender['decade'] = pop_data_by_gender['date'].
dt.year
pop_data_by_gender['decade'] = round((pop_data_by_
gender['decade'] / 10), 0) * 10
pop_data_by_gender.groupby(['gender', 'decade']).mean()
```

This produces the following summary table:

Out[83]:

		population aged 100+ (000)
gender	**decade**	
female	**1950**	21.600000
	1960	15.818182
	1970	18.444444
	1980	36.545455
	1990	74.000000
	2000	127.545455
	2010	240.888889
	2020	386.166667
male	**1950**	7.600000
	1960	5.181818
	1970	5.777778
	1980	10.454545
	1990	18.333333
	2000	28.818182
	2010	58.222222
	2020	103.166667

Figure 10.10 – Summary of 100+ population by gender and decade

Note that the result of using `.groupby()` on multiple columns is a pandas **multi-index**. The multi-index was introduced in *Chapter 5, Data Selection – DataFrames*. You can see that females account for more of the increases than males.

Other important pandas methods

Earlier, we discussed the correlation between variables and showed the `.corr()` method. A closely related statistical method is to compute a covariance matrix for multiple variables. Pandas supports this directly with the `.cov()` method. Returning to our car dataset from *Chapter 9, Training, Validation, and Test Splits*, here, `StandardScaler()` is used to first scale the data, then the `.cov()` method is used. As before with `.corr()`, the `.cov()` method returns a `Numpy` array, so we need to restore the columns and row names. Also, we drop the car name column as we can't perform calculations on that:

```
from sklearn.preprocessing import StandardScaler
scaler = StandardScaler()
car_data = pd.read_csv('Datasets/auto-mpg.data.csv')
scaled_data = scaler.fit_transform(car_data.iloc[:, :-1])
scaled_data = pd.DataFrame(scaled_data).cov()
scaled_data.columns = car_data.columns[:-1]
scaled_data.set_index(car_data.columns[:-1], inplace = True)
scaled_data
```

This results in the following output:

Out[25]:

	mpg	cyl	disp	hp	weight	accel	my
mpg	1.002558	-0.779606	-0.807186	-0.780418	-0.834373	0.424411	0.582026
cyl	-0.779606	1.002558	0.953255	0.845139	0.899823	-0.505974	-0.346531
disp	-0.807186	0.953255	1.002558	0.899552	0.935381	-0.545191	-0.370801
hp	-0.780418	0.845139	0.899552	1.002558	0.866749	-0.690958	-0.417426
weight	-0.834373	0.899823	0.935381	0.866749	1.002558	-0.417905	-0.309910
accel	0.424411	-0.505974	-0.545191	-0.690958	-0.417905	1.002558	0.291059
my	0.582026	-0.346531	-0.370801	-0.417426	-0.309910	0.291059	1.002558

Figure 10.11 – Covariance matrix for the car data

As evident, the correlations are as we saw before – cylinders, displacement, horsepower, or weight all decrease mileage, but newer models have better mileage, and oddly, mileage increases with acceleration. (The latter we theorized was due to acceleration being highly correlated to weight; as we can see, they are negatively correlated, meaning less weight means more acceleration.) Note that the `.cov()` method returns the column variances on the diagonal – here, they are close to `1.0` because we scaled to a standard deviation of `1` before using `.cov()`.

Windowing functions

There is one more important group of pandas methods that is very useful in data modeling: **windowing functions**. Earlier, when using `.resample()`, the data was aggregated in non-overlapping regions; data points falling in the 1950s could not also be counted in the 1960s, for example. There are many cases where you would like to aggregate in a window of time or points and move that window smoothly along the data. That's, in essence, what windowing functions can do. Let's begin by looking at the `.rolling()` function and comparing to the `.resample()` method we used before.

You are given data on the unemployment rate by month for the United States for the last 30 years (adapted from `https://data.bls.gov/timeseries/LNS14000000`). Here, you read in the data, and use `.rolling()` and `.resample()` to change from 1 month to a 3-month period:

```
emp_data = pd.read_csv('Datasets/US_unemployment_by_month.csv')
emp_data['date'] = pd.to_datetime(emp_data['date'], format =
'%m/%d/%Y')
rolling = emp_data.rolling(on = 'date', window = '90d').mean()
samples = emp_data.resample(on = 'date', rule = '90d', label =
'right').mean()
```

Note that in the `.rolling()` method, the width of the window is specified by `window`, while in `.resample()`, the period is specified by `rule`. (We did not use `rule =` before; here, we make it explicit to contrast to `window =`.) Also, each of the pandas methods has defaults for how to align the computed interval to the labels. In the case of `.rolling()`, the default is to use the '`right`' or most recent label, while for `.resample()` it is '`left`', so `label = 'right'` was specified in `.resample()` to make the methods comparable.

Now, each series is plotted using different colors and symbols. To see more detail in some points, `.set_xrange()` is used, and we use `.to_datetime()` to easily convert human readable dates to values for the axis limits:

```
fig, ax = plt.subplots(figsize = (13, 9))
ax.scatter(rolling['date'], rolling['unemployment'],
          marker = 'o', color = 'red',
           label = 'rolling unemployment 90 days')
ax.scatter(samples.index, samples['unemployment'],
          marker = '+', color = 'black', s = 300,
           label = 'resampled unemployment 90 days')
ax.set_xlim(pd.to_datetime('1/1/2001'),
            pd.to_datetime('12/31/2002'))
ax.legend()
plt.show()
```

The preceding snippet will lead to the following output:

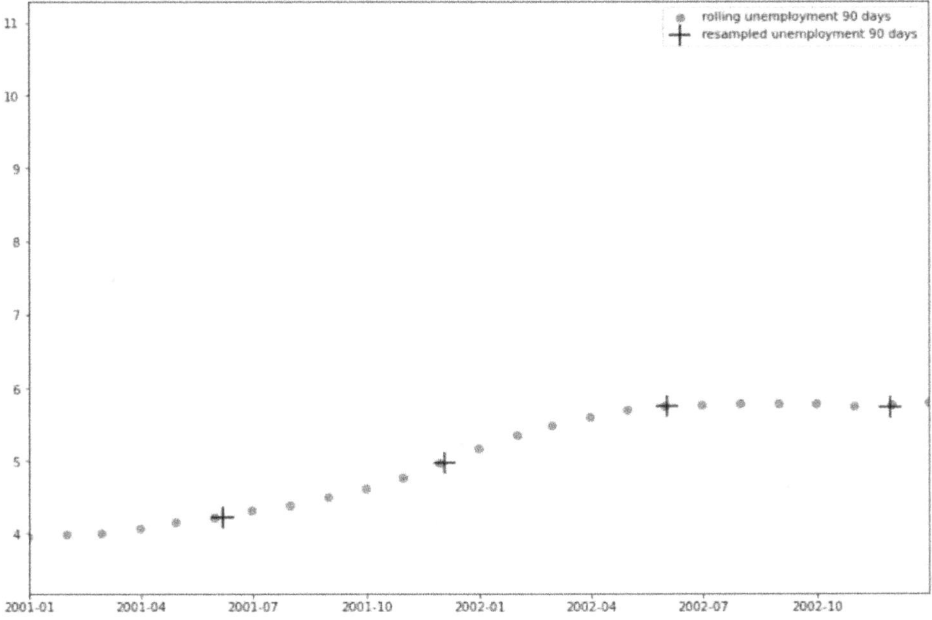

Figure 10.12 – Comparing calculating means using .rolling() and .resample()

You can see from *Figure 10.12* that the first series, using `.rolling()`, has values every month, while the second series, using `.resample()`, has values every sixth month.

Another windowing method in pandas is `.expanding()`. There are cases where you would like some summary value up until the current time, and to see that summary over time. An example would be if you were analyzing sales data and were asked to summarize cumulative sales by month. By using the `.expanding()` method, this is exactly what you can accomplish. Suppose you have data for total vehicle sales for many years (original data from `https://fred.stlouisfed.org/series/TOTALSA`) and are interested in the trend of cumulative sales. Here, the data is read in, and the date is converted, as we have done previously. Then, the `'cumulative'` column is created using the `.expanding()` method and `.sum()` as the aggregation function, and the pandas `.plot()` method is used again to make a simple visualization:

```
veh_sales = pd.read_csv('Datasets/TOTALSA.csv')
veh_sales['DATE'] = pd.to_datetime(veh_sales['DATE'], format =
'%Y-%m-%d')
veh_sales['cumulative'] = veh_sales['TOTALSA'].expanding().
sum()
veh_sales.plot('DATE', 'cumulative')
```

Upon running this code, you will see an output as follows:

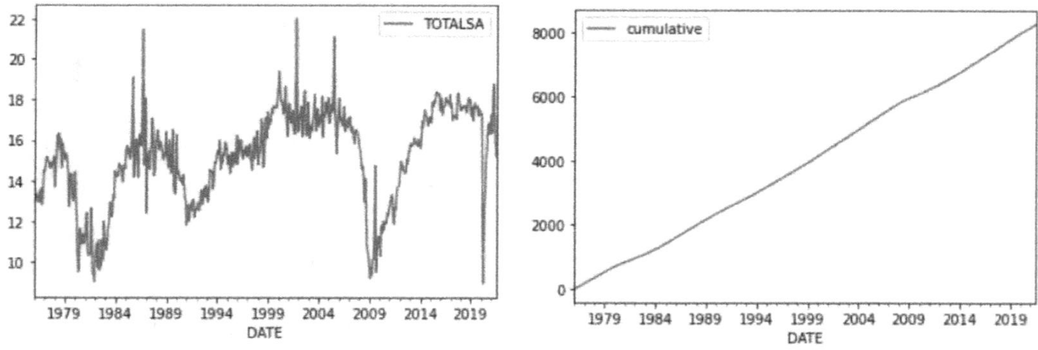

Figure 10.13 – Raw and cumulative vehicle sales using .expanding().sum()

In *Figure 10.13*, you see that the data is quite variable over time, but in the bigger picture of cumulative sales, things look much more constant. If you look carefully, you can see slight decreases in slope in the cumulative curve on the bottom at the times the raw curve has large dips.

Windowing methods

One additional feature of the window functions in pandas we will highlight here is the ability to define the windowing method. When using `.rolling()` with the `.sum()` or `.mean()` method, `win_type` can be specified in the `.rolling()` call. In the default usage we've seen so far, `win_type` is `None` (not specified, but equivalent to `win_type = None`) and the samples are equally weighted, so the sum and mean are the normally expected values. However, there are many other weightings available (see `https://pandas.pydata.org/pandas-docs/stable/reference/api/pandas.Series.rolling.html` for the list used in pandas, and `https://docs.scipy.org/doc/scipy/reference/signal.windows.html#module-scipy.signal.windows` for the definitions which come from SciPy). This provides the capability to use weighted sums or means, such as exponential weighting. Note that the SciPy window types are designed around signal analysis and filter design and, as such, move a weighted window over the data, which by default is centered on the window. This can be confusing if what is desired is something where the weights decay towards older times, as in time series data modeling, where it's not uncommon to choose a weighting that puts more emphasis on more recent values.

As an example of using a windowing method, here, we use `.rolling()` as before to compute 90-day averages of the vehicle sales data, and we apply it again but use `win_type = 'exponential'`. To do the latter, we have to make a couple of changes, one being that we have to specify the number of intervals instead of being able to use a time (so we use `window = 3` instead of `window = '90d'`, as the data is on 1-month intervals), because the underlying SciPy function does not understand dates. Another is we have to pass additional parameters for the exponential function in the aggregation function, so we use `.mean(tau = 1, sym = False, center = 3)`. Each SciPy function has a particular set of parameters; in the case of exponential, we want to override the aforementioned default of centering (`center = 3`, or the right edge, which also requires `sym = False`), and `tau = 1` determines how fast the function decays over time. You can work on the code and compare it with the one on GitHub: `https://github.com/PacktWorkshops/The-Pandas-Workshop/blob/master/Chapter10/Examples.ipynb`.

The effect of the previous code is to make the weighted averages more like the most recent points, which, in this case, means they will track sudden changes better. Here, we plot both series for comparison:

```
fig, ax = plt.subplots(figsize = (11, 9))
ax.plot(veh_sales['DATE'], veh_sales['TOTALSA'],
        color = 'black', linestyle = 'dashed', label = 'raw
sales')
```

```
ax.plot(base_forecast.index.shift(1, 'D'), base_
forecast['TOTALSA'],
        color = 'red', linestyle = 'dotted', label = 'naive
forecast')
ax.plot(weighted_forecast.index.shift(1, 'D'), weighted_
forecast['TOTALSA'],
        color = 'blue', linestyle = (0, (5, 10)), label =
'weighted forecast')
ax.set_xlim(pd.to_datetime('6/1/2019'), pd.to_
datetime('6/30/2020'))
ax.legend()
plt.show()
```

This produces the following:

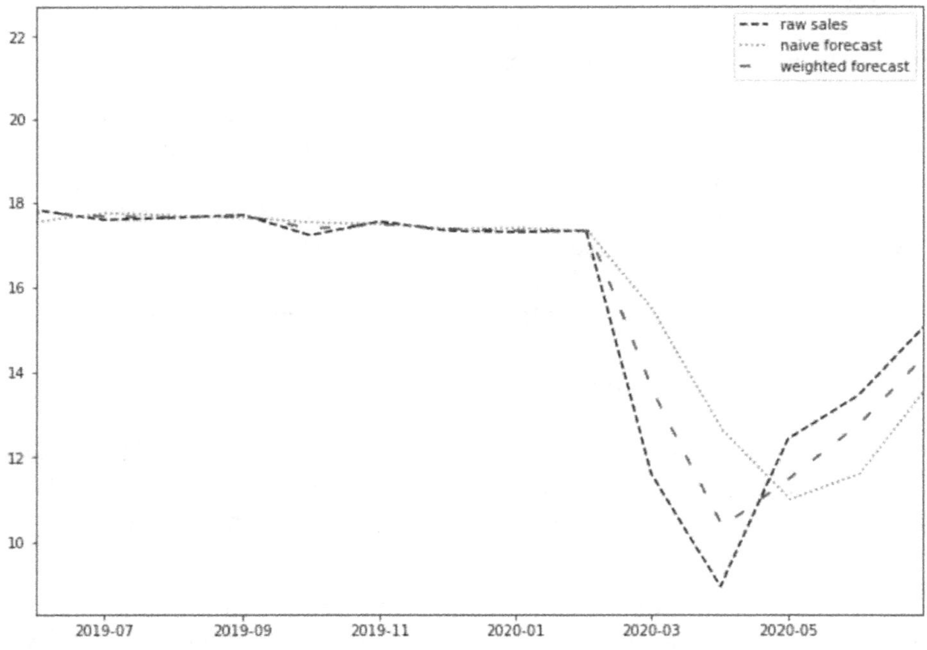

Figure 10.14 – Rolling .mean() versus exponentially weighted .mean()

Pandas also offers the `.ewm()` function as an alternative to `.rolling()`; the `.ewm()` function uses exponential decay as opposed to the SciPy filter windows, and, as such, might be more intuitive. In the comparison here, the original data is plotted, then a SciPy exponential function is used as before, and finally, the pandas `.ewm()` method is used. In `.ewm()`, we specify the span as half the window size to get results consistent with the SciPy method; however, you can specify the function using several different parameters depending on your application. Here, using a smaller span will narrow the time window. You can check out the code here: `https://github.com/PacktWorkshops/The-Pandas-Workshop/blob/master/Chapter10/Examples.ipynb`.

This results in the following output:

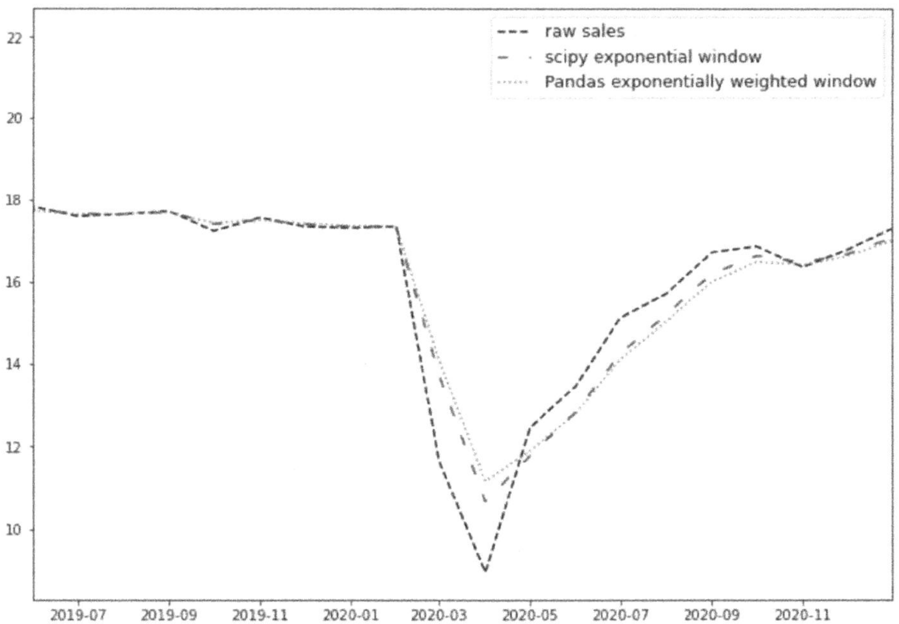

Figure 10.15 – Comparison of various moving averages to the original time series data

In *Figure 10.15*, you can see that both the smoothed series lag the actual series somewhat, which is expected due to averaging in some past data points. You can also see that `.ewm()` is nearly equivalent to the SciPy `'exponential'` window, but easier to specify, so you might prefer it for these use cases.

Smoothing data

The methods we have used so far for downsampling (`.resample()`) and applying calculations on rolling windows (`.rolling`), as well as `.ewm()`, can all be considered smoothing. Smoothing refers to replacing original values in a series with some sort of combination of nearby values, which has the effect of *smoothing* out short-term variations but retaining the overall behavior. As you have seen already, some care must be taken to ensure the smoothed data is aligned (in time, for time series data) as expected. Different methods in Python have different parameters and defaults so it's important to check the documentation.

A possible use case is to suppose you are analyzing some signal data, and you have reason to think the underlying signal might be **periodic** (such as a sine wave) but the data you have is very noisy. Although there are more advanced signal processing methods that could be used, an initial step might be to apply smoothing and see what the smoothed signal looks like. Here, we will approach this in reverse to illustrate this case. First, construct a (clean, not noisy) sine wave over a period of a year:

1. Recall that datetimes in Python have resolution to nanoseconds, so the *60 * 60 * 1e9 * 24*, which converts to days (*60 seconds/min * 60 min/hour * 1e9 nS/second * 24 hrs/day*), and the period creates times every 30 minutes. Then, in the sine function, convert to radians using *2 * np.pi * time*, and divide by 93 days as the recurring period:

```
times = pd.to_datetime(np.arange(0, 60*60*1e9*24*365, 
60*60*1e9*0.5),
                       origin = '2020-01-01')
data = np.sin(2 * np.pi * 
              times.values.astype(float) / 1e9 / 93 / 60 
/ 60/ 24)
fig, ax = plt.subplots(figsize = (11, 9))
ax.plot(times, data)
plt.show()
```

Running this code gives the following output:

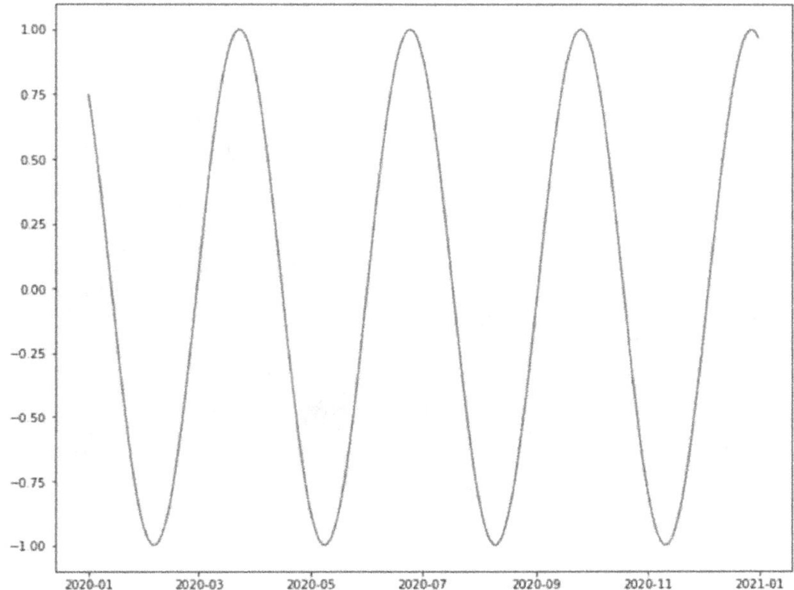

Figure 10.16 – A sine function repeating every 93 days

2. Now, create some noise to add to the data, by using the NumPy `.random.normal()` method, which samples from a normal (**Gaussian**) distribution. Here, the `.random.normal()` method is called with a mean of `0`, a standard deviation of `5`, and the number of samples equal to the length of the data. Note that this is a lot of noise compared to the original signal, but you hope to be able to recover most of the original. A line plot is created to compare the two series:

```
noisy_data = data + np.random.normal(0, 5, len(times))
fig, ax = plt.subplots(figsize = (11, 9))
ax.plot(times, noisy_data,
        linewidth = 0.5,
        label = 'noisy data')
ax.plot(times, data,
        linestyle = 'dashed',
        label = 'original data')
ax.legend()
plt.show()
```

This code produces the following. Note that the noise is more or less evenly distributed above and below the original values:

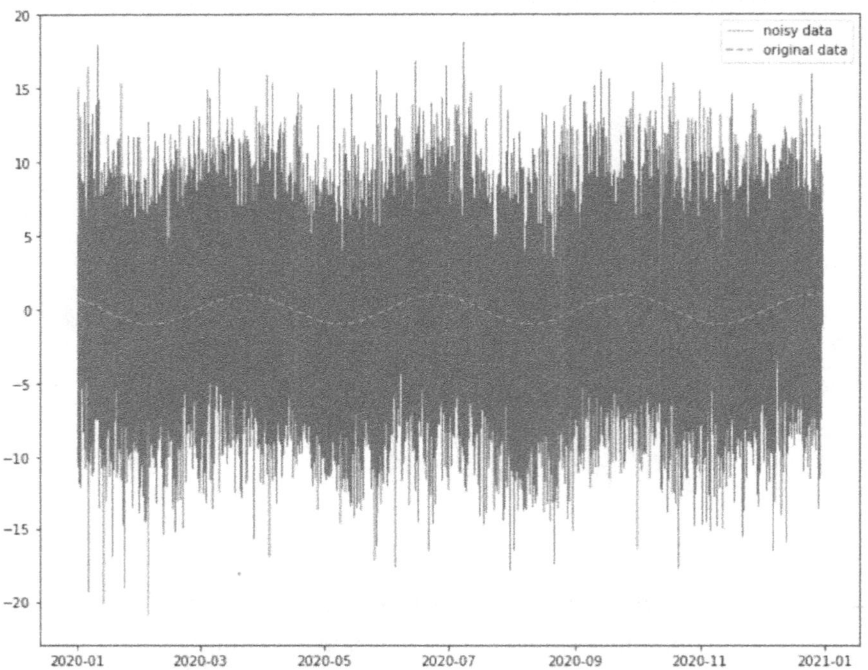

Figure 10.17 – Comparison of noisy data to original signal

3. You should inspect the data before jumping into analysis (even though you know what it looks like, since you generated it yourself). Here, histograms are generated for the original and noisy data. Note the use of the `alpha` parameter in matplotlib. This is the opacity (so smaller is more transparent) and makes the overlapping areas more easily distinguished:

```
fig, ax = plt.subplots(figsize = (11, 8))
ax.hist(noisy_data, bins = 50, label = 'noisy data',
       hatch = '///', alpha = 0.25)
ax.hist(data, bins = 50, label = 'original data',
       hatch = '+', alpha = 0.5)
ax.legend()
plt.show()
```

Running the previous code produces the following:

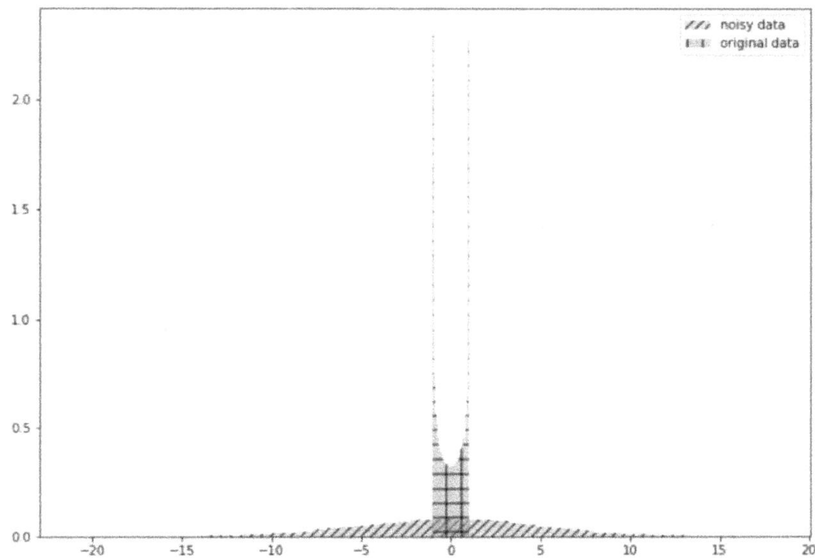

Figure 10.18 – Comparing the distributions of the original and noisy data

In *Figure 10.18*, note that while the sine wave distribution is bounded sharply by -1 to 1, the addition of the noise *smears* the noisy values out to a bit past -15 to 15, and the sharp peaks at -1 and 1 are gone.

4. Now, we return to the scenario that the noisy data is what you received, and you are looking into the possibility that there is a clear underlying repeating pattern. Here, the `.rolling()` method is used with varying window sizes. You can think of a larger window size as applying more smoothing, so in the following, you can get a feel for how much smoothing might be needed to see the true nature of the signals:

```
fig, ax = plt.subplots(figsize = (11, 8))
smoothing = [1, 100, 2000]
hatches = ['//', '|', '.']
for i in range(len(smoothing)):
    smooth = smoothing[i]
    hatch = hatches[i]
    ax.hist(pd.Series(noisy_data).rolling(window =
smooth, center = True).mean(),
            density = True, bins = 50,
            hatch = hatch,
            label = 'smoothing = ' + str(smooth),
```

```
                alpha = 0.5)
ax.set_xlim(-15, 15) // ax.legend()
plt.show()
```

This code snippet produces the following:

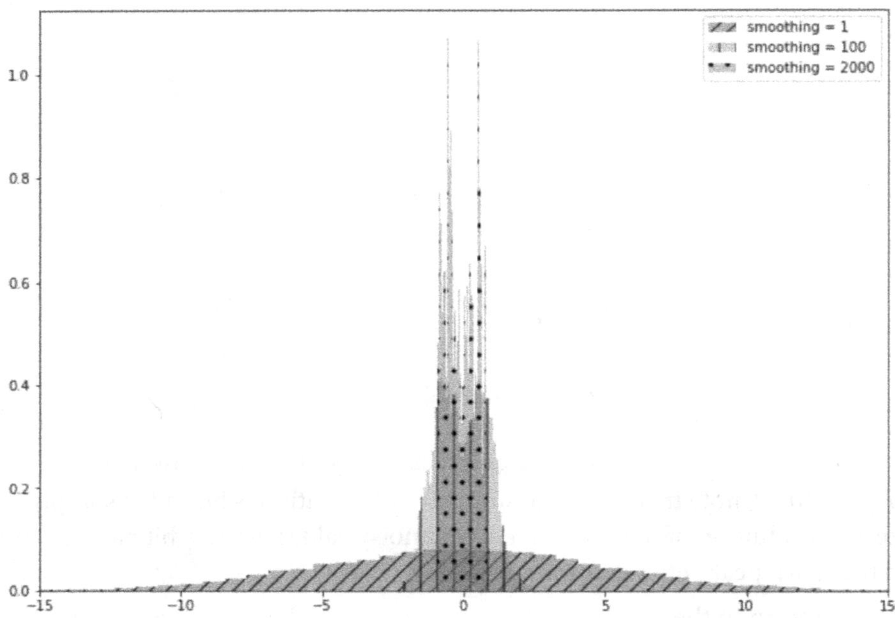

Figure 10.19 – Distribution of data with different levels of smoothing

Note that in *Figure 10.19*, the distribution with no smoothing (`smoothing = 1`, which just returns the original values) is the same as in *Figure 10.18*, but also note that as the smoothing increases, the original distribution somewhat reappears. However, even smoothing over 2,000 periods still shows evidence of noise.

5. Using this, you decide to compare the smoothed and original time series. Here, a line plot is made with smoothing over 100 or 2,000 periods:

```
smoothing = [100, 2000]
fig, ax = plt.subplots(figsize = (11, 8))
for smooth in smoothing:
    ax.plot(times, pd.Series(noisy_data).rolling(window =
smooth, center = True).mean(),
               label = 'smoothed noisy data @ ' +
str(smooth),
```

```
            linewidth = smooth / 500)
ax.plot(times, data,
        label = 'original data',
        linestyle = 'dashed',
        linewidth = 2)
ax.legend()
plt.show()
```

This produces the following comparison plot:

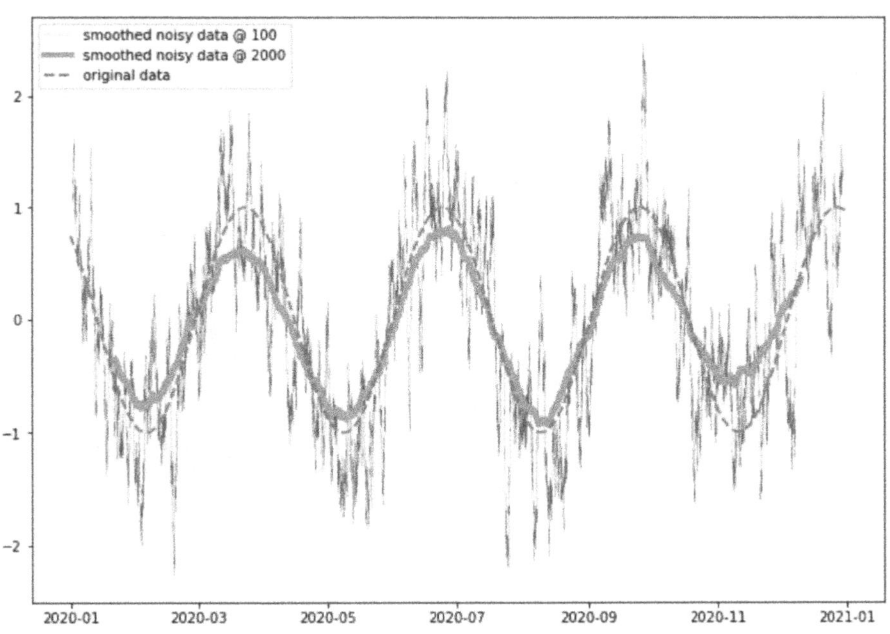

Figure 10.20 – The original data and two levels of smoothing

Of course, in the scenario, you don't have the original, so you would look at the data more broadly.

6. As a final step, you plot the data smoothed over 2,000 periods versus time for the entire dataset, with the noisy data re-labeled as `'received signal'` and the `y` range limited to focus on the smoothed series:

```
fig, ax = plt.subplots(figsize = (11, 8))
ax.plot(times, noisy_data,
        label = 'received signal',
        linewidth = 0.5
```

```
        alpha = 0.25)
ax.plot(times, pd.Series(noisy_data).rolling(window =
2000,
                                                center =
True).mean(),
        label = 'smoothed data @ 2000',
        linestyle = 'dashed')
ax.set_ylim(-2, 2)
ax.legend()
plt.show()
```

This produces the following output:

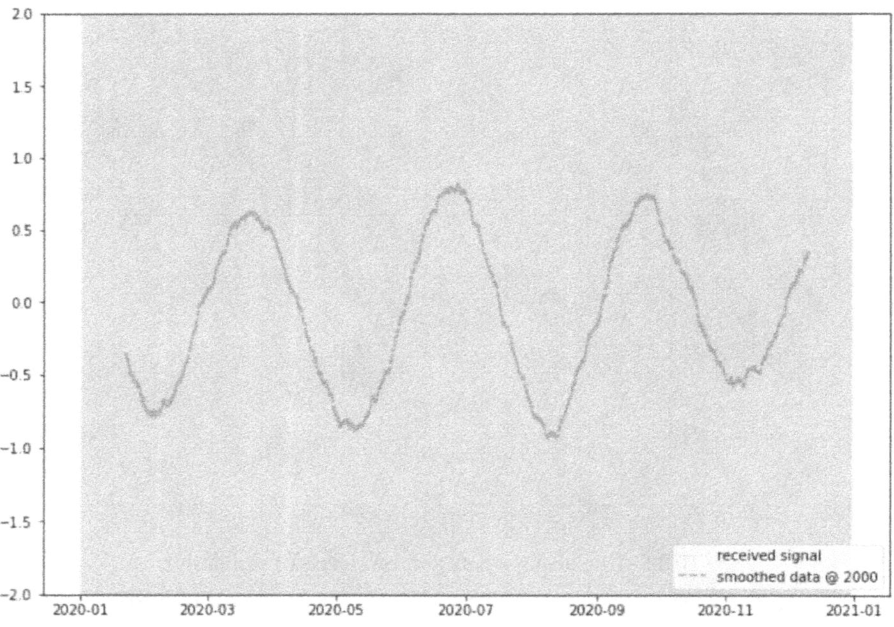

Figure 10.21 – The smoothed signal detail over the noisy data

Looking at *Figure 10.21*, imagine you are observing signals from space in the search for extra-terrestrial intelligent life. If you looked at the noisy data, you might see nothing at all. Yet, applying a simple smoothing in pandas can reveal a good reconstruction of the key underlying signal.

Smoothing over time intervals can be useful in cases other than noise reduction. Suppose you are trying to estimate the total number of exoplanets (planets outside our solar system) in the universe. You have data from NASA (original source `https://exoplanetarchive.ipac.caltech.edu/cgi-bin/TblView/nph-tblView?app=ExoTbls&config=PS`) with all the published exoplanet discoveries to date:

1. Here, you read the data in and look at the structure of the table:

```
exoplanets = pd.read_csv('Datasets/exo_planet_reporting.
csv')
exoplanets
```

Out[500]:

	date	num_recorded	cumulative_recorded
0	1/1/1992	2	2
1	4/1/1994	1	3
2	11/1/1995	1	4
3	1/1/1997	3	7
4	7/1/1997	2	9
...	...	...	...
234	4/1/2021	2	4418
235	5/1/2021	28	4446
236	6/1/2021	6	4452
237	7/1/2021	16	4468
238	8/1/2021	4	4472

239 rows × 3 columns

Figure 10.22 – Exoplanets reported in the literature by date

2. As you have done before, the following code converts the dates to datetimes, then creates a bar plot of reports by year using the `.groupby()` method you saw in the *Pandas modeling tools* section, and a line plot of cumulative reports:

```
exoplanets['date'] = pd.to_datetime(exoplanets['date'],
format = '%m/%d/%Y')
exoplanets['year'] = exoplanets['date'].dt.year
yearly_totals = exoplanets[['year', 'num_recorded']].
```

```
groupby('year').sum()
fig, ax = plt.subplots(1, 2, figsize = (15, 9))
ax[0].bar(yearly_totals.index, yearly_totals['num_
recorded'])
ax[0].set_title('yearly published new exoplanents')
ax[1].plot(exoplanets['date'], exoplanets['cumulative_
recorded'])
ax[1].set_title('cumulative expolanets published')
plt.show()
```

This results in the following output:

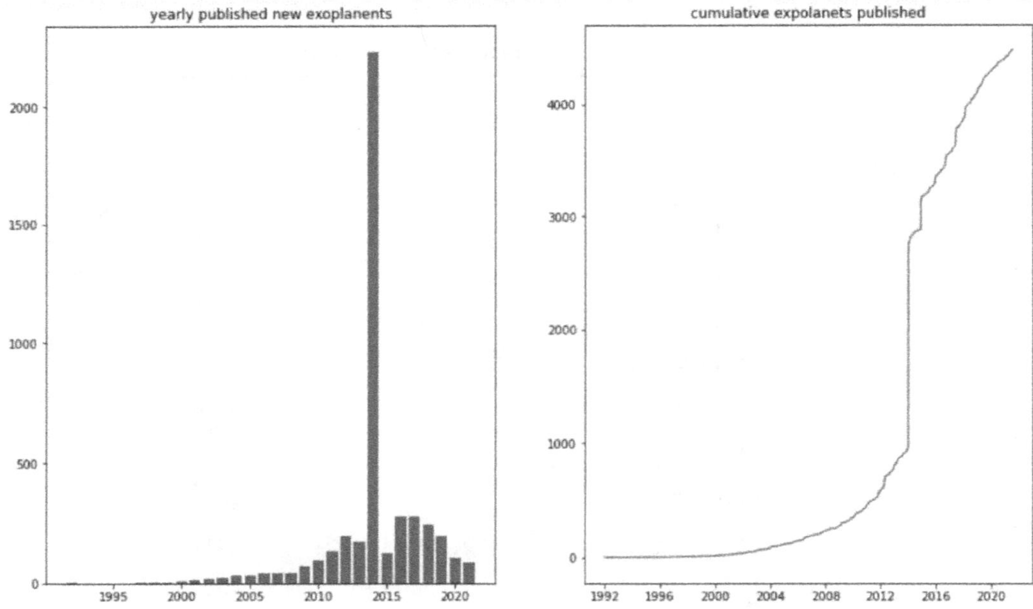

Figure 10.23 – Exoplanet discoveries by year and cumulative discoveries.

The large spike in 2014 is a result of the Kepler telescope, which began making exoplanet discoveries in 2011 (which were reported in scientific literature in 2014). This also causes the jump in the cumulative curve.

3. To make your estimate, you need to estimate the discoveries per year and combine that with the area searched. The left side of *Figure 10.23* gives the values in a given year, but you would like something that is more continuous versus time. To accomplish that, you use `.rolling()` with a window size of 365 days and apply that to the data after setting the index to the dates for `.rolling()` to operate on.

This then re-computes the number per year at every date, instead of only one value per year:

```
exoplanets.set_index('date', drop = True, inplace = True)
fig, ax = plt.subplots(figsize = (11,8))
ax.plot(exoplanets.index,
        exoplanets.rolling(window = '365d').sum()['num_
recorded'],
       label = 'average per year')
ax.legend()
plt.show()
```

This produces the following line chart:

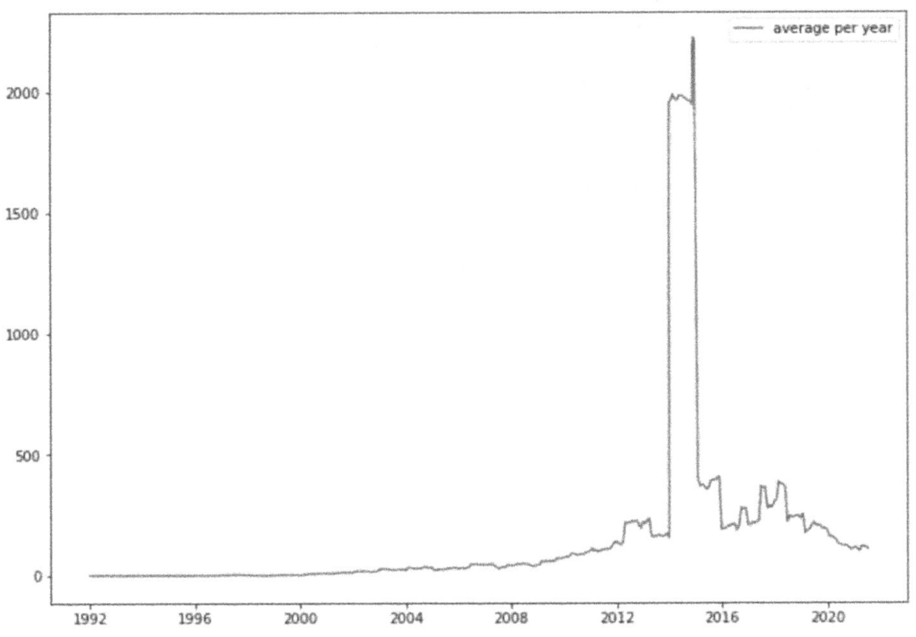

Figure 10.24 – Continuous values of the exoplanets reported per year

Figure 10.24 is similar to the left side of *Figure 10.23*, but the data is more granular. It appears from this data that reports peaked around 2018 and have been declining. Accounting for future missions, you plan to use this curve to estimate discoveries in the future. In this section, you have learned about the use of smoothing methods to find insights within datasets. Another common application is to try to use data to predict a future event or trend. We now turn to the challenge of prediction.

Predicting future values of time series

You have seen how smoothing can be used to uncover important information in a series that might be hidden by noise. It might be tempting to think that smoothing is a very easy data modeling method, so why not use it to make predictions? The issue that arises is, in many cases, the process of smoothing data and aligning it to the original series means you are using information for any given point in the smoothed series that includes future values. Therefore, using such values as predictions is an example of data leakage, discussed in *Chapter 9, Data Modeling – Preprocessing* in the *Avoiding information leakage* section.

Suppose you are again analyzing the SPX index data you saw in *Chapter 9, Data Modeling – Preprocessing*:

1. Here, you read the data, convert the dates to datetimes, and make a simple plot over a limited time range:

```
SPX = pd.read_csv('Datasets/spx.csv')
SPX['date'] = pd.to_datetime(SPX['date'], format =
'%d-%b-%y')
SPX.set_index('date', drop = True, inplace = True)
fig, ax = plt.subplots(figsize = (11, 8))
ax.plot(SPX.index, SPX['close'])
ax.set_xlim(pd.to_datetime('2016-01-01'), pd.to_
datetime('2018-07-01'))
ax.set_ylim(1500, 3000)
plt.show()
```

Running this gives the following output:

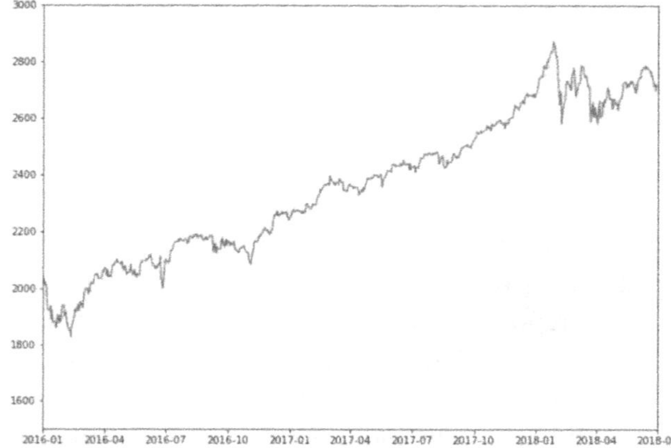

Figure 10.25 – SPX index values

2. You can see there are many small variations in the data, and you decide to smooth the data over 90 days and compare to the original. Here, you use `.rolling()` to smooth the data and make a similar plot of *Figure 10.25* with the original and the smoothed series:

```
fig, ax = plt.subplots(figsize = (11, 8))
ax.plot(SPX.index, SPX['close'], label = 'raw index')
ax.plot(SPX.index, SPX.rolling(window = 90).mean(), label
= 'smoothed @ 90d')
ax.set_xlim(pd.to_datetime('2016-01-01'), pd.to_
datetime('2018-07-01'))
ax.set_ylim(1500, 3000)
ax.legend()
plt.show()
```

Running this code will give the following output:

Figure 10.26 – Comparison of original to smoothed SPX

3. In *Figure 10.26*, you see a significant time shift between the smoothed series and the original. This is due to the fact that the default for `.rolling()` is to use the date at the right (latest time) of the window as the label. In this case, that clearly isn't what you want, so you re-create the plot, but use the `center = True` option in the `.rolling()` method:

```
fig, ax = plt.subplots(figsize = (11, 8))
ax.plot(SPX.index, SPX['close'], label = 'raw index')
ax.plot(SPX.index, SPX.rolling(window = 90, center =
True).mean(),
        label = 'smoothed @ 90d')
ax.set_xlim(pd.to_datetime('2016-01-01'), pd.to_
datetime('2018-07-01'))
ax.set_ylim(1500, 3000)
ax.legend()
plt.show()
```

This produces the following result:

Figure 10.27 – Smoothed SPX centered on original dates

In *Figure 10.27*, the smoothed data is now aligned to the original, but if you inspect the right-most dates, there are no values for the smoothed series. So, as-is, we could not use the smoothed data to predict future values (we don't even have values for all the past data). However, you saw earlier that you could use a weighted window to make the computed values more like the most recent, and we can shift the labels (dates) using the pandas `.shift()` method. This is a very simplistic prediction approach and is generally only useful for very short periods.

4. In the code here, first, a **Lambda function** is defined called `linear_window`, which just takes the previous values of a `z` series, and computes a weighted sum, where the weights begin as `1/w`, where `w` is the window size, and the weights increase to `1`. This has the effect that the sum is most affected by the most recent values. The pandas `.rolling()` method is used with `window = w`, and the aggregation function is replaced by the `.apply()` method. In `.apply()`, the function is passed, and `raw = False` tells pandas to send the data in as a pandas Series (raw means a raw NumPy array). Any function that takes in a series and returns a value can be used with `.apply()` after `.rolling()`.

   An x series is computed as a simple combination of sine and cosine functions, and the data used to compute the rolling series excludes the last x value. The last value of the rolling series is then used as the prediction of the next x value. The data is plotted for comparison:

```
linear_window = (lambda z: np.sum(z[len(z)-w:len(z)] *
                                  (np.arange(1, w + 1)/
(w))) /
                        (np.sum(np.arange(1, w + 1)/(w))))
length = 51
x = pd.Series(np.sin(2 * np.pi * np.arange(length) / 13)
+
              np.cos(2 * np.pi * np.arange(length) / 11))
w = 3
weighted = x[:-1].rolling(window = w).apply(linear_
window, raw = False)
fig, ax = plt.subplots(figsize = (11, 8))
ax.plot(x.index, x,
        marker = 'o', label = 'raw', color = 'blue')
ax.scatter(list(weighted.index + 1)[-1], weighted.iloc[-
1],
           marker = 's', s = 50,
           label = 'weighted prediction 1 period ahead',
```

```
color = 'red')
ax.legend()
plt.show()
```

This will result in the following output:

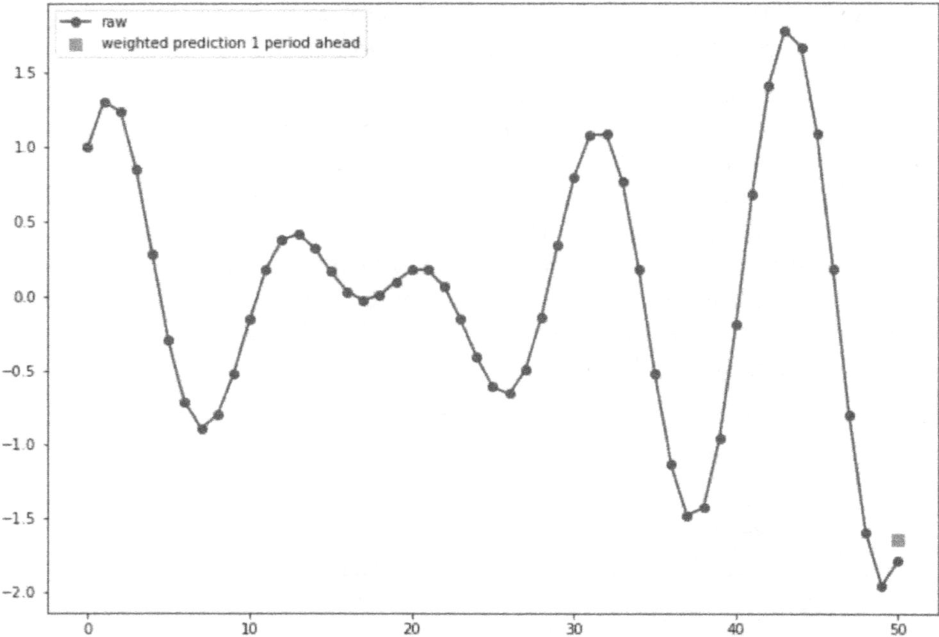

Figure 10.28 – Predicting the next value with a weighted rolling window

You see in *Figure 10.28* that the prediction isn't too far from the actual value. However, it's unlikely this simple method would work if we tried to predict much further, and it also depends a lot on the nature of the data. Nonetheless, using `.apply()` with `.rolling()` and defining Lambda functions for your use case can be powerful and easy to implement. You should now be aware of a variety of approaches to smoothing as well as some use cases. In the next exercise, you will use smoothing to investigate some data and discover underlying patterns.

Exercise 10.01 – Smoothing data to discover patterns

You decide to revisit the SPX index data and use some closer analysis and smoothing to see whether there are some smaller patterns in the data on shorter time scales. The goal is to see whether there are periods of weekly patterns:

1. For this exercise, all you will need is the `pandas` library and matplotlib. Load them in the first cell of the notebook:

```
import pandas as pd
import matplotlib.pyplot as plt
```

2. Load the SPX data from the `spx.csv` file into a DataFrame called `stock_data`:

```
stock_data = pd.read_csv('Datasets\\spx.csv')
stock_data.date = pd.to_datetime(stock_data.date)
stock_data.head()
```

The output should be as follows:

Out[3]:

	date	close
0	1986-01-02	209.59
1	1986-01-03	210.88
2	1986-01-06	210.65
3	1986-01-07	213.80
4	1986-01-08	207.97

Figure 10.29 – SPX closing price data

3. Now, visualize the raw data as a scatter plot. To see some detail, narrow the date range from *2013-1-10* to *2014-9-30*:

```
subset = stock_data.loc[(stock_data.date > '2013-09-30')
&
                        (stock_data.date < '2014-10-01'),
:]
fig, ax = plt.subplots(figsize = (9, 7))
ax.scatter(subset.date, subset.close,
        color = 'blue', s = 4)
ax.set_title('SPX closing price performance', fontsize =
16)
```

```
ax.set_ylabel('closing price', fontsize = 14)
ax.tick_params(labelsize = 12)
plt.show()
```

The output should be as follows:

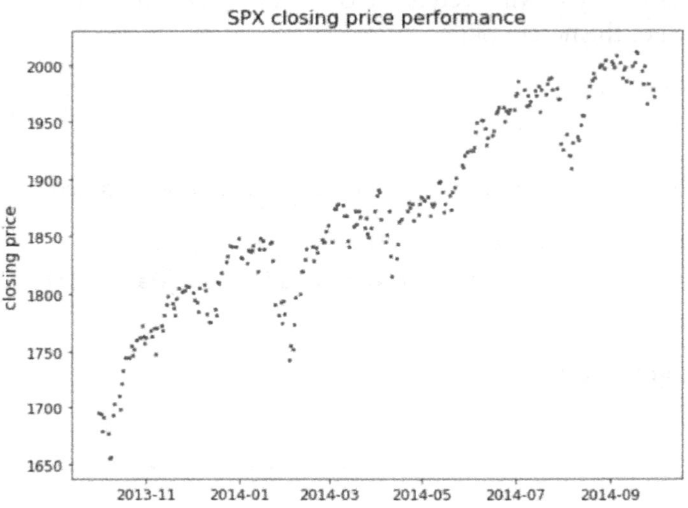

Figure 10.30 – SPX data detail

4. In science, it's frowned upon to plot individual data points connected by lines that imply that the data is continuous (such as data from an analog sensor). However, in data science, our goal is to understand and model the data, so let's replot the daily data with a line plot to see what patterns emerge. Considering that stock trading might well have a weekly cycle, add some vertical lines on weekly intervals, starting on *Monday, 2014-4-13* for 12 weeks:

```
subset = stock_data.loc[(stock_data.date > '2013-09-30')
&
                        (stock_data.date < '2014-10-01'),
:]
fig, ax = plt.subplots(figsize = (11, 8))
ax.plot(subset.date, subset.close,
        color = 'blue', lw = 1)
ax.set_title('SPX closing price performance', fontsize =
16)
ax.set_ylabel('closing price', fontsize = 14)
ax.tick_params(labelsize = 12)
```

```
for date in pd.date_range('2014-04-13', periods = 12,
freq = '1W'):
    ax.axvline(date, ymin = 0, ymax = 1, color = 'black',
lw = 0.25)
plt.show()
```

The resulting chart should look as follows:

Figure 10.31 – SPX data plotted as a line plot and highlighting some weekly intervals

The time span highlighted by the vertical lines is interesting. It might be that many of the apparent dips are after the weekend. That might be interesting, but you want to remove some of the *weekly* noise and see what emerges.

5. Now, suppose that you are interested in more substantial patterns, and you want to filter out the weekly noise. For this, use the pandas `.rolling()` method to smooth the data over a 7-day period and add vertical lines every 3 months as guides:

```
subset_smooth = subset['close'].rolling(window = 7, min_
periods = 0, center = True).mean()
fig, ax = plt.subplots(figsize = (9, 7))
ax.plot(subset.date, subset_smooth,
        color = 'red',
        lw = 1)
ax.set_title('SPX closing price performance', fontsize =
```

```
16)
ax.set_ylabel('closing price', fontsize = 14)
ax.tick_params(labelsize = 12)
for date in pd.date_range('2013-10-31', periods = 4,
freq= '3M'):
    ax.axvline(date, ymin = 0, ymax = 1, color = 'black',
lw = 0.25)
plt.show()
```

The resulting plot should look as follows:

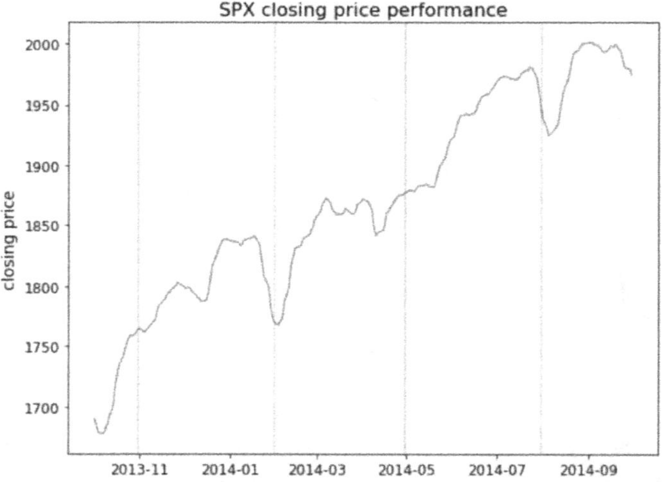

Figure 10.32 – Smoothed SPX data

As far as patterns go, this may be inconclusive, but without the short-duration noise, you can focus on where big changes have occurred and analyze those to see whether some external events correlate. If you found some external factors accounting for some of the longer-term patterns, you could consider using that in a model to predict in the future.

Now that you have practiced the methods in this chapter, the next activity will test some of the skills in this chapter, as well as those from *Chapter 9*, *Data Modeling – Preprocessing*.

> **Note**
>
> You can find the code for this exercise here: `https://github.com/PacktWorkshops/The-Pandas-Workshop/blob/master/Chapter10/Exercise10.01.ipynb`.

Activity 10.01 - Normalizing and smoothing data

Suppose you are an analyst in a financial advisory firm. Your manager has given three stock symbols to you and requested your input on how they may be correlated with their price behavior. You are provided a `stocks.csv` data file, which contains the symbols, closing prices, trading volumes, and a sentiment indicator (some view of the quality of the stocks, but you are not told the exact definition). Your initial goal here is to determine whether all three stocks show similar market characteristics or not, and if any or all of them do, make an initial visualization using smoothing. The long-term goal is to try to build some predictive models, so you will split the data into train and test sets. As it is time series, it's important to split on time, not randomly. For this activity, all you will need is the `pandas` library, a scaling module from `sklearn`, and `matplotlib`. Load them in the first cell of the notebook:

```
ximport pandas as pd
from sklearn.preprocessing import StandardScaler
import matplotlib.pyplot as plt
```

1. Use the `stocks.csv` dataset.
2. Inspect `.dtypes` and convert dates to pandas `datetime` if needed.
3. Split the data into `train` and `test` sets based on the date, keeping the last 3 months as the `test` set.
4. Generate a scatter plot that shows the prices over time of different stock symbols and identifies the `train` and `test` splits.
5. You will find that the initial scatter plot isn't very informative, because different symbols have very different pricing, so some are compressed at the bottom of the *y* axis. Plot a histogram of the price distribution for each symbol separately, and use enough bins to see the detail.
6. Since you see the prices have very different ranges, you need to scale each symbol separately. Use sklearn's `StandardScaler` to scale the original price and volume data by symbol, storing each symbol as a new DataFrame in a list, with `scalers` as another list.
7. Plot the `train/test` data as before, using a loop over the symbols.
8. You can now see that two of the symbols are similar in trend and one differs. Replot the two similar symbols but apply smoothing of 14 days and compare whether the two stocks are behaving the same way over the period from 2017-09 forward.

The result should indicate some potential similarities between two of the symbols' price behavior, which motivates some additional market research to look for reasons why they behave this way.

> **Note**
> You can find the solution for this activity in the Appendix.

Summary

In this chapter, you built on the topics of independent and dependent variables, splitting data into train/validation/test splits for modeling and providing unbiased estimates of model performance. Here, you learned a range of basic data modeling methods using resampling (up and downsampling data frequency) and rolling window approaches to smoothing and estimating. You began your detailed investigation of data modeling with pandas tools for smoothing and resampling data, and some particular capabilities to handle time series. Importantly, you saw that smoothing methods can highlight patterns in very noisy data and that smoothing can be non-uniform in time, such as using `.ewm()` or a custom weighting function. With these foundational methods in hand, the next chapter will conclude data modeling with a deeper exploration of linear regression and then non-linear and powerful modeling methods, using Random Forest as a regression model.

11
Data Modeling - Regression Modeling

In this final chapter on data modeling, you will learn details about linear regression using the `sklearn` library's `LinearRegression` method, and non-linear regression modeling using the `sklearn` library's `RandomForestRegressor` method. As you learn more about these methods, you will also learn details about measuring model performance using measures such as the sum of square error and root mean squared error, as well as powerful visual methods, including constructing histograms of model errors and other plotting methods.

By the end of this chapter, you will have brought together all you have learned about data modeling and be ready to address a wide range of business and technical data challenges.

This chapter covers the following topics:

- An introduction to regression modeling
- Exploring regression modeling
- Model diagnostics
- Activity 11.01 – Implementing multiple regression

An introduction to regression modeling

The term **regression**, as used in regression analysis and linear regression, was first used by Sir Francis Galton and appeared in 1886 in his essay in *Anthropological Miscellanea, Regression towards Mediocrity in Hereditary Stature* (see `https://galton.org/essays/1880-1889/galton-1886-jaigi-regression-stature.pdf`, where he says he had already made an address in Aberdeen that had been published in *Nature in September* (`www.nature.com`)). We will visit arguably his most famous dataset in the next section, but in modern usage, regression can mean a large number of statistical methods to estimate a function that describes a set of data. In some cases, and especially in current data science practice, such models can also be used to make predictions on new data. Essentially, if data is of a continuous nature (meaning the data values can be any value between some range, or even unbounded), then regression methods are used to fit mathematical functions to the data. Thus, although there may be a range from simple (such as linear regression) to complex (such as artificial neural networks) methods used to accomplish regression, they all have features in common with the core subjects of this chapter.

Exploring regression modeling

You have already used regression models in *Chapter 9* and *Chapter 10*. Here, we will go deeper into regression modeling and compare linear and non-linear models for data modeling. A famous early example of regression analysis was produced by Sir Francis Galton, who lived in England from 1822 to 1911. Among many activities, Galton collected data on the heights of fathers and mothers and their adult children. It is notable that today, the data would be considered biased, as the sample was most likely from more affluent families that had access to better nutrition and living conditions than the average for the time in England. Nonetheless, the data serves as a good introduction to regression:

1. Here, we load a simplified version of the data (adapted from the original) into a pandas DataFrame and plot the heights of all the children and the fathers:

```
galton_heights = pd.read_csv('Datasets/galton.csv')
galton_heights.head()
```

This produces the following:

Out[17]:

	ht_father	ht_child
0	78.5	73.2
1	78.5	69.2
2	78.5	69.0
3	78.5	69.0
4	75.5	73.5

Figure 11.1 – Galton's height data

2. The next code snippet produces a simple scatter plot to visualize the data:

```
fig, ax = plt.subplots(figsize = (11, 8))
ax.scatter(galton_heights.ht_father,
           galton_heights.ht_child)
ax.set_xlabel('height of father (inches)')
ax.set_ylabel('adult child height (inches)')
ax.set_xlim(60, 80)
ax.set_ylim(50, 85)
plt.show()
```

This produces the following plot:

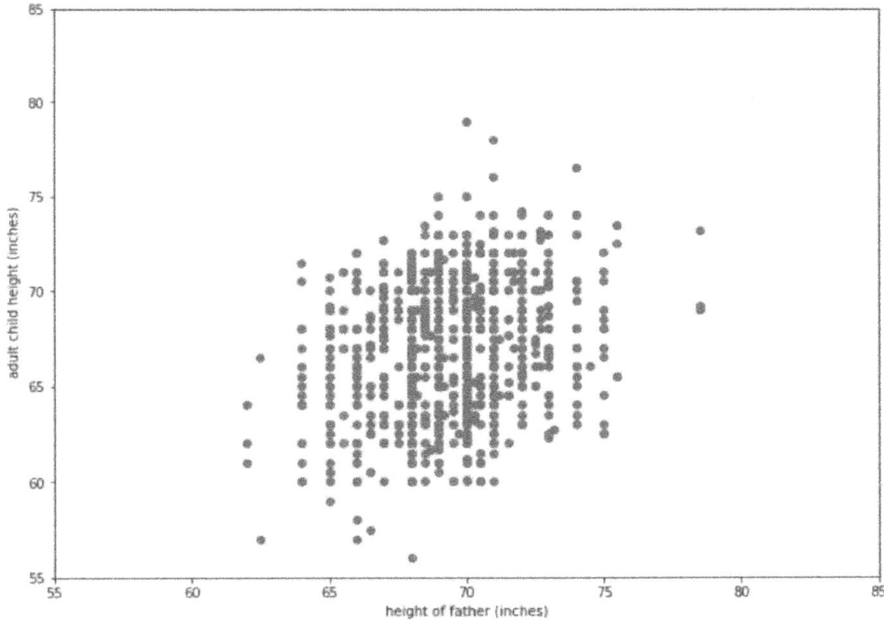

Figure 11.2 – The heights of the children are positively correlated to the heights of their fathers

You can see that there appears to be a positive correlation – the taller the father, on average, the taller the children. Of course, there is a lot of scatter – there are likely many factors that influence the height besides the height of the father.

3. An obvious way to analyze the data in *Figure 11.2* is to try to find a line that is a "good" or "best" fit for the data. As a starting point, without any math, you can try the `y = x` model. Such a line is a diagonal in *Figure 11.2,* since the x and y axes are of the same scale and range. The following code adds the diagonal to the previous plot:

```
fig, ax = plt.subplots(figsize = (11, 8))
ax.scatter(galton_heights.ht_father,
           galton_heights.ht_child)
ax.plot([0, 100], [0, 100], color = 'green', linestyle =
'--')
ax.set_xlabel('height of father (inches)')
ax.set_ylabel('adult child height (inches)')
ax.set_xlim(55, 85)
ax.set_ylim(55, 85)
plt.show()
```

This produces the following plot:

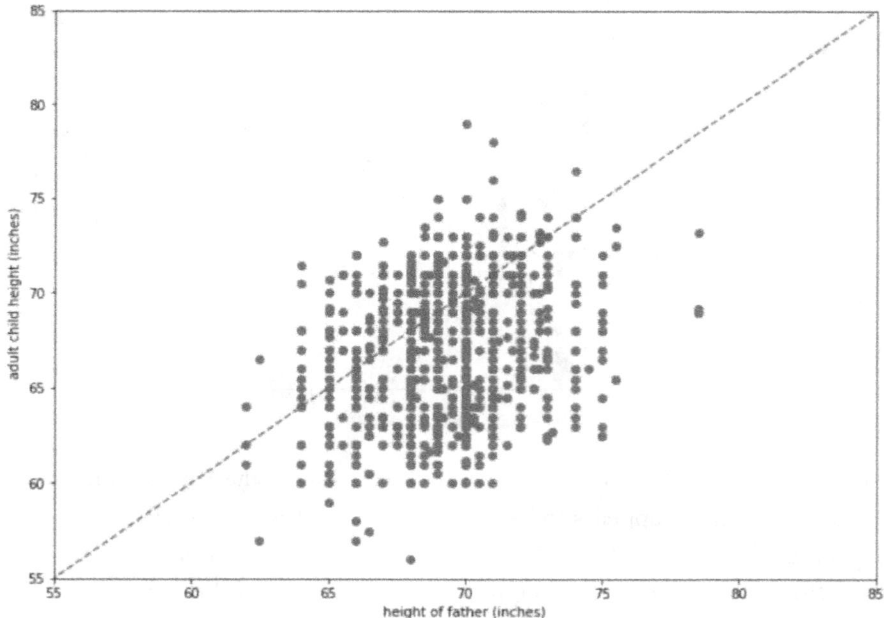

Figure 11.3 – Galton's data with the y = x model plotted as a line over the data

You may note by inspecting *Figure 11.3* that this simple model seems a little biased – more of the actual child heights are below the line than above it.

4. A way to characterize the errors such a model makes is to compute and plot the **residuals**. The term *residual* is used to mean the difference between the model and the data – literally, what is left after using the model to make a prediction. One way to define a "good" model is to ask that the residuals be symmetrically distributed and centered on 0 (note that there are many more rigorous statistical model evaluation criteria – the goal here is intuition, not full mathematical rigor). The next code snippet analyzes the residuals by first creating a new column in the DataFrame with the predictions, and then plotting a histogram of the difference between the predicted and actual heights:

```
galton_heights['naive_pred'] =b galton_heights['ht_
father']
naive_res = galton_heights['naive_pred'] - galton_
heights['ht_child']
fig, ax = plt.subplots(figsize = (11, 8))
ax.hist(naive_res, bins = 20)
ax.set_title('Distribution of errors\npredicted minus
actual height of child')
plt.show()
```

This generates the following chart:

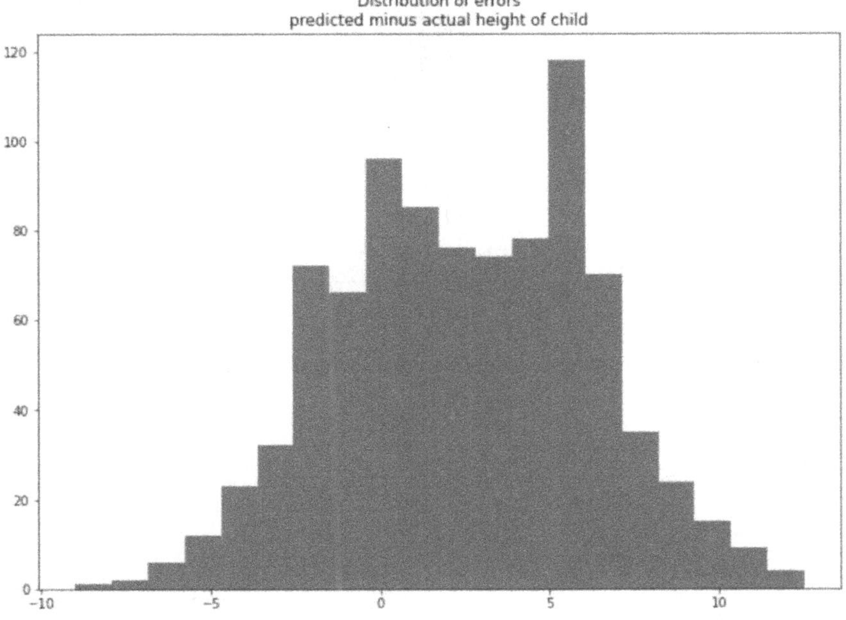

Figure 11.4 – The residuals (errors) using the simple y = x model of the child heights

The histogram in *Figure 11.4* is a powerful analysis tool. You can see that the average error is above zero (on average, the model predicts too large a height), and the distribution is more or less symmetric, with some deviations.

5. In the next section, *Linear models*, we will explore regression modeling in more detail, and inspect residuals and other properties of the models. As a preview, the following code uses the `sklearn LinearRegression()` method to fit a model to the data, and then plots the model on the data points, along with the previous 'naïve' model. You can check out the code here: `https://github.com/PacktWorkshops/The-Pandas-Workshop/blob/master/Chapter11/Examples11.ipynb`.

This generates the following comparison plot:

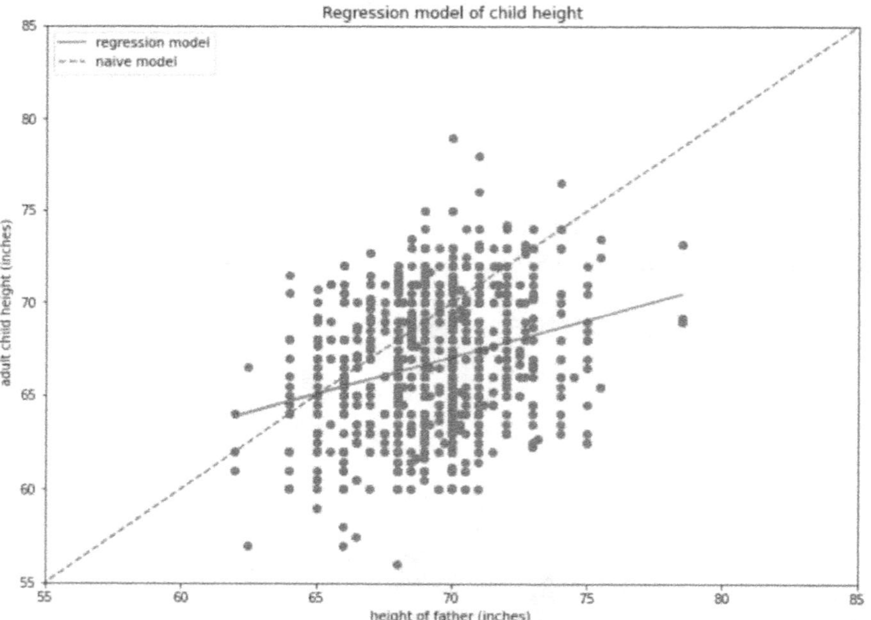

Figure 11.5 – A regression model versus a naïve model of child heights

6. It's clear that the regression model is quite a bit different from the naïve y = x model. Inspecting the residual can give a more quantified assessment. Here, the code computes the predictions and residuals as before, and plots the distribution of the residuals:

```
galton_heights['OLS_pred'] = \
    linear_model.predict(np.array(galton_heights.ht_
father).reshape(-1, 1))
OLS_res = galton_heights['OLS_pred'] - galton_
heights['ht_child']
fig, ax = plt.subplots(figsize = (11, 8))
ax.hist(OLS_res, bins = 20)
ax.set_title('Distribution of errors\npredicted minus
actual height of child')
plt.show()
```

This produces the new histogram, as follows:

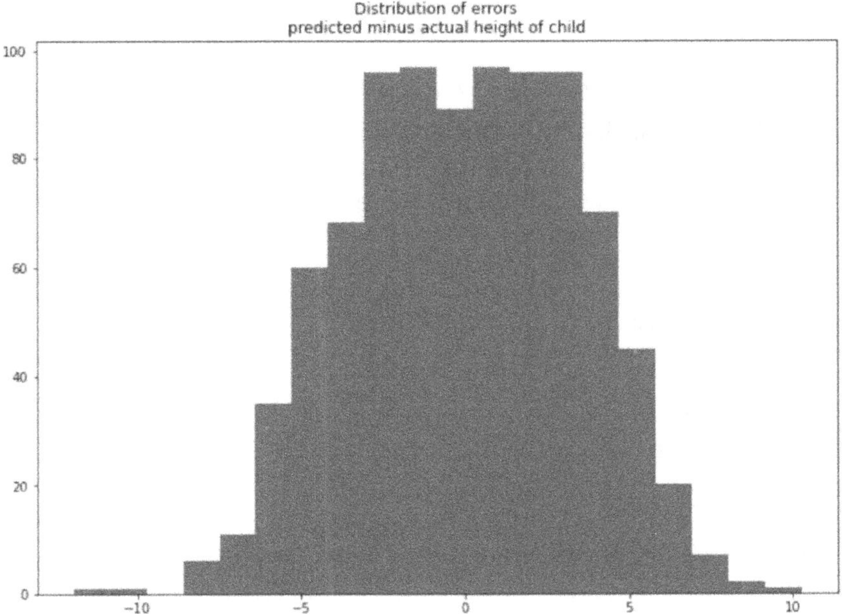

Figure 11.6 – Residuals using the regression model

You can see that the distribution in *Figure 11.6* is more centered than in *Figure 11.4*.

7. A measure of the overall goodness of the fit is the sum of squared errors or, sometimes, the root mean squared error, which is just the square root of the average squared error. These values can be computed directly from the residuals by squaring, or summing, or squaring and summing, dividing by the number of points, and taking the square root. The following code does that manually for the naïve errors and the regression errors to compare them:

```
In [31]:
naive_SSE = np.sum((galton_heights.naive_pred - galton_heights.ht_child)**2)
OLS_SSE = np.sum((galton_heights.OLS_pred - galton_heights.ht_child)**2)
naive_RMSE = np.sqrt(naive_SSE / galton_heights.shape[0])
OLS_RMSE = np.sqrt(OLS_SSE / galton_heights.shape[0])
print('naive model gives:\n',
      'SSE = ', naive_SSE.round(3), '\n',
      'RMSE = ', naive_RMSE.round(3), '\n',
      'regression model gives:\n',
      'SSE = ', OLS_SSE.round(3), '\n',
      'RMSE = ', OLS_RMSE.round(3))

naive model gives:
 SSE =  18104.76
 RMSE =  4.49
 regression model gives:
 SSE =  10641.987
 RMSE =  3.442
```

Figure 11.7 – Measuring the overall fit

A nice feature of the RMSE is that it is in the units of the variable, so the preceding results tell you that the error using the regression model is around 25% less, or an inch less of an error, than the naïve model. By all measures, you would say that the regression model is "better" but certainly not a great model.

With this introduction to regression, residuals, and residual histograms, you are ready to look into linear models more closely.

Using linear models

Linear models can be surprisingly effective in many cases; however, in the car mileage case, introduced in *Chapter 9* in the *Training, validation, and test splits of data* section, we could see some possibly non-linear dependencies in the data. Here, we reproduce *Figure 9.11* with the pairwise scatter plots for the car data. The following code is also the same as in *Chapter 9*, but we're not reproducing the code for the plotting function (refer to *Chapter 9* or the `Examples.ipynb` file for this chapter for that):

```
import pandas as pd
import numpy as np
my_data = pd.read_csv('datasets/auto-mpg.data.csv')
plot_corr_grid(my_data, variables = list(my_data.columns)[:-1])
```

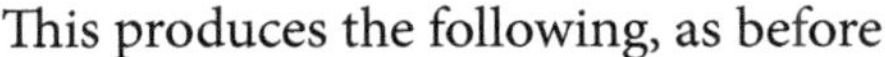

This produces the following, as before:

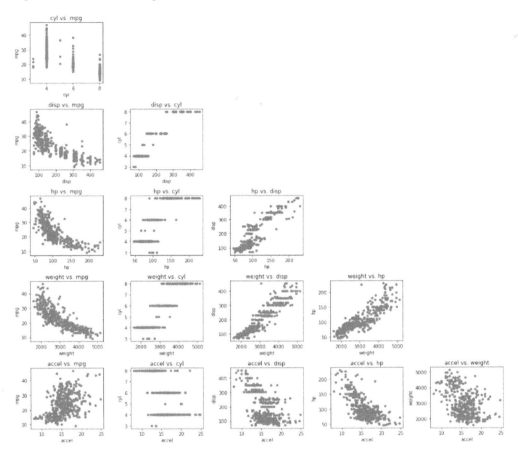

Figure 11.8 – The same charts as Figure 9.11

In *Figure 11.8*, we can see that `displacement`, `horsepower`, and `weight` appear to have non-linear relationships with the `mpg` target variable. There are a few approaches we can take to accommodate these relationships and still use a linear model. The most common is to add new columns that are powers of the original variables (such as `hp**2`) or to apply a transform to the data, such as taking the logarithm.

To illustrate how a transformation of the independent data can improve a model, here you inspect the `weight` variable:

```
fig, ax = plt.subplots(figsize = (11, 8))
ax.hist(my_data.weight, bins = 30) // plt.show()
```

This produces the following histogram:

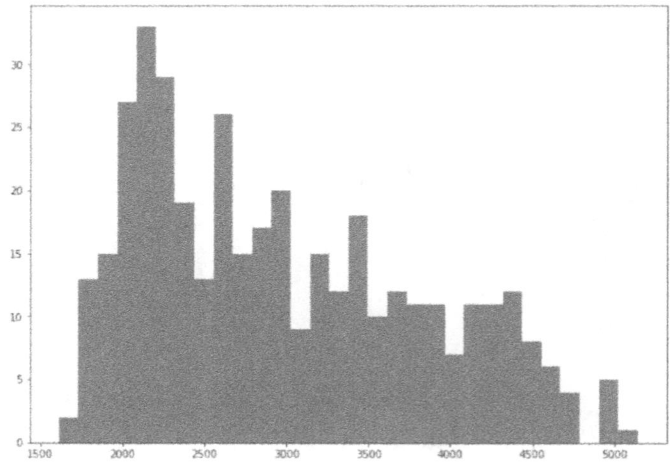

Figure 11.9 – The histogram of the weights in the car data

Now, repeat the same process, but apply `np.log()` to the weight values before plotting:

```
fig, ax = plt.subplots(figsize = (11, 8))
ax.hist(np.log(my_data.weight), bins = 30) \\ plt.show()
```

This produces another histogram, as shown here:

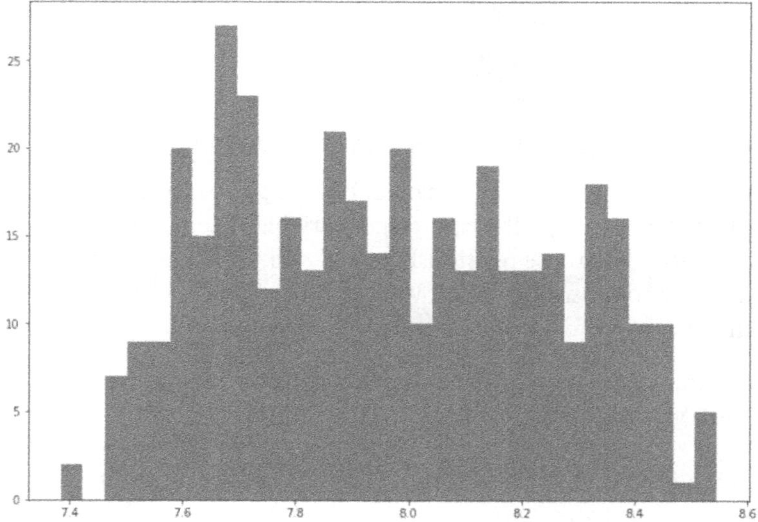

Figure 11.10 – A histogram of car weights after a log transform

Comparing *Figure 11.10* to *Figure 11.9*, you can see that the raw data is skewed, and the transformed data is more symmetric. Now, plot mpg versus `log(weight)` and compare it to *Figure 11.8*:

```
fig, ax = plt.subplots(figsize = (11, 8))
ax.scatter(np.log(my_data.weight), my_data.mpg)
plt.show()
```

This produces the following plot:

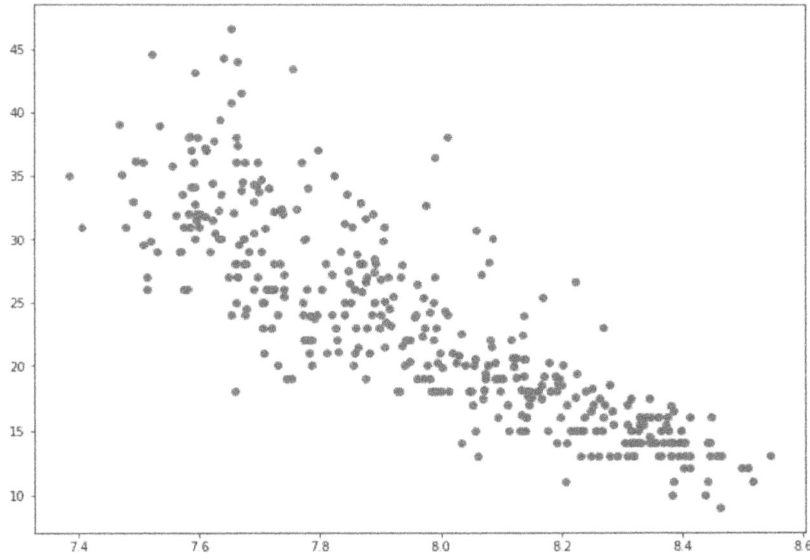

Figure 11.11 – A scatter plot of mpg versus log (weight) for the car data

Although the pattern in *Figure 11.11* isn't perfectly linear, it has decidedly less curvature than the original in *Figure 11.8*. This can make it easier to use in a linear model. Next, we will test that.

Now, you apply `log()` transforms to the variables with the most curvature in *Figure 11.8*, displacement, horsepower, and weight, and then fit a linear model. In this case, some data is held back to test the model in a more rigorous way, as seen in *Chapter 10*:

```
np.random.seed(42)
train = np.random.choice(my_data.shape[0], int(0.7 * my_data.
shape[0]))
validation = [i for i in range(my_data.shape[0]) if i not in
train]
```

```
X = my_data.iloc[train, 1:-1]
X['disp'] = np.log(X['disp'])
X['hp'] = np.log(X['hp'])
X['weight'] = np.log(X['weight'])
log_scaler = StandardScaler()
X = log_scaler.fit_transform(X)
y = np.reshape(np.array(my_data.loc[train, 'mpg']), (-1, 1))
log_lin_model = OLS()
my_model = log_lin_model.fit(X, y)
print('R2 score is ', my_model.score(X, y))
print('model coefficients:\n', my_model.coef_,
      '\nintercept: ', my_model.intercept_)
RMSE = mean_squared_error(y, my_model.predict(X), squared =
False)
print('the root mean square error is ', RMSE)
```

Running this code results in the following output:

```
R2 score is  0.8259169101408546
model coefficients:
 [ 1.02477193 -1.50441625 -2.06014001 -3.61709086 -0.49864781  2.81397581]
intercept:  23.34270072992701
the root mean square error is  3.298247031404574
```

Figure 11.12 – The result of using the log() transform on some variables of the car data in a linear model

Note that we repeated the manual splitting used in *Chapter 9* for consistency, but you have now learned how to use the `sklearn .train_test_split()` method for this purpose. If you recall from earlier, our RMSE on the original data was 3.24, so we have achieved a significant improvement by transforming non-linear predictors.

Exercise 11.1 – Linear regression

Here, you are working as an analyst for the Minneapolis Public Works, developing models for traffic on a key highway between Minneapolis and Saint Paul, Minnesota. You are provided initially with simple data (data from the UCI data repository, originally from `https://archive.ics.uci.edu/ml/datasets/Metro+Interstate+Traffic+Volume`) comprised of date-times and hourly traffic volume counts. The first step is to construct a simple linear regression model to begin to get an understanding of the data. Here, you will load the data and use the `sklearn LinearRegression()` method to build a model, and look at the model versus actual, and the residuals (errors).

1. For this exercise, all you will need is the pandas and numpy libraries, a module from sklearn, and matplotlib. Load them in the first cell of the notebook:

```
import pandas as pd
import numpy as np
from sklearn.linear_model import LinearRegresson as OLS
import matplotlib.pyplot as plt
```

2. Read the traffic_date.csv file into a DataFrame called my_data and use .head() to look at the first few rows:

```
my_data = pd.read_csv('Datasets\\traffic_date.csv')
my_data.head()
```

This should produce the following:

Out[2]:

	date_time	traffic_volume
0	10/2/2012 9:00	5545
1	10/2/2012 10:00	4516
2	10/2/2012 11:00	4767
3	10/2/2012 12:00	5026
4	10/2/2012 13:00	4918

Figure 11.13 – The traffic data

3. Convert the date_time column to a pandas datetime:

```
my_data.date_time = pd.to_datetime(my_data.date_time)
```

4. Take a subset of the data from September 1, 2018 going forward and plot it with respect to time:

```
traffic_subset = my_data.loc[my_data.date_time > '2018-
08-31', :]
traffic_subset.reset_index(drop = True, inplace = True)
fig, ax = plt.subplots(figsize = (11, 8))
ax.plot(traffic_subset.date_time, traffic_subset.traffic_
volume)
plt.xticks(rotation = 90, size = 12)
plt.yticks(size = 14)
plt.show()
```

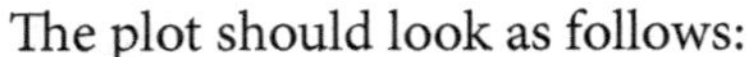
The plot should look as follows:

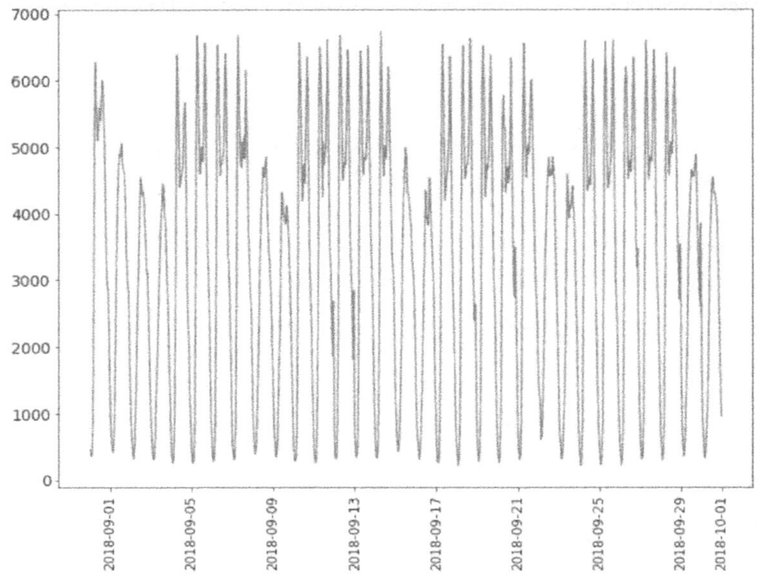

Figure 11.14 – The most recent month of the traffic data

5. Use the `sklearn` `LinearRegression` method to fit a linear model to the data over this time range, using the date-time as `X` and traffic volume as the `Y` variable:

```
lin_model = OLS()
model_X = \
    (np.reshape(np.array((traffic_subset.date_time -
                          traffic_subset.date_time[0]).
                         astype(np.int64)), (-1, 1)))
model_y = \
    np.array(traffic_subset['traffic_volume']).reshape(-
1, 1)
my_model = lin_model.fit(model_X, model_y)
print(my_model.intercept_)
print(my_model.coef_)
```

The output should be as follows:

```
[3079.01901131]
[[1.78032483e-13]]
```

Figure 11.15 – The intercept and slope of the ordinary least squares model

Note that the slope is very small. Given the apparent period nature of the data, you don't expect this to be a good model but instead something to consider if there is a trend over time. The small slope indicates a slight increase over time.

6. Replot the data and add the regression line to the plot:

```
fig, ax = plt.subplots(figsize = (11, 8))
ax.plot(traffic_subset.date_time, traffic_subset.traffic_
volume)
ax.plot(traffic_subset.date_time, my_model.
predict(model_X))
plt.xticks(rotation = 90, size = 12)
plt.yticks(size = 14)
plt.show()
```

The plot should appear as shown here:

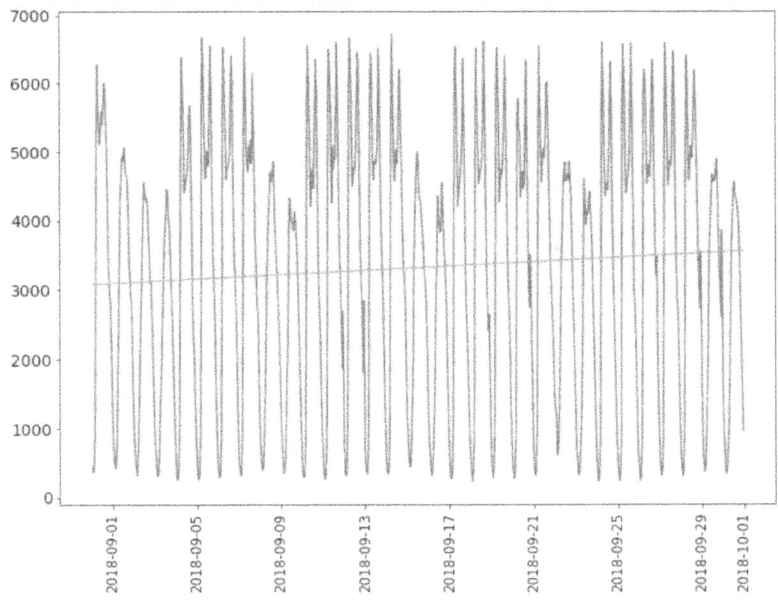

Figure 11.16 – A simple regression model of traffic versus time

It's very evident that a linear model using time as the independent variable is not sufficient. However, it gives you a good sense that the data is varying, likely during the course of every day, and you need to somehow account for that.

7. To finish this example, calculate the residuals and plot a histogram:

```
residuals = my_model.predict(model_X)[:, 0] - traffic_
subset.traffic_volume
fig, ax = plt.subplots(figsize = (9, 5))
ax.hist(residuals, bins = 50)
plt.show()
```

The data should appear as follows:

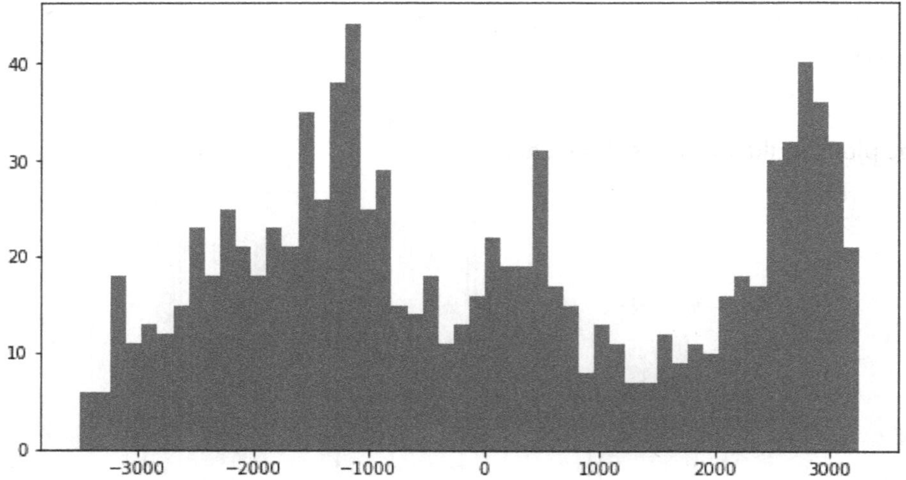

Figure 11.17 – A histogram of the residuals

We can see there are three dominant modes in the errors, one near the mean (a residual near 0), one much lower, and one much higher. Generally, you want the residuals to look more like a normal distribution, and *Figure 11.16* is far from that. It's evident that a simple linear model is not sufficient to model this data. Let's look at non-linear models as a possible improvement.

> **Note**
>
> You can find the code for this exercise here: `https://github.com/PacktWorkshops/The-Pandas-Workshop/tree/master/Chapter11/Activity11_01`.

Non-linear models

There are many other models available in `sklearn`, including non-linear models. As with the other `sklearn` methods we've already seen, pandas can integrate easily with these methods, allowing you to work with `DataFrames` and build powerful models. Let's look at the `sklearn` `RandomForestRegressor` method and compare it to the linear model, returning to the car `mpg` data. Here, you import `RandomForestRegressor` and then use it with default parameters to fit the same data as before:

```
from sklearn.ensemble import RandomForestRegressor
RF_model = RandomForestRegressor(random_state = 42)
X = my_data.iloc[train, 1:-1]
y = my_data.loc[train, 'mpg']
my_RF_model = RF_model.fit(X, y)
print('R2 score is ', my_RF_model.score(X, y))
RMSE = mean_squared_error(y, my_RF_model.predict(X), squared =
False)
print('the root mean square error is ', RMSE)
```

You should see the following output:

```
R2 score is  0.9790825512446402
the root mean square error is  1.143297473599363
```

Figure 11.18 – The resulting modeling car data with a RandomForest regression model

It's immediately evident that you are now accounting for nearly all the variation in the `mpg` data, and our RMSE error has dropped by a factor of about 3. Let's look at some additional ways to evaluate the model performance.

Model diagnostics

So far, you have seen some metrics such as R2 and RMSE to measure model performance. Also, graphical methods have been introduced to inspect the errors in predictions (the residuals). In addition to what you've learned by plotting residuals to investigate the quality of a model, in regression, there are a couple more powerful and important methods you can use.

Comparing predicted and actual values

In *Figure 11.15*, the prediction using simple linear regression was plotted on the same time series chart as the data. While this is very informative, another way to look at the model is to plot the predicted values versus the actual ones. In such a plot, if the scales are the same for `x` and `y`, then "perfect" predictions lie on a diagonal line. This makes it easy to see by inspection if there are trends at, for example, low or high values.

Here, the predictions for the linear model (using the log-transformed data) and the Random Forest model are plotted together against the actual values:

```
fig, ax = plt.subplots(figsize = (8, 8))
ax.scatter(y, my_model.predict(X_log),
           marker = 'o', s = 100,
           alpha = 0.35, facecolor = 'None', color = 'blue',
           label = 'linear regression, train data (log
transform)')
ax.scatter(y, my_RF_model.predict(X),
           marker = '^', s = 50,
           alpha = 0.35, facecolor = 'None', color = 'red',
           label = 'Random Forest, train data')ax.set_xlim(0,
50)
ax.set_ylim(0, 50)
ax.plot([0, 50], [0, 50], color = 'black')
ax.set_xlabel('actual mpg', fontsize = 14)
ax.set_ylabel('predicted mpg', fontsize = 14)
ax.legend(fontsize = 12)
plt.show()
```

This gives the following comparison:

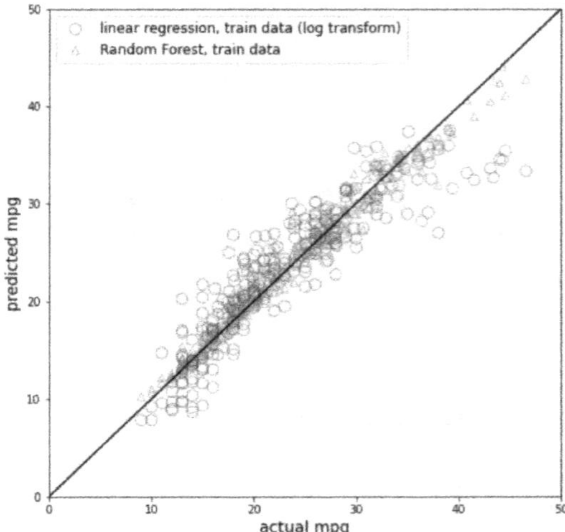

Figure 11.19 – Predictions using linear regression and random forest

You can see in *Figure 11.18* that the predictions from the Random Forest model are closer to the actual value than the predictions using linear regression. You can also see a trend in the linear regression predictions – they are low for both low and high actual values of `mpg`. For the Random Forest model, the predictions are nearer the "perfect" diagonal but trend to low at the high values of `mpg`.

Using the Q-Q plot

Another way to look at regression model performance is to consider how well the residuals conform to a normal distribution. Ideally, the residuals are random and normally distributed. In addition to looking at the histogram of residuals, a so-called **Q-Q plot** charts the predicted values versus the actual ones in terms of **quantiles**, which are just bins sampling the normal distribution. You may have used **Z-scores** in the past – a Z-score tells you how many standard deviations from the mean is a sample from a normal distribution. The Q-Q plot charts the observed Z-score for the actual residual distribution versus what is expected for a normal distribution. The power of this method is that it shows very clearly when the residuals are not conforming to the expected distribution, which can be hard to tell by visual inspection of the residual histogram. The `scipy` library provides a simplified version of a method to create the Q-Q plot called `probplot()`.

Here, `probplot()` is used to create the Q-Q plot, and the residual histogram is plotted alongside:

```
residuals = my_model.predict(X_log) - y
fig, ax = plt.subplots(1, 2, figsize = (11, 6))
probplot(residuals, plot = ax[0])
ax[1].hist(residuals, bins = 50)
ax[1].set_title('residuals')plt.show()
```

This generates the following chart:

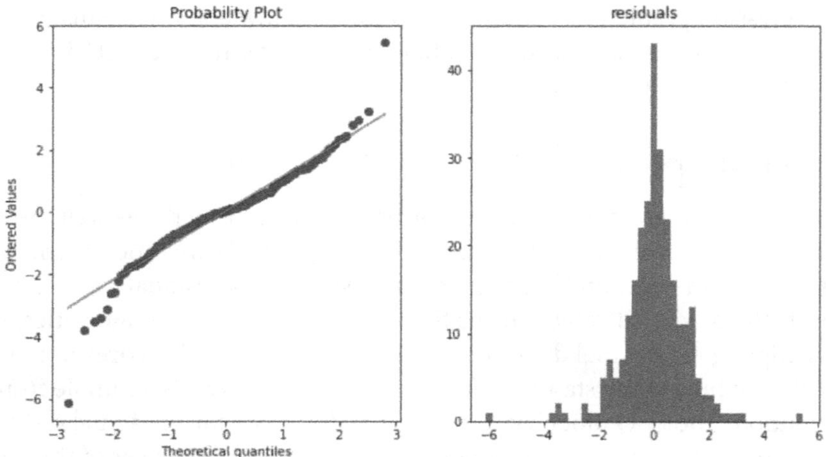

Figure 11.20 – The Q-Q plot and residual histogram for the Random Forest model of mpg

If you were inspecting the residual histogram, you might conclude that the results were really good. However, the Q-Q plot shows that the residuals beyond +/- 2 `mpg` are larger/smaller than ideal.

Continuing with the car data, we seem to have achieved a much better result. Before we celebrate too much, let's look at the validation data:

```
X_val = my_data.iloc[validation, 1:-1]
y_val = my_data.loc[validation, 'mpg']
val_pred = my_RF_model.predict(X_val)
val_RMSE = mean_squared_error(val_pred, y_val, squared = False)
print('the validation RMSE is ', val_RMSE)
```

```
the validation RMSE is  2.2932374329163605
```

Figure 11.21 – Validation RMSE for the car data using the RandomForest model

This result is an improvement over the 3.53 `mpg` error we had before but not nearly as good as indicated by the initial fit. This is a possible indication of **overfitting**. When modeling data with machine learning models, it is possible for a model to be powerful enough to "memorize" the training data but make larger errors on the unseen data. There are various strategies to mitigate overfitting that depend on the model type and the data. In general, most methods involve imposing some constraints on the model parameters, which results in somewhat poorer performance on the train data but better performance on new, unseen data. The code on GitHub summarizes where things stand now, and then we will attempt to improve the validation performance:

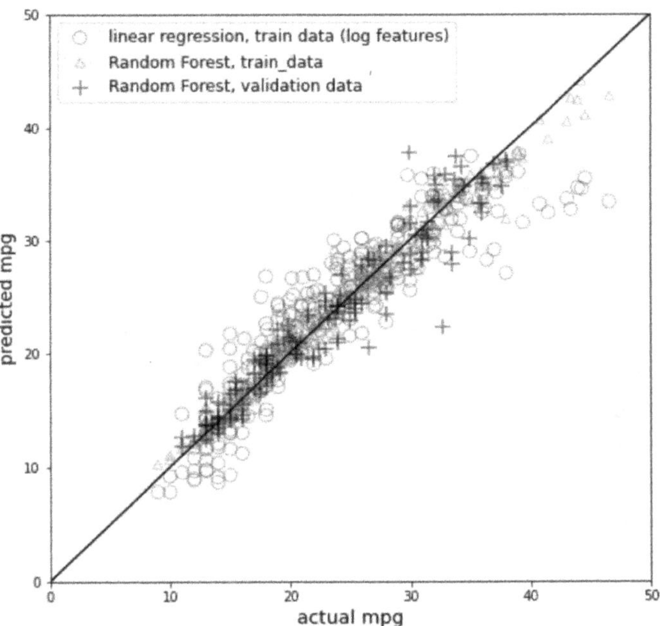

Figure 11.22 – A summary of predictions with the Random Forest and linear regression models

In *Figure 11.21*, you can see the scatter of the validation predictions is larger than for the training predictions. Now, we will introduce an approach to improve upon that.

For the `sklearn` `RandomForestRegressor` method, there are many parameters we can adjust that change the performance of the model. A common strategy is to search over a range of the most important parameters and choose models that provide good validation data performance (that is, attempt to optimize the model configuration for the validation data while training the model only on the training data). There is a range of methods in `sklearn` within the `model_selection` class; importantly, `GridSearchCV` allows us to set up a dictionary of parameters over which we want to search and then automatically try to optimize the model.

As we mentioned earlier, this optimization leaks information about the validation set into the model. However, for our purposes here, let's see how we might optimize for our validation data. In the following code, we load `GridSearchCV` to do the optimization search and then `PredefinedSplit`, which is used to force the search to use our validation split. Without the fixed split, `GridSearchCV` will perform multiple splits on the data randomly, which in general is a good strategy. Here, since we have a defined validation split, this method optimizes for that data. In each iteration of the grid search, the data is split, the training data is used to fit the model with the given set of parameters from the grid, and then the score for the validation data is recorded and used to select the best model. There are additional parameters available in the `RandomForestRegressor()` method; here, the choices are those that typically have a significant impact on such a model. Note that since the final parameters are a dictionary, `pprint` (Pretty Print) is loaded and used to print the best parameters:

```
from sklearn.model_selection import GridSearchCV as GSCV
from sklearn.model_selection import PredefinedSplit
import pprint
my_val_index = [-1 if i in validation else 0 for i in my_data.
index]
my_val = PredefinedSplit(test_fold = my_val_index)
X_grid = my_data.iloc[:, 1:-1]
y_grid = my_data.loc[:, 'mpg']
RF_grid = {'n_estimators': [900, 1100, 1300],
           'criterion' : ['mae', 'mse'],
           'min_samples_leaf' : [1, 2, 3],
           'max_features' : [2, 3, 4],
           'max_depth' : [15, 17, 19],
           'min_samples_split' : [2, 3, 4]}
best_model = GSCV(RandomForestRegressor(random_state = 42),
                  param_grid = RF_grid,
                  cv = my_val,
                  verbose = 1,
                  n_jobs = -1).fit(X_grid, y_grid)
print('best model:')
pprint.pprint(best_model.best_params_)
print(mean_squared_error(best_model.predict(X_val), y_val,
squared = False))
```

This will result in the following output:

```
Fitting 1 folds for each of 486 candidates, totalling 486 fits
best model:
{'criterion': 'mse',
 'max_depth': 15,
 'max_features': 4,
 'min_samples_leaf': 2,
 'min_samples_split': 2,
 'n_estimators': 900}
1.1678655741624786
```

Figure 11.23 – The output of running GridSearchCV using the RandomForest regression model

The RMSE for the validation data is significantly lower than before. Now, we visualize the results, using the code here: `https://github.com/PacktWorkshops/The-Pandas-Workshop/blob/master/Chapter11/Examples11.ipynb`.

This produces the following plot:

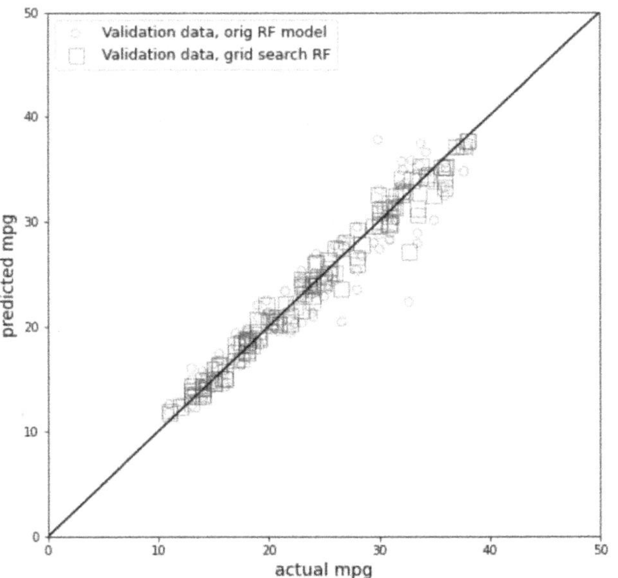

Figure 11.24 – The validation predictions with the original RF model and the optimized model

If you inspect *Figure 11.23*, you can see that the squares, which are the optimized validation results, are closer to the ideal line than the circles from the original `RandomForest` regression model.

This concludes our discussion on data modeling. We've reviewed the fundamentals of inspecting data for correlation, scaling data, and using both linear and non-linear models. We've used the `sklearn` methods for making predictive models and showed the basic optimization and validation approach, using the `sklearn` methods for model selection. A great aspect of combining `pandas` with `sklearn` is that once you have your cleaned, tabular data in a `pandas` `DataFrame`, the way it is consumed by a wide range of models is nearly identical. Therefore, if you start by getting your data into `pandas` and doing the preprocessing there, you can then rapidly experiment and optimize models to support your particular business case.

Exercise 11.02 – Multiple regression and non-linear models

In this exercise, you are again working as an analyst for the Minneapolis Public Works and are asked to revisit the traffic model. New data has been made available – in particular, information about weather and holidays, which may be useful to build a better model.

Load a `DataFrame` of multiple `X` variables and one `y`. Plot `y` versus each `X` and consider whether there are any potential non-linear relationships. Use the `sklearn` `LinearRegression` method to fit a multiple regression model. Compare the model to the data and inspect the residuals. Add a new `X` variable that is a non-linear function of an existing `X` and use `LinearRegression` again to fit a new model. Compare the model to the data and inspect the residuals. Fit a non-linear model using the `sklearn` `RandomForestRegressor` method. Compare the final model to the data, inspect the residuals, and plot the predicted values versus the actual values:

1. For this exercise, you will need the `pandas` and `numpy` libraries, three modules from `sklearn`, the `scipy` `probplot` method, and `matplotlib`. Load them in the first cell of the notebook:

```
import pandas as pd
import numpy as np
from sklearn.linear_model import LinearRegression as OLS
from sklearn.ensemble import RandomForestRegressor
from sklearn.preprocessing import StandardScaler
from scipy.stats import probplot
import matplotlib.pyplot as plt
```

2. Read the `Metro_Interstate_Traffic_Volume.csv` file into a `DataFrame` called `my_data`. We know there is a lot of data; let's focus on the last month by dropping data older than August 31, 2018. Reset the index after dropping the older data, and print out `head()` and `shape`:

```
my_data = pd.read_csv('Datasets\\Metro_Interstate_
Traffic_Volume.csv')
my_data = my_data.loc[my_data.date_time > '2018-08-31',
:]
my_data.reset_index(drop = True, inplace = True)
print(my_data.head())
print(my_data.shape)
```

This should produce the following:

```
  holiday    temp  rain_1h  snow_1h  clouds_all  weather_main  \
0    None  294.76     0.25      0.0          75          Rain
1    None  294.61     0.25      0.0          75          Rain
2    None  294.54     0.25      0.0          90          Rain
3    None  294.54     0.25      0.0          90  Thunderstorm
4    None  294.04     1.40      0.0          90          Rain

      weather_description            date_time  traffic_volume
0              light rain  2018-08-31 00:00:00             764
1              light rain  2018-08-31 01:00:00             456
2              light rain  2018-08-31 02:00:00             358
3  proximity thunderstorm  2018-08-31 02:00:00             358
4           moderate rain  2018-08-31 03:00:00             378
(968, 9)
```

Figure 11.25 – The traffic data

3. Convert the `date_time` column to a `pandas` `datetime` and create a new column, converting the time offset from the start to an integer seconds:

```
my_data.date_time = pd.to_datetime(my_data.date_time)
my_data['int_time'] = (my_data.date_time - my_data.date_
time[0]).astype(np.int64) / 1e9
```

Note that the `1e9` divisor is due to the fact that the `timedelta` returned by the subtraction has units of nanoseconds.

4. Collect the categorical variables in a list and then plot a bar plot of the traffic volume versus each value in each category. Use the method to plot the charts in a grid seen in other examples. You can check out the code here: `https://github.com/PacktWorkshops/The-Pandas-Workshop/blob/master/Chapter11/Exercise11_02/Exercise11.02.ipynb`.

The output should be as follows:

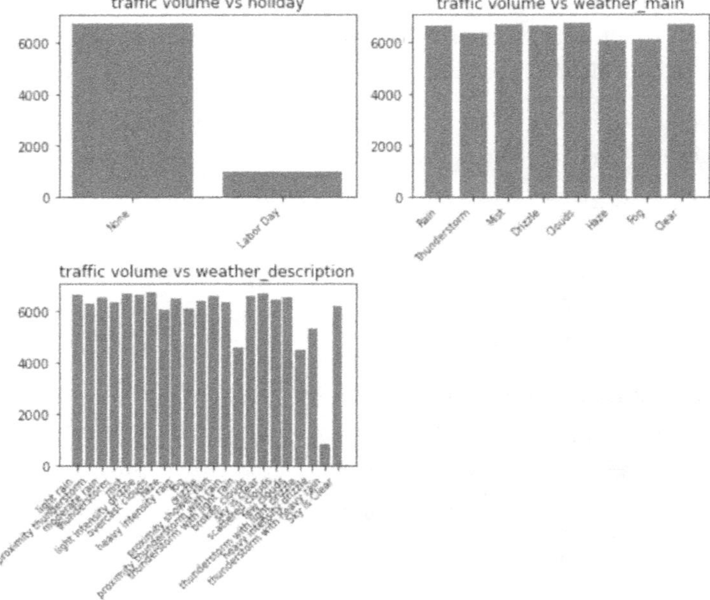

Figure 11.26 – The traffic volume versus the various categorical variables

5. Do the same for the numeric variables; exclude the target (`traffic_volume`) and the date/time variables from the list.

   The output should look as follows:

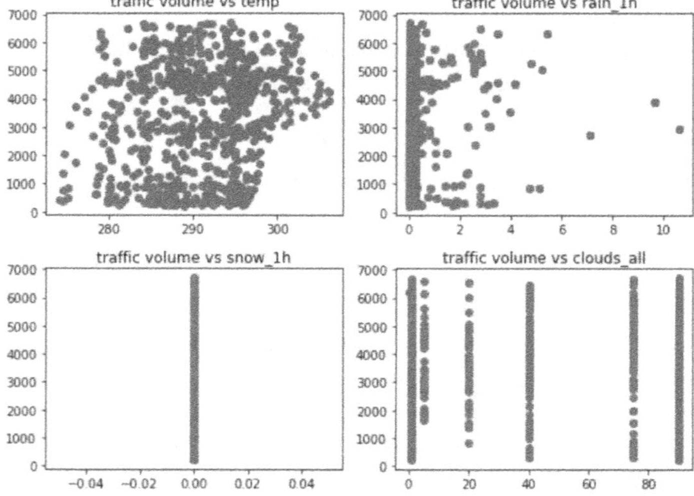

Figure 11.27 – The traffic volume versus the numeric variables

6. We see that the `snow` variable isn't providing any information. Drop the `snow_1h` column, replace the categorical variables with dummy variables, and display `head()` for the resulting data:

```
num_vars = [num_vars[i]
             for i in range(len(num_vars))
            if num_vars[i] != 'snow_1h']
model_data = pd.concat([my_data.loc[:, num_vars + ['int_
time']],
                        pd.get_dummies(my_data.loc[:,
cat_vars])],
                       axis = 1)
model_data.head()
```

The result should be as follows:

Out[46]:

	temp	rain_1h	clouds_all	int_time	holiday_Labor Day	holiday_None	weather_main_Clear	weather_main_Clouds	weather_main_Drizzle	weather_main_Fog	...
0	294.76	0.25	75	0.0	0	1	0	0	0	0	...
1	294.61	0.25	75	3600.0	0	1	0	0	0	0	...
2	294.54	0.25	90	7200.0	0	1	0	0	0	0	...
3	294.54	0.25	90	7200.0	0	1	0	0	0	0	...
4	294.04	1.40	90	10800.0	0	1	0	0	0	0	...

5 rows × 36 columns

Figure 11.28 – The updated data for modeling

7. Fit a linear regression model to predict traffic volume, and print out the coefficients, the intercept, and the R2 score:

```
y = my_data.traffic_volume
mls_model = OLS()
mls_model.fit(model_data, y)
print(mls_model.coef_)
print(mls_model.intercept_)
print(mls_model.score(model_data, y))
```

The result should be as follows:

```
[ 8.27936773e+01  1.23720714e+01  8.25196608e+00  4.17553926e-04
 -6.97245578e+02  6.97245577e+02  1.08786701e+03  2.46037578e+01
  3.06313660e+02 -2.57506459e+02 -2.98144206e+02 -2.09165922e+02
  4.80612430e+01 -7.02029087e+02  2.52845633e+03  2.38788198e+02
 -4.06271139e+02  7.09318270e+01 -2.57506459e+02 -2.98144206e+02
  1.19834748e+03  1.50177878e+02 -4.85762684e+02 -6.24466385e+02
 -2.09165922e+02 -2.45963831e+02 -3.88937455e+02  7.68313580e+02
  7.45203887e+02  2.46047307e+03  1.03821187e+02 -1.44058932e+03
 -1.16769699e+03 -2.59254554e+03 -6.03660704e+02  4.56197190e+02]
-22174.31595560447
0.12478430548635289
```

Figure 11.29 – The coefficients and R2 score for the initial linear model

8. Generate a plot of the traffic data and overlay the predictions:

```
fig, ax = plt.subplots(figsize = (11, 8))
ax.plot(my_data.date_time, my_data.traffic_volume)
ax.plot(my_data.date_time, mls_model.predict(model_data))
plt.xticks(rotation = 90)
plt.show()
```

The plot should look as follows:

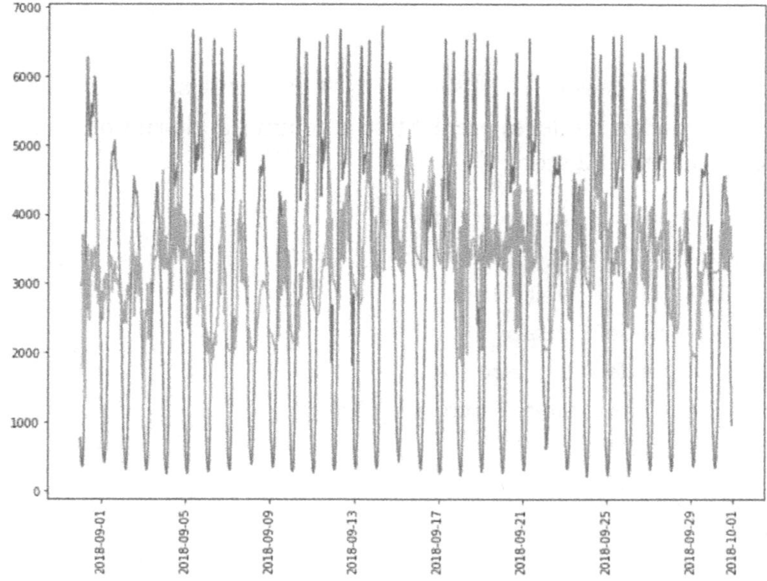

Figure 11.30 – The initial fit results with linear regression

Note that, although this is not a great model, using the additional variables has captured some of the behavior missed by the naïve linear model used earlier.

9. Plot the residuals for this fit:

```
residuals = pd.Series(mls_model.predict(model_data) -
                      my_data.traffic_volume)
fig, ax = plt.subplots()
ax.hist(residuals, bins = 50)
plt.show()
```

The plot should look as follows:

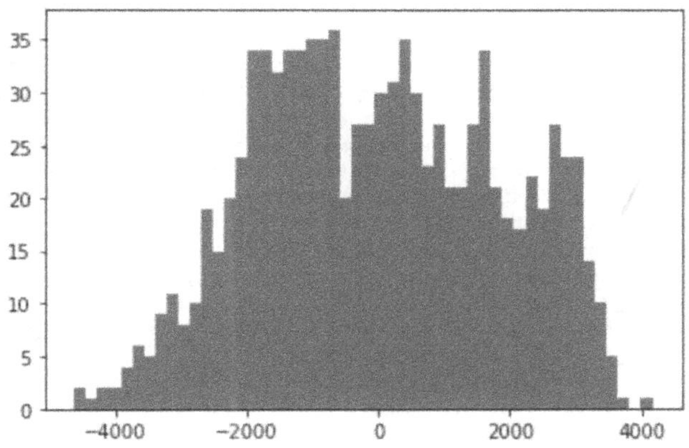

Figure 11.31 – Residuals from the initial model fit

We can see this is already an improvement over our naïve linear model versus time from the previous exercise.

10. It's evident adding in the environmental factors has improved the model, but there is a very clear cycle that appears to be daily. This seems intuitively correct for traffic data, so let's add a feature that can account for that. A method to provide a repeating cycle to a linear model is to add a sinusoidal feature. The general formula is as follows:

$$f(t) = a_1 * \sin\left(\frac{2 * \pi * t}{period}\right) + a_2 * \cos\left(\frac{2 * \pi * t}{period}\right)$$

Here, a_1 and a_2 are unknown coefficients that, taken together, account for any time offset (or "phase" – this is a mathematical property that uses both the sin and cos terms). We can add two variables, one as the sin of time and the other as the cos of time, and let the regression model fit a_1 *and* a_2 along with the other coefficients.

Add a column for the sin and the cos. We assume from looking at the charts that the period is 1 day. Since our time is in seconds, we'll have to divide by 24 * 60 * 60 to convert to days:

```
model_data['sin_t'] = np.sin(2 * np.pi * model_data.int_
time / 24 * 60 * 60)
model_data['cos_t'] = np.cos(2 * np.pi * model_data.int_
time / 24 * 60 * 60)
model_data.columns
```

The columns output should be as follows, with the two added columns last:

```
Out[52]: Index(['temp', 'rain_1h', 'clouds_all', 'int_time', 'holiday_Labor Day',
            'holiday_None', 'weather_main_Clear', 'weather_main_Clouds',
            'weather_main_Drizzle', 'weather_main_Fog', 'weather_main_Haze',
            'weather_main_Mist', 'weather_main_Rain', 'weather_main_Thunderstorm',
            'weather_description_Sky is Clear', 'weather_description_broken clouds',
            'weather_description_drizzle', 'weather_description_few clouds',
            'weather_description_fog', 'weather_description_haze',
            'weather_description_heavy intensity drizzle',
            'weather_description_heavy intensity rain',
            'weather_description_light intensity drizzle',
            'weather_description_light rain', 'weather_description_mist',
            'weather_description_moderate rain',
            'weather_description_overcast clouds',
            'weather_description_proximity shower rain',
            'weather_description_proximity thunderstorm',
            'weather_description_proximity thunderstorm with rain',
            'weather_description_scattered clouds',
            'weather_description_sky is clear', 'weather_description_thunderstorm',
            'weather_description_thunderstorm with heavy rain',
            'weather_description_thunderstorm with light drizzle',
            'weather_description_thunderstorm with light rain', 'sin_t', 'cos_t'],
           dtype='object')
```

Figure 11.32 – The model data columns after adding sin and cos

11. Now, fit the model again, using `LinearRegression`, and print out the coefficients, intercept, and score:

```
mls_model = OLS()
mls_model.fit(model_data, y)
print(mls_model.coef_)
print(mls_model.intercept_)
print(mls_model.score(model_data, y))
```

The result should be as follows:

```
[ 8.28107060e+01  1.26203181e+01  8.24244476e+00  4.16119259e-04
 -6.98263168e+02  6.97891031e+02  1.08771818e+03  2.45839013e+01
  3.08964701e+02 -2.56895343e+02 -2.97414530e+02 -2.09558127e+02
  4.82579100e+01 -7.04325158e+02  2.52967857e+03  2.39476255e+02
 -4.08640308e+02  7.01252031e+01 -2.56859474e+02 -2.97365101e+02
  1.20821386e+03  1.49254877e+02 -4.89857091e+02 -6.24935205e+02
 -2.09245319e+02 -2.46088298e+02 -3.89358352e+02  7.67961945e+02
  7.47634289e+02  2.46226663e+03  1.02939561e+02 -1.44040378e+03
 -1.16767842e+03 -2.59893840e+03 -6.05853914e+02  4.58041745e+02
 -4.44455705e+07 -2.34378006e+01]
-22156.158917458004
0.12479213645488285
```

Figure 11.33 – The updated model performance

We see only a small improvement in R2 (the last value, obtained using the `.score()` method). Note that the coefficient of the sin term is very large compared to the cos term; this indicates that there isn't much time offset.

12. Now that we have features to explicitly account for the periodic nature of the data, let's fit the data with a non-linear model. Scale the data using `StandardScaler()` and then, fit with `RandomForestRegressor()`:

```
scaler = StandardScaler()
scaled_model_X = scaler.fit_transform(model_data)
RF = RandomForestRegressor(random_state = 42)
RF = RF.fit(scaled_model_X, my_data.traffic_volume)
RF.score(scaled_model_X, my_data.traffic_volume)
```

This should produce the following:

```
Out[56]: 0.9481262295271707
```

Figure 11.34 – The R2 for the RandomForest model

You achieved an improvement of 0.83 in the R2, increasing by 0.95. Next, inspect the results.

13. Plot a histogram of the residuals:

```
residuals = pd.Series(RF.predict(scaled_model_X) -
                      my_data.traffic_volume)
fig, ax = plt.subplots(figsize = (11, 8))
ax.hist(residuals, bins = 50)
plt.show()
```

This produces a nice looking result, as follows:

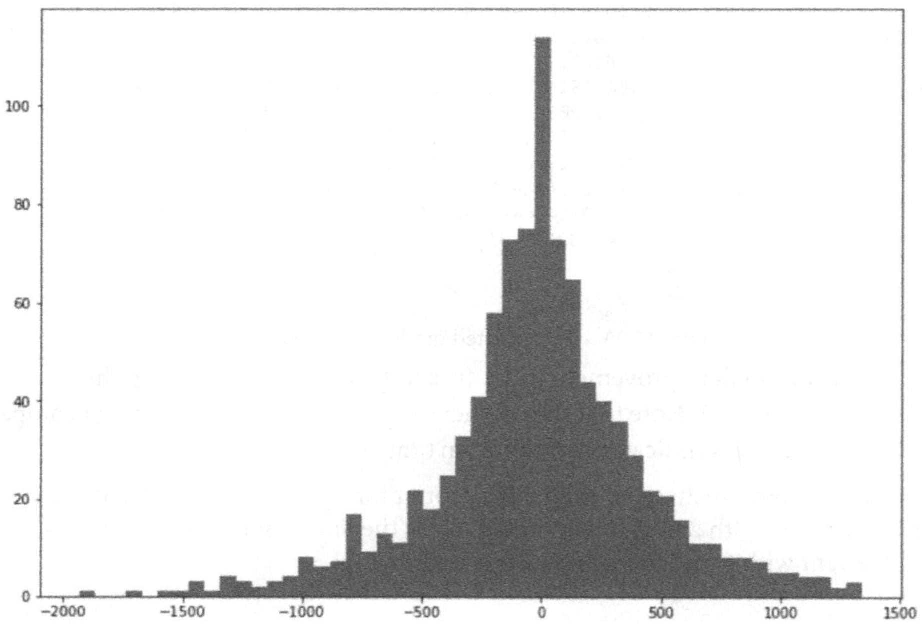

Figure 11.35 – The residuals from the RandomForest model fit

14. Now, as before, plot the traffic data and overlay the model predictions:

```
fig, ax = plt.subplots(figsize = (11, 7.5))
ax.plot(my_data.date_time, my_data.traffic_volume,
        label = 'actual traffic')
ax.plot(my_data.date_time, RF.predict(scaled_model_X),
        label = 'predicted traffic')
ax.legend(fontsize = 12)
ax.set_ylabel('Cars per hour', fontsize = 14)
ax.tick_params(labelsize = 12)
ax.set_title('Prediction of traffic volume using weather
data',
             fontsize = 16)
plt.xticks(rotation = 90)
plt.show()
```

This provides the following result:

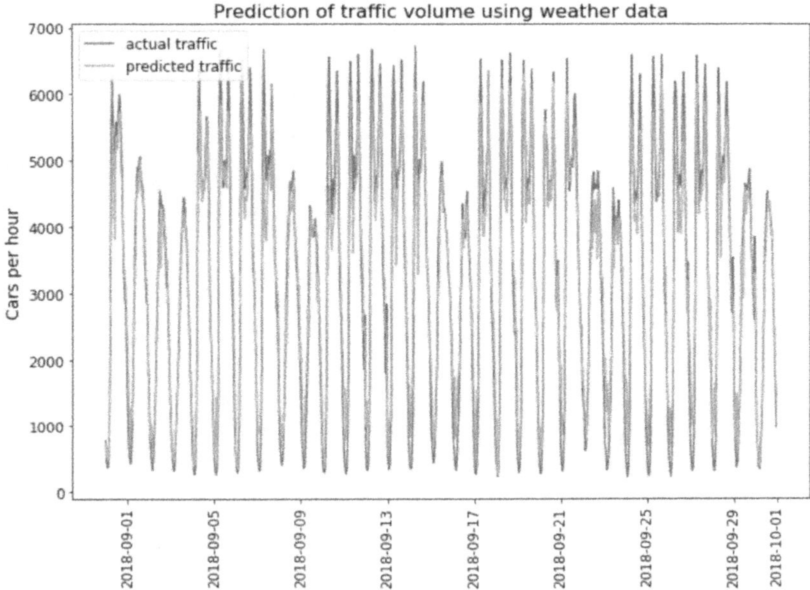

Figure 11.36 – The Random Forest model predictions on top of the traffic data

15. As a final step, when using regression models, a very good and intuitive visualization is to plot the predicted values versus the actual values in a scatterplot where the x and y scales are the same. In such a plot, perfect predictions lie directly on the diagonal, and we can see very easily how well the model performs and whether it behaves differently over different values of the predictions. Make such a plot for these results:

```
fig, ax = plt.subplots(figsize = (11, 11))
ax.scatter(my_data.traffic_volume,
           RF.predict(scaled_model_X))
ax.plot([0, 7000], [0, 7000], color = 'black', lw = 1)
ax.set_xlabel('Actual cars/hour', fontsize = 14)
ax.set_ylabel('Predicted cars/hour', fontsize = 14)
ax.set_title('Model performance\nRandom Forest
regression', fontsize = 16)
plt.tick_params(labelsize = 12)
plt.show()
```

This produces the following plot:

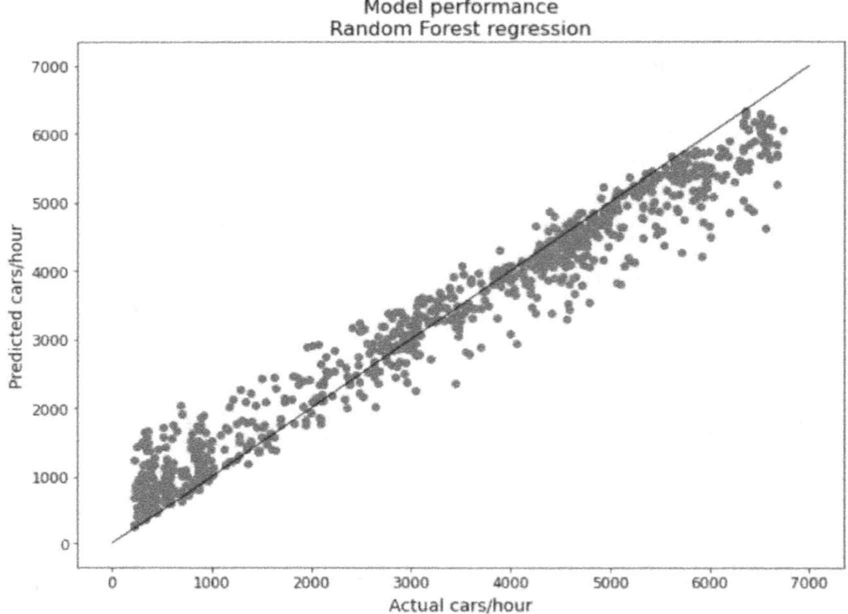

Figure 11.37 – A plot of the predicted values versus the actual values for the traffic data

Note an interesting aspect of this chart. The predictions at low values are too high and smoothly transition to under-predicting at high values. This is a behavior of some tree-based models. We can try to optimize the model to see whether it improves, but sometimes you can still see this behavior. There are methods to apply corrections to such results and improve the final model, but those are beyond the scope of what we are doing here.

16. As a final step, generate the Q-Q plot for the residuals:

```
residuals = RF.predict(scaled_model_X) - my_data.traffic_
volume
probplot(residuals, plot = plt)
plt.show()
```

This should generate the following result:

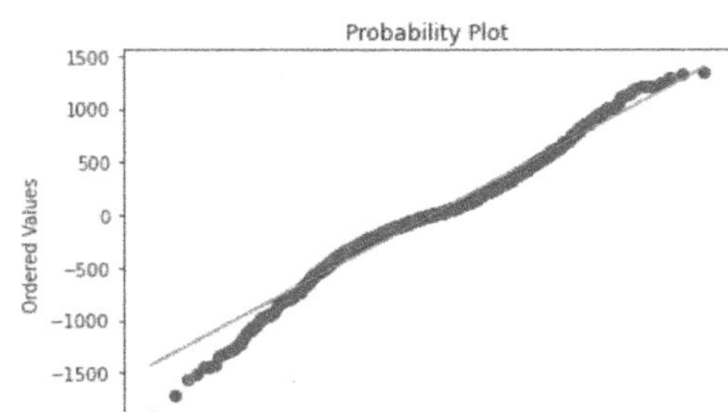

Figure 11.38 – A Q-Q plot of the residuals for the RF model

You can see in *Figure 11.37* that the predictions below the actual traffic volume are the largest deviations. Since you know from the previous step that those occur at high traffic volumes, that would be an area to investigate further to understand whether there are specific causes of the higher volumes.

In this exercise, you have reinforced all the methods learned so far for applying regression models to complex data, including using feature engineering (adding sin and cos) to achieve a better model. The final activity will challenge all those skills on a complex dataset.

Activity 11.01 – Multiple regression with non-linear models

As part of a research effort to improve metallic-oxide semiconductor sensors for the toxic gas carbon monoxide (CO), you are asked to investigate models of the sensor response for an array of sensors. You will review the data, perform some feature engineering for non-linear features, and then compare a baseline linear regression approach to a random forest model:

1. For this exercise, you will need the `pandas` and `numpy` libraries, and three modules from `sklearn`, `matplotlib`, and `seaborn`. Load them in the first cell of the notebook:

```
import pandas as pd
import numpy as np
from sklearn.linear_model import LinearRegression as OLS
from sklearn.ensemble import RandomForestRegressor
from sklearn.preprocessing import StandardScaler
import matplotlib.pyplot as plt
import seaborn as sns
```

2. As we have done before, create a `utility` function to plot a grid of histograms after being given the data, which variables to plot, the rows and columns of the grid, and how many bins. Similarly, create a utility function that allows you to plot a list of variables as scatter plots against a given x variable, also after being given the rows and columns of the grid.

3. Now, load the `CO_sensors.csv` file (the original data from `https://archive.ics.uci.edu/ml/datasets/Gas+sensor+array+temperature+modulation`) into a `DataFrame` called `my_data`.

   This should produce the following:

	Time (s)	CO (ppm)	Humidity (%r.h.)	Temperature (C)	Flow rate (mL/min)	Heater voltage (V)	R1 (MOhm)	R2 (MOhm)	R3 (MOhm)	R4 (MOhm)	R5 (MOhm)	R6 (MOhm)	R7 (MOhm)	R8 (MOhm)	R9 (MOhm)	R10 (MOhm)	F (MOh
0	0.000	0.0	49.21	26.38	247.2771	0.1994	0.5114	0.5863	0.5716	1.9386	1.1669	0.7103	0.5541	51.0146	40.8079	47.8748	4.6
1	0.311	0.0	49.21	26.38	243.3618	0.7158	0.0626	0.1586	0.1161	0.1347	0.1385	0.1545	0.1307	0.1935	0.1341	0.1773	0.1
2	0.620	0.0	49.21	26.38	242.4944	0.8840	0.0654	0.1496	0.1075	0.1076	0.1131	0.1363	0.1188	0.1195	0.1049	0.1289	0.1
3	0.930	0.0	49.21	26.38	241.6242	0.8932	0.0722	0.1444	0.1074	0.1032	0.1106	0.1306	0.1190	0.1125	0.1014	0.1232	0.1
4	1.238	0.0	49.21	26.38	240.8151	0.8974	0.0767	0.1417	0.1098	0.1025	0.1116	0.1284	0.1208	0.1111	0.1008	0.1226	0.1

Figure 11.39 – The CO sensor data

4. Use `.describe().T` to inspect the data further.

5. Use the histogram grid `utility` function to plot histograms of all columns, except `Time (s)`.

   The output should appear as follows:

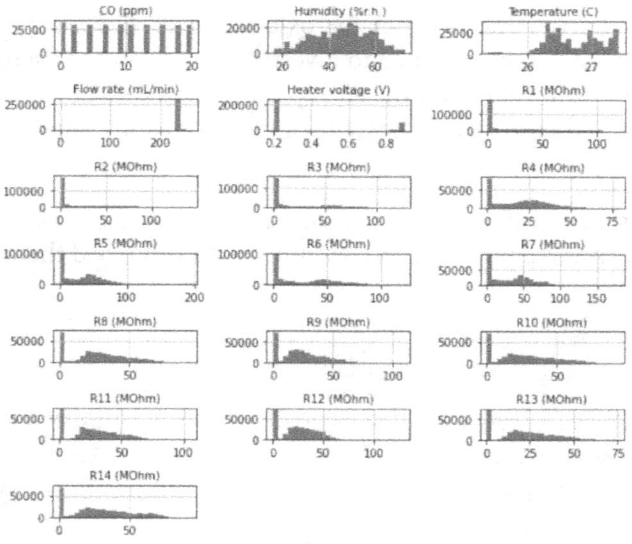

Figure 11.40 – Histograms of the sensor data

6. Use `seaborn` to generate a `pairplot` of the first five columns (excluding the sensor readings).

   The plot should look as follows (depending on how you defined your function and options):

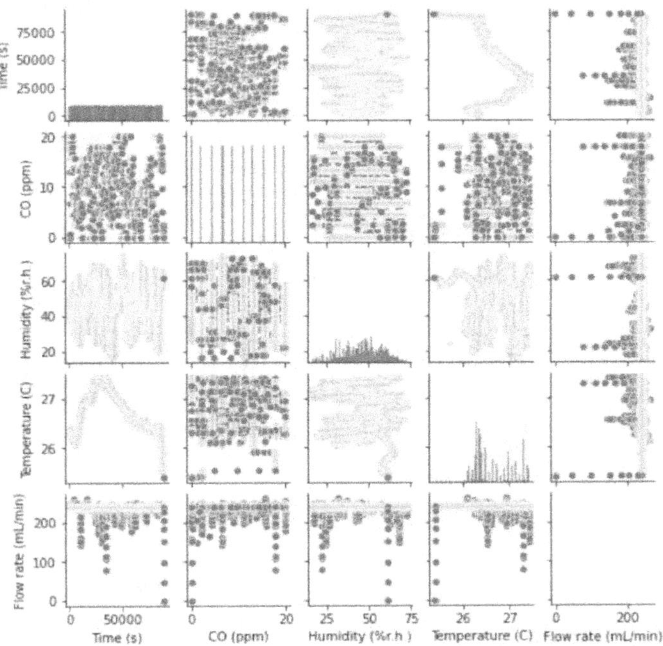

Figure 11.41 – A pairplot of the sensor data

7. Use the scatter plot grid `utility` function to plot all the sensor data versus time. The result should look as follows (depending on how you defined your function and options):

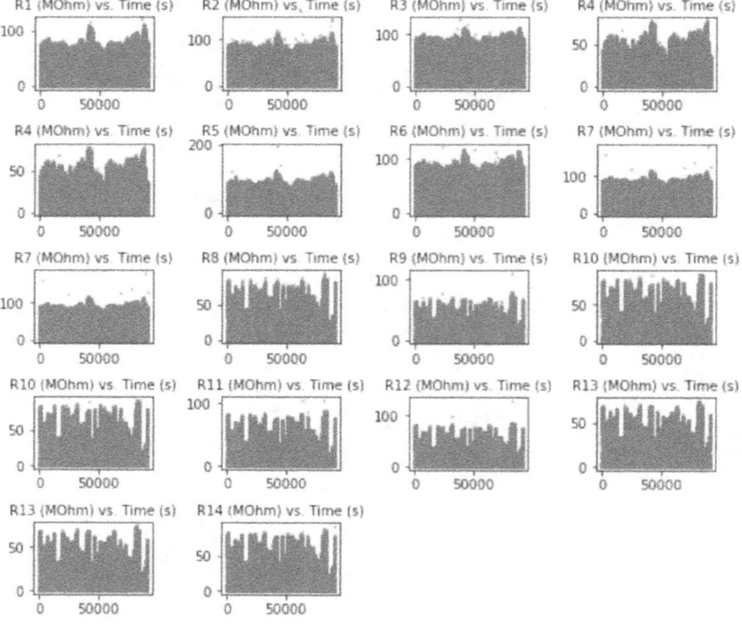

Figure 11.42 – The sensor traces versus time

8. It's difficult to tell whether there is a time dependency or a periodic component. Zoom in on **R13** over the time from **40000** to **45000** seconds. This detail should look as follows:

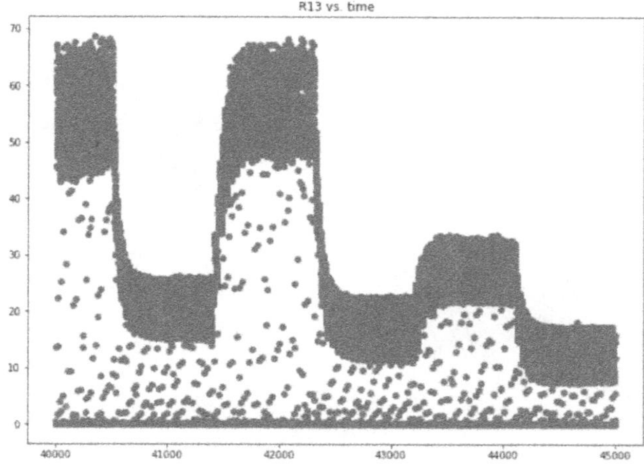

Figure 11.43 – Detail of one of the sensor traces

You can now see that the tests appear to comprise step functions of various sizes. This shows that the `time` variable is arbitrary and not useful for modeling the `CO` response. We can also see that there are a significant number of values that deviate from the steps, which may be due to the humidity variations, measurement errors, or some other issues. These may limit how well we can model the results.

9. Investigate the relationship of the changes in **R13** with the **CO** and **Humidity** values during one step change – for example, from about **41250** to **42500**. Plot the **R13** values using the `.plot()` method in `matplotlib`, and overlay the **CO** and **Humidity** values as line plots on the same plot.

   The resulting plot should look something like the following (depending on your choices):

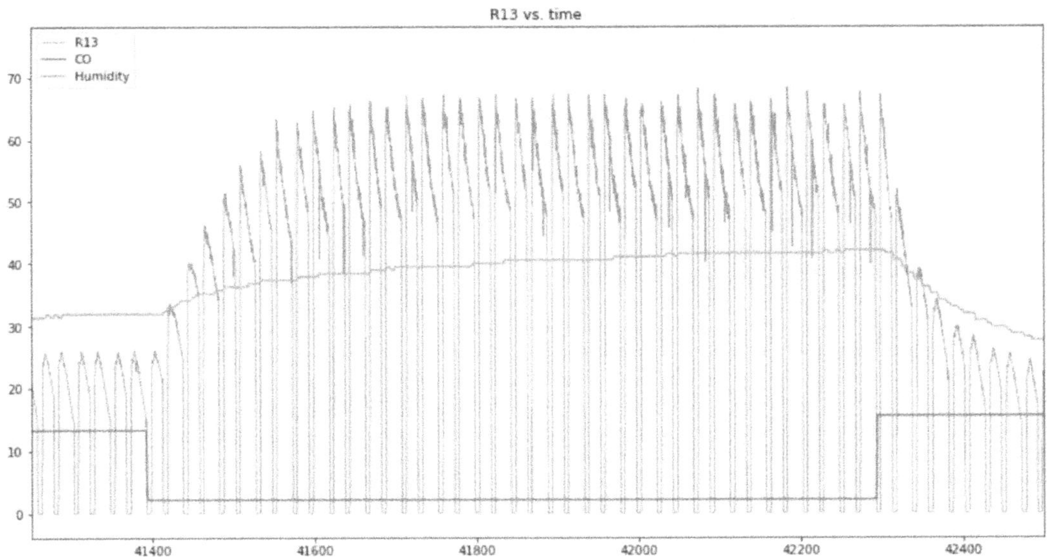

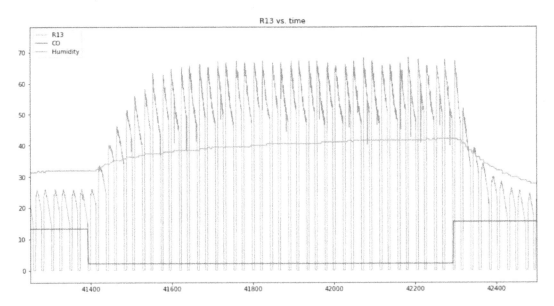

Figure 11.44 – R13, CO, and Humidity versus time

As you saw in the detail, there are a series of step changes for both **CO** and **Humidity**, resulting in changes in the resistance values. However, there are evident time lags involved, as shown by the curved traces of **Humidity**. In addition, **R13** seems to spike, then fall, and have intervening periods where the value appears to be **0** and at a steady state. Perhaps this is a function of the electronics, but further investigation would be required to be sure.

10. Now, use `seaborn` to plot a correlation `heatmap` for the sensor columns.

    The heatmap should look as follows:

```
Out[9]: <matplotlib.axes._subplots.AxesSubplot at 0x18d58f87908>
```

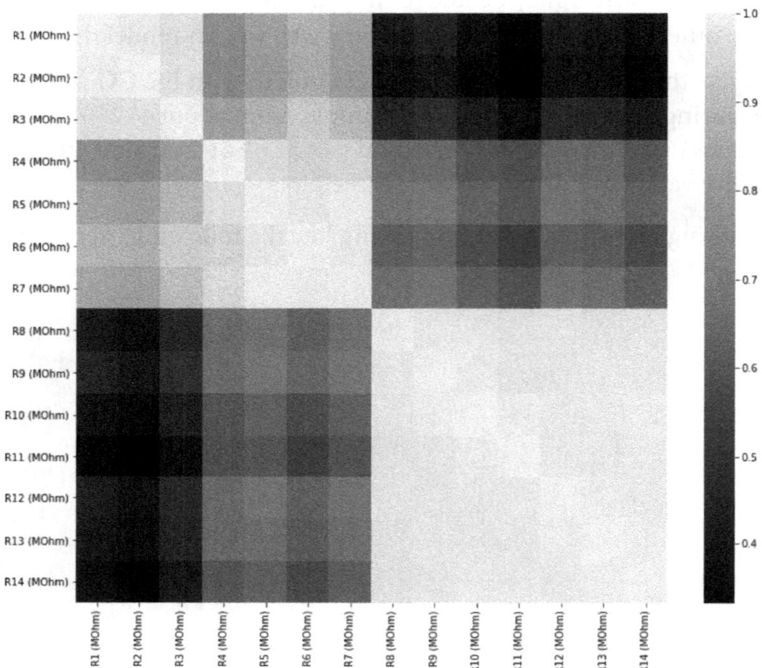

Figure 11.45 – Correlations among the sensors

    Note that there are two or three groups in the plot. The last seven sensors are all highly correlated with one another. The first three are as well, as are the next four.

11. The data description for this data (see `https://archive.ics.uci.edu/ml/datasets/Gas+sensor+array+temperature+modulation` and the provided text file in the `Datasets` directory) says there are two kinds of sensors: *"Figaro Engineering (7 units of TGS 3870-A04) and FIS (7 units of SB-500-12)."* Now, it is apparent that **R1** to **R7** are one kind of sensor, and **R8** to **R14** are the other kind. The data was collected to evaluate the performance of the sensors measuring CO at various conditions of temperature and humidity. In particular, the humidity is taken to be an "uncontrolled variable," and during the tests, random levels of humidity were imposed. In the field, the humidity would not be controlled or measured, which impacts the interpretation of data, especially for low levels of CO. The sensors output is reported as the resistance in MOhms, which are the main independent variables with which to predict CO. Temperature and the voltage applied to the sensor heater are also available.

Investigate the behavior of the sensors versus CO and humidity. Use the pandas `.corr()` method to generate the correlation matrix, and then use the first two rows of the result to make a `barplot` of the sensor correlations versus **CO** and **Humidity** respectively.

The resulting plots should look as follows:

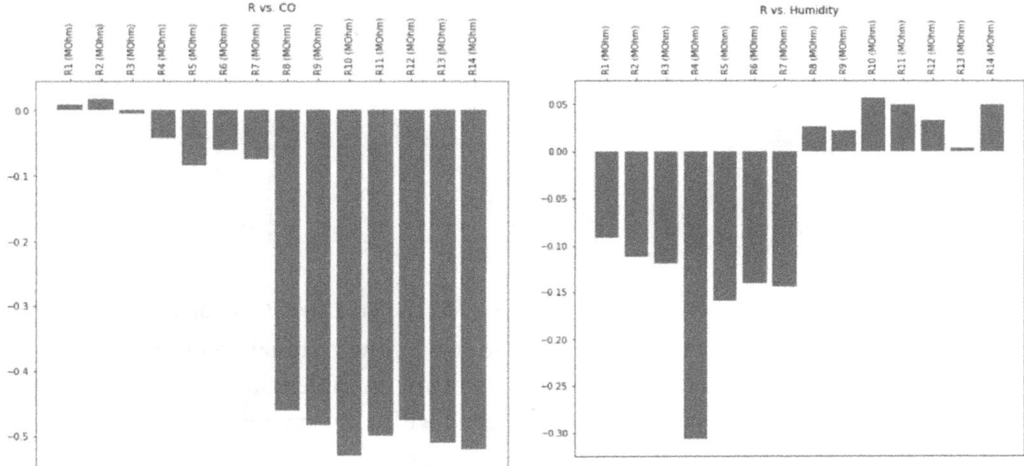

Figure 11.46 – Correlation of the sensor outputs to CO (left) and Humidity (right)

You see that while all the sensors are to measure the CO, they have markedly different behavior, depending on which of the two types we are measuring. From the problem description, it is evident that the sensors are impacted by humidity, but in the application, humidity is an uncontrolled and possibly unknown value. Hopefully, the different sensor behaviors can provide humidity information to a model and enable good predictions.

12. Apply a `sqrt()` transform to each of the sensor columns (since there are 0 or near- zero values, a log transform would not be appropriate) and add the columns to the dataset.
13. For the initial model, drop `Time`, `Humidity`, and `CO` from the `X` data. Use `CO` as the y data. Use `LinearRegression` to fit a model and plot the residuals, as well as the predicted versus actual values.

The output should look as follows:

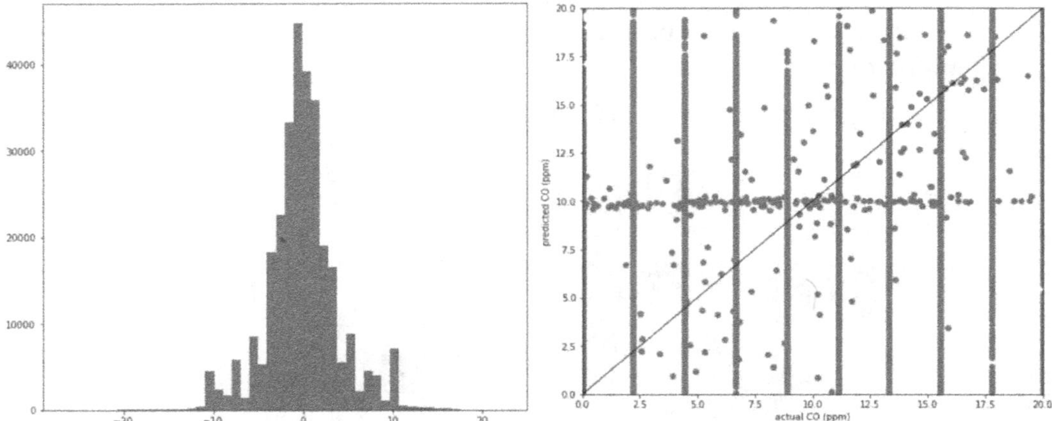

Figure 11.47 – The initial linear model – residuals (left) and predicted versus actual (right)

This model produces unbiased results, as shown by the residuals centered around 0, but from the second plot, we can see that there are multiple issues. There are groups of incorrect predictions at various levels, along with a clump near the middle of the predicted **CO** readings of **10** ppm. This result clearly is not acceptable.

14. Scale the data with `StandardScaler()` and then fit a `RandomForestRegressor()` method to the model. Plot the residuals and the predicted versus actual values.

    The plots should look as follows:

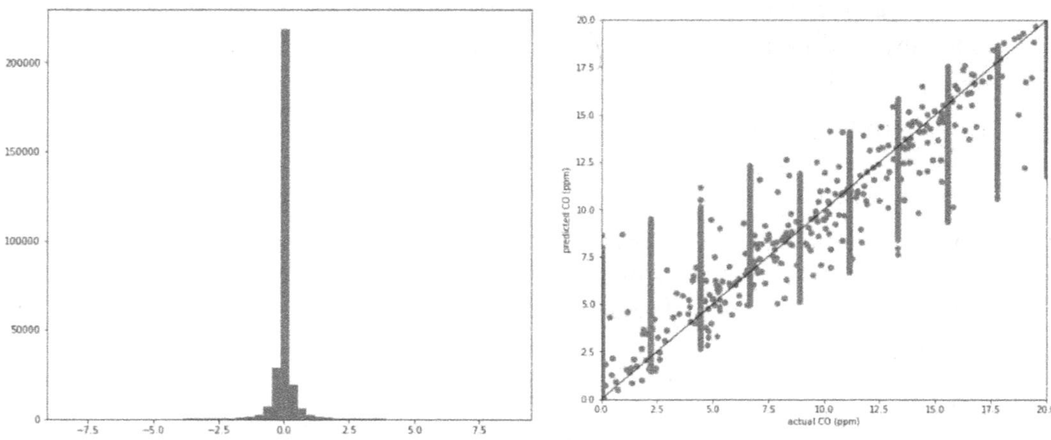

Figure 11.48 – The results of the RandomForestRegressor() fit

Although it is evident that the Random Forest model has reduced the residuals, the vertical groupings are still present, which is not a satisfactory result. Reviewing *Figure 11.43*, note that although the **CO** values are shown and nearly constant, there is a lag time in the humidity and sensor resistance values. A possible approach would be to average readings. A simple test of this idea is to group by the **CO** values and take the mean sensor values, and the model with those. In addition, it seems reasonable to filter out the regions where the resistance values drop to low values, as those seem anomalous.

15. Create a dataset, filtering out all the rows where a sensor resistance value drops to a low value, say 0.1. Then, group by `CO (ppm)` and aggregate as the mean values. Build a Random Forest model using the sensor mean resistances and the **CO** group values. Also, refit a linear regression model to this data. Plot the predicted values versus the actual values for both results.

    The resulting plot should look something like the following:

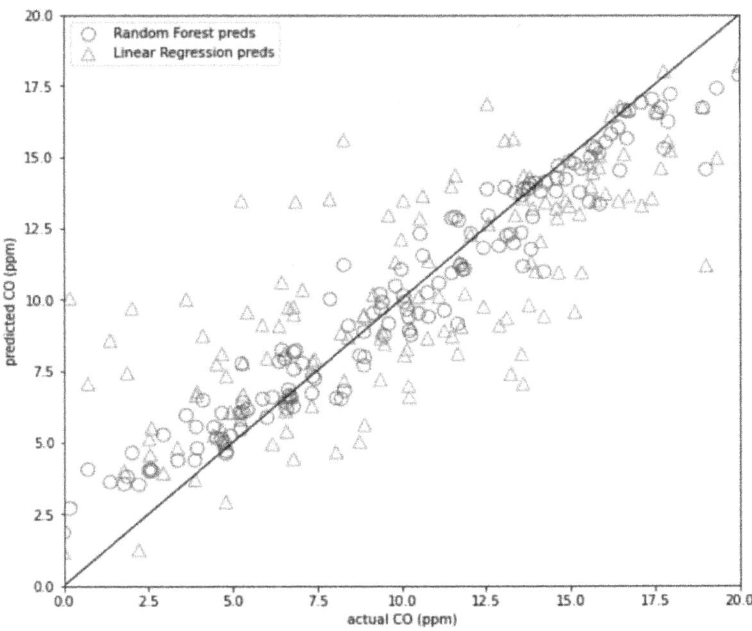

Figure 11.49 – The results using the grouped CO values and mean sensor values

These results are much better. This would require more discussion with expert stakeholders to confirm this approach, but it is a promising direction to obtain a calibration for the sensors. Note that the linear regression model cannot fit a lot of the data nearly as well. Also, note that the vertical scatter in the Random Forest predictions might be an indicator of noise caused by the random humidity values. This can be investigated further by building another model including **Humidity** as an independent variable.

> **Note**
>
> The solution for this activity can be found in the *Appendix*.

Summary

In this final chapter on *data modeling*, we have covered a wide range of topics on regression as a way to model data and make predictions. You learned how to make linear regression models as well as non-linear models, and ways to properly prepare data for such modeling. Metrics such as the **Sum of Squared Errors** (**SSE**) and the **Root Mean Squared Error** (**RMSE**) were introduced to assess the quality of models fitting data. In addition, visual techniques such as inspecting the histogram of residuals, Q-Q plots, and plotting predicted values versus actual values were shown to be important and easily used tools to determine the quality of a model.

You learned that even with simple linear models, some modest feature engineering such as transforming independent variables (the square root or log, for example) can improve results, at the cost of making it difficult to interpret the model coefficients. The common case of time series data with periodic features (such as daily or weekly) was addressed, adding new features such as sin/cos of time.

In addition to introducing the Random Forest model as a powerful non-linear regression technique, the power of the `sklearn` library's grid search cross-validation was demonstrated as a way to further improve such models by optimizing hyperparameters. As a final reinforcement of your learning, you solved a complex sensor array-modeling task using everything you have learned here.

You are now well-equipped to approach many technical and business data challenges. Nearly everything in this chapter and the two preceding ones applies equally to many classification problems, as well as a wide range of regression ones. You should be confident to make initial explorations, visualizations, and models that you may encounter in business or other work. We will now start diving into advanced usecases for pandas, starting with time series.

Part 4 – Additional Use Cases for pandas

This section covers some use cases for pandas, especially time series. We will learn about the various capabilities of pandas time series, some of the use cases, along with exercises that you can work on. We will close the book with various real-world case studies that utilize pandas as a key part of data processing and modeling.

This section contains the following chapters:

12 Using Time in pandas

In this chapter, you will focus entirely on one kind of data that is, on one hand, quite common in data analysis but, on the other hand, requires a number of special considerations – time series. **pandas** provides many methods specifically for working with time series. Here, you will learn about time-aware data types (such as `datetime` and `Timedelta`). In *Chapter 13, Exploring Time Series*, you will learn how to use these in the index to enable advanced capabilities such as resampling to different time intervals, interpolating, and modeling as a function of time.

You will cover the following topics as you work through this chapter:

- What are datetimes?
- Activity 12.01 – understanding power usage
- Datetime math operations

Introduction to time series

Time series data is nearly ubiquitous but can be a pain point in many analyses. For example, suppose you are asked to forecast sales for a retail store and are given daily sales figures for the last 6 months. When you review the data, you realize the store is usually open 5 days a week but sometimes has sales on Saturdays and even some Sundays. This makes most weekend days have missing values, and the time interval of the data is inconsistent. Also, when you consider estimating a monthly forecast, you realize months are of different lengths and have varying numbers of sales days. As simple and obvious as the issues are, they create a number of issues in analyzing and modeling the data over time.

The machine learning literature and popular articles are heavily biased toward classification problems, with little mention of time series. Yet much of the data we deal with is time series or at least starts out that way. Time series is a general term used to refer to data that is naturally ordered by time. For example, tweets arrive as a stream of timestamped data. Similarly, store transactions or online credit card transactions are time series. The log streams from data centers are time series.

It's important to note that, unlike tabular data in classification problems, time series data is **ordered**. In tabular data, random samples are shuffled before being used in a model. In time series, the order matters and we generally want to preserve it. The temporal relationship of events is critical; we can only recognize unusual server traffic if we analyze the sequence of data compared to *normal* use periods. The time sequence of store transactions can be compared day to day and over longer periods to anticipate high-demand periods for inventory and staff planning. The examples are endless.

pandas has a wide range of features to work with time series data. In the pandas documentation, it is noted that pandas time series objects are based on NumPy `datetime64` and `timedelta64` object types. pandas consolidates some useful methods from libraries such as `scikit.timeseries` (so much so that pandas will eventually absorb this library), and adds a lot of additional functionality used for working with time series data. In this chapter, we'll introduce some of the more important capabilities and review how to deal with timestamps in data. The key to understanding how time series differs from other pandas data structures is that pandas provides a couple of additional object types, namely `Timestamp` and `Timedelta`, as well as `Period`; we will review these beginning in the next section, *What are datetimes?*. Also, recall the importance of the index in pandas; for many operations with time series, we will make `index` one of these new object types, instead of working exclusively with integers or labels, as we did in the previous chapters till now. Making `index` a timestamp, for example, enables new functionality to simplify manipulating time series. Let's get started by understanding datetimes.

What are datetimes?

You probably already understand that in the computer memory, all numeric information is represented as ones and zeros, so at the most basic level, there isn't anything special about dates or times. However, when working with real data in business and technical projects, we tend to think about time or dates in their own units, differently from other numbers. Time is most often thought of as hours, minutes, or seconds, and dates are usually years, months, and days. Other common patterns are the weekdays, day of the week, business days, and quarters. We often group data into **bins** of days, weeks, months, or quarters. Within these bins, there might be data every second, minute, hour, or on some other or even random period. Because it is natural to think of dates and time of day together, Python in general, and pandas in particular, provides objects to make it easy to work this way. The most fundamental time component in pandas is `Timestamp`, and it is equivalent to `Datetime` in Python, provided by the `datetime` package. `Timestamp` is used as the index for pandas time series data types, as we'll see a bit later. pandas provides the `Timestamp` method to convert various types of data into timestamps. Here, we convert a string in a familiar date format into a pandas `timestamp`:

```
my_timestamp = pd.Timestamp('12-25-2020')
my_timestamp
```

This code snippet produces the following output:

```
Out[3]: Timestamp('2020-12-25 00:00:00')
```

Figure 12.1 – pandas timestamp object converted from a string in a date format

It's intuitive that `Timestamp` consists of year, month, day, hour, minute, and second. Since we did not provide any time information, `Timestamp()` assumes that the time portion is `00:00:00`. The fact that `Timestamp` has these components is what makes pandas times series operations so flexible. To illustrate that pandas is already simplifying things for us, here, we import Python `datetime`, and use it to accomplish the same conversion:

```
from datetime import datetime
my_datetime = datetime.strptime('12-25-2020','%m-%d-%Y')
my_datetime
```

This produces the following output:

```
Out[9]: datetime.datetime(2020, 12, 25, 0, 0)
```

Figure 12.2 – Accomplishing the string to datetime conversion using the datetime module

You can see that, in the case of using the `datetime` method, we have to provide a date format for Python to decode the string. The two resulting objects have very similar methods available to them.

Attributes of datetime objects

You might recall that since everything in Python is an object, we can use the `dir()` method to see what methods are attached to an object. In the code here, we use the `dir()` method on both of the preceding code snippets to return the methods attached to them. As we are mainly interested in the methods for routine coding, we use the `.startswith()` Python method (introduced in *Chapter 6*, *Data Selection*) in a list comprehension to remove the methods that have `'_'` in the first character (which denote a range of special behaviors in Python, such as ignoring such methods in some imports or allowing you to use reserved class names, all of which are outside of our scope here). We then combine them into one DataFrame using the pandas `.merge()` method, which performs a SQL-like merge, and by using `how = 'outer'`, we get the items from both sources, using the pandas `.fillna()` method to put `'-'` where there is not a match. You can check out the code here: `www.github.com/PacktWorkshops/The-Pandas-Workshop/blob/master/Chapter12/Examples.ipynb`.

To expand on the list comprehensions, they use `with dir(object)`, which is a method that produces a list of all the attached methods, then we add `i for i in dir()` so that we can iterate over the returned list. Finally, we use the `if` statement with `i.startswith("_")`, which says search for the underscore at the beginning of the item. Finally, we merge using `"outer"`, which includes all results from both DataFrames, and will return `NaN` if there is not a match, and the pandas `.fillna("-")` method replaces `NaN` with `"-"`. This effectively filters the results of `dir()` for *normal* methods we would use in our code. The `methods` DataFrame looks as follows (we have formatted it to combine the three printed groups):

	TS	DT		TS	DT		TS	DT
0	asm8	-	25	is_month_end	-	50	time	time
1	astimezone	astimezone	26	is_month_start	-	51	timestamp	timestamp
2	ceil	-	27	is_quarter_end	-	52	timetuple	timetuple
3	combine	combine	28	is_quarter_start	-	53	timetz	timetz
4	ctime	ctime	29	is_year_end	-	54	to_datetime64	-
5	date	date	30	is_year_start	-	55	to_julian_date	-
6	day	day	31	isocalendar	isocalendar	56	to_numpy	-
7	day_name	-	32	isoformat	isoformat	57	to_period	-
8	day_of_week	-	33	isoweekday	isoweekday	58	to_pydatetime	-
9	day_of_year	-	34	max	max	59	today	today
10	dayofweek	-	35	microsecond	microsecond	60	toordinal	toordinal
11	dayofyear	-	36	min	min	61	tz	-
12	days_in_month	-	37	minute	minute	62	tz_convert	-
13	daysinmonth	-	38	month	month	63	tz_localize	-
14	dst	dst	39	month_name	-	64	tzinfo	tzinfo
15	floor	-	40	nanosecond	-	65	tzname	tzname
16	fold	fold	41	normalize	-	66	utcfromtimestamp	utcfromtimestamp
17	freq	-	42	now	now	67	utcnow	utcnow
18	freqstr	-	43	quarter	-	68	utcoffset	utcoffset
19	fromisocalendar	fromisocalendar	44	replace	replace	69	utctimetuple	utctimetuple
20	fromisoformat	fromisoformat	45	resolution	resolution	70	value	-
21	fromordinal	fromordinal	46	round	-	71	week	-
22	fromtimestamp	fromtimestamp	47	second	second	72	weekday	weekday
23	hour	hour	48	strftime	strftime	73	weekofyear	-
24	is_leap_year	-	49	strptime	strptime	74	year	year

Figure 12.3 – Comparison of the time series methods and datetime methods

What you can notice from *Figure 12.3* is that there are `NaN` values in the `datetime` columns, because pandas provides more methods in `Timeseries` than `datetime`, for example, `dayofweek` and `dayofyear`. You might also notice that there are methods in both columns, for example, `microsecond`. Recall we mentioned that the pandas `Timeseries` objects are based on NumPy, and it turns out the underlying resolution of the `Timeseries` objects is nanoseconds. You may never need such fine-grained time, but it is actually available. It's good to know this, because in machine learning or data science, we will need to convert `Timestamps` or `datetimes` to integer values, and we will make use of the fact that they are stored in units of nanoseconds. Let's modify our source `datetime` string to see some of these attributes.

> **Note**
> For the remainder of this chapter, unless there could be confusion, we'll consider Timestamp and datetime to be equivalent.

Here, in the string, we include values for the hours, minutes, and seconds, and include values after the decimal point in the seconds down to nanosecond resolution. We then use the methods from our `methods` DataFrame to display the time using some formatting and a bit of math to resolve A.M. and P.M.:

```
my_time = pd.Timestamp('2020-12-25 15:05:09.001234987')
print(my_time.hour - 12, ('AM' if my_time.hour < 12 else 'PM'),
      my_time.minute, 'minutes',
      my_time.second, 'seconds',
      my_time.microsecond, 'microseconds',
      my_time.nanosecond, 'nanoseconds')
```

This code snippet creates the following output, which should intuitively be what you expect:

```
3 PM 5 minutes 9 seconds 1234 microseconds 987 nanoseconds
```

Figure 12.4 – Formatted time portion of the timestamp stored in my_time

pandas also provides the `.to_datetime()` method to convert objects to timestamps. Here, we first store `datetime` as a string object, then use the `.to_datetime()` method to convert to a `Timestamp` object:

```
string_date = '2020-07-31 13:51'
TS_date = pd.to_datetime(string_date)
TS_date
```

This produces the following:

```
Out[7]: Timestamp('2020-07-31 13:51:00')
```

Figure 12.5 – A timestamp created using the pandas .to_datetime() method

This completes the introduction to timestamps and prepares you to explore the range of methods pandas provides to manipulate time data in data from various sources, which may be incomplete, on periods we don't want to use, or that require other transformations. In the following exercise, you will work with some datetime information stored in a `.csv` file to explore a dataset.

Exercise 12.01 – working with datetime

In this exercise, as an analyst in an automotive company, you are given a `.csv` file containing data for a long-term test of an automobile engine. Your initial goal is to determine whether the power output is varying over the course of the test, such as decreasing over time or varying unusually. You will need to convert a date read in as a string to a timestamp, then create some columns from that for use in your analysis and create a plot to see the power output over time.

> **Note**
>
> You can find the code for this exercise here: `https://github.com/PacktWorkshops/The-Pandas-Workshop/tree/master/Chapter12/Exercise12_01`.

Perform the following steps to complete the exercise:

1. Create a `Chapter12` directory for all the exercises of this chapter. In the `Chapter12` directory, create an `Exercise12_01` directory.
2. Open your Terminal (macOS or Linux) or Command Prompt (Windows), navigate to the `Chapter12` directory, and type `jupyter notebook`. The Jupyter Notebook window should open.

   Select the `Exercise12_01` directory, which will change the Jupyter working directory to that folder, then click **New** | **Python 3** to create a new `Python 3` notebook, as shown in the following screenshot:

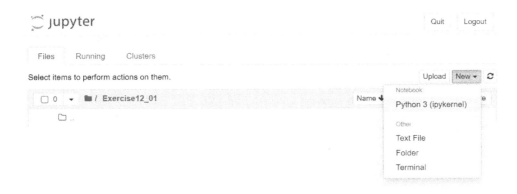

Figure 12.6 – Creating a new Python 3 Jupyter notebook

3. For this exercise, all you will need is the `pandas` library. Load it in the first cell of the notebook:

```
import pandas as pd
```

4. Read the `engine_power.csv` file into a pandas DataFrame and use the `.head()` method to inspect the first few lines:

```
engine_power = pd.read_csv('../datasets/engine_power.
csv')
engine_power.head()
```

> **Note**
> Please change the path of the dataset file (highlighted) based on where you have downloaded it on your system

The output should look as follows:

Out[14]:

	datetime	power
0	2020-1-1 00:00	221.403465
1	2020-1-1 01:55	327.370592
2	2020-1-1 03:50	223.272440
3	2020-1-1 04:04	328.380592
4	2020-1-1 05:45	329.109239

Figure 12.7 – The engine_power dataset

5. Print the first value in the `datetime` column, as well as printing its type, so you understand what the data type is. You know you want to have a `Timestamp` type, so this step informs what transformation will be needed:

```
print(engine_power.datetime[0])
print(type(engine_power.datetime[0]))
```

The output should be as follows:

```
2020-1-1 00:00
<class 'str'>
```

Figure 12.8 – The first value in the datetime column shows the data is stored as strings

You see the values are strings, so you need to convert those to timestamps.

6. Use the pandas `.to_datetime()` method to convert the `datetime` column to `Timestamps`:

```
engine_power['datetime'] = pd.to_datetime(engine_
power['datetime'])
engine_power
```

You should see an output as follows:

Out[9]:

	datetime	power
0	2020-01-01 00:00:00	221.403465
1	2020-01-01 01:55:00	327.370592
2	2020-01-01 03:50:00	223.272440
3	2020-01-01 04:04:00	328.380592
4	2020-01-01 05:45:00	329.109239
...	...	...
2114	2020-04-15 11:02:00	131.620792
2115	2020-04-15 11:16:00	8.703348
2116	2020-04-15 12:43:00	23.701833
2117	2020-04-15 12:57:00	110.785479
2118	2020-04-15 13:12:00	22.869297

2119 rows × 2 columns

Figure 12.9 – The engine_power data after converting the datetime column to Timestamps

7. Verify the result by using `type()` on the first element in the `datetime` column:

```
type(engine_power.datetime[0])
```

This should produce the following:

```
Out[10]:  pandas._libs.tslibs.timestamps.Timestamp
```

Figure 12.10 – The datetime column has been converted to pandas Timestamps

8. Now, you would like a column with the month as an integer. Recall earlier, we listed out all the methods available for the `Timestamp` objects, and one of those is `.month`. This method returns the calendar month of a timestamp. You need to iterate through all the items in the `datetime` column using the `.month` method on each one, so use a list comprehension to do that:

```
engine_power['month'] = [engine_power.datetime[i].month
                          for i in engine_power.index]
engine_power
```

This should produce the following result:

Out[11]:

	datetime	power	month
0	2020-01-01 00:00:00	221.403465	1
1	2020-01-01 01:55:00	327.370592	1
2	2020-01-01 03:50:00	223.272440	1
3	2020-01-01 04:04:00	328.380592	1
4	2020-01-01 05:45:00	329.109239	1
...	...	...	...
2114	2020-04-15 11:02:00	131.620792	4
2115	2020-04-15 11:16:00	8.703348	4
2116	2020-04-15 12:43:00	23.701833	4
2117	2020-04-15 12:57:00	110.785479	4
2118	2020-04-15 13:12:00	22.869297	4

2119 rows × 3 columns

Figure 12.11 – The engine_power data with a new column having integer month values

9. Another method provided by pandas is the `.day` method, which returns the day of the month. Use that to create a `day_of_month` column in the `engine_data` DataFrame. Print the DataFrame to confirm the addition:

```
engine_power['day_of_month'] = [engine_power.datetime[i].
day
        for i in engine_power.index]
print(engine_power)
```

This produces the following:

```
                datetime       power  month  day_of_month
0    2020-01-01 00:00:00  221.403466      1             1
1    2020-01-01 01:55:00  327.370592      1             1
2    2020-01-01 03:50:00  223.272440      1             1
3    2020-01-01 04:04:00  328.380592      1             1
4    2020-01-01 05:45:00  329.109239      1             1
...                  ...         ...    ...           ...
2114 2020-04-15 11:02:00  131.620792      4            15
2115 2020-04-15 11:16:00    8.703348      4            15
2116 2020-04-15 12:43:00   23.701833      4            15
2117 2020-04-15 12:57:00  110.785479      4            15
2118 2020-04-15 13:12:00   22.869297      4            15

[2119 rows x 4 columns]
```

Figure 12.12 – The updated DataFrame with day_of_month

10. Verify all days of the month are represented by printing the unique values that are in the `day_of_month` column:

```
print(engine_power.day_of_month.unique())
```

The result should be as follows:

```
[ 1  2  3  4  5  6  7  8  9 10 11 12 13 14 15 16 17 18 19 20 21 22 23 24
 25 26 27 28 29 30 31]
```

Figure 12.13 – Days of month include all values from 1 to 31

11. Now, verify that all the days of the week are represented using the `.weekday()` method in a list comprehension and the `.unique()` method as in the previous step:

```
pd.Series([engine_power.datetime[i].weekday()
           for i in engine_power.index]).unique()
```

This produces the following:

```
Out[9]: array([2, 3, 4, 5, 6, 0, 1], dtype=int64)
```

Figure 12.14 – The data including every day of the week

Although not a complete review, it appears that there are indeed daily values and no missing days. Therefore, you can make some summary plots by month and day to investigate the variation over time.

12. Use the `.groupby()` method on the `month` column to compute the average of power for each month, and the pandas `.plot()` method to make a simple chart to inspect the data:

```
(engine_power[['power', 'month']]. groupby(['month']).
mean()).plot()
```

This should produce the following:

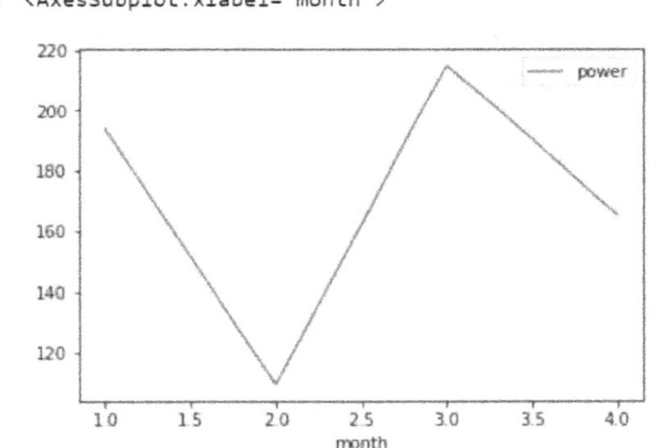

Figure 12.15 – The monthly values of average power from the engine test

13. As there is a large difference between the months, use another `.groupby()` method on `month` and `day_of_month` and plot again to see more detail.

This produces the following chart:

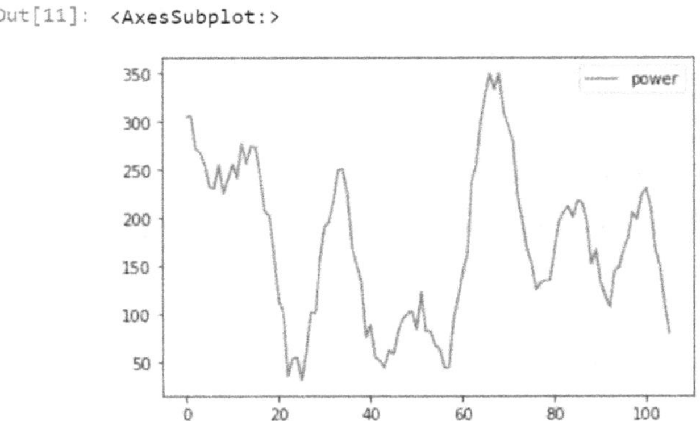

Figure 12.16 – The daily averages of engine power

You conclude from the two plots that there is significant variation over time without a clear trend. Your next step is to schedule a meeting with the stakeholders to understand whether there are known reasons for the variation or whether there are problems with this test.

We've now seen the basics of pandas `Timestamps` and `datetimes`, so let's dive a bit deeper into creating and manipulating these objects.

Creating and manipulating datetime objects/time series

To begin exploring pandas time series capabilities, let's create a time series. We saw in *Exercise 12.01* that we could create a series of timestamps by reading in some dates as strings and converting them. pandas provides several methods to generate sequences of timestamps. Here, we introduce the `.date_range()` method to generate a sequence of dates. In this method, we will need to define the start and end, and a `freq` argument, which tells pandas what period we want to fill from start to end. pandas again makes things easy, as we can specify intuitive values to `freq` such as `M` (month) and D (day), and here, we use `W` with the suffix `-MON`, meaning weeks starting on Mondays:

```
dates = pd.date_range(start = '2012-01-01',
                      end = '2019-12-31',
                      freq = 'W-MON')
print(type(dates))
```

This generates the following:

```
Out[14]: pandas.core.indexes.datetimes.DatetimeIndex
```

Figure 12.17 – Generating a range of dates using date_range

Although the result is an index, we can still print the `dates` object and inspect the contents:

```
print(dates)
```

This produces the following output:

```
DatetimeIndex(['2012-01-02', '2012-01-09', '2012-01-16', '2012-01-23',
               '2012-01-30', '2012-02-06', '2012-02-13', '2012-02-20',
               '2012-02-27', '2012-03-05',
               ...
               '2019-10-28', '2019-11-04', '2019-11-11', '2019-11-18',
               '2019-11-25', '2019-12-02', '2019-12-09', '2019-12-16',
               '2019-12-23', '2019-12-30'],
              dtype='datetime64[ns]', length=418, freq='W-MON')
```

Figure 12.18 – The values in the dates object

You see that the values start on `2012-01-02` instead of `2012-01-01`, which was specified, and similarly end on `2019-12-30` instead of `2019-12-31`. The reason is the specification of the frequency as `W-MON` means the result will have a period of 1 week with the weeks beginning on Mondays, and the first Monday in 2012 is the second, not the first of January, and the last Monday in 2019 is the thirtieth of December.

We can perform operations on the date values intuitively, such as subtraction, and we obtain datetime-like objects as a result called `TimeDelta`. In particular, a common pattern is to compute offsets, such as subtracting out the starting date. Let's inspect the type of the result when we subtract one of the values in our `dates` series from another:

```
type(dates[1] - dates[0])
```

This produces the following results:

```
Out[22]: pandas._libs.tslibs.timedeltas.Timedelta
```

Figure 12.19 – The result of subtracting two dates in pandas is a Timedelta

The `Timedelta` object is exactly what it sounds like; it stores the difference between two times. Several of the methods available for `Timedelta` are the same as `Timestamp` (recall, you can always see them using `dir()`), but importantly, a method provided for `Timedelta` not available in `Timestamp` is `.totalseconds()`. This makes intuitive sense in that a difference in times should be available in time units.

Here, we subtract the first date from all the dates so that we are starting at `0` and use the `.totalseconds()` method on the result to get seconds as the difference and divide twice by 60 (first for minutes per hour, then seconds per minute) to get values in minutes. Since the `.total_seconds()` method returns numeric data (no longer `Timedelta` or `datetime`), we can operate on the result in values as we can with any numeric object. To illustrate, a simple expression is applied to values (the minutes) and stored in the `values` variable to get a computed result (for example, if we had an expression for growth of some value versus time, we could be computing it this way):

```
values = pd.Series([(dates[i] - dates[0]).total_seconds()
                    for i in range(len(dates))]) / 60 / 60 #
convert to hours
values = 1037.65 + values**1.5 / 1000
```

Now, we use the familiar `DataFrame` constructor to combine the dates and values into a DataFrame:

```
time_series = pd.DataFrame({'date' : dates,
                            'value' : values})
time_series.head()
```

This produces the following output:

Out[12]:

	date	value
0	2012-01-02	1037.650000
1	2012-01-09	1039.827529
2	2012-01-16	1043.808982
3	2012-01-23	1048.964772
4	2012-01-30	1055.070231

Figure 12.20 – The time_series DataFrame constructed from the series of weekly dates

We said in the *What are datetimes?* section that `datetime` objects in pandas have underlying numeric data with a resolution of nanoseconds (10^{-9} seconds). Let's look at that in a bit more detail. Here, we use the pandas `.astype()` method to convert `datetimes` directly to integers then to days by dividing by 10^9 ($1e^9$) to convert from nanoseconds to seconds, then dividing by 60 twice to convert to hours, and finally, dividing by 24 to get days. We subtract the resulting day value for the first date from all the dates so that our days start at zero:

```
time_series['int_date'] = (dates.astype(int)/1e9/60/60/24 -
                           dates.astype(int)[0]/1e9/60/60/24)
time_series
```

This produces the following update to our DataFrame:

Out[13]:

	date	value	int_date
0	2012-01-02	1037.650000	0.0
1	2012-01-09	1039.827529	7.0
2	2012-01-16	1043.808982	14.0
3	2012-01-23	1048.964772	21.0
4	2012-01-30	1055.070231	28.0
...	...	...	...
413	2019-12-02	19313.980376	2891.0
414	2019-12-09	19380.399465	2898.0
415	2019-12-16	19446.898818	2905.0
416	2019-12-23	19513.478340	2912.0
417	2019-12-30	19580.137933	2919.0

418 rows × 3 columns

Figure 12.21 – The updated time_series DataFrame with the integer days as a new column, int_date

Time periods in pandas

pandas also provides the idea of a period in conjunction with timestamps. Here, we use the `.period_range()` method to generate a sequence of dates, similarly to how we used `.date_range`. The result is a unique pandas object, `PeriodIndex`, where the values are periods. This index type can be used in some business situations where things logically accumulate in certain periods (such as weeks, months, or quarters):

```
time_periods = pd.period_range('2020-01-01', periods = 13, freq
= 'W-MON') \ time_periods
```

This produces the following:

```
Out[51]: PeriodIndex(['2019-12-31/2020-01-06', '2020-01-07/2020-01-13',
                      '2020-01-14/2020-01-20', '2020-01-21/2020-01-27',
                      '2020-01-28/2020-02-03', '2020-02-04/2020-02-10',
                      '2020-02-11/2020-02-17', '2020-02-18/2020-02-24',
                      '2020-02-25/2020-03-02', '2020-03-03/2020-03-09',
                      '2020-03-10/2020-03-16', '2020-03-17/2020-03-23',
                      '2020-03-24/2020-03-30'],
                     dtype='period[W-MON]')
```

Figure 12.22 – A series of date ranges generated using pandas .period_range()

Note that each value is a date range. In this case, we specified `periods` and `freq` as `start`, and pandas aligns the `end` values with the `freq` argument, so these periods end on Mondays. Although potentially confusing, `period_range` has some useful properties, including being able to get the start and end of the periods, and easily shift periods with simple math. Here, we print out the start and end of all the periods:

```
print(time_periods.start_time, '\n', time_periods.end_time)
```

This produces the following:

```
DatetimeIndex(['2019-12-31', '2020-01-07', '2020-01-14', '2020-01-21',
               '2020-01-28', '2020-02-04', '2020-02-11', '2020-02-18',
               '2020-02-25', '2020-03-03', '2020-03-10', '2020-03-17',
               '2020-03-24'],
              dtype='datetime64[ns]', freq='W-TUE')
 DatetimeIndex(['2020-01-06 23:59:59.999999999',
               '2020-01-13 23:59:59.999999999',
               '2020-01-20 23:59:59.999999999',
               '2020-01-27 23:59:59.999999999',
               '2020-02-03 23:59:59.999999999',
               '2020-02-10 23:59:59.999999999',
               '2020-02-17 23:59:59.999999999',
               '2020-02-24 23:59:59.999999999',
               '2020-03-02 23:59:59.999999999',
               '2020-03-09 23:59:59.999999999',
               '2020-03-16 23:59:59.999999999',
               '2020-03-23 23:59:59.999999999',
               '2020-03-30 23:59:59.999999999'],
              dtype='datetime64[ns]', freq=None)
```

Figure 12.23 – The start and end times of each period

So, you can see that each period starts at the beginning of the day (all the 0s for the times aren't shown by default), and the end time is the very last nanosecond of the end day. If we needed to shift everything two periods, we just add 2:

```
time_periods + 2
```

This produces the following. Note that each period has been shifted by two periods:

```
Out[57]: PeriodIndex(['2020-01-14/2020-01-20', '2020-01-21/2020-01-27',
                      '2020-01-28/2020-02-03', '2020-02-04/2020-02-10',
                      '2020-02-11/2020-02-17', '2020-02-18/2020-02-24',
                      '2020-02-25/2020-03-02', '2020-03-03/2020-03-09',
                      '2020-03-10/2020-03-16', '2020-03-17/2020-03-23',
                      '2020-03-24/2020-03-30', '2020-03-31/2020-04-06',
                      '2020-04-07/2020-04-13'],
                     dtype='period[W-MON]')
```

Figure 12.24 – Adding two (periods) to every period on time_periods

Information in pandas time-aware objects

Because objects such as datetimes in pandas have properties and methods associated with them just as with all Python objects, pandas methods can easily extract and manipulate many kinds of time-based information. In particular, pandas provides many methods to make manipulating datetimes easy. Here, we use `.daysinmonth` with `.unique()` to see the number of days in the various months in our data:

```
print(dates.daysinmonth.unique())
```

```
Int64Index([31, 29, 30, 28], dtype='int64')
```

Figure 12.25 – Using pandas .daysinmonth on the dates

It's not uncommon in business situations to need to understand quarters, weeks, and other business periods. Here, we show how pandas simplifies our work by providing `.isquarterend`, which returns a Boolean `True` or `False` result, and we use that as a Boolean `index` to get just the quarter end days in our data:

```
print(dates[dates.is_quarter_end])
```

```
DatetimeIndex(['2012-12-31', '2013-09-30', '2014-03-31', '2014-06-30',
               '2018-12-31', '2019-09-30'],
              dtype='datetime64[ns]', freq=None)
```

Figure 12.26 – The dates that are quarter end dates in the dates data

We can retrieve the week numbers using the pandas `.isocalendar()` method, which by itself returns a *tuple* containing the year, week number, and weekday. Here, we subset that with the `.week` attribute to get the week numbers:

```
print(dates.isocalendar().week)
```

```
2012-01-02     1
2012-01-09     2
2012-01-16     3
2012-01-23     4
2012-01-30     5
              ..
2019-12-02    49
2019-12-09    50
2019-12-16    51
2019-12-23    52
2019-12-30     1
Freq: W-MON, Name: week, Length: 418, dtype: UInt32
```

Figure 12.27 – The week numbers for the dates data

Similarly, there are methods associated with timedeltas. Here, we create a timedelta and store it in `time_diff`:

```
time_diff = (dates[33] - dates[0])
print(type(time_diff))
```

This produces the following output:

```
<class 'pandas._libs.tslibs.timedeltas.Timedelta'>
```

Figure 12.28 – The result of subtracting two datetimes in pandas is a timedelta

Note that the subtraction of two of the dates creates a `Timedelta` object, as we mentioned earlier. Here, we inspect the `.seconds` attribute of `time_diff`:

```
print(time_diff.seconds)
```

This gives us the following output:

```
0
```

You may be surprised initially by this result. The reason the `.seconds` attribute is `0` is that this attribute is the value of the seconds in the timedelta, not the total seconds of the difference. We can clarify this by printing the actual `time_diff` variable and the value:

```
print(time_diff, " equals ", time_diff.value, "nanoseconds")
```

```
231 days 00:00:00  equals  19958400000000000 nanoseconds
```

Figure 12.29 – The contents of the time_diff variable and the value

You can see the underlying value is still in nanoseconds, and that you can use the `.total_seconds()` method to get seconds directly:

```
print(time_diff.total_seconds())
```

```
19958400.0
```

Figure 12.30 – The total seconds of difference between the two dates we subtracted

Note that the total seconds have fewer zeroes, as the value in the previous figure is in nanoseconds. Also, as a check, if you divide this result by (24 * 60 * 60) to convert to days, you get exactly 231 days, as shown in *Figure 12.30*.

Exercise 12.02 – math with datetimes

In this exercise, we will continue working with the engine test data. As the data scientist on the project, you realize you may want to analyze the data over different time scales from the start of the test. You decide to create features and compute seconds from the start and days from the start.

> **Note**
> You can find the code for this exercise here: `https://github.com/PacktWorkshops/The-Pandas-Workshop/tree/master/Chapter12/Exercise12_02`.

Perform the following steps to complete the exercise:

1. In the `Chapter12` directory, create the `Exercise12_02` directory.
2. Open your Terminal (macOS or Linux) or Command Prompt (Windows), navigate to the `Chapter12` directory, and type `jupyter notebook`. Jupyter Notebook should open up.
3. Select the `Exercise12_02` directory, which will change the Jupyter working directory to that folder, then click **New | Python 3** to create a new `Python 3` notebook, as shown in the following screenshot:

Figure 12.31 – Creating a new Python 3 Jupyter notebook

4. For this exercise, all you will need is the `pandas` library. Load it in the first cell of the notebook:

```
import pandas as pd
```

5. Read the `engine_power.csv` file into a pandas DataFrame:

```
engine_power = pd.read_csv('../datasets/engine_power.
csv')
```

> **Note**
>
> Please change the path of the dataset file (highlighted) based on where you have downloaded it on your system.

6. Use the `.head()` method to view the contents of the DataFrame:

```
engine_power.head()
```

Out[10]:

	datetime	power
0	2020-1-1 00:00	221.403466
1	2020-1-1 01:55	327.370592
2	2020-1-1 03:50	223.272440
3	2020-1-1 04:04	328.380592
4	2020-1-1 05:45	329.109239

Figure 12.32 – The engine_power DataFrame

7. Use `.dtypes` to inspect the data types for `engine_power`:

```
engine_power.dtypes
```

```
Out[11]: datetime     object
         power       float64
         dtype: object
```

Figure 12.33 – The data types for the datetime and power columns

8. Convert the `datetime` column to `datetime` using pandas `.to_datetime()`. Verify the result by using `.dtypes`:

```
engine_power.loc['datetime'] = pd.to_datetime(engine_
power['datetime'])
engine_power.dtypes
```

This should produce the following:

```
Out[5]: datetime    datetime64[ns]
        power              float64
        dtype: object
```

Figure 12.34 – The updated types in engine_power

9. Use the `.total_seconds()` method on the difference of the dates from the starting dates to create a column containing seconds from the start:

```
engine_power['sec_from_start'] = \
[(engine_power.loc[i, 'datetime'] -
  engine_power.loc[0, 'datetime']).total_seconds()
 for i in engine_power.index] \ engine_power
```

Out[37]:

	datetime	power	sec_from_start
0	2020-01-01 00:00:00	221.403465	0.0
1	2020-01-01 01:55:00	327.370592	6900.0
2	2020-01-01 03:50:00	223.272440	13800.0
3	2020-01-01 04:04:00	328.380592	14640.0
4	2020-01-01 05:45:00	329.109239	20700.0
...	...	...	...
2114	2020-04-15 11:02:00	131.620792	9111720.0
2115	2020-04-15 11:16:00	8.703348	9112560.0
2116	2020-04-15 12:43:00	23.701833	9117780.0
2117	2020-04-15 12:57:00	110.785479	9118620.0
2118	2020-04-15 13:12:00	22.869297	9119520.0

2119 rows × 3 columns

Figure 12.35 – The engine_power data with the seconds from start added as a new column

You see, as expected, that your new values start at `0`.

10. You decide you would be better served with a column containing the days from start, but as a decimal value since there are obviously many data points per day. Use the difference of the values of the starting date to all the dates and directly calculate the `days_from_start` column:

```
engine_power['days_from_start'] = \
(engine_power['datetime'].values -
 engine_power['datetime'][0].value).astype(float) / (24 *
60 * 60 * 1e9)
engine_power
```

This gives you the following updated DataFrame:

Out[33]:

	datetime	power	sec_from_start	days_from_start
0	2020-01-01 00:00:00	221.403465	0.0	0.000000
1	2020-01-01 01:55:00	327.370592	6900.0	0.079861
2	2020-01-01 03:50:00	223.272440	13800.0	0.159722
3	2020-01-01 04:04:00	328.380592	14640.0	0.169444
4	2020-01-01 05:45:00	329.109239	20700.0	0.239583
...	...	...	...	...
2114	2020-04-15 11:02:00	131.620792	9111720.0	105.459722
2115	2020-04-15 11:16:00	8.703348	9112560.0	105.469444
2116	2020-04-15 12:43:00	23.701833	9117780.0	105.529861
2117	2020-04-15 12:57:00	110.785479	9118620.0	105.539583
2118	2020-04-15 13:12:00	22.869297	9119520.0	105.550000

2119 rows × 4 columns

Figure 12.36 – The updated engine_data DataFrame with days from the start as decimal values

You now have the data needed to analyze the power variation versus time in days and seconds. With the data in this form, it will be easy to create visualizations with different resolutions, do comparisons on different timescales, and do other time-based analyses.

We showed here that using the Python `datetime` module, we can create `datetime` objects, supplying a format string to tell `datetime` how to decode the string. Let's look at formatting and how it works with converting to time objects in more detail.

Timestamp formats

We have seen all of the dates mentioned previously with a fixed format of yyyy-mm-dd. But, there are many other ways of representing dates, which sometimes causes ambiguity if we don't specify the format used beforehand. For example, 2020-03-09 would be interpreted as the ninth of March, 2020, when the format is yyyy-mm-dd, whereas the same data will be interpreted as the third of September, 2020, when the format is yyyy-dd-mm (yet another common format).

Such cases cause ambiguity with pandas and datetime packages as well. To counter this, datetime packages use a format parameter to interpret the date, as shown here:

```
datetime.strptime('09-30-2020','%m-%d-%Y')
datetime.datetime(2020, 9, 30, 0, 0)
```

pandas also provides a `.to_datetime()` method that can use formats. The format strings are taken from the `datetime` module, and you can review all the possible string elements at the official Python documentation by clicking this link: `https://docs.python.org/3/library/datetime.html`. Some of the common ones are shown in *Figure 12.37*. pandas also follows the same notations for the format as the datetime package:

string	usage	examples
%a	abbreviated weekday	Mon, Wed
%A	full weekday	Sunday, Monday
%w	numeric weekday, Sunday = 0	0, 1
%d	zero-padded day of month	07, 29
%b	abbreviated month	Jan, Mar
%B	full month	February, September
%m	zero-padded month	01, 07, 11
%f	zero-padded microsecond	012989, 000002
%Y	numeric year with century	2020, 1987
%H	zero-padded hour on 24-hour clock	00, 23
%I	zero-padded hour	01, 11
%p	A.M. or P.M.	A.M., P.M.
%M	zero-padded minutes	23, 59
%S	zero-padded seconds	00, 13

Figure 12.37 – Some common date and time formatting strings

To illustrate, in the following code, we create the same timestamp in pandas using two different strings and show how we can also format when we print out using the `.strftime()` method in pandas:

```
num_date = pd.to_datetime('12-20-2020 13:57:03.13',
                          format = "%m-%d-%Y %H:%M:%S.%f")
print(num_date)
```

This produces the following:

```
2020-12-20 13:57:03.130000 2020-12-20 13:57:03
December 20, 2020 01:57:03 PM
```

Figure 12.38 – Creating a timestamp for 12/20/2020 at 1:57 P.M. using a format string

Here, we create the same date but start with a human-readable string. We regenerate the human-readable string in the second `print` statement using the appropriate format string:

```
text_date = pd.to_datetime('December 20, 2020 1:57:03 PM',
                           format = '%B %d, %Y %I:%M:%S %p')
print(text_date)
print(text_date.strftime(format = '%B %d, %Y %I:%M:%S %p'))
```

This produces the following:

```
2020-12-20 13:57:03
December 20, 2020 01:57:03 PM
```

Figure 12.39 – The datetime from a human-readable string and back

Note that in the format strings, everything is literal except the `%` operator, which tells pandas that what follows is a format instruction. So, `%m-%d-%Y` tells pandas to expect *a zero-padded month followed by a dash, followed by a zero-padded day, followed by a dash, followed by a 4-digit year with the century*. In the second example, `text_date`, we see that spaces also need to be included to inform pandas of the format, so `%B %d %Y` means *a full month name followed by a space, followed by a zero-padded day, followed by a comma, then a space, followed by a 4-digit year with the century.*

Datetime localization

pandas can also handle time zones. The methods beginning with `.tz` have to do with time zone awareness. In the following, we first make pandas aware that `text_date` is in the US Mountain time zone, then we print it using `.tz_convert()` to change it to Tokyo time. Notice the difference in the date and the time (A.M. versus P.M.):

```
text_date = text_date.tz_localize('US/Mountain')
print(text_date)
print(text_date.tz_convert('Asia/Tokyo').
      strftime(format = '%B %d, %Y %I:%M:%S %p'))
```

This produces the following output:

```
2020-12-20 13:57:03-07:00
December 21, 2020 05:57:03 AM
```

Figure 12.40 – Printing December 20, 1:57 PM in US Mountain time in the Tokyo time zone

You can see that utilizing formats and time zones supports developing applications that can work in multiple locations in a time zone-aware fashion.

Limits on timestamps

Since pandas `Timestamps` are stored in nanosecond resolution, there is a limit to the date/time ranges that can be stored in a 64-bit integer internal representation. Here, we use the `.min` and `.max` attributes to display the limits:

```
print(pd.Timestamp.min)
print(pd.Timestamp.max)
```

This produces the following:

```
1677-09-21 00:12:43.145225
2262-04-11 23:47:16.854775807
```

Figure 12.41 – The limits of pandas timestamps

So, if you are working with a database that includes dates before the late fifteenth century, you could run into an error. Similarly, if your dates extend out more than about 240 years from now, an error could occur. There are a variety of packages available that provide additional capability to work with datetimes. As an example, here we import the arrow package, get a datetime for the date 1/1/1970, and show that the `0` time is 1/1/1970 by getting the integer value and that we can get integer values for dates beyond what pandas can support by getting the integer value for the date `2475-03-07 07:11:23`. The negative is that, in this case, the resolution is now seconds instead of nanoseconds:

```
import arrow
print(arrow.get('1970-01-01 00:00:00'))
print(arrow.get('1970-01-01 00:00:07').int_timestamp)
print(arrow.get('2475-03-07 07:11:23').int_timestamp)
```

This produces the following output:

```
1970-01-01T00:00:00+00:00
7
15941949083
```

Other options include the `datetime` package, which supports microsecond resolution. Now that you have some familiarity with `datetime` objects, you are ready to explore some common and useful operations using them, such as converting to integers to use in a model.

Activity 12.01 – understanding power usage

In this activity, as a data analyst involved in an energy conservation study, you are provided with data collected from a home in France over several years, which includes frequent measurements of total power usage as well as submeters that isolate the kitchen, laundry room, and some heating and air conditioning use. You have been asked to look at kitchen energy use and understand the time of day trends. The data is from the UCI repository (`https://archive.ics.uci.edu/ml/datasets/Individual+household+electric+power+consumption`):

1. For this activity, all you will need is the `pandas` and `numpy` libraries. Load them in the first cell of the notebook:

    ```
    import pandas as pd
    import numpy as np
    ```

2. Read in the `household_power_consumption.csv` data from the `Datasets` directory, and list the first few rows:

Out[5]:

	Date	Time	Global_active_power	Global_reactive_power	Voltage	Global_intensity	Sub_metering_1	Sub_metering_2	Sub_metering_3
0	1/8/2008	00:00:00	0.500	0.226	239.750	2.400	0.000	0.000	1.0
1	1/8/2008	00:01:00	0.482	0.224	240.340	2.200	0.000	0.000	1.0
2	1/8/2008	00:02:00	0.502	0.234	241.680	2.400	0.000	0.000	0.0
3	1/8/2008	00:03:00	0.556	0.228	241.750	2.600	0.000	0.000	1.0
4	1/8/2008	00:04:00	0.854	0.342	241.550	4.000	0.000	1.000	7.0

Figure 12.42 – The household_power_consumption.csv data

3. You should inspect the data types of the columns, and further investigate whether there are non-numeric values. If so, correct them by converting to `NA` values and then filling them by interpolation.
4. Make a quick visualization to understand the timeframe of the data. Your plan is to identify a year with complete data and focus on that:

```
Out[10]: <AxesSubplot:xlabel='Date'>
```

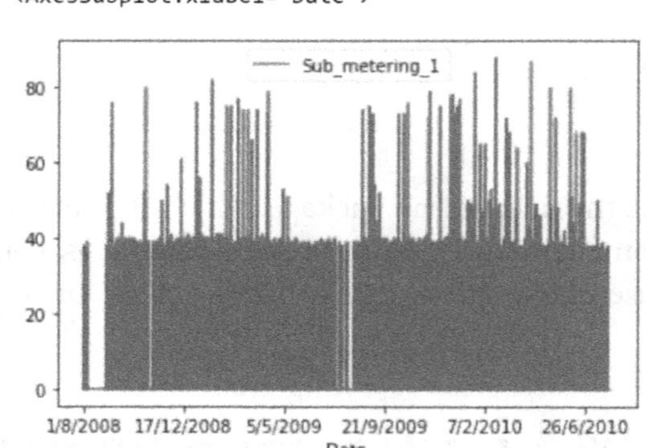

Figure 12.43 – Sub_metering_1 data versus date. The year 2009 is a full year

5. Using the year you identified, create a new DataFrame having `Date`, `Time`, and `Kitchen_power_use`. Note that `Sub_metering_1` is the kitchen:

```
Out[13]:
```

	Date	Time	Kitchen_power_use
1074636	1/1/2009	00:00:00	0.0
1074637	1/1/2009	00:01:00	0.0
1074638	1/1/2009	00:02:00	0.0
1074639	1/1/2009	00:03:00	0.0
1074640	1/1/2009	00:04:00	0.0

Figure 12.44 – The new DataFrame containing kitchen power use

6. `Date` and `Time` are strings; combine them on each row, then convert the combined string to a datetime, and store that in a new column called `timestamp`. Keep in mind the European format of the original date strings:

```
Out[12]:
```

	Date	Time	Kitchen_power_use	timestamp
1074636	1/1/2009	00:00:00	0.0	2009-01-01 00:00:00
1074637	1/1/2009	00:01:00	0.0	2009-01-01 00:01:00
1074638	1/1/2009	00:02:00	0.0	2009-01-01 00:02:00
1074639	1/1/2009	00:03:00	0.0	2009-01-01 00:03:00
1074640	1/1/2009	00:04:00	0.0	2009-01-01 00:04:00

Figure 12.45 – Creation of a timestamp

7. Create an `hour` and `date` column using methods on the `timestamp` column to represent the hour of the day and the date in a standard format:

Out[34]:

	Date	Time	Kitchen_power_use	timestamp	hour	date
1074636	1/1/2009	00:00:00	0.0	2009-01-01 00:00:00	0	2009-01-01
1074637	1/1/2009	00:01:00	0.0	2009-01-01 00:01:00	0	2009-01-01
1074638	1/1/2009	00:02:00	0.0	2009-01-01 00:02:00	0	2009-01-01
1074639	1/1/2009	00:03:00	0.0	2009-01-01 00:03:00	0	2009-01-01
1074640	1/1/2009	00:04:00	0.0	2009-01-01 00:04:00	0	2009-01-01

Figure 12.46 – Addition of the hour and new date column

8. Group the data by `date` and `hour` aggregating `Kitchen_power_use`:

Out[55]:

	date	hour	Kitchen_power_use
20	2009-01-01	20	0.0
21	2009-01-01	21	0.0
22	2009-01-01	22	0.0
23	2009-01-01	23	0.0
24	2009-01-02	0	0.0
25	2009-01-02	1	0.0
26	2009-01-02	2	0.0
27	2009-01-02	3	0.0

Figure 12.47 – Kitchen power use by hour for each day

9. For January, aggregate the data by hour and make a bar plot by hour of kitchen energy use:

Out[50]: <AxesSubplot:xlabel='hour'>

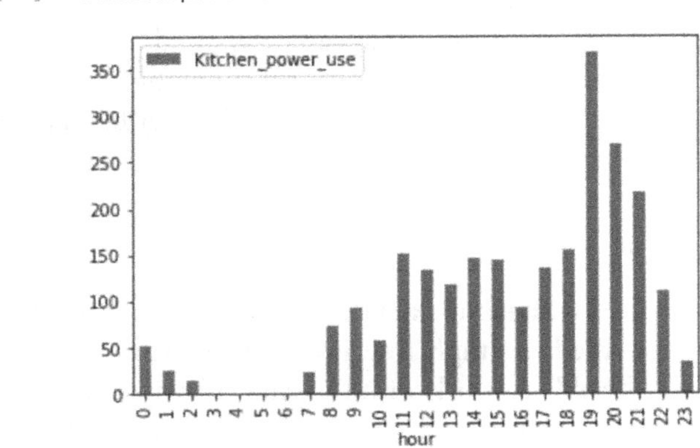

Figure 12.48 – Kitchen energy use by hour for January, 2009

10. You see that usage seems to begin around breakfast time, continue through the day with a peak at dinner time, then trail off. Make a similar plot for the entire year to compare:

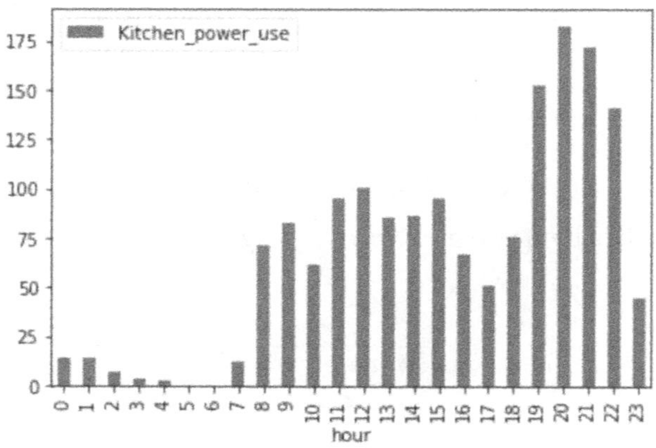

Figure 12.49 – Kitchen energy use by hour for all of 2019

> **Note**
> You can find the solution for this activity in the *Appendix*.

In the next section, you will learn more about methods on datetimes and how you can do mathematical operations on them to obtain desired results.

Datetime math operations

We have shown already some operations that can be applied to `datetime/Timestamp` objects. In this section, we will go a little deeper and show the use of the `origin` parameter, which is useful if converting dates in some integer formats (such as might come from Excel).

Date ranges

Suppose you were provided with some data that is daily temperatures from 1/1/2019 to 6/30/2020 but it is provided as simply a series of values without the corresponding dates. You would like to be able to work with the data corresponding to actual dates, for example, to look for repeating patterns or seasonal variations. You can add the dates easily using the pandas `date_range` method:

1. Here, you create a `temperatures` series beginning with just an integer series, using the NumPy `sin()` function and a period of 180 days to generate variation over time, and adding noise to represent the hypothetical data. The first step is to create the integer series:

```
x_values = pd.Series(range(1, 548))
```

2. Next, use the integer series with `np.sin()` and a period of 180 days to create `sin_series`:

```
import numpy as np \ period = 180
sin_series = np.sin(2 * np.pi * x_values / period) * 5
```

3. Finally, simulate some noise in the data using `np.random.normal`, and add the noise to `sin_series`:

```
noise = 65 + np.random.normal(0, 3, 547)
temperatures = sin_series + noise
```

4. Now, make a simple plot to see the result:

```
temperatures.plot()
```

```
Out[33]: <AxesSubplot:>
```

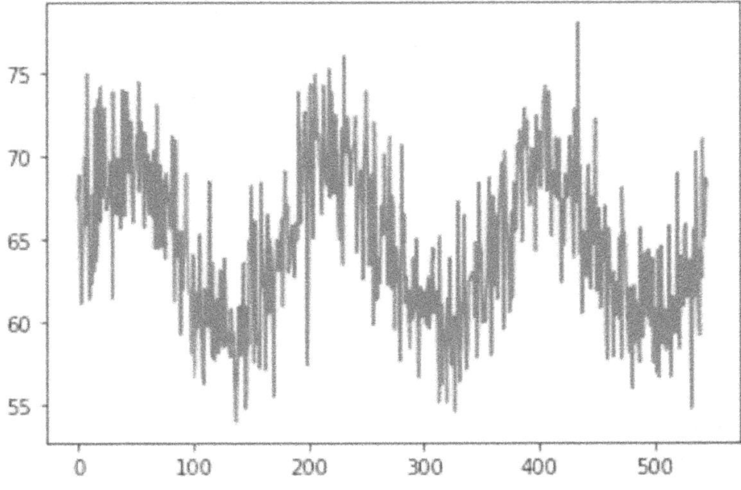

Figure 12.50 – The synthetic temperature data

5. Now, you want to add the dates. You can use the pandas `.date_range()` method, which can take date strings as was used before, to create sequential dates. Note that `.date_range()` can take a third parameter for the number of periods to create, and a fourth, the frequency, to specify the size of the steps. Here, use defaults that generate sequential days. Then, combine the dates with the temperatures using a `DataFrame` constructor, and again, use the pandas `.plot()` method, this time specifying the `x` and `y` variables explicitly:

```
dates = pd.date_range('2019-01-01', '2020-06-30')
temp_data = pd.DataFrame({'date' : dates,'temperature' : 
temperatures})
temp_data.plot(x = 'date', y = 'temperature')
```

```
Out[34]: <AxesSubplot:xlabel='date'>
```

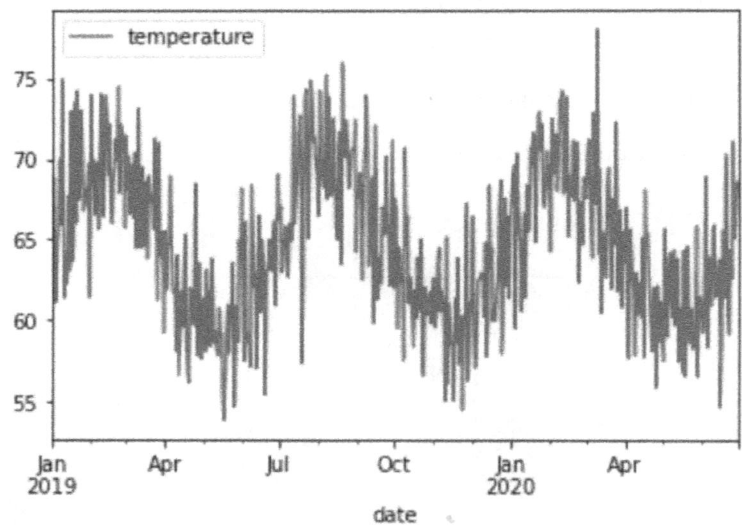

Figure 12.51 – The temperature data with a date axis

Timedeltas, offsets, and differences

Now, suppose you found that the temperatures were measured every day at 11:30 A.M. and you wanted to reflect this in the timestamps. First, let's inspect the starting date; here, we simply print the first three elements of the `date` column using a list comprehension to iterate:

```
_ = [print(temp_data['date'][i]) for i in range(3)]
```

This shows that the values created earlier are `Timestamps`, and the time portions are all `0`:

```
2019-01-01 00:00:00
2019-01-02 00:00:00
2019-01-03 00:00:00
```

Figure 12.52 – The first three date values in the temp_data DataFrame

pandas provides the `Timedelta` object type, which makes it easy to perform simple math on `Timestamps`. Here, we create a variable offset using `pd.Timedelta()`, specifying the value as `11.5`, and the units as hours. We then simply add that to the `date` column, and again, list out the first three values:

```
offset = pd.Timedelta(11.5, unit = 'h')
temp_data['date'] = temp_data['date'] + offset
_ = [print(temp_data['date'][i]) for i in range(3)]
```

This produces the expected output:

```
2019-01-01 11:30:00
2019-01-02 11:30:00
2019-01-03 11:30:00
```

Figure 12.53 – The updated timestamps are now at 11:30 A.M.

A wide range of units is allowed, making using `Timedelta` very intuitive:

'W', 'D', 'T', 'S', 'L', 'U', or 'N'
'days' or 'day'
'hours', 'hour', 'hr', or 'h'
'minutes', 'minute', 'min', or 'm'
'seconds', 'second', or 'sec'
'milliseconds', 'millisecond', 'millis', or 'milli'
'microseconds', 'microsecond', 'micros', or 'micro'
'nanoseconds', 'nanosecond', 'nanos', 'nano', or 'ns'

Figure 12.54 – The available units values for pd.Timedelta()

In the first line, `W` is for weeks, `D` is for days, `T` is for minutes, `S` is for seconds, `L` is for milliseconds, `U` is for microseconds, and `N` is for nanoseconds.

Although you can't add two `datetime` objects directly in pandas, you can subtract them, and the result is a timedelta. In addition to the value we see when we list the object, `Timedelta` objects have a number of attributes, such as days or seconds. Here, we take the difference between two datetimes and then look at some of the attributes of the result:

```
time_difference = (pd.to_datetime('2019-01-11') -
                   pd.to_datetime('2019-01-04'))
print(time_difference.days)
print(time_difference.seconds)
```

This code produces the following outputs:

```
7
0
```

The result of `7` days is obvious, but you might question the `0` seconds. The key is to understand that attributes are the values you would see if you printed the timedelta as *days:hours:minutes:seconds:milliseconds:microseconds:nanoseconds*. Since, in this case, we subtracted 2 days with `0` time values, the values of everything less than days are exactly `0`. Here, we make two datetimes with resolution down to minutes and look at the attributes again:

```
time_difference = (pd.to_datetime('2019-01-11 13:57:03') -
                   pd.to_datetime('2019-01-04 14:31:47'))
print(time_difference.days)
print(time_difference.seconds)
```

This produces a slightly different result than before:

```
6
84316
```

The days are now `6`, because the time we put on `2019-01-04` is later than the time on `2019-01-11`. The additional difference is `84316` seconds (which is 23 hours, 25 minutes, and 16 seconds). The `Timedelta` attributes do not have any options between days and seconds, so all the time less than an even day shows up in the seconds. It is not unusual in data science to want to know the exact difference between two datetimes, when a model includes datetime information as a predictor. Using `Timedelta` objects makes this easy. In the next section, you will learn how pandas understands business concepts such as business days and makes it simpler to work with real data.

Date offsets

pandas provides a wide range of features for working with dates in time series, making it powerful in business analysis and financial calculations. As a simple example, consider the case that many businesses define financial quarters to be 13 weeks. We can create a date range for the first quarter of 2021, using the `.date_range()` method, setting the start as `2021-01-01` then specifying the period to be `13` (weeks) times `7` (days per week) and a frequency of `1` day:

```
first_quarter = pd.date_range('2021-01-01 00:00:00',
                              periods = 13 * 7, freq = '1d')
first_quarter
```

This produces the following:

```
Out[33]: DatetimeIndex(['2021-01-01', '2021-01-02', '2021-01-03', '2021-01-04',
               '2021-01-05', '2021-01-06', '2021-01-07', '2021-01-08',
               '2021-01-09', '2021-01-10', '2021-01-11', '2021-01-12',
               '2021-01-13', '2021-01-14', '2021-01-15', '2021-01-16',
               '2021-01-17', '2021-01-18', '2021-01-19', '2021-01-20',
               '2021-01-21', '2021-01-22', '2021-01-23', '2021-01-24',
               '2021-01-25', '2021-01-26', '2021-01-27', '2021-01-28',
               '2021-01-29', '2021-01-30', '2021-01-31', '2021-02-01',
               '2021-02-02', '2021-02-03', '2021-02-04', '2021-02-05',
               '2021-02-06', '2021-02-07', '2021-02-08', '2021-02-09',
               '2021-02-10', '2021-02-11', '2021-02-12', '2021-02-13',
               '2021-02-14', '2021-02-15', '2021-02-16', '2021-02-17',
               '2021-02-18', '2021-02-19', '2021-02-20', '2021-02-21',
               '2021-02-22', '2021-02-23', '2021-02-24', '2021-02-25',
               '2021-02-26', '2021-02-27', '2021-02-28', '2021-03-01',
               '2021-03-02', '2021-03-03', '2021-03-04', '2021-03-05',
               '2021-03-06', '2021-03-07', '2021-03-08', '2021-03-09',
               '2021-03-10', '2021-03-11', '2021-03-12', '2021-03-13',
               '2021-03-14', '2021-03-15', '2021-03-16', '2021-03-17',
               '2021-03-18', '2021-03-19', '2021-03-20', '2021-03-21',
               '2021-03-22', '2021-03-23', '2021-03-24', '2021-03-25',
               '2021-03-26', '2021-03-27', '2021-03-28', '2021-03-29',
               '2021-03-30', '2021-03-31', '2021-04-01'],
              dtype='datetime64[ns]', freq='D')
```

Figure 12.55 – A date range for the first financial quarter (13 weeks) of 2021

Note that this quarter includes `2021-01-04`. But, you have been asked to do an analysis focusing on the business days; in this case, we define business days simply to exclude weekends. pandas provides a frequency of `'B'`, which means business day. Here, we modify the previous snippet in two ways – we make the periods `13` (weeks) times `5` (weekdays per week), and we change the frequency to `'B'`:

```
first_quarter_bus = pd.date_range('2021-01-01 00:00:00',
                                  periods = 13 * 5, freq = 'B')
first_quarter_bus
```

This produces the following, which contains only weekdays:

```
Out[43]: DatetimeIndex(['2021-01-01', '2021-01-04', '2021-01-05', '2021-01-06',
                        '2021-01-07', '2021-01-08', '2021-01-11', '2021-01-12',
                        '2021-01-13', '2021-01-14', '2021-01-15', '2021-01-18',
                        '2021-01-19', '2021-01-20', '2021-01-21', '2021-01-22',
                        '2021-01-25', '2021-01-26', '2021-01-27', '2021-01-28',
                        '2021-01-29', '2021-02-01', '2021-02-02', '2021-02-03',
                        '2021-02-04', '2021-02-05', '2021-02-08', '2021-02-09',
                        '2021-02-10', '2021-02-11', '2021-02-12', '2021-02-15',
                        '2021-02-16', '2021-02-17', '2021-02-18', '2021-02-19',
                        '2021-02-22', '2021-02-23', '2021-02-24', '2021-02-25',
                        '2021-02-26', '2021-03-01', '2021-03-02', '2021-03-03',
                        '2021-03-04', '2021-03-05', '2021-03-08', '2021-03-09',
                        '2021-03-10', '2021-03-11', '2021-03-12', '2021-03-15',
                        '2021-03-16', '2021-03-17', '2021-03-18', '2021-03-19',
                        '2021-03-22', '2021-03-23', '2021-03-24', '2021-03-25',
                        '2021-03-26', '2021-03-29', '2021-03-30', '2021-03-31',
                        '2021-04-01'],
                       dtype='datetime64[ns]', freq='B')
```

Figure 12.56 – The first quarter of 2021, but only weekdays (business days)

Now, suppose you also want the corresponding second quarter to do some order booking comparisons. The idea of business days is also available in pandas date offsets. Here, we use the `.BusinessDay()`, method, which is provided in pandas `tseries.offsets`, to add `13` (weeks) times `5` (business days per week) to the first business quarter:

```
offset = pd.tseries.offsets.BusinessDay(13 * 5)
second_quarter_bus = first_quarter_bus + offset
second_quarter_bus
```

```
Out[44]: DatetimeIndex(['2021-04-02', '2021-04-05', '2021-04-06', '2021-04-07',
                        '2021-04-08', '2021-04-09', '2021-04-12', '2021-04-13',
                        '2021-04-14', '2021-04-15', '2021-04-16', '2021-04-19',
                        '2021-04-20', '2021-04-21', '2021-04-22', '2021-04-23',
                        '2021-04-26', '2021-04-27', '2021-04-28', '2021-04-29',
                        '2021-04-30', '2021-05-03', '2021-05-04', '2021-05-05',
                        '2021-05-06', '2021-05-07', '2021-05-10', '2021-05-11',
                        '2021-05-12', '2021-05-13', '2021-05-14', '2021-05-17',
                        '2021-05-18', '2021-05-19', '2021-05-20', '2021-05-21',
                        '2021-05-24', '2021-05-25', '2021-05-26', '2021-05-27',
                        '2021-05-28', '2021-05-31', '2021-06-01', '2021-06-02',
                        '2021-06-03', '2021-06-04', '2021-06-07', '2021-06-08',
                        '2021-06-09', '2021-06-10', '2021-06-11', '2021-06-14',
                        '2021-06-15', '2021-06-16', '2021-06-17', '2021-06-18',
                        '2021-06-21', '2021-06-22', '2021-06-23', '2021-06-24',
                        '2021-06-25', '2021-06-28', '2021-06-29', '2021-06-30',
                        '2021-07-01'],
                       dtype='datetime64[ns]', freq=None)
```

Figure 12.57 – The second business quarter of 2021

Note that if you make frequent use of the offsets, it can be convenient to import them all, as we show here:

```
from pandas.tseries.offsets import *
```

After the preceding import, there is a range of methods available, including `BusinessHour`, `CustomBusinessDay`, and `MonthEnd`. You can find all the available offset methods at the official documentation by clicking this link: `https://pandas.pydata.org/pandas-docs/stable/reference/offset_frequency.html?highlight=offset`.

Suppose you are a business analyst and are asked to compute some metrics as of the last day of each month. Here, we apply `MonthEnd()` to `second_quarter_bus`. In the code here, adding the `MonthEnd()` offset to all the dates in the `second_quarter_bus` dates shifts each one to the corresponding end of month. The `0` parameter is the month offset; it is possible to compute end days for months in another period by specifying an offset. The `.unique()` method simply drops all the duplicate dates generated by shifting each day in our list to the end of the month:

```
(second_quarter_bus + MonthEnd(0)).unique()
```

This produces a date index with the desired dates:

```
Out[52]: DatetimeIndex(['2021-04-30', '2021-05-31', '2021-06-30', '2021-07-31'],
         dtype='datetime64[ns]', freq=None)
```

Figure 12.58 – The month-end dates for the months in the second business quarter

Note that using `MonthEnd(0)` and `MonthEnd(1)` produces the same offset, equivalently shifting to the end of the current month. Other offsets move the results farther, as illustrated here:

```
(second_quarter_bus + MonthEnd(3)).unique()
```

This produces the following:

```
Out[53]: DatetimeIndex(['2021-06-30', '2021-07-31', '2021-08-31', '2021-09-30'],
         dtype='datetime64[ns]', freq=None)
```

Figure 12.59 – The values using MonthEnd offset by three

You've seen now that pandas can work with date series and understand weekdays and other business concepts. The `CustomBusinessDay()` method, in particular, allows accounting for holidays, which might be important depending on the business context. Let's reinforce all these ideas with an exercise.

Exercise 12.03 – timedeltas and date offsets

In this exercise, you are a business analyst for a retail clothing and accessories online store and are required to prepare a revenue summary by month for a report to management. The data is a subset of an online orders dataset from the UCI repository (`https://archive.ics.uci.edu/ml/datasets/Online+Retail+II`).

> **Note**
>
> You can find the code for this exercise here: `https://github.com/PacktWorkshops/The-Pandas-Workshop/tree/master/Chapter12/Exercise12_03`.

Perform the following steps to complete the exercise:

1. In the `Chapter12` directory, create the `Exercise12_03` directory.
2. Open your Terminal (macOS or Linux) or Command Prompt (Windows), navigate to the `Chapter12` directory, and type `jupyter notebook`. Jupyter Notebook should open up.

   Select the `Exercise09_03` directory, which will change the Jupyter working directory to that folder, then click **New** | **Python 3** to create a new `Python 3` notebook, as shown in the following screenshot:

Figure 12.60 – Creating a new Python 3 Jupyter notebook

3. For this exercise, all you need is the `pandas` library. Load it in the first cell of the notebook:

```
import pandas as pd
```

4. Read the `online_retail_II.csv` file into a pandas DataFrame:

```
retail_sales = pd.read_csv('../datasets/online_retail_
II.csv')
retail_sales.head()
```

> **Note**
> Please change the path of the dataset file (highlighted) based on where you have downloaded it on your system.

This should produce the following:

Out[2]:

	Invoice	StockCode	Description	Quantity	InvoiceDate	Price	Customer ID	Country
0	539993	22386	JUMBO BAG PINK POLKADOT	10	1/4/2011 10:00	1.95	13313.0	United Kingdom
1	539993	21499	BLUE POLKADOT WRAP	25	1/4/2011 10:00	0.42	13313.0	United Kingdom
2	539993	21498	RED RETROSPOT WRAP	25	1/4/2011 10:00	0.42	13313.0	United Kingdom
3	539993	22379	RECYCLING BAG RETROSPOT	5	1/4/2011 10:00	2.10	13313.0	United Kingdom
4	539993	20718	RED RETROSPOT SHOPPER BAG	10	1/4/2011 10:00	1.25	13313.0	United Kingdom

Figure 12.61 – The retail sales data

5. Convert the `InvoiceDate` column to a datetime using pandas `.to_datetime()`, then inspect the earliest and latest values of `InvoiceDate`:

```
retail_sales['InvoiceDate'] = pd.to_datetime(retail_
sales['InvoiceDate'])
print('start: ', retail_sales.InvoiceDate.min(),
      '\nend: ', retail_sales.InvoiceDate.max())
```

This should produce the following:

```
start:  2011-01-04 10:00:00
end:  2011-06-30 20:08:00
```

Figure 12.62 – The start and end datetimes of the retail sales data

6. Now, use `pd.tseries.offsets.MonthEnd(0)` to convert `InvoiceDate` to the month-end dates. Use the `.dt.date` attribute to select just the date, ignoring the timestamps. Inspect the DataFrame again:

```
retail_sales['InvoiceDate'] = (retail_
sales['InvoiceDate'] +
                               pd.tseries.offsets.
MonthEnd(0)).dt.date
retail_sales.head()
```

This produces the following result:

Out[4]:

	Invoice	StockCode	Description	Quantity	InvoiceDate	Price	Customer ID	Country
0	539993	22386	JUMBO BAG PINK POLKADOT	10	2011-01-31	1.95	13313.0	United Kingdom
1	539993	21499	BLUE POLKADOT WRAP	25	2011-01-31	0.42	13313.0	United Kingdom
2	539993	21498	RED RETROSPOT WRAP	25	2011-01-31	0.42	13313.0	United Kingdom
3	539993	22379	RECYCLING BAG RETROSPOT	5	2011-01-31	2.10	13313.0	United Kingdom
4	539993	20718	RED RETROSPOT SHOPPER BAG	10	2011-01-31	1.25	13313.0	United Kingdom

Figure 12.63 – The retail_sales data with InvoiceDate converted to MonthEnd dates

7. Compute the revenue by line as `Quantity` times `Price` and save in a new column called `Revenue`:

```
retail_sales['Revenue'] = (retail_sales.loc[:,
'Quantity'] *
                           retail_sales.loc[:, 'Price'])
retail_sales.head()
```

This produces the following:

Out[5]:

	Invoice	StockCode	Description	Quantity	InvoiceDate	Price	Customer ID	Country	Revenue
0	539993	22386	JUMBO BAG PINK POLKADOT	10	2011-01-31	1.95	13313.0	United Kingdom	19.5
1	539993	21499	BLUE POLKADOT WRAP	25	2011-01-31	0.42	13313.0	United Kingdom	10.5
2	539993	21498	RED RETROSPOT WRAP	25	2011-01-31	0.42	13313.0	United Kingdom	10.5
3	539993	22379	RECYCLING BAG RETROSPOT	5	2011-01-31	2.10	13313.0	United Kingdom	10.5
4	539993	20718	RED RETROSPOT SHOPPER BAG	10	2011-01-31	1.25	13313.0	United Kingdom	12.5

Figure 12.64 – The retail_sales DataFrame updated with Revenue

8. Now, use `groupby` on the `InvoiceDate` and `Revenue` columns, grouping by `InvoiceDate`, and use `.sum()` to aggregate the monthly sums:

```
sales_by_month = \
  retail_sales[['InvoiceDate', 'Revenue']].
groupby('InvoiceDate').sum()
sales_by_month
```

This produces the following result:

Out[7]:

InvoiceDate	Revenue
2011-01-31	560000.260
2011-02-28	498062.650
2011-03-31	683267.080
2011-04-30	493207.121
2011-05-31	723333.510
2011-06-30	691123.120

Figure 12.65 – The revenue by month

In this exercise, you converted dates read as strings to `datetimes` as you have done before, then used the pandas `MonthEnd` offset method to change all the dates to the month ending dates, which allowed you to group by the dates and get monthly totals. You are now ready to dive deeper into some advanced time series manipulations provided by pandas.

Summary

In this chapter, you've mastered many core methods provided by pandas that make working with and analyzing time series data easy. Starting with the enhancements provided by pandas for the `Timestamp` data types, you saw that we can use time-aware methods to interpolate missing values or resample a time series to a higher or lower frequency (period). These methods are all very common patterns in business analysis and data science, and you are now well equipped to analyze most data that may come your way. This chapter prepares you for the next one, *Chapter 13*, *Exploring Time Series*, where you will learn advanced methods for working with time as in an index, culminating with building a time series model.

13 Exploring Time Series

Here, you will learn how to use time data in the index to enable advanced capabilities such as resampling to different time intervals, interpolating, and modeling as a function of time. In *Chapter 11*, *Data Modeling – Regression Modeling*, you learned how to use multiple regressions as a powerful data modeling approach, and by the end of this chapter, you will be able to use regression with time series as well.

You will cover the following topics as you work through this chapter:

- The time series as an index
- Resampling, grouping, and aggregation by time
- Activity 13.01 – Creating a time series model

The time series as an index

In many of the examples so far, we have had a **column** in a DataFrame containing dates or `datetime` information, and we've manipulated that. In many cases, when we want to perform operations on time-stamped data, it is simpler and more natural to have a time-based index. In general, you may want to consider time series to refer to a data structure with a time-based index and one or more columns of data. Let's explore a bit more what we can do with such a time series.

Time series periods/frequencies

We've seen the use of the pandas `.date_range()` method to generate a sequence of dates. The method is intuitive; we simply provide the `start`, `end`, and optional frequency (`freq`) arguments. The latter is the key to a lot of the convenience provided by pandas. The `freq` argument can take many values, and we've summarized them here.

freq string	meaning	freq string	meaning
B	business day frequency	Q	quarter end frequency
C	custom business day frequency	BQ	business quarter end frequency
D	calendar day frequency	QS	quarter start frequency
W	weekly frequency	BQS	business quarter start frequency
M	month end frequency	A, Y	year end frequency
SM	semi-month end frequency	BA, BY	business year end frequency
	(15th and end of month)	AS, YS	year start frequency
BM	business month end frequency	BAS, BYS	business year start frequency
CBM	custom business month	BH	business hour frequency
MS	month start frequency	H	hourly frequency
	end frequency	T, min	minutely frequency
SMS	semi-month start frequency	S	secondly frequency
	(1st and 15th)	L, ms	milliseconds
BMS	business month start frequency	U, us	microseconds
CBMS	custom business month	N	nanoseconds
	start frequency		

Figure 13.1 – The possible values and meanings of the freq argument for date_range()

Notice several convenience `freq` arguments are defined, such as `SMS` for semi-month-start (dates starting on the first and fifteenth of the month) and others.

In some situations, it makes more sense to focus on a time period between data points, versus the explicit times or dates. Pandas provides the `.Period()` method to define a time period and a time point to which periods are aligned. Here, we define a period aligned to end on the first Monday of 2019 (`2019-01-07`):

```
W1_2019 = pd.Period('2019', freq = 'W-MON')
W1_2019
```

```
Out[90]: Period('2019-01-01/2019-01-07', 'W-MON')
```

Figure 13.2 – A period defined to be one week, ending on the first Monday in 2019

The convenience of the `Periods` object in pandas is that we can then add them in **period units**, as shown here, where we add one:

```
W1_2019 + 1
```

This produces a period one week later than the previous:

```
Out[91]: Period('2019-01-08/2019-01-14', 'W-MON')
```

Figure 13.3 – A period, one period later than the previous one

Notice that the **anchoring**, as pandas calls the use of the suffixes such as `-MON`, can be a little confusing with periods, as they anchor to the *end* of the period, not the start.

Shifting, lagging, and converting frequency

We've seen examples of using pandas `Timedelta` before, and now we are in a position to see how to use it to shift a time series. Here, we first define `Timedelta`, then we apply it to a date range:

```
Weeks_2020 = pd.date_range('2020-01-01', '2020-12-31', freq =
'W')
print(Weeks_2020[:6])
shift = pd.Timedelta('6 days')
print(Weeks_2020 + shift)
```

This produces the following:

```
DatetimeIndex(['2020-01-05', '2020-01-12', '2020-01-19', '2020-01-26',
               '2020-02-02', '2020-02-09'],
              dtype='datetime64[ns]', freq='W-SUN')
DatetimeIndex(['2020-01-11', '2020-01-18', '2020-01-25', '2020-02-01',
               '2020-02-08', '2020-02-15'],
              dtype='datetime64[ns]', freq=None)
```

Figure 13.4 – Shifting a time series by 6 days using a Timedelta and adding to the original series

Note that, in this case, we used `freq = 'W'` with no suffix, which defaults to Sunday, so the original series is aligned to the first Sunday in 2020.

As a final example, let's look at a frequency that might be appropriate in another setting. Suppose we were given some sensor data and told that it was collected every nanosecond. We could generate a time series to represent this. Here, we use `.date_range()` with a frequency of `'n'`, which is *nanoseconds*. That seems a little odd but recall that although we refer to date ranges, it is a general method operating on the underlying nanosecond resolution time objects. For convenience, we subtract the starting datetime from the range, which returns `Timedelta` objects, and we can use the `.total_seconds()` method to extract real-valued times. Note here that the end of the range is 0.00001 seconds, which is 10,000 nanoseconds; the resulting series has 10,001 elements.

```
sensor_times = (pd.date_range('00:00:00.0',
                              '00:00:00.00001',
                              freq = 'n') -
                pd.to_datetime('00:00:00.0')).total_seconds()
sensor_times
```

This produces the following:

```
Out[141]: Float64Index([                   0.0,                  1e-09,
                                         2e-09, 3.0000000000000004e-09,
                                         4e-09,                  5e-09,
                         6.000000000000001e-09,  7.000000000000001e-09,
                                         8e-09,  9.000000000000001e-09,
              ...
                         9.991000000000001e-06,              9.992e-06,
                                     9.993e-06,  9.994000000000001e-06,
                                     9.995e-06,              9.996e-06,
                         9.997000000000001e-06,              9.998e-06,
                                     9.999e-06,                  1e-05],
             dtype='float64', length=10001)
```

Figure 13.5 – Real-valued time-index with nanosecond resolution

Keep in mind that it would be possible to multiply by *10^9* to convert nanoseconds to integers, which is a typical pattern if we are to use such times in modeling. Also note that, due to the numerical precision, some of the values are not exact; we could address this with rounding or some other method if it were important. Another common use of time/date information is to group or aggregate data (such as by day or week) or convert it to another frequency (such as from days to weeks, or vice-versa). You will address these methods next.

Resampling, grouping, and aggregation by time

We have now covered many of the components of time series and the great convenience offered by pandas to work with time-stamped data. As we mentioned in the last section, most of the time you will think of a time series as a time-based index and one or more columns of data. Now, let's take that structure as a starting point and then move on to introduce some advanced capabilities in pandas.

Using the resample method

Suppose you were given 6,000 readings of a sensor dataset, and the sample rate was 10 Hz or 10 times per second. We can make a simulated series like this as follows. We can start as we did in the last section, creating a sequence of timestamps. Using an end time of 9:59.9 and a frequency of 100 ms (milliseconds) generates the correct number of points (6,000) on the correct interval (10 per second = 100 ms):

```
sensor_times = ((pd.date_range('00:00:00', '00:09:59.9', freq = 
'100ms')) -
                pd.to_datetime('00:00:00')).total_seconds()
sensor_times
```

This produces the following:

```
Out[125]: TimedeltaIndex([       '0 days 00:00:00', '0 days 00:00:00.100000',
                          '0 days 00:00:00.200000', '0 days 00:00:00.300000',
                          '0 days 00:00:00.400000', '0 days 00:00:00.500000',
                          '0 days 00:00:00.600000', '0 days 00:00:00.700000',
                          '0 days 00:00:00.800000', '0 days 00:00:00.900000',
                          ...
                                 '0 days 00:09:59', '0 days 00:09:59.100000',
                          '0 days 00:09:59.200000', '0 days 00:09:59.300000',
                          '0 days 00:09:59.400000', '0 days 00:09:59.500000',
                          '0 days 00:09:59.600000', '0 days 00:09:59.700000',
                          '0 days 00:09:59.800000', '0 days 00:09:59.900000'],
                         dtype='timedelta64[ns]', length=6000, freq=None)
```

Figure 13.6 – A sequence of timestamps every 0.1 second

Now, we simulate some data using `numpy.random.normal()`:

```
raw_data = (np.sin(2 * np.pi * np.arange(6000) / 500) +
            np.random.normal(5, 5, 6000))
```

We then construct a DataFrame using the times as the index:

```
sensor_data = pd.DataFrame({'data' : raw_data},
                            index = (sensor_times))
sensor_data
```

This gives us our simulated sensor data as follows:

Out[127]:

	data
0 days 00:00:00	10.692253
0 days 00:00:00.100000	11.023602
0 days 00:00:00.200000	-4.541415
0 days 00:00:00.300000	1.843395
0 days 00:00:00.400000	0.011138
...	...
0 days 00:09:59.500000	4.563188
0 days 00:09:59.600000	12.157702
0 days 00:09:59.700000	12.994389
0 days 00:09:59.800000	9.201197
0 days 00:09:59.900000	5.269805

6000 rows × 1 columns

Figure 13.7 – Simulated sensor data every 0.1 second

Note that the data values will vary due to the random numbers generated by NumPy. We can perform a quick visualization using the pandas `.plot()` method:

```
sensor_data.plot()
```

Out[128]: <AxesSubplot:>

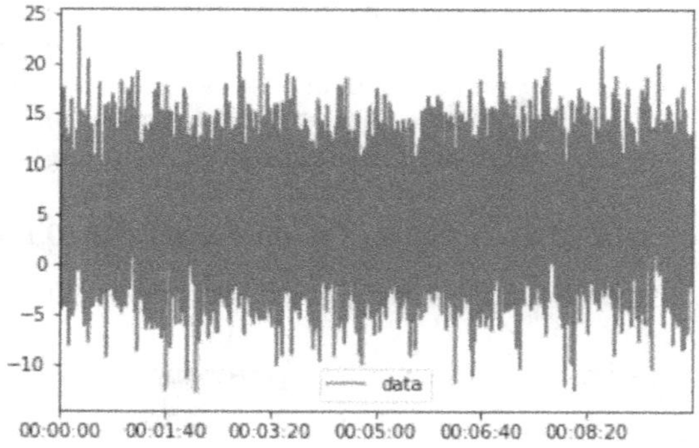

Figure 13.8 – The simulated sensor data

We see we have a lot of noise in the data. We can use the pandas `.resample()` method to address this.

Sampling in time series refers to various ways to use the existing timestamps and generate new timestamps at different intervals. **Downsampling** means converting to longer intervals, and **upsampling** means converting to shorter intervals. Note that new data values must be computed at the new timestamps. In pandas, this is done with an aggregation function applied to the result of the `.resample()` method, such as `.mean()` or an interpolation method. Sampling can be used to increase or decrease the time-granularity of data for analysis purposes, as well as a way to align two datasets to common timestamps so they can be merged or manipulated together. In the data in *Figure 13.8*, we can downsample to try to remove some noise. Generally, downsampling loses some information compared to the original data, so it is possible to downsample and then upsample back to the original frequency, but with a loss of information.

Resampling takes a rule that essentially says *construct a new series with the period specified in the rule*. In our case, let's try *downsampling* from a period of 0.1 seconds to a period of 5 seconds. The `.resample()` method requires some kind of *aggregation* function to define the downsampling. Here, we use `.mean()`, and `.resample()` applies that function to all values included in each of the new intervals. In this case, that is 50 samples averaged per interval:

```
sensor_data_smooth = sensor_data.resample('5000ms').mean()
sensor_data_smooth
```

This gives the following output:

Out[192]:

	data
0 days 00:00:00	4.595379
0 days 00:00:05	5.667126
0 days 00:00:10	6.863734
0 days 00:00:15	4.523539
0 days 00:00:20	4.402065
...	...
0 days 00:09:35	5.221181
0 days 00:09:40	3.371805
0 days 00:09:45	3.632050
0 days 00:09:50	5.272400
0 days 00:09:55	5.361699

120 rows × 1 columns

Figure 13.9 – The resampled sensor data, every 5 seconds

Note how `.resample()` automatically uses the time-based index to do the operation. We can plot the resampled data to see the effect:

```
sensor_data_smooth.plot()
```

This gives the following plot:

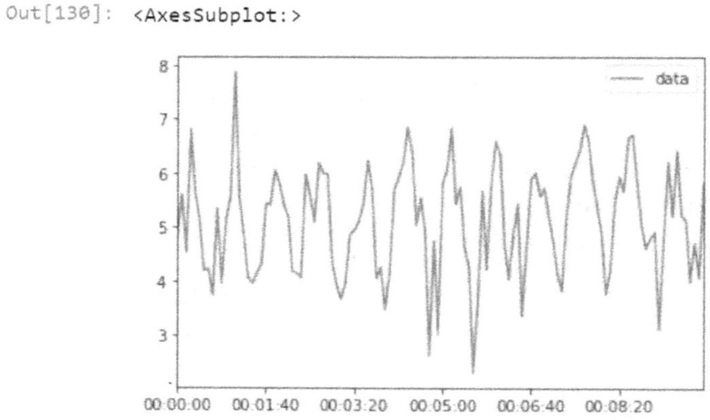

Figure 13.10 – The simulated sensor data smoothed using .resample()

We can now see the periodic nature of the data (we created it with a `sine` function).

The `.resample()` method can also *upsample*. The approach looks exactly the same, we just specify a rule that is a *higher* frequency period than the data. Here, we upsample the smoothed data to a period of 1 second. However, to do this, we will have gaps in the data, so instead of aggregating, we need a *filling* function. Here, we use `.interpolate()`. The `.interpolate()` method by default simply connects the nearest points with a line to estimate the missing values:

```
sensor_data_1s = sensor_data_smooth.resample('1s').
interpolate()
sensor_data_1s.plot()
```

This produces a nearly identical plot since we are linearly interpolating between the existing points:

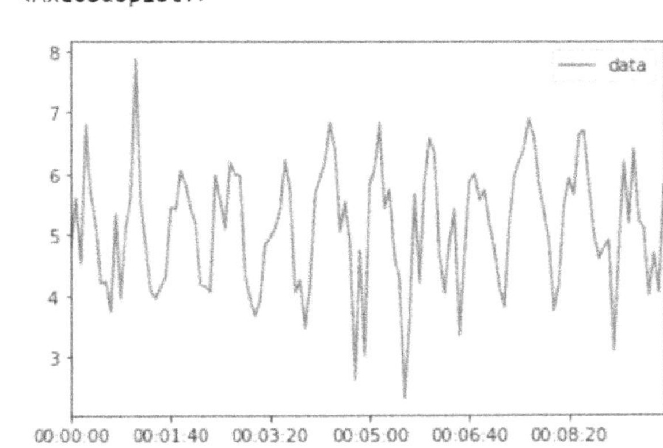

Figure 13.11 – The upsampled data

You can see that in *Figure 13.11*, the upsampled (interpolated) data looks the same as *Figure 13.10*. In this case, it makes sense, as *Figure 13.10* is the smoothed data, so upsampling just connects the points.

Exercise 13.01 – Aggregating and resampling

In this exercise, we will use the same subset of an online orders dataset we used in *Exercise 3* in *Chapter 12*. The data include sales for 36 countries and over 3,400 products. In this case, as the business analyst for corporate, you have been asked to compute hourly total sales. To accomplish this, you will resample and aggregate the data into one-hour periods.

> **Note**
>
> You can find the code for this exercise here: `https://github.com/PacktWorkshops/The-Pandas-Workshop/tree/master/Chapter13/Exercise13_01`.

Perform the following steps to complete the exercise:

1. In the `Chapter12` directory, create the `Exercise12_04` directory.
2. Open your Terminal (macOS or Linux) or Command Prompt (Windows), navigate to the `Chapter12` directory, and type `jupyter notebook`. The Jupyter notebook should open up.

3. Select the `Exercise12_04` directory, which will change the Jupyter working directory to that folder, then click **New | Python 3** to create a new `Python 3` notebook, as shown in the following screenshot:

Figure 13.12 – Creating a new Python 3 Jupyter notebook

4. For this exercise, all you will need is the `pandas` library. Load it in the first cell of the notebook:

```
import pandas as pd
```

5. Read the `online_retail_II.csv` file into a pandas DataFrame:

```
retail_sales = pd.read_csv('../datasets/online_retail_
II.csv')
retail_sales.head()
```

> **Note**
> Please change the path of the dataset file (highlighted) based on where you have downloaded it on your system.

This should produce the following:

Out[2]:

	Invoice	StockCode	Description	Quantity	InvoiceDate	Price	Customer ID	Country
0	539993	22386	JUMBO BAG PINK POLKADOT	10	1/4/2011 10:00	1.95	13313.0	United Kingdom
1	539993	21499	BLUE POLKADOT WRAP	25	1/4/2011 10:00	0.42	13313.0	United Kingdom
2	539993	21498	RED RETROSPOT WRAP	25	1/4/2011 10:00	0.42	13313.0	United Kingdom
3	539993	22379	RECYCLING BAG RETROSPOT	5	1/4/2011 10:00	2.10	13313.0	United Kingdom
4	539993	20718	RED RETROSPOT SHOPPER BAG	10	1/4/2011 10:00	1.25	13313.0	United Kingdom

Figure 13.13 – The retail sales data

6. Convert the `InvoiceDate` column to `datetime` using the pandas `.to_datetime()` method:

```
retail_sales['InvoiceDate'] = pd.to_datetime(retail_
sales['InvoiceDate'])
```

7. Make the index of `retail_sales` the `InvoiceDate` column, so we can use methods such as `.resample()`:

```
retail_sales.set_index('InvoiceDate', inplace = True)
retail_sales.head()
```

This produces the following result:

Out[27]:

	Invoice	StockCode	Description	Quantity	Price	Customer ID	Country
InvoiceDate							
2011-01-04 10:00:00	539993	22386	JUMBO BAG PINK POLKADOT	10	1.95	13313.0	United Kingdom
2011-01-04 10:00:00	539993	21499	BLUE POLKADOT WRAP	25	0.42	13313.0	United Kingdom
2011-01-04 10:00:00	539993	21498	RED RETROSPOT WRAP	25	0.42	13313.0	United Kingdom
2011-01-04 10:00:00	539993	22379	RECYCLING BAG RETROSPOT	5	2.10	13313.0	United Kingdom
2011-01-04 10:00:00	539993	20718	RED RETROSPOT SHOPPER BAG	10	1.25	13313.0	United Kingdom

Figure 13.14 – The retail_sales data with the InvoiceDate index

8. Compute the revenue by line as `Quantity` times `Price` and save in a new column, `Revenue`:

```
retail_sales['Revenue'] = retail_sales['Quantity'] *
retail_sales['Price']
```

9. Quickly visualize the data as-is by plotting the newly created `Revenue` column, for the first two weeks of data (before `2011-01-15`):

```
retail_sales.loc[retail_sales.index < pd.to_
datetime('2011-01-15'), 'Revenue'].plot()
```

```
Out[17]: <AxesSubplot:xlabel='InvoiceDate'>
```

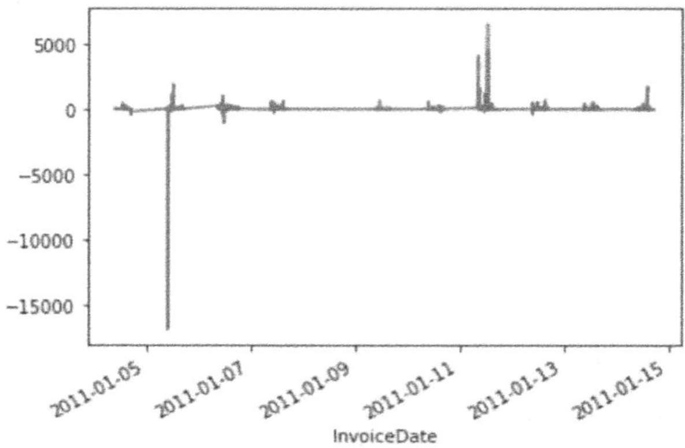

Figure 13.15 – The raw revenue figures

10. Use the `.resample()` method to downsample the data to 1-hour periods, and aggregate using `.sum()` to get the total per period. Note that you are ignoring `Countries` and `Products`:

```
retail_sales = retail_sales['Revenue'].resample('1h').
sum()
retail_sales
```

This produces the following output:

```
Out[29]: InvoiceDate
         2011-01-04 10:00:00    1696.12
         2011-01-04 11:00:00    1462.48
         2011-01-04 12:00:00    2223.33
         2011-01-04 13:00:00    5627.52
         2011-01-04 14:00:00    2785.46
                                  ...
         2011-06-30 16:00:00    1321.58
         2011-06-30 17:00:00    1539.94
         2011-06-30 18:00:00    1144.65
         2011-06-30 19:00:00     816.17
         2011-06-30 20:00:00     203.86
         Freq: H, Name: Revenue, Length: 4259, dtype: float64
```

Figure 13.16 – The retail_sales data downsampled to 1-hour periods

11. Now that the data is on 1-hour intervals, you would like to see whether there are some patterns. Make a quick visualization using the pandas `.plot()` method and limit the plot to `datetimes` before `2011-01-15`:

    ```
    retail_sales[retail_sales.index < pd.to_
    datetime('2011-01-15')].plot()
    ```

    This produces the following result:

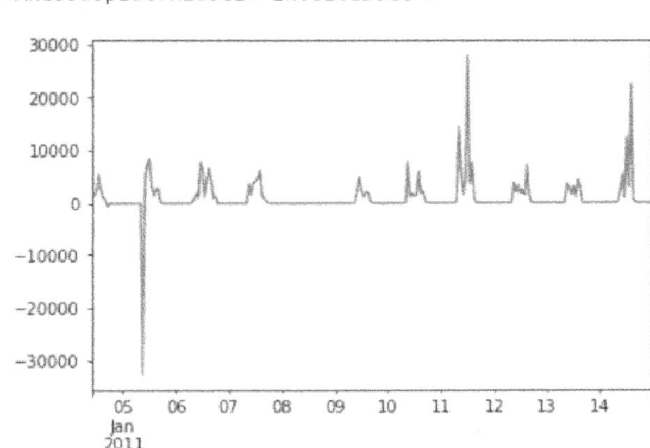

Figure 13.17 – The hourly revenue up to 2015-01-01

You can see in the chart that there are daily cycles of sales and a gap at the weekend. There are also apparently a lot of returns in January (the large negative spike). pandas makes it easy to gain such insights from your data. In this exercise, you saw how to use resampling and aggregation to convert orders data to a natural time period and look for patterns. We'll wrap up this chapter by looking at some other methods we can use to process time series data, namely the so-called **windowing** operations in pandas.

Windowing operations with the rolling method

Sometimes, instead of putting data into fixed time buckets, you may want a continuous transformation, such as a moving average. Pandas makes such transformations easy with the `.window()` method. Returning to our simulated data from earlier, let's recreate the original series as before:

```
sensor_data = pd.DataFrame({'data' : raw_data},
                           index = (sensor_times))
sensor_data.plot()
```

This reproduces the same plot from earlier:

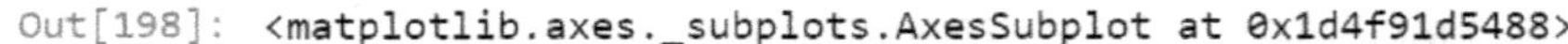

```
Out[198]: <matplotlib.axes._subplots.AxesSubplot at 0x1d4f91d5488>
```

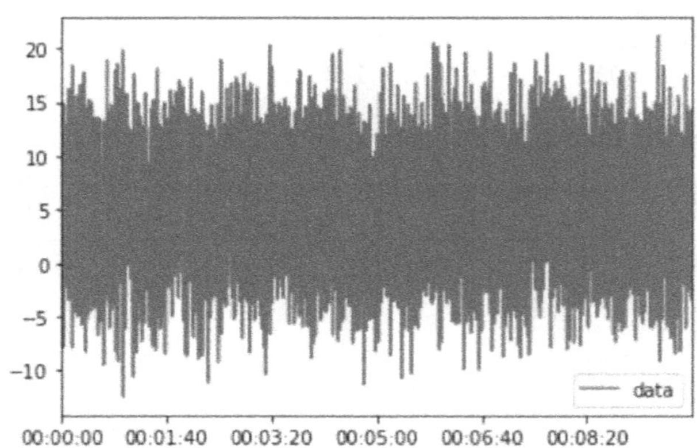

Figure 13.18 – The simulated sensor_data

Previously, we used `.resample()` with `.mean()` to smooth the data. However, that puts each data point in a fixed bin. Instead, we can use `.rolling()` with a window size to define a moving window that *slides* over the data and compute a function at every point across the window size. Before applying `.window()`, let's look at a simple example of how it works. Suppose your kids operate a lemonade stand in July. You help them keep track of daily receipts. We simulate the data as follows:

```
lemonade_income = \
    pd.DataFrame({'date' : pd.date_range('07-1-2021',
                                          '07-30-2021'),
                  'receipts' : pd.Series([50, 75,
                                          25, 33,
                                          17, 6,
                                          57]).sample(30,
                                                      replace =
True,
                                                      random_
state = 6)})
lemonade_income.set_index('date', drop = True, inplace = True)
print(lemonade_income)
lemonade_income.plot()
```

This produces the following:

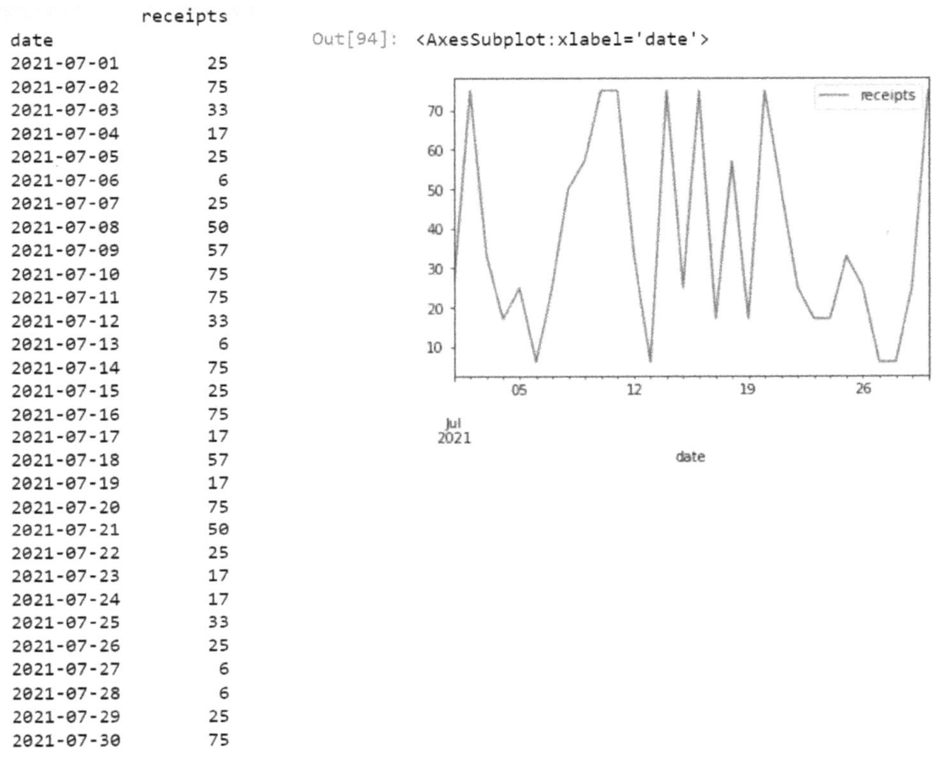

date	receipts
2021-07-01	25
2021-07-02	75
2021-07-03	33
2021-07-04	17
2021-07-05	25
2021-07-06	6
2021-07-07	25
2021-07-08	50
2021-07-09	57
2021-07-10	75
2021-07-11	75
2021-07-12	33
2021-07-13	6
2021-07-14	75
2021-07-15	25
2021-07-16	75
2021-07-17	17
2021-07-18	57
2021-07-19	17
2021-07-20	75
2021-07-21	50
2021-07-22	25
2021-07-23	17
2021-07-24	17
2021-07-25	33
2021-07-26	25
2021-07-27	6
2021-07-28	6
2021-07-29	25
2021-07-30	75

Figure 13.19 – Lemonade stand sales data

Now, suppose you and the kids are interested in how things varied over time. You, being a data analyst, suggest that using a moving average for a week would give a good picture. A moving average is like sliding a window over the data and averaging whatever is inside the window at each step. This process is illustrated here:

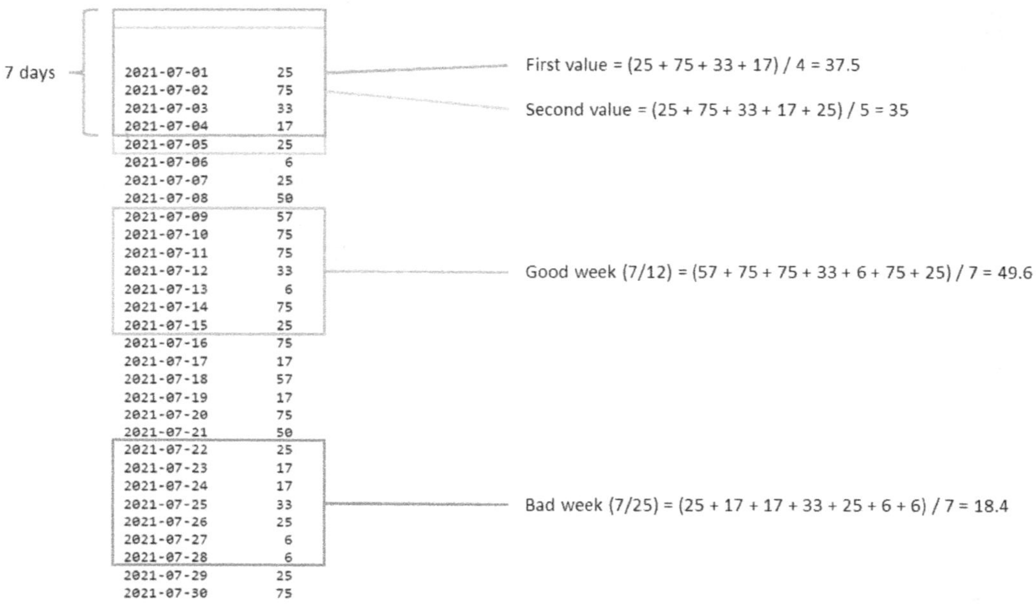

Figure 13.20 – Calculating moving averages

Note that we start with the window extending before our data. Taking averages accounts for the different number of days in any given window. You can see the values start in the 30s, and in the middle of the month, they reach nearly 50 and seem to have declined towards the end. You can accomplish this with the `.rolling()` function in pandas with a window of 7 days, as follows:

```
lemonade_income.rolling(window = 7,
                        center = True,
                        min_periods = 0).mean().plot()
```

This produces the following:

```
Out[95]: <AxesSubplot:xlabel='date'>
```

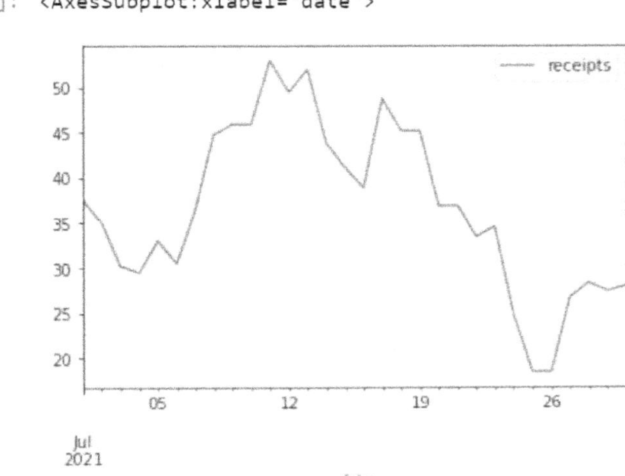

Figure 13.21 – 7-day moving average lemonade sales

It seems sales were, on average, better in the middle of the month. Perhaps the fourth of July US holiday had an impact, as well as back-to-school activities at the end of the month.

Notice that `.rolling()` generates a new value at every point of the original data, instead of reducing the number of values. Here, we use the `window` parameter with a window size of 100 points. We specify `center = True` to tell pandas to return the value at the *center* of the window; the default would be the value at the *right* (end) of the window. We also use `min_periods = 0` so that pandas will calculate values at either end of the data, even though there aren't enough points for the first and last 50 windows. The default is `min_periods` equal to the window, so pandas will put `NaN` in the points where the minimum cannot be met. Using `min_periods = 0` ensures every point has a value. Other values could be used depending on the use case. As with `.resample()`, we need to specify a function to aggregate data in the window, and here we use `.mean()`:

```
sensor_data.rolling(window = 100,
                    center = True,
                    min_periods = 0).mean().plot()
```

This produces the following:

```
Out[204]: <matplotlib.axes._subplots.AxesSubplot at 0x1d4fa69cd88>
```

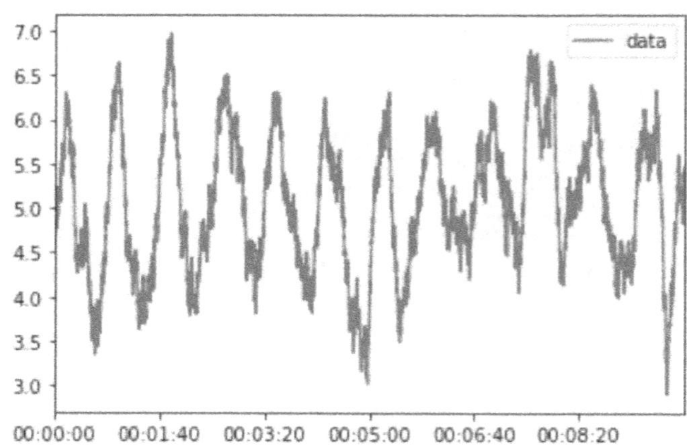

Figure 13.22 – The simulated sensor_data smoothed using a 100-period moving average

Note how this plot is similar to the one obtained using `.resample()` but has more points. Using `.rolling()`, we can also specify whether the intervals are *closed* or *open* on each side; in other words, is the first and/or last point in the interval included. Note that if we specify `both` (`closed = 'both'`), then we will be *reusing* the endpoints in the previous/next interval. The default is `right`, which means using the last (right-most) point in the interval, but not the first point. We can also specify a column to use to determine the intervals if there is a datetime-like column in the data we want to use instead of the index.

One important distinction using `.window()` is we can also specify the `win_type` parameter. The `win_type` takes string values as specified by the `scipy.signal` methods. The default is to evenly weigh all the points in the window. However, `scipy.signal` provides many other options (see the official documentation at `https://docs.scipy.org/doc/scipy/reference/signal.windows.html#module-scipy.signal.windows`) such as `triang` (triangular weighting) or `cosine` (cosine function weighting). Many of these are useful in **signal processing** applications. Here, we zoom in to a smaller range and compare the default to the triangular window:

```
fig, ax = plt.subplots()
ax.plot(sensor_data.iloc[:200, :].rolling(window = 100,
                                          win_type = None).
mean(),
        label = 'default',
        color = 'black')
```

```
ax.plot(sensor_data.iloc[:200, :].rolling(window = 100,
                                           win_type = 'triang').
mean(),
        label = 'triangular',
        color = 'red')
ax.legend()
plt.show()
```

This generates the following plot:

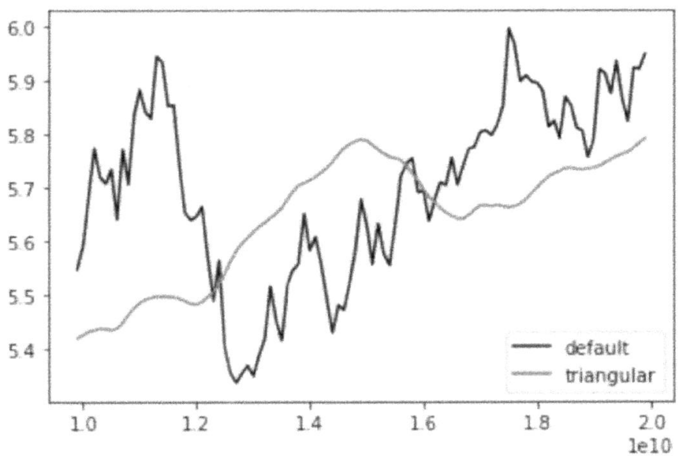

Figure 13.23 – Comparison of a triangular window to the default when computing a 100-point moving average

Because the triangular window weights the center points more than the other, it has smoothed out some of the information compared to a standard moving window. You might wonder how the choice of a 100-point window size was made. Here, we first define a utility function that can plot out a grid of moving window plots, given some data and a list of window sizes. You can find the code here: `https://github.com/PacktWorkshops/The-Pandas-Workshop/blob/master/Chapter13/Examples.ipynb`.

Now, you can explore the effect of the window size in a single line of code:

```
plot_rolling_grid(sensor_data, windows = [10, 50, 100, 500])
```

This produces the following:

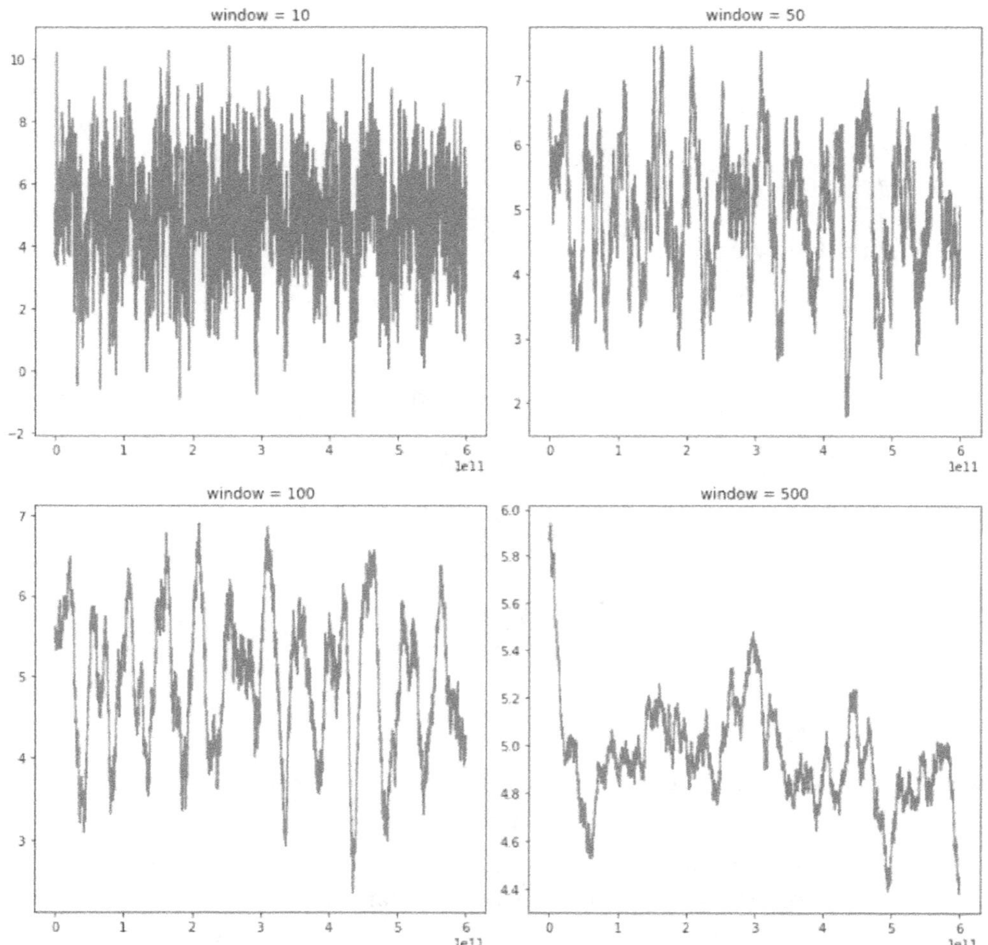

Figure 13.24 – The effect of window size on the smoothing of the sensor data

You can see that around 100, the periodic nature of the data becomes more apparent. Of interest, using even more smoothing shows that after the initial data, there is an apparent downward shift in the average values, which was not obvious previously.

Activity 13.01 – Creating a time series model

In this activity, as a data analyst for a bike-share startup, you are provided with a dataset that has hourly unit rentals for a bike-share business. You are tasked to create a very simple model to predict the rentals one week in advance. Here, you will use linear regression from scikit-learn, which you saw in *Chapter 11*, *Data Modeling –Regression*:

1. For this activity, you will need the `pandas` library, the `matplotlib.pyplot` library, and the `sklearn.linear_model.LinearRegression` module. Load them in the first cell of the notebook:

```
import pandas as pd
import matplotlib.pyplot as plt
from sklearn.linear_model import LinearRegression
```

2. Read in the `bike_share.csv` data from the `Datasets` directory, and list the first few rows:

Out[91]:

	date	hour	rentals
0	1/1/2011	0	16
1	1/1/2011	1	40
2	1/1/2011	2	32
3	1/1/2011	3	13
4	1/1/2011	4	1

Figure 13.25 – The bike_share.csv data

3. You need to create a `datetime` index. Construct a new datetime-valued column as a combination of the date and the hour and make it the index:

Out[137]:

	date	hour	rentals	date_time
date_time				
2011-01-01 00:00:00	1/1/2011	0	16	1/1/2011 00:00:00
2011-01-01 01:00:00	1/1/2011	1	40	1/1/2011 01:00:00
2011-01-01 02:00:00	1/1/2011	2	32	1/1/2011 02:00:00
2011-01-01 03:00:00	1/1/2011	3	13	1/1/2011 03:00:00
2011-01-01 04:00:00	1/1/2011	4	1	1/1/2011 04:00:00
...	...	...	...	...
2012-12-31 19:00:00	12/31/2012	19	119	12/31/2012 19:00:00
2012-12-31 20:00:00	12/31/2012	20	89	12/31/2012 20:00:00
2012-12-31 21:00:00	12/31/2012	21	90	12/31/2012 21:00:00
2012-12-31 22:00:00	12/31/2012	22	61	12/31/2012 22:00:00
2012-12-31 23:00:00	12/31/2012	23	49	12/31/2012 23:00:00

17379 rows × 4 columns

Figure 13.26 – Addition of a datetime column and index

4. Generate a simple plot of the data versus time:

```
Out[94]: <matplotlib.axes._subplots.AxesSubplot at 0x25969241208>
```

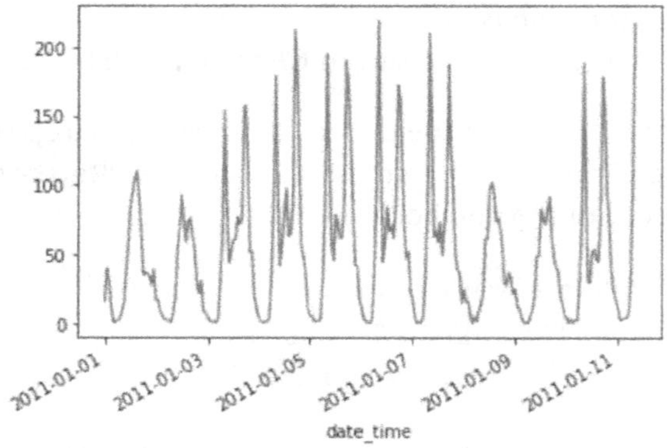

Figure 13.27 – Simple time series plot

5. Using the index and the `rentals` column, *downsample* the data to 1-day intervals. You want total rentals per day, so choose the appropriate aggregation function:

```
Out[95]:
```

date_time	rentals
2011-01-01	985
2011-01-02	801
2011-01-03	1349
2011-01-04	1562
2011-01-05	1600
2011-01-06	1606
2011-01-07	1510
2011-01-08	959
2011-01-09	822
2011-01-10	1321
2011-01-11	1263
2011-01-12	1162
2011-01-13	1406
2011-01-14	1421

Figure 13.28 – Total rentals per day

6. Generate a simple plot of the first 8 weeks (56 days) of the resampled data:

```
Out[86]: <matplotlib.axes._subplots.AxesSubplot at 0x25968f5dac8>
```

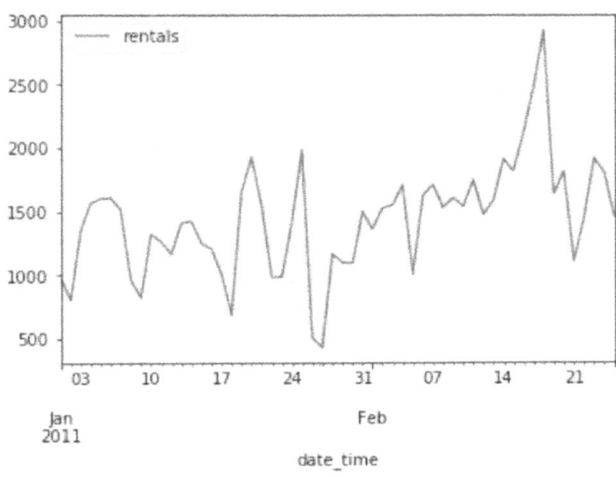

Figure 13.29 – Plot of total rentals per day

7. You should notice there are ups and downs that seem to be on about a 7-day cycle. Explore this by plotting the data versus itself, shifted appropriately:

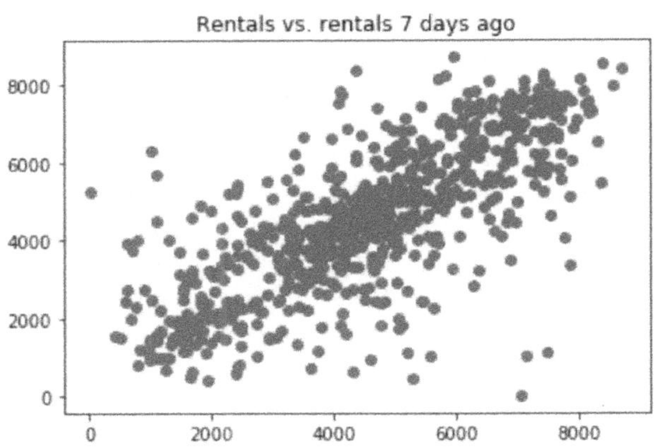

Figure 13.30 – Rentals per day versus the same values a week earlier

8. Preparing to apply linear regression to the data in *Figure 13.31*, create a new column with shifted data:

Out[109]:

date_time	rentals	lagged_rentals
2011-01-01	985	NaN
2011-01-02	801	NaN
2011-01-03	1349	NaN
2011-01-04	1562	NaN
2011-01-05	1600	NaN
...	...	...
2012-12-27	2114	4128.0
2012-12-28	3095	3623.0
2012-12-29	1341	1749.0
2012-12-30	1796	1787.0
2012-12-31	2729	920.0

731 rows × 2 columns

Figure 13.31 – Adding a lagged column to the data

9. Now, use the `LinearRegression` module to fit a linear model, using the lagged data as the `X` data, and the actual rentals as the `Y` data:

```
R2 is  0.5145071365683822  using:
             rentals  lagged_rentals
date_time
2011-01-08       959           985.0
2011-01-09       822           801.0
2011-01-10      1321          1349.0
2011-01-11      1263          1562.0
2011-01-12      1162          1600.0
```

Figure 13.32 – Result of the LinearRegression() model

10. Plot the predicted values versus the actual values and compare them to an *ideal* prediction:

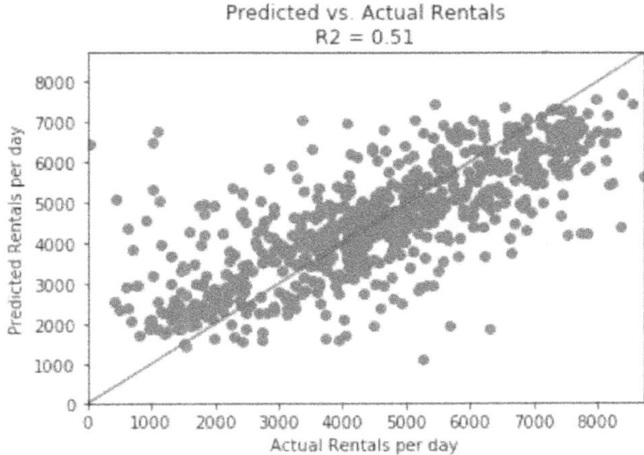

Figure 13.33 – Predicted versus actual values using simple linear regression

> **Note**
>
> You can find the solution for this activity in the Appendix.

Summary

In this chapter, starting with the enhancements provided by pandas for the `Timestamp` data types, you saw we can use time-aware methods to interpolate missing values or resample a time series to a higher or lower frequency (period). You used a linear regression model using a lagged series of data, which was enabled by making a `datetime` index from some text data. You are now prepared to analyze tabular data as well as order time series data, and carry out many transformations to find information hidden in complex data, which we will do in the next chapter.

14
Applying pandas Data Processing for Case Studies

So far in this book, we have progressively learned different data processing techniques using pandas, such as working with different types of data structures, accessing data from multiple sources, data cleaning, data transformation, visualization, code optimization, and finally, data modeling. This chapter aims to harness all these techniques you have learned so far, in analyzing four different case studies. The different case studies you will work through in this chapter will expose you to the different ways data needs to be preprocessed to be workable and help you see how good preparation is the key to good analysis. By the end of this chapter, you will have reinforced your understanding of all the data processing techniques you learned in this book by applying them to four case studies.

This chapter covers the following topics:

- Introduction to the case studies and datasets
- Recap of the preprocessing steps
- Activity 14.01 – analyzing air quality data

Introduction to the case studies and datasets

Data cleaning and preparation usually take up to 80% of the time in a data analytics life cycle. Transactional datasets can have multiple failure modes, some of the prominent ones being missing data points, incompatible formats, variability in data types, incorrect spellings in data, and unwanted characters and white spaces in data.

These are just some examples of how data can be messy. The success of a data analyst will depend on how well they are able to traverse these quagmires of messy data and transform the data into the required format. A sure-shot way to be adept at this all-too-important process is to get hands-on experience with multiple real-world datasets. In this chapter, you will analyze four different datasets, with each analysis focusing on different facets of data wrangling. The following list offers a snapshot of the datasets we will be dealing with in this chapter and the different techniques we will be applying to work on this dataset:

- German climate dataset

  Working through this dataset will help you hone your skills in creating a new dataset from multiple sources. You will combine datasets for three climate parameters (precipitation, vapor pressure, and sunshine) to create a new dataset. Once the new dataset is created, you will implement different methods, such as changing data formats and merging datasets, along with the grouping and aggregation of datasets. After creating a consolidated dataset, you will also do some exploratory analysis and try to answer questions on climate.

- Earnings dataset

  This dataset is the aggregation of different parameters related to working adults. Using this, you will be implementing some of the visualization techniques that were covered in *Chapter 8, Data Visualization*. You will explore the data using different transformation skills, such as aggregation and grouping, and then answer questions by looking at the visualizations.

- Bus trajectory dataset

  This dataset contains multiple files that contain the route information of public bus services. Using this dataset, you will apply some interesting techniques for preprocessing, such as geolocation extraction from latitude and longitude information. After preprocessing the data, you will answer different questions on the service quality levels of the bus services.

- Air quality dataset

  This dataset contains different parameters pertaining to air quality. You will use this dataset in an activity where you will be implementing different pandas preprocessing steps in addition to steps for thevisualization of data. The preprocessing and visualizations will be used to answer some questions on air quality.

Recap of the preprocessing steps

Unlike the previous chapters, in this chapter, we will only be reinforcing the skills that were taught in the previous chapters. This will be in the form of various exercises and an activity.

This section will help you recap some of the important preprocessing steps covered in this book so far and also go through some techniques that will be used in the exercises:

1. Reading CSV files

   ```
   pd.read_csv('file path' , delimiter=';')
   ```

   As you may recall, the `pd.read_csv` function is used to read the data from a CSV file available at the specified path.

2. Recasting data

   One of the most frequent transformation steps is changing the format from wide format to long format. For example, the following figure shows some data in wide format. You can see that the data for each month is spread across the columns:

	Jahr	Jan	Feb	Mrz	Apr	Mai	Jun	Jul	Aug	Sep	Okt	Nov	Dez
0	1951	48.0	49.0	98.0	61.0	23.0	34.0	44.0	146.0	64.0	89.0	47.0	72.0

   Figure 14.1 – Wide format data

   Often, when we have to preprocess data, we need data of months one after the other for ease of slice and dice. This is called long format, as depicted in the following figure:

	Jahr	variable	value
0	1951	Jan	48.0
1	1952	Jan	100.0
2	1953	Jan	100.0
3	1954	Jan	31.0
4	1955	Jan	21.0

   Figure 14.2 – Long format data

You can see that the data has been transformed into just three columns, with the year, month, and value for the month spread across three separate columns.

3. A good utility function to transform the data is the `melt` function in pandas. The pseudocode for this function is as follows:

```
pd.melt(data, id_vars = ['var1'], value_vars=
['val1','val2'])
```

The parameters used for this function are as follows:

- `data`: This is the DataFrame to be recast.
- `id_vars`: This is the column name on which the conversion is to be done.
- `value_vars`: This is where you can list the unique values of `id_variable`, which has to be recast. In the previous example, each year from the `Jahr` column has a unique value, which are the names of the months, `Jan`, `Feb`, and so on; these will form `value_vars`.

4. Merging DataFrames

pandas provides an intuitive method called `merge` to join different DataFrame objects. The syntax of the `merge` operation is as follows:

```
pd.merge(dat1,dat2,how='inner',on=['variable1','varia-
ble2'])
```

Here, `pd.merge` is used for merging two DataFrames, `dat1` and `dat2`. The following parameters are used for the `merge` operation:

- `how`: This defines how the `merge` operation needs to happen. There are different ways merging can be done, as can be seen from the following figure:

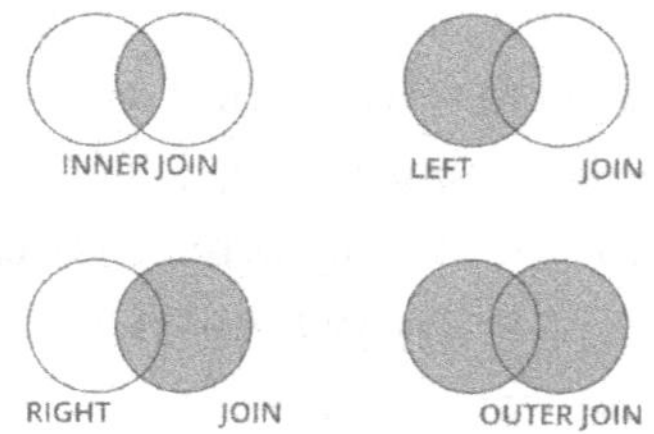

Figure 14.3 – Venn diagram for merging data

- `inner`: For inner joins, only the common records of both the DataFrames will be joined together. Note that joining is always done with respect to one or more variables.

- `left`: In left join, all the records of the DataFrame on the left will be retained. Only the relevant data of the right DataFrame that is common to the left one will be added to the combined DataFrame. If the right DataFrame does not have any data corresponding to the joining variable(s) of the left DataFrame, it will be filled with `NA` values.
- `right`: Right join is a mirror image of left join. In this case, the records of the right DataFrame will be retained and the records of the left one, common with the right, will be added to the combined DataFrame.
- `outer`: In outer join, all the records of the left and right will be present in the combined DataFrame. The common records will be merged on the selected variable.
- `on`: To merge two DataFrames, it is important to have a common variable(s) present for both the DataFrames. This is defined using the `on` parameter. In the preceding pseudocode, the merging is being done on the `variable1` and `variable2` variables.

5. Imputing missing values using `DataFrame.interpolate()`

   Missing data is a problem you will encounter quite often in the data analytics life cycle. Using datasets with missing data necessitates imputing values in place of `NaN` values. The type of imputing strategy to be adopted depends on the dataset and the domain.

   For example, consider that a dataset contains climate-related data, and assume that there is a linear trend for different variables across the months. In such a scenario, you can adopt a linear imputing method called interpolation. You will be using the `DataFrame.interpolate()` method in the upcoming exercises. The pseudocode for this method is as follows:

```
Df.interpolate(method ='linear',limit_direction ='both')
```

   In this method, `Df` is the DataFrame for which the null values are being imputed. The `linear` parameter implies a linear interpolation of missing values. There are other methods too, such as `spline` and `quadratic`, and the choice of method depends on the distribution of data and your understanding of the data generation process.

   The second parameter, `limit_direction`, gives an indication of the direction that the code must look for when interpolating. For example, assume that a dataset contains 10 records and there are null values in the first, second, and fourth records. When interpolating with the `both` direction, the third row, which has values, will interpolate both forward to fill the fourth row and also backward to fill the first and second rows.

6. Converting continuous data to ordinal by binning

   There will be many instances when you have to convert continuous data into categorical data. The best strategy in this scenario is to convert the continuous data into different bins. For example, consider that you have continuous data between 0 and 100; you can convert these data points into five specific groups, that is, 0-20, 20-40, 40-60, 60-80, and 80-100. Data points falling under each range can be treated as one category, for example, 0-20 as category 1, 20-40 as category 2, and so on. This is called binning. pandas has a good function called `pd.cut()` for converting data into bins. The pseudocode for this is as follows:

   ```
   pd.cut(data['variable'], bins=[0, 20, 40, 60, 80,100])
   ```

   Here, the variable that must be converted to categorial is specified, and then the bins are defined. The output will be categorical data based on the bins we have defined.

7. Extracting geolocation information

   You will be using a library called `geopy.geocoders` and a method called `Nominatim` to extract geographical locations from latitude and longitude information. To work with this package, you'll need to first initialize the API using the following pseudocode:

   ```
   geolocator = Nominatim(user_agent="geoapiExercises")
   ```

   Here, `Nominatim` is the package that extracts the geolocations, given some latitude-longitude information. When initializing the `Nominatim` object, you need to provide an application name such as `geoapiExercises` so that the library API sends the geolocation identifiers to your application name. After creating this variable called `geolocator`, the relevant information can be extracted using a method as follows:

   ```
   geolocator.reverse(coordinates)
   ```

   Here, `coordinates` contains the latitude and longitude information of the location.

The next section briefly covers the dataset containing information about German climate data, which will be used in the following exercise.

Preprocessing the German climate data

Over the next few exercises, you will be looking at three different datasets related to climate in order to answer some climate-related questions. You will use different preprocessing techniques in the exercises.

The datasets contain German climate data, and there are three files related to precipitation, sunshine, and vapor pressure.

> **Note**
>
> The raw data files have been sourced from the following links:
>
> **Precipitation data**: `https://opendata.dwd.de/climate_environment/CDC/observations_global/CLIMAT/monthly/qc/precipitation_total/historical/01001_195101_201712.txt`
>
> **Vapor pressure data**: `https://opendata.dwd.de/climate_environment/CDC/observations_global/CLIMAT/monthly/qc/vapour_pressure/historical/98836_196801_201712.txt`
>
> **Sunshine data**: `https://opendata.dwd.de/climate_environment/CDC/observations_global/CLIMAT/monthly/qc/sunshine_duration/historical/98836_197803_201612.txt`

These datasets are available at the GitHub link of this book and can be downloaded from there: `https://github.com/PacktWorkshops/The-Pandas-Workshop/`. Before you begin the exercises, make sure to download the data and keep it in your local drive in a folder named `data`.

Exercise 14.01 – preprocessing the German climate data

In this exercise, you will read three different files containing German climate data and convert the format from wide format to long format.

Perform the following steps to complete this exercise:

1. Open a new Jupyter notebook and then rename it `Exercise 14.01`.
2. Import the pandas library, as follows:

```
import pandas as pd
```

3. Define the path to the files. Note that you will have to give the path to the folder in which you have downloaded the raw files:

```
precipitation_dataPath = '/content/drive/MyDrive/Packt_
Colab/pandas_chapter11/chapter11/01001_195101_201412.txt'
vapor_dataPath = '/content/drive/MyDrive/Packt_Colab/
pandas_chapter11/chapter11/98836_196801_201712.txt'
sunshine_dataPath = '/content/drive/MyDrive/Packt_Colab/
pandas_chapter11/chapter11/98836_197803_201612.txt'
```

4. Read the files using the `pd.read_csv` function, as follows:

```
precipitation_data = pd.read_csv(precipitation_
dataPath,delimiter=';')
vapor_data = pd.read_csv(vapor_dataPath,delimiter=';')
sunshine_data = pd.read_csv(sunshine_
dataPath,delimiter=';')
```

5. Print out the top five rows for the precipitation data, as follows:

```
precipitation_data.head()
```

You should get the following output:

	Jahr	Jan	Feb	Mrz	Apr	Mai	Jun	Jul	Aug	Sep	Okt	Nov	Dez
0	1951	48.0	49.0	98.0	61.0	23.0	34.0	44.0	146.0	64.0	89.0	47.0	72.0
1	1952	100.0	24.0	28.0	23.0	16.0	35.0	25.0	19.0	59.0	105.0	56.0	80.0
2	1953	100.0	102.0	50.0	86.0	15.0	16.0	2.0	31.0	113.0	91.0	124.0	127.0
3	1954	31.0	58.0	39.0	50.0	20.0	26.0	65.0	34.0	53.0	90.0	135.0	40.0
4	1955	21.0	10.0	70.0	75.0	3.0	2.0	53.0	105.0	126.0	30.0	103.0	74.0

Figure 14.4 – Top five rows of the precipitation data

6. Now, take a look at the top five rows of the vapor data:

```
vapor_data.head()
```

You should get the following output:

	Jahr	Jan	Feb	Mrz	Apr	Mai	Jun	Jul	Aug	Sep	Okt	Nov	Dez
0	1968	26.4	26.4	28.1	28.1	29.8	29.8	29.8	29.8	29.8	29.8	28.1	28.1
1	1969	26.4	26.4	28.1	29.8	29.8	29.8	29.8	NaN	29.8	29.8	29.8	29.8
2	1970	28.1	28.1	28.1	29.9	29.8	29.9	28.1	29.8	29.8	29.8	29.8	NaN
3	1971	26.4	26.1	26.4	NaN	29.8	28.1	29.8	28.1	NaN	NaN	28.0	NaN
4	1972	NaN	NaN	NaN	NaN	NaN	NaN	NaN	28.1	NaN	NaN	NaN	28.0

Figure 14.5 – Top five rows of the vapor data

7. Similarly, look at the top five rows of the third dataset:

```
sunshine_data.head()
```

You should get the following output:

	Jahr	Jan	Feb	Mrz	Apr	Mai	Jun	Jul	Aug	Sep	Okt	Nov	Dez
0	1978	NaN	NaN	257.0	NaN	NaN	NaN	NaN	170.0	NaN	NaN	209.0	245.0
1	1979	253.0	NaN	NaN	228.0	207.0	NaN	169.0	NaN	NaN	NaN	NaN	NaN
2	1980	NaN	230.0	249.0	232.0	NaN	NaN	213.0	195.0	195.0	195.0	NaN	197.0
3	1981	NaN	212.0	244.0	NaN	206.0	NaN	193.0	NaN	168.0	148.0	234.0	261.0
4	1982	199.0	174.0	NaN	NaN	212.0	153.0	214.0	NaN	180.0	197.0	271.0	268.0

Figure 14.6 – Top five rows for the sunshine data

8. As you can see, the datasets are in wide format. Each year has a record with the monthly data across the columns. However, after combining the three datasets, this wide format would not be ideal for further processing. To resolve this, you must convert the data into long format, where there will be only three columns, the first corresponding to the year, the second relating to the month, and the last one having values corresponding to the year and month.

To convert the dataset from wide format to long format, we will use `pd.melt`, as follows:

```
precipitation_data = pd.melt(precipitation_data,\
                             id_vars = ['Jahr'],\
                             value_vars=['Jan','Feb',\
                                         'Mrz','Apr',\
                                         'Mai','Jun',\
                                         'Jul','Aug',\
                                         'Sep','Okt',
                                         'Nov','Dez'])


precipitation_data.head()
```

You should get the following output:

	Jahr	variable	value
0	1951	Jan	48.0
1	1952	Jan	100.0
2	1953	Jan	100.0
3	1954	Jan	31.0
4	1955	Jan	21.0

Figure 14.7 – Precipitation data in long format

You can see how the format of the data has changed. The months are listed under the column named `variable` and the corresponding values appear under the `value` column.

9. Now, rename the `value` column to `Precipitation`, using the following code:

```
precipitation_data = precipitation_data.rename\
                    (columns={"value": "Precipitation"})
```

You will get the following output:

	Jahr	variable	Precipitation
0	1951	Jan	48.0
1	1952	Jan	100.0
2	1953	Jan	100.0
3	1954	Jan	31.0
4	1955	Jan	21.0

Figure 14.8 – Output after renaming the value column to Precipitation

10. Similarly, convert the other two datasets into long format; start with the vapor dataset:

```
vapor_data = pd.melt(vapor_data,\
                     id_vars = ['Jahr'],\
                     value_vars= ['Jan','Feb',\
                                  'Mrz','Apr',\
                                  'Mai','Jun',\
                                  'Jul','Aug',\
                                  'Sep','Okt',\
```

```
                                    'Nov','Dez'])

vapor_data = vapor_data.rename\
          (columns={"value": "Vapour_Pressure"})

vapor_data.head()
```

You'll get the following output:

	Jahr	variable	Vapour_Pressure
0	1968	Jan	26.4
1	1969	Jan	26.4
2	1970	Jan	28.1
3	1971	Jan	26.4
4	1972	Jan	NaN

Figure 14.9 – Top five rows of the vapor pressure data

11. Perform the same step for the sunshine dataset:

```
sunshine_data = pd.melt(sunshine_data,\
                        id_vars = ['Jahr'],\
                        value_vars= ['Jan','Feb',\
                                     'Mrz','Apr',\
                                     'Mai','Jun',\
                                     'Jul','Aug',\
                                     'Sep','Okt',\
                                     'Nov','Dez'])

sunshine_data = \
sunshine_data.rename(columns={"value": "Sun_shine"})

sunshine_data.head()
```

You'll get the following output:

	Jahr	variable	Sun_shine
0	1978	Jan	NaN
1	1979	Jan	253.0
2	1980	Jan	NaN
3	1981	Jan	NaN
4	1982	Jan	199.0

Figure 14.10 – Top five rows of the sunshine dataset

This concludes the first exercise. In this exercise, you loaded the data from the `data` folder and then preprocessed the data by converting the wide format into long format. In the next exercise, you will use the same three DataFrames that were created in this exercise for further processing.

Exercise 14.02 – merging DataFrames and renaming variables

In this exercise, you will merge the three DataFrames from the previous exercise into one DataFrame. You will also rename some of the variables in this exercise. Since there is a dependency on the data generated in the last exercise, use the same Jupyter notebook used for the last exercise:

1. First, merge the different datasets together. Start by merging the first two and then, on this combined DataFrame, merge the third one. The columns that the merging must happen on are `Jahr` and `variable`, as it is important to know the year and month that is common to both the DataFrames. Do an inner merge, which means that only the common variables within the two DataFrames will be considered; any other joining method will introduce `NA` values in the combined DataFrame, which is not desired:

```
conDf = pd.merge(precipitation_data,vapor_data,\
                 how='inner',\
                 on=['Jahr','variable'])

conDf.head()
```

You should get the following output:

	Jahr	variable	Precipitation	Vapour_Pressure
0	1968	Jan	49.0	26.4
1	1969	Jan	19.0	26.4
2	1970	Jan	19.0	28.1
3	1971	Jan	51.0	26.4
4	1972	Jan	50.0	NaN

Figure 14.11 – Merged DataFrame with precipitation and vapor pressure data

You may notice that the years in the precipitation dataset start from 1951; however, for the vapor pressure dataset, the years only start from 1968. The inner merge has considered only those values common to both datasets and, therefore, starts from 1968.

2. Now, merge the third DataFrame with the combined DataFrame:

```
conDf = pd.merge(conDf,\
                 sunshine_data,\
                 how='inner',on=['Jahr','variable'])
conDf.head()
```

You should get the following output:

	Jahr	variable	Precipitation	Vapour_Pressure	Sun_shine
0	1978	Jan	62.0	28.1	NaN
1	1979	Jan	61.0	28.1	253.0
2	1980	Jan	60.0	NaN	NaN
3	1981	Jan	78.0	28.1	NaN
4	1982	Jan	59.0	28.0	199.0

Figure 14.12 – Consolidated DataFrame

From the output, you can see that the data from all three datasets is neatly stacked in separate columns, which is a great format for performing further wrangling steps.

3. If you noticed, the `month` variables and the `year` variable are in the German format. Convert them to the corresponding English names using the `map` function. You only need to convert the months where the German name is different from the English name. Add the following code for this:

```
months = {'Mrz':'Mar','Mai':'May','Okt':'Oct','Dez':'
Dec'}

conDf['variable'] = \
conDf['variable'].map(months).fillna(conDf['variable'])
conDf
```

You should get the following output:

	Jahr	variable	Precipitation	Vapour_Pressure	Sun_shine
0	1978	Jan	62.0	28.1	NaN
1	1979	Jan	61.0	28.1	253.0
2	1980	Jan	60.0	NaN	NaN
3	1981	Jan	78.0	28.1	NaN
4	1982	Jan	59.0	28.0	199.0
...	...	...	...	...	...
415	2012	Dec	42.0	29.8	220.0
416	2013	Dec	51.0	30.6	217.0
417	2014	Dec	70.0	30.2	234.0
418	2015	Dec	40.0	29.9	274.0
419	2016	Dec	63.0	29.9	229.0

Figure 14.13 – Consolidated DataFrame with variable names changed

You can now see that the months are changed to their English names.

4. Now, rename the `Jahr` variable to `Year`, as follows:

```
conDf = conDf.rename(columns={"Jahr": "Year"})
conDf.head()
```

You should get the following output:

	Year	variable	Precipitation	Vapour_Pressure	Sun_shine
0	1978	Jan	62.0	28.1	NaN
1	1979	Jan	61.0	28.1	253.0
2	1980	Jan	60.0	NaN	NaN
3	1981	Jan	78.0	28.1	NaN
4	1982	Jan	59.0	28.0	199.0

Figure 14.14 – Consolidated DataFrame with variable names changed

In this exercise, you practiced the merge methods and also renamed the variables to the corresponding English names. In the next exercise, you will do further preprocessing and answer some questions about the data.

Exercise 14.03 – data interpolation and answering questions after data preprocessing

In the previous exercise, you combined three datasets into a single DataFrame. In this exercise, we will take this combined DataFrame from the previous one and do further processing, such as imputing the null values and aggregating the data to answer some questions.

From this dataset, you need to find out what the average vapor pressure is for the month of January across all the years. Again, since there is a dependency on the data generated in the previous exercise, you will continue using the same notebook we used in it:

1. Up until now, the data was sorted and arranged in a way that all the months were consolidated. To have all the months of the same year one after the other, you need to first sort with respect to the year and then the month. However, sorting with respect to the month might be problematic, as the default sort will be done based on the alphabetical order of months, which will result in April and August occupying the first and second spots. To overcome this, you can first introduce a new column called `months`, copy the existing months into the new column, and then map each month according to its numerical order; that is, `Jan` will be mapped to `1`, `Feb` to `2`, and so on. Add the following code for this:

```
conDf['months'] = conDf['variable']
conDf.head()
```

You should get the following output:

	Year	variable	Precipitation	Vapour_Pressure	Sun_shine	months
0	1978	Jan	62.0	28.1	NaN	Jan
1	1979	Jan	61.0	28.1	253.0	Jan
2	1980	Jan	60.0	NaN	NaN	Jan
3	1981	Jan	78.0	28.1	NaN	Jan
4	1982	Jan	59.0	28.0	199.0	Jan

Figure 14.15 – DataFrame after adding a new variable

2. Having created the new column, now create a dictionary to map the month to the corresponding numerical order, as follows:

```
monthsMap = {'Jan':1,'Feb':2,\
             'Mar':3,'Apr':4,\
             'May':5,'Jun':6,\
             'Jul':7,'Aug':8,\
             'Sep':9,'Oct':10,\
             'Nov':11,'Dec':12}
```

3. Using the dictionary, map the months in the new column to their corresponding numerical equivalents, as follows:

```
conDf['months'] =\
conDf['months'].map(monthsMap).fillna(conDf['months'])
conDf
```

You should see the following output:

	Year	variable	Precipitation	Vapour_Pressure	Sun_shine	months
0	1978	Jan	62.0	28.1	NaN	1
1	1979	Jan	61.0	28.1	253.0	1
2	1980	Jan	60.0	NaN	NaN	1
3	1981	Jan	78.0	28.1	NaN	1
4	1982	Jan	59.0	28.0	199.0	1
...	...	...	...	...	...	...
391	2010	Dec	70.0	NaN	219.0	12
392	2011	Dec	58.0	31.1	188.0	12
393	2012	Dec	42.0	29.8	220.0	12
394	2013	Dec	51.0	30.6	217.0	12
395	2014	Dec	70.0	30.2	234.0	12

Figure 14.16 – DataFrame after mapping

4. Now that the months are mapped to their numerical equivalents, it's easy to sort them the way you want. Add the following code for sorting the months:

```
newCondf = conDf.sort_values(by=['Year','months'])
newCondf
```

The preceding snippet results in the following output:

	Year	variable	Precipitation	Vapour_Pressure	Sun_shine	months
0	1978	Jan	62.0	28.1	NaN	1
33	1978	Feb	25.0	NaN	NaN	2
66	1978	Mar	45.0	28.1	257.0	3
99	1978	Apr	32.0	26.4	NaN	4
132	1978	May	77.0	NaN	NaN	5
...	...	...	...	...	...	...
263	2014	Aug	36.0	31.4	216.0	8
296	2014	Sep	66.0	31.0	188.0	9
329	2014	Oct	82.0	31.7	186.0	10
362	2014	Nov	46.0	31.4	235.0	11
395	2014	Dec	70.0	30.2	234.0	12

Figure 14.17 – DataFrame after sorting

From the output, we can see that the sorting is done year-wise and also in the right order of the months.

5. In the new dataset, you can see that there are quite a few `NaN` values. Impute missing values using the `df.interpolate()` method, assuming that there is a linear trend for different variables across the months:

```
cleanDf1 = newCondf.interpolate(method ='linear',\
                                  limit_direction ='both')
cleanDf1
```

You should get the following output:

	Year	variable	Precipitation	Vapour_Pressure	Sun_shine	months
0	1978	Jan	62.0	28.100000	257.0	1
33	1978	Feb	25.0	28.100000	257.0	2
66	1978	Mar	45.0	28.100000	257.0	3
99	1978	Apr	32.0	26.400000	239.6	4
132	1978	May	77.0	26.683333	222.2	5
...	...	...	...	...	...	...
263	2014	Aug	36.0	31.400000	216.0	8
296	2014	Sep	66.0	31.000000	188.0	9
329	2014	Oct	82.0	31.700000	186.0	10
362	2014	Nov	46.0	31.400000	235.0	11
395	2014	Dec	70.0	30.200000	234.0	12

Figure 14.18 – DataFrame after interpolation

As seen from the output, all the null values are filled using the `.interpolate()` method.

6. Now that you have a clean dataset, you can find out what the average vapor pressure is for the month of January across all the years. To answer this question, group the dataset on the `variable` column and then do a mean aggregation for the `vapor_pressure` column and filter the month of January from the result, as follows:

```
Q1 = cleanDf1.groupby\
(['variable'])['Vapour_Pressure'].agg('mean').loc['Jan']
Q1
```

You should get the following output:

```
28.966044056953148
```

You can answer many more questions like this from the dataset. The key takeaway from the preceding exercises is the strategy that was adopted to combine multiple datasets, doing some basic cleaning up by changing variable names and then finally imputing missing values. Actions such as these are very common when dealing with datasets in real life. All the methods you have learned in this book will come in handy to address different scenarios. The important factor is to decide which method to use for various scenarios. That comes with a lot of practice with different datasets.

Next, you will look at another dataset and apply some data processing skills we have learned in the book.

Exercise 14.04 – using data visualizations to answer questions

As seen in *Chapter 8, Data Visualization*, plots and graphs are some of the most effective tools for data analytics. In this section, you will apply your visualization skills to a different dataset, which is the adult earnings dataset.

> **Note**
>
> The dataset being used here has been sourced from the UCI Machine Learning library and the original link for this is `http://archive.ics.uci.edu/ml/datasets/Adult`.
>
> This dataset is also kept in the GitHub repository of this book. You can download it and place it in the `data` folder of your local drive.

The dataset contains different variables related to working adults, such as age, occupation, gender, and educational qualifications, along with an indicator for whether the individual earns more than $50,000 per year or not. We will use our analysis and visualization skills to answer the following questions:

- *Does education have an influence on earning capacity?*
- *Is there any relationship between the hours worked and the earnings?*

Follow these steps to complete this exercise:

1. Open a new Jupyter notebook and name it `Exercise 11.04`.
2. Import the `pandas` library:

```
import pandas as pd
```

3. Define the paths for the dataset to read the dataset using pandas. You need to define the path of the folder where you have placed the file:

```
filePath = '/content/drive/MyDrive/Packt_Colab/pandas_
chapter11/chapter11/adult.data'
```

4. Load the data using the `read_csv()` function:

```
data = pd.read_csv(filePath,delimiter=',',header=None)
data.head()
```

You should get the following output:

	0	1	2	3	4	5	6	7	8	9	10	11	12	13	14
0	39	State-gov	77516	Bachelors	13	Never-married	Adm-clerical	Not-in-family	White	Male	2174	0	40	United-States	<=50K
1	50	Self-emp-not-inc	83311	Bachelors	13	Married-civ-spouse	Exec-managerial	Husband	White	Male	0	0	13	United-States	<=50K
2	38	Private	215646	HS-grad	9	Divorced	Handlers-cleaners	Not-in-family	White	Male	0	0	40	United-States	<=50K
3	53	Private	234721	11th	7	Married-civ-spouse	Handlers-cleaners	Husband	Black	Male	0	0	40	United-States	<=50K
4	28	Private	338409	Bachelors	13	Married-civ-spouse	Prof-specialty	Wife	Black	Female	0	0	40	Cuba	<=50K

Figure 14.19 – Earnings dataset

5. Summarize the dataset using the `.info()` method:

```
data.info()
```

You should get the following output:

```
<class 'pandas.core.frame.DataFrame'>
RangeIndex: 32561 entries, 0 to 32560
Data columns (total 15 columns):
0     32561 non-null int64
1     32561 non-null object
2     32561 non-null int64
3     32561 non-null object
4     32561 non-null int64
5     32561 non-null object
6     32561 non-null object
7     32561 non-null object
8     32561 non-null object
9     32561 non-null object
10    32561 non-null int64
11    32561 non-null int64
12    32561 non-null int64
13    32561 non-null object
14    32561 non-null object
dtypes: int64(6), object(9)
memory usage: 3.7+ MB
```

Figure 14.20 – Summary of the dataset

6. If you look at the data, you can observe that it does not have any headers, which you would have to add manually. Besides, there are data points related to race and gender, which are sensitive data. It would be prudent to eliminate the sensitive data points before analysis. Along with the sensitive data, we will also eliminate some of the continuous data that has no relevance to our analysis:

```
data = data.drop([2,4,8,9],axis=1)
```

7. Now, add headers to the data and then print the head of the data:

```
data.columns = ['age','workclass',\
                'education','marital-status',\
                'occupation','relationship',\
                'capital-gain','capital-loss',\
                'hours-per-week','native-
country','earning']
data.head()
```

You should get the following output:

	age	workclass	education	marital-status	occupation	relationship	capital-gain	capital-loss	hours-per-week	native-country	earning
0	39	State-gov	Bachelors	Never-married	Adm-clerical	Not-in-family	2174	0	40	United-States	<=50K
1	50	Self-emp-not-inc	Bachelors	Married-civ-spouse	Exec-managerial	Husband	0	0	13	United-States	<=50K
2	38	Private	HS-grad	Divorced	Handlers-cleaners	Not-in-family	0	0	40	United-States	<=50K
3	53	Private	11th	Married-civ-spouse	Handlers-cleaners	Husband	0	0	40	United-States	<=50K
4	28	Private	Bachelors	Married-civ-spouse	Prof-specialty	Wife	0	0	40	Cuba	<=50K

Figure 14.21 – Top five rows of the earnings dataset after adding headers and dropping some columns

The dataset is now ready for analysis, and you can answer some questions about it.

The first question to try to answer is, *does education have an influence on earning capacity?* To answer this question, you will have to find the proportion of people in each education group that earns greater than $50,000. This can be achieved by taking the following steps:

I. Aggregate the records based on education and then find the total count of people under each education group.

II. Filter the records where the earning is greater than $50,000, group it by education, and find the count.

III. Divide the first DataFrame with the second one to find the proportion of people.

IV. Plot the education level with respect to the earnings.

5. First, find the total number of people under each education category:

```
Q1_1 = data.groupby(['education'])['earning'].
agg('count')
Q1_1
```

You should get the following output:

```
education
 10th             933
 11th            1175
 12th             433
 1st-4th          168
 5th-6th          333
 7th-8th          646
 9th              514
 Assoc-acdm      1067
 Assoc-voc       1382
 Bachelors       5355
 Doctorate        413
 HS-grad        10501
 Masters         1723
 Preschool         51
 Prof-school      576
 Some-college    7291
Name: earning, dtype: int64
```

Figure 14.22 – Number of people under each education category

In this step, you first grouped the records in the `education` column and then counted the number of earning records within each group.

6. Next, filter the records based on people who are earning more than $50,000, aggregate this data based on education, and find the count under each category. Add the following code for this:

```
Q1_2 = data[data['earning'] == ' >50K'].groupby\
                                       (['education'])\
                                       ['earning'].
agg('count')
Q1_2
```

You should get the following output:

```
education
 10th              62
 11th              60
 12th              33
 1st-4th            6
 5th-6th           16
 7th-8th           40
 9th               27
 Assoc-acdm       265
 Assoc-voc        361
 Bachelors       2221
 Doctorate        306
 HS-grad         1675
 Masters          959
 Prof-school      423
 Some-college    1387
Name: earning, dtype: int64
```

Figure 14.23 – Number of people earning above $50,000 under each education category

7. Now create a new DataFrame by dividing the second DataFrame by the first one and multiplying it by 100, to get the proportion:

```
Q1_3 = pd.DataFrame((Q1_2 / Q1_1) * 100)
Q1_3.head()
```

You should see the following output:

	earning
education	
10th	6.645230
11th	5.106383
12th	7.621247
1st-4th	3.571429
5th-6th	4.804805

Figure 14.24 – DataFrame after calculations

8. Now, rename the `earning` column to `Proportion`:

```
Q1_3.columns = ['Proportion']
Q1_3.head()
```

The output will now look like the following:

	Proportion
education	
10th	6.645230
11th	5.106383
12th	7.621247
1st-4th	3.571429
5th-6th	4.804805

Figure 14.25 – Renaming the column

9. Round off the decimals in the `Proportion` column to two decimal places using the `.round()` method:

```
Q1_3 = Q1_3.round({'Proportion': 2})
Q1_3.head()
```

You should get the following output:

	Proportion
education	
10th	6.65
11th	5.11
12th	7.62
1st-4th	3.57
5th-6th	4.80

Figure 14.26 – DataFrame after rounding off

You can see that the data has been rounded off to two digits.

10. Create the first plot, a line plot, to visualize the trend in the proportion of people earning more than $50,000 in each education category:

```
Q1_3.plot.line(y='Proportion',rot=90,\
               title='Earning proportion with Education')
```

You should get the following output:

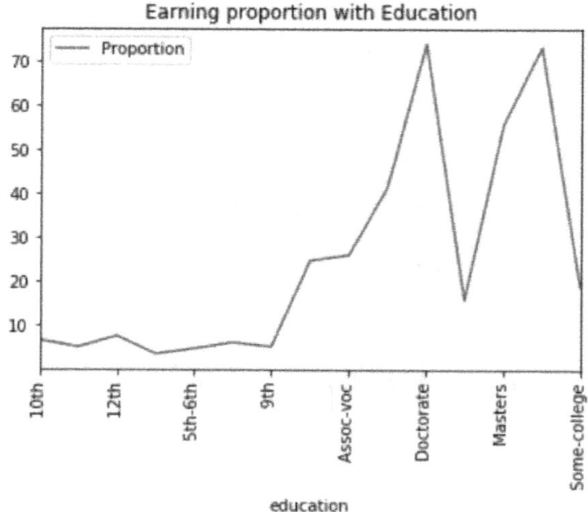

Figure 14.27 – Plot on earning potential

In the preceding snippet, the `plot.line()` function was used to create the plot. The proportion is given as the *y* axis, the index of the DataFrame, which is the education category, is the *x* axis, and `rot = 90` is used to rotate the axis labels vertically.

However, there is a big problem with the plot. You can see that there is no trend in the plot. What could be the reason?

Well, the obvious reason is the `education` category has not been kept in an increasing or decreasing order of the classes. Ideally, the `1st - 4th` education category has to first be followed by `5th - 6th` and so on. Only then will you be able to identify any trends within the data. For this, sort the education category in the logical order of education:

```
Q1_3 = Q1_3.reindex(index = [' 1st-4th',' 5th-6th',\
                             ' 7th-8th',' 9th',' 10th',\
                             ' 11th', ' 12th','
HS-grad',\
                             ' Some-college',' Assoc-
acdm',\
                             ' Assoc-voc',' Bachelors',\
                             ' Masters',' Prof-school',\
                             ' Doctorate'])
Q1_3.head()
```

You should get the following output:

	Proportion
education	
1st-4th	3.57
5th-6th	4.80
7th-8th	6.19
9th	5.25
10th	6.65

Figure 14.28 – DataFrame sorted in the order of education

From the output, you can see that the indexes are arranged in increasing order of education. Now, plot the data again, as follows:

```
propPlot = Q1_3.plot.line(y='Proportion',\
                          rot=90,\
                          title =\
                          'Earning proportion with 
Education')

propPlot.set_xlabel("Education")
propPlot.set_ylabel("Proportion")
```

You should get the following output:

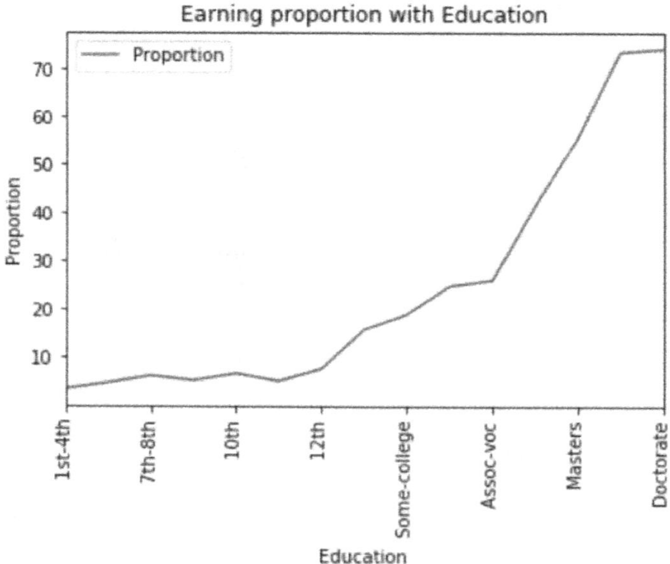

Figure 14.29 – Reordered plot on earning potential

You can see that there is an increasing trend in the proportion of people who earn more than $50,000 as the level of education increases.

In this exercise, you used your skills in aggregating datasets to get a consolidated dataset. You then used your visualization skills to answer a question about earning potential. The next exercise continues with the same dataset to answer the second question, to see whether there is a relationship between hours worked and earnings.

Exercise 14.05 – using data visualizations to answer questions

In the previous exercise, you preprocessed the adult earnings data and saw the relationship between the level of education and earning capacity. In this exercise, you will analyze whether there is any relationship between the hours worked and the earnings. For this exercise, there is a dependency on the datasets created in the previous one, so continue this exercise in the same notebook:

1. Start the analysis by looking at the variable under consideration, `hours-per-week`:

```
data['hours-per-week'].describe()
```

You should get the following output:

```
count    32561.000000
mean        40.437456
std         12.347429
min          1.000000
25%         40.000000
50%         40.000000
75%         45.000000
max         99.000000
Name: hours-per-week, dtype: float64
```

Figure 14.30 – Statistics for the hours-per-week variable

2. From the summary of the data, it can be seen that the data is continuous, not categorical. Just as you did in the previous exercise, you will need to group the records to find the proportion. Look at how many unique values there are to group:

```
len(set(data['hours-per-week']))
```

You should get the following output:

```
94
```

From the output, we can see that there are `94` unique values for hours-per-week. It wouldn't make sense to aggregate under each of these unique values. The best strategy in this scenario is to convert the continuous data into categorical data. For this, create five specific groups: `0-20`, `20-40`, `40-60`, `60-80`, and `80-100`. Define the bins with which you can group the data by adding the following code:

```
cut_bins = [0, 20, 40, 60, 80, 100]
```

3. Next, convert the continuous data into categories based on the bins, using the `pd.cut()` function:

```
data['cut_hours'] = pd.cut(data['hours-per-week'],\
                           bins=cut_bins)
data.head()
```

You should get the following output:

	age	workclass	education	marital-status	occupation	relationship	capital-gain	capital-loss	hours-per-week	native-country	earning	cut_hours
0	39	State-gov	Bachelors	Never-married	Adm-clerical	Not-in-family	2174	0	40	United-States	<=50K	(20, 40]
1	50	Self-emp-not-inc	Bachelors	Married-civ-spouse	Exec-managerial	Husband	0	0	13	United-States	<=50K	(0, 20]
2	38	Private	HS-grad	Divorced	Handlers-cleaners	Not-in-family	0	0	40	United-States	<=50K	(20, 40]
3	53	Private	11th	Married-civ-spouse	Handlers-cleaners	Husband	0	0	40	United-States	<=50K	(20, 40]
4	28	Private	Bachelors	Married-civ-spouse	Prof-specialty	Wife	0	0	40	Cuba	<=50K	(20, 40]

Figure 14.31 – DataFrame after binning

From the output, you can see that the `cut_hours` column represents the `hours-per-week` data as categorical data as per the bins that were defined. Once the bins are formed, it is easier to group on those bins and get the aggregated desired values.

4. Group the earnings data as per the bins and then find the total records under each bin:

```
Q2_1 = data.groupby(['cut_hours'])['earning'].
agg('count')
Q2_1
```

You should get the following output:

```
cut_hours
(0, 20]        2928
(20, 40]      20052
(40, 60]       8471
(60, 80]        902
(80, 100]       208
Name: earning, dtype: int64
```

Figure 14.32 – Result after groupby

5. Similar to the steps in the previous exercise, filter the records for those who earn more than $50,000, group them according to the bins, and find the counts. After this, divide the second DataFrame by the first and multiply by 100 to get the proportion:

```
Q2_2 =\
data[data['earning'] == ' >50K'].groupby(['cut_hours'])\
```

```
                                ['earning'].agg('count')

Q2_3 = pd.DataFrame(Q2_2/Q2_1 * 100)
Q2_3
```

The preceding snippet will result in the following output:

cut_hours	earning
(0, 20]	6.659836
(20, 40]	18.900858
(40, 60]	40.750797
(60, 80]	37.804878
(80, 100]	30.288462

Figure 14.33 – Result after filtering and division

6. Rename the column and round the values to 2 decimal places:

```
Q2_3.columns = ['Proportion']

Q2_3 = Q2_3.round({'Proportion': 2})
Q2_3
```

You should get the following output:

cut_hours	Proportion
(0, 20]	6.66
(20, 40]	18.90
(40, 60]	40.75
(60, 80]	37.80
(80, 100]	30.29

Figure 14.34 – DataFrame after renaming and rounding off

7. Having found the proportions, it's now time to plot the data. Use a bar plot for the plotting:

```
hoursPlot = Q2_3.plot.bar(y='Proportion', rot=90,\
title = 'Earning proportion with Hours of work')

hoursPlot.set_xlabel("Hours of work")
hoursPlot.set_ylabel("Proportion")
```

You should get the following output:

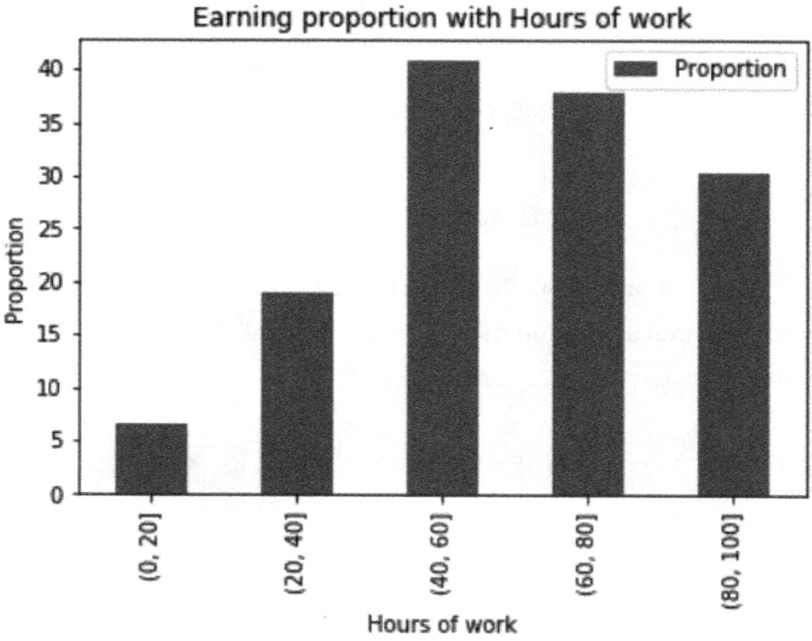

Figure 14.35 – Final data plot for the earning statistic

From the plot, you can see that there is a rising trend and after peaking at the `40-60` bin, the proportion dips. This means that there is no linear relation between hours of work and earning proportion. The people who earn the highest are the ones who work between 40 and 60 hours a week.

In this exercise, you saw how plots can be effectively used to find answers to questions on data. The biggest takeaway is how pandas can be used to transform the data for plotting effectively. In the next exercise, you will extract information about a location and use it to answer different questions.

Exercise 14.06 – analyzing data on bus trajectories

Imagine you are working as the data analyst for the public administration department of your city. The department of transportation in your city wants to improve the services offered by the buses plying in different suburbs of the city. To that end, they have decided to look at the data related to the different routes and get a feel of some of the service parameters of the bus service. The transport department is interested in finding answers to three questions to get a feel of some of the service parameters:

- *Which suburb has the highest number of commencing points of bus routes?*
- *What are the time bands when the highest number of routes commence?*
- *How does the frequency of bus services during peak hours vary across weekdays?*

For this exercise, you will work with two datasets related to the bus trajectories for different municipalities in Brazil. The first dataset has the variables related to the trajectories of buses in different municipalities. This dataset also has different attributes related to the route of the bus, such as the speed, total distance traveled, capacity utilization of the bus, and weather. The second dataset has the latitude and longitude details of each of the trajectories.

> **Note**
>
> The datasets have been sourced from the following link:
>
> `https://archive.ics.uci.edu/ml/machine-learning-databases/00354/GPS%20Trajectory.rar`
>
> The `.rar` files can be found in the GitHub repository of this book and have two datasets, `go_track_tracks.csv` and `go_track_trackspoints.csv`. Download the `.rar` file, extract it, and keep the two files in the `data` folder of your local drive.

You will use your knowledge of pandas to combine these datasets, preprocess them, and then answer the three questions. The following steps will help you to complete this exercise:

1. Open a new Jupyter notebook and import the required libraries:

```
import pandas as pd
from dateutil.parser import parse
```

2. Define the paths for both datasets and read them using pandas. Please note that you will have to provide the path where you have saved the two files instead of the paths in this step:

```
filePath1 = '/content/drive/MyDrive/Packt_Colab/'\
            'pandas_chapter11/chapter11/go_track_tracks.
csv'
filePath2 = '/content/drive/MyDrive/Packt_Colab/'\
            'pandas_chapter11/chapter11/go_track_
trackspoints.csv'
```

3. Read the first file:

```
data1 = pd.read_csv(filePath1,delimiter=",")
data1.head()
```

You should get the following output:

	id	id_android	speed	time	distance	rating	rating_bus	rating_weather	car_or_bus	linha
0	1	0	19.210586	0.138049	2.652	3	0	0	1	NaN
1	2	0	30.848229	0.171485	5.290	3	0	0	1	NaN
2	3	1	13.560101	0.067699	0.918	3	0	0	2	NaN
3	4	1	19.766679	0.389544	7.700	3	0	0	2	NaN
4	8	0	25.807401	0.154801	3.995	2	0	0	1	NaN

Figure 14.36 – The first five values of the dataset

4. Read the second dataset:

```
data2 = pd.read_csv(filePath2,delimiter=",")
data2.head()
```

You should see the following output:

	id	latitude	longitude	track_id	time
0	1	-10.939341	-37.062742	1	2014-09-13 07:24:32
1	2	-10.939341	-37.062742	1	2014-09-13 07:24:37
2	3	-10.939324	-37.062765	1	2014-09-13 07:24:42
3	4	-10.939211	-37.062843	1	2014-09-13 07:24:47
4	5	-10.938939	-37.062879	1	2014-09-13 07:24:53

Figure 14.37 – The first five values of the second dataset

5. Merge the two datasets on the trajectory ID. The `id` column name in the first dataset and the `track_id` column of the second DataFrame represent the IDs for the trajectories traversed by the buses. Join both these datasets on the trajectories to consolidate the travel details of each bus along each trajectory. Since you want only the relevant data from dataset 2, which is present in dataset 1, use the `left` join for the merge operation:

```
data = pd.merge(data1,\
                data2,\
                left_on='id',\
                right_on="track_id",\
                how="left")

data.head()
```

You will get the following output:

	id_x	id_android	speed	time_x	distance	rating	rating_bus	rating_weather	car_or_bus	linha	id_y	latitude	longitude	track_id	time_y
0	1	0	19.210586	0.138049	2.652	3	0	0	1	NaN	1	-10.939341	-37.062742	1	2014-09-13 07:24:32
1	1	0	19.210586	0.138049	2.652	3	0	0	1	NaN	2	-10.939341	-37.062742	1	2014-09-13 07:24:37
2	1	0	19.210586	0.138049	2.652	3	0	0	1	NaN	3	-10.939324	-37.062765	1	2014-09-13 07:24:42
3	1	0	19.210586	0.138049	2.652	3	0	0	1	NaN	4	-10.939211	-37.062843	1	2014-09-13 07:24:47
4	1	0	19.210586	0.138049	2.652	3	0	0	1	NaN	5	-10.938939	-37.062879	1	2014-09-13 07:24:53

Figure 14.38 – DataFrame after left join

From the output, you can see that the `id_x` column is the trajectory ID and the `id_y` column is a unique ID for each point in a trajectory. With the merge operation, you have got the different routes the bus will take and each point within the route is identified with its corresponding latitude-longitude coordinates. You will be using this information to explore geolocations of different trajectories.

6. Get the geolocations of the starting point of a route. For this, remove all duplicates under each trajectory and take only the first point, using the `drop_duplicates()` function, as follows:

```
df = data.drop_duplicates(subset='id_x', keep="first")
df.head()
```

You should get the following output:

	id_x	id_android	speed	time_x	distance	rating	rating_bus	rating_weather	car_or_bus	linha	id_y	latitude	longitude	track_id	time_y
0	1	0	19.210586	0.138049	2.652	3	0	0	1	NaN	1	-10.939341	-37.062742	1	2014-09-13 07:24:32
90	2	0	30.848229	0.171485	5.290	3	0	0	1	NaN	91	-10.939439	-37.062428	2	2014-09-13 13:37:54
203	3	1	13.560101	0.067699	0.918	3	0	0	2	NaN	204	-10.903162	-37.048294	3	2014-09-17 05:09:23
226	4	1	19.766679	0.389544	7.700	3	0	0	2	NaN	227	-10.908893	-37.052372	4	2014-09-17 05:09:23
355	8	0	25.807401	0.154801	3.995	2	0	0	1	NaN	564	-10.943777	-37.052344	8	2014-09-26 15:26:53

Figure 14.39 – DataFrame after removing duplicates

7. Next, extract the locations corresponding to the starting points of each trajectory. Use the `reverse(coordinates)` method of the `geopy` package to get the locations. First, initialize the `geopy` package. When initializing the `Nominatim` object, provide an application name such as `geoapiExercises` so that the library API sends the geolocation identifiers to the application name:

```
from geopy.geocoders import Nominatim
geolocator = Nominatim(user_agent="geoapiExercises")
```

8. Next, start an iterative loop to add the location details against the latitude-longitude information:

```
df['Suburb'] = 'NA'

for i in range(len(df)):
    lat = df.iloc[i]['latitude']
    long = df.iloc[i]['longitude']
    coordinates = str(lat) + ',' + str(long)
    location = geolocator.reverse(coordinates)
    suburb = location.raw['address'].get('suburb', '')
    df['Suburb'].iloc[i] = suburb

df.head()
```

You should get the following output:

l_x	id_android	speed	time_x	distance	rating	rating_bus	rating_weather	car_or_bus	linha	id_y	latitude	longitude	track_id	time_y	Suburb
1	0	19.210586	0.138049	2.652	3	0	0	1	NaN	1	-10.939341	-37.062742	1	2014-09-13 07:24:32	Grageru
2	0	30.848229	0.171485	5.290	3	0	0	1	NaN	91	-10.939439	-37.062428	2	2014-09-13 13:37:54	Grageru
3	1	13.560101	0.067699	0.918	3	0	0	2	NaN	204	-10.903162	-37.048294	3	2014-09-17 05:09:23	Industrial
4	1	19.766679	0.389544	7.700	3	0	0	2	NaN	227	-10.908893	-37.052372	4	2014-09-17 05:09:23	Centro
8	0	25.807401	0.154801	3.995	2	0	0	1	NaN	564	-10.943777	-37.052344	8	2014-09-26 15:26:53	Jardins

Figure 14.40 – DataFrame after the addition of locations

From the output, you can see that all the locations are extracted for the starting point of each trajectory.

9. Next, use the `date` column and then extract information for the weekday, day, month, and hour. Use the `parse()` function to extract this information, as follows:

```
df['Parse_date'] = df['time_y'].apply(lambda x: parse(x))

# Parsing the weekday
df['Weekday'] = df['Parse_date']\
                .apply(lambda x: x.weekday())

# Parsing the Day
df['Day'] = df['Parse_date']\
            .apply(lambda x: x.strftime("%A"))

# Parsing the Month
df['Month'] = df['Parse_date']\
              .apply(lambda x: x.strftime("%B"))

# Parsing the Time
df['StartHour'] = df['Parse_date']\
                  .apply(lambda x: x.strftime("%H"))

df.head()
```

You will see an output as follows:

rating_bus	rating_weather	car_or_bus	linha	...	latitude	longitude	track_id	time_y	Suburb	Parse_date	Weekday	Day	Month	StartHour
0	0	1	NaN	...	-10.939341	-37.062742	1	2014-09-13 07:24:32	Grageru	2014-09-13 07:24:32	5	Saturday	September	07
0	0	1	NaN	...	-10.939439	-37.062428	2	2014-09-13 13:37:54	Grageru	2014-09-13 13:37:54	5	Saturday	September	13
0	0	2	NaN	...	-10.903162	-37.048294	3	2014-09-17 05:09:23	Industrial	2014-09-17 05:09:23	2	Wednesday	September	05
0	0	2	NaN	...	-10.908893	-37.052372	4	2014-09-17 05:09:23	Centro	2014-09-17 05:09:23	2	Wednesday	September	05
0	0	1	NaN	...	-10.943777	-37.052344	8	2014-09-26 15:26:53	Jardins	2014-09-26 15:26:53	4	Friday	September	15

Figure 14.41 – Parsing the date data

You can see that the relevant date-related information is added to the dataset. You can now answer some questions on the dataset using the information we have extracted.

10. **Question 1: Which suburb has the most number of routes commencing?**

    To answer this question, group the data based on the suburb and get the count of the records, and then sort the data in descending order:

```
Q1_1 = df.groupby(['Suburb'])['Suburb']\
         .agg('count').sort_values(ascending=False)
Q1_1.head()
```

    The preceding snippet will lead to the following output:

```
Suburb
Industrial      49
São José        12
Coroa do Meio   11
Jabutiana        9
Centro           9
Name: Suburb, dtype: int64
```

    Figure 14.42 – Answer output for question 1

    From the output, you can see that the `Industrial` suburb has the highest number of buses starting from the location.

11. **Question 2: What are the time bands when the highest number of routes commence?**

    You will perform multiple steps to answer this question. As a first step, aggregate the `StartHour` column into multiple bins of hours. For this, convert the hours to numeric form:

```
df['StartHour'] = pd.to_numeric(df['StartHour'])
```

Next, you need to convert the time to different bins. In making the bins, just assume a typical work cycle when people go to work and in the evening return from work. Based on that, you can use the following bins. Note that you may choose different bins you want based on the granularity you want to have:

```
cut_bins = [0, 6, 10,15 ,20,23]
```

Next, cut the data based on the bins we defined:

```
df['cut_hours'] = pd.cut(df['StartHour'], bins=cut_bins)
df.head()
```

The output for this should be as follows:

rating_bus	rating_weather	car_or_bus	linha	...	longitude	track_id	time_y	Suburb	Parse_date	Weekday	Day	Month	StartHour	cut_hours
0	0	1	NaN	...	-37.062742	1	2014-09-13 07:24:32	Grageru	2014-09-13 07:24:32	5	Saturday	September	7	(6, 10]
0	0	1	NaN	...	-37.062428	2	2014-09-13 13:37:54	Grageru	2014-09-13 13:37:54	5	Saturday	September	13	(10, 15]
0	0	2	NaN	...	-37.048294	3	2014-09-17 05:09:23	Industrial	2014-09-17 05:09:23	2	Wednesday	September	5	(0, 6]
0	0	2	NaN	...	-37.052372	4	2014-09-17 05:09:23	Centro	2014-09-17 05:09:23	2	Wednesday	September	5	(0, 6]
0	0	1	NaN	...	-37.052344	8	2014-09-26 15:26:53	Jardins	2014-09-26 15:26:53	4	Friday	September	15	(10, 15]

Figure 14.43 – Dataframe after processing

Finally, group the data as per the bins:

```
Q2_1 = df.groupby(['cut_hours'])['Suburb'].agg('count')
Q2_1
```

You should get the following output:

```
cut_hours
(0, 6]      22
(6, 10]     59
(10, 15]    49
(15, 20]    28
(20, 23]     4
```

Figure 14.44 – Answer output for question 2

From the data, you can see that the frequency is the highest for the time band between 6 and 10 in the morning.

12. **Question 3: How does the frequency of bus services during peak hours vary across weekdays?**

    First, filter out all records in the 6 to 10-hour time band.

    ```
    Q3_1 = df[pd.arrays.IntervalArray(df['cut_hours'])\
              .overlaps(pd.Interval(6, 10))]
    ```

    On the filtered data, aggregate on weekdays, as follows:

    ```
    Q3_2 = Q3_1.groupby(['Day'])['Suburb'].agg('count')
    Q3_2
    ```

    You should see the following output:

    ```
    Day
    Friday        8
    Monday        6
    Saturday     15
    Thursday     13
    Tuesday       6
    Wednesday    11
    ```

    Figure 14.45 – Output for data after filtering specific days

    Re-index the days of the week from Monday to Saturday, as follows:

    ```
    Q3_2 = Q3_2.reindex(index = ['Monday',\
                                 'Tuesday',\
                                 'Wednesday',\
                                 'Thursday',\
                                 'Friday',\
                                 'Saturday'])
    Q3_2
    ```

    You should get the following output:

    ```
    Day
    Monday        6
    Tuesday       6
    Wednesday    11
    Thursday     13
    Friday        8
    Saturday     15
    ```

    Figure 14.46 – Output after rearranging weekdays

13. Plot the values using a line plot and see the distribution:

```
dayPlot = Q3_2.plot.line(y='Day',rot=90,\
title = 'Frequency of service on week days')

dayPlot.set_xlabel("Week Day")
dayPlot.set_ylabel("Frequency")
```

You will see the following output:

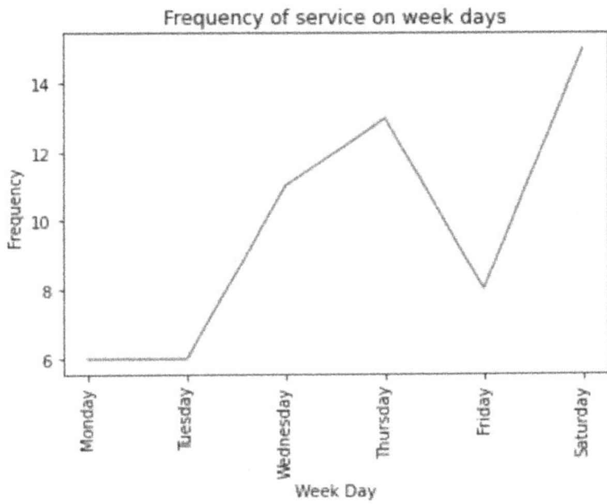

Figure 14.47 – Final plot of the processed data

From the plot, you can see that Saturday has the highest frequency of service for the time band from 6 A.M. to 10 A.M.

In this exercise, you extracted information from time objects and then answered various questions about the data.

Activity 14.01 – analyzing air quality data

Consider that you're working as a data analyst for your city's municipality. The Department for the Environment needs your help in getting answers to some questions related to emissions. The following are the questions the department wants answers to:

- *Which day of the week has the highest NO2(GT) emissions?*
- *At what time of the day are NMHC(GT) emissions highest?*

- *Which month has the lowest CO(GT) emissions?*

> **Note**
> The emissions dataset has been sourced from the following link:
> `https://archive.ics.uci.edu/ml/machine-learning-databases/00360/`
> You can find the dataset in the GitHub repository for this book. Download the data, unzip the data, and then load the CSV file in a `data` folder of your local machine. The department needs the answers through good visualizations.

The following steps will help you complete this activity:

1. Open a new Jupyter notebook.
2. Download the data and then read the data using pandas.
3. Drop unknown columns and rows with `NA` values.
4. Extract the different attributes from the `date` column.
5. Use the different methods you have learned in this chapter to answer all the questions.
6. Use suitable plotting methods to get relevant charts.

For the first question, you should get a plot like the following:

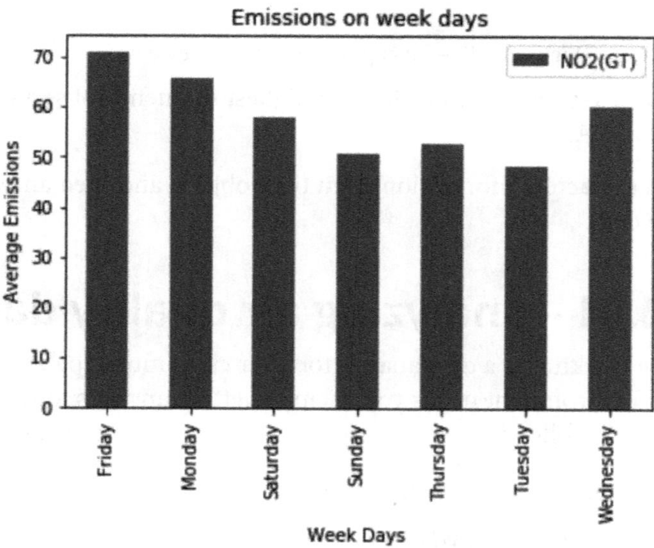

Figure 14.48 – Final expected output for the first question

Summary

In this final chapter, you got hands-on practice with different data processing tasks done on real-world datasets. In the first dataset, you explored different methods of data processing. Some of the key methods implemented were for converting from wide format to long format, merging two DataFrames, and imputing missing data using the `interpolate` method.

With the second dataset, you practiced preprocessing tasks before plotting, such as grouping and aggregation, and converting continuous data into categorical data using binning. You also answered questions about the data using line plots and bar charts.

Using the third dataset, you extracted geolocations from latitude and longitude information. After extracting geolocation information, you also answered some questions on the service level of bus routes.

Finally, with the fourth dataset, we used different methods to preprocess data to build a classification model. You should now be able to confidently tackle most data science problems that come your way using pandas!

15 Appendix

Solution 1.1

Perform the following steps to complete the activity:

1. Open a new Jupyter notebook and select the `Pandas_Workshop` kernel.
2. Import `pandas` and `random` into your notebook:

```
import pandas as pd
import random
```

3. Load the DataFrame for `Store1`:

```
Store1 = pd.read_csv('../Datasets/Store1.csv')
Store1.head()
```

> **Note**
>
> Please change the path of the dataset file (highlighted) based on where you have downloaded it in your system.

You should get the following output:

	Months	Grocery_sales	Stationary_sales
0	Jan	16	57
1	Jan	44	139
2	Jan	15	85
3	Jan	59	8
4	Jan	36	106

Figure 15.1 – Sales figures of the first store

4. Similarly, load the DataFrame for `Store2`:

```
Store2 = pd.read_csv('../Datasets/Store2.csv')
Store2.head()
```

> **Note**
> Please change the path of the dataset file (highlighted) based on where you have downloaded it in your system.

You should get the following output:

	Months	Grocery_sales	Stationary_sales
0	Jan	36	84
1	Jan	51	63
2	Jan	17	71
3	Jan	48	65
4	Jan	57	66

Figure 15.2 – Sales figures of the second store

Answering question 1: Which store has greater sales for the quarter?

5. Find the total number of sales for `Store 1` by selecting only the sales values and then summing them successively:

```
Sales_store1 = (Store1[['Grocery_sales','Stationary_
sales']].sum(axis=1)).sum()
Sales_store1
```

You should get the following value:

```
6082
```

Similarly, get the total number of sales for `Store2`:

```
Sales_store2 = (Store2[['Grocery_sales','Stationary_
sales']].sum(axis=1)).sum()
Sales_store2
```

You should get the following value:

```
5847
```

From both the values, you can see that `Store1` has higher sales.

Answering question 2: Which store has the highest sales for grocery products?

6. Take the grocery sales column for `Store1` and sum the values, as follows:

```
# Grocery sales for store 1
Sales_store1 = Store1['Grocery_sales'].sum()
Sales_store1
```

You should get the following value:

```
2097
```

Similarly, take the grocery sales column for `Store2` and sum the values:

```
# Grocery sales for store 2
Sales_store2 = Store2['Grocery_sales'].sum()
Sales_store2
```

You should get the following value:

```
2696
```

From the results, you can see that `Store2` has higher grocery sales.

Answering question 3: Which store has the highest sales for March?

7. For `Store1`, group the sales transactions based on month and then aggregate the values to get monthly sales figures:

```
Sales_store1 = Store1.groupby(['Months'])[['Grocery_
sales','Stationary_sales']].agg('sum').agg('sum',axis =
1)


Sales_store1
```

You should get the following output:

```
Months
Feb      1744
Jan      1467
Mar      2871
dtype: int64
```

From the output, you can see that the number of sales for March is `2871`.

Similarly, for `Store2`, find the values for March:

```
Sales_store2 = Store2.groupby(['Months'])[['Grocery_
sales','Stationary_sales']].agg('sum').agg('sum',axis =
1)
Sales_store2
```

You should get the following values:

```
Months
Feb      2050
Jan      1419
Mar      2378
dtype: int64
```

From the results, you can see that `Store1` has higher sales for March.

Answering question 4: For how many days were the sales values for stationery greater in store 1 than store 2?

8. To answer this question, first, use the `df.gt()` function to find out the number of `stationary` sales for both `Store1` and `Store2`:

```
Store1_stationary_greater = Store1[Store1['Stationary_
sales'].gt(Store2['Stationary_sales'],axis=0)]
Store1_stationary_greater
```

You should see the following output. You can ignore any warnings you get as part of the output.

	Months	Grocery_sales	Stationary_sales
1	Jan	44	139
2	Jan	15	85
4	Jan	36	106
5	Jan	27	136
6	Jan	74	116
7	Jan	63	142
8	Jan	65	129
9	Jan	12	138
10	Feb	34	112
11	Feb	73	100
12	Feb	45	135
13	Feb	31	13

Figure 15.3 – Store1 sales (output truncated for brevity)

9. To find the total number of days, use the `len()` function on the output of the preceding step:

```
len(Store1_stationary_greater)
```

You should get the following value:

```
35
```

From the values, you can see that there are `35` days where `Store1` stationery sales are higher than that of `Store2`.

Solution 2.1

Perform the following steps to complete the activity:

1. Import the `pandas` library:

```
import pandas as pd
```

2. Read the `US_GDP.csv` file from the `Datasets` directory into a DataFrame named `GDP_data`. The data is stored in two columns, `date` and `GDP`, and the date is read in (by default) as the `object` type. The goal of this activity is to first convert the `date` column into a timestamp and then set this column as the index. Finally, save the updated dataset to a new file:

```
fname = '../Datasets/US_GDP.csv'
GDP_data = pd.read_csv(fname)
```

> **Note**
> Please change the path of the dataset file (highlighted) based on where you have downloaded it in your system. You can download the file from `The-Pandas-Workshop/US_GDP.csv at master · PacktWorkshops/The-Pandas-Workshop · GitHub.`

3. Display the head of `GDP_data` so that you can see the format of the data in the file:

```
GDP_data.head()
```

The output should look as follows:

Out[2]:

	date	GDP
0	2017-03-31	19190.4
1	2017-06-30	19356.6
2	2017-09-30	19611.7
3	2017-12-31	19918.9
4	2018-03-31	20163.2

Figure 15.4 – First five rows of GDP_data

4. Inspect the object types of `GDP_data`, in particular, the `date` column:

```
GDP_data.dtypes
```

The output should be as follows:

```
Out[3]: date      object
        GDP      float64
        dtype: object
```

Figure 15.5 – Data types of GDP_data

5. Use the `pd.to_datetime()` method to convert the `date` column into a timestamp:

```
GDP_data['date'] = pd.to_datetime(GDP_data['date'])
```

6. Use the `.set_index()` method to replace the index with the `date` column. Be sure to use `inplace = True` so the result is applied to the existing DataFrame, and `drop = True` to remove the `date` column after it is used for the index. Use `.head()` to confirm the changes:

```
GDP_data.set_index('date', inplace = True)
GDP_data.head()
```

The output should be as follows:

Out[4]:

date	GDP
2017-03-31	19190.4
2017-06-30	19356.6
2017-09-30	19611.7
2017-12-31	19918.9
2018-03-31	20163.2

Figure 15.6 – GDP_data after using the date for the index

7. Use the `.to_csv()` method to save the file to a new `.csv` file named `US_GDP_date_index.csv`. Remember to change the path to the dataset (highlighted), as required:

```
GDP_data.to_csv('../Datasets/US_GDP_date_index.csv')
```

Solution 3.1

Perform the following steps to complete the activity:

1. Load the `pandas` and `sqlite3` libraries in the first cell of the notebook:

```
import pandas as pd
import sqlite3
```

2. Get the list of tables present in `supply_company.db`. The database can be downloaded from `https://github.com/PacktWorkshops/The-Pandas-Workshop/blob/master/Chapter03/datasets/supply_company.db`:

```
tables = pd.read_sql("SELECT name FROM sqlite_master
WHERE type = 'table'",
                     sqlite3.connect('../datasets/supply_
company.db'))
tables
```

Note

Please change the path of the dataset file (highlighted) based on where you have downloaded it on your system.

This produces the following result:

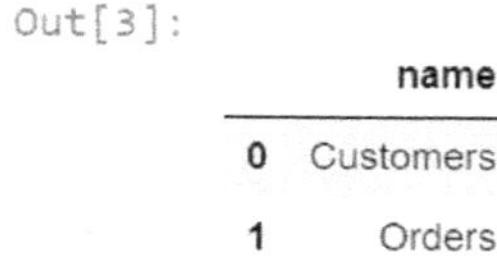

Out[3]:

	name
0	Customers
1	Orders

Figure 15.7 – List of tables in supply_company.db

3. Use a pandas SQL method to load the table containing the orders into a DataFrame, as follows:

```
orders = pd.read_sql("select * from Orders",\
                     sqlite3.connect("../datasets/supply_
company.db"))
orders
```

This produces the following output (truncated to save space—you should get 28 rows):

Out[3]:

	index	Customer_Number	date	item	qty	price	amount
0	0	25058	10/19/2020	354161666	62	91.50	5673.14
1	1	25058	11/10/2020	1129038342	38	79.79	3032.14
2	2	26069	11/23/2020	421919566	40	55.67	2226.76
3	3	26069	12/22/2020	1156861472	54	80.30	4336.03
4	4	26858	11/30/2020	936049686	64	45.37	2903.99
5	5	26858	12/9/2020	458515506	54	15.55	839.51
6	6	26858	11/6/2020	937462037	83	44.92	3728.20

Figure 15.8 – The Orders table from supply_company.db (truncated)

4. The following code will retrieve the ID of the customer with the largest purchases in the data. `.groupby()` aggregates the data by the customer ID, and `.sum()` tells pandas to do the aggregation by summing values. This has the effect that if a customer has more than one order, the total amounts will be summed. Note that `['amount']` is simply indexing just the `amount` column, as you saw in *Chapter 2, Data Structures*. The `.sort_values(ascending = False)` method sorts the largest value in `amount` to the first position, and `index[0]` then returns the customer ID since `.groupby()` makes the grouping argument the index:

```
largest_cust= \
    orders.groupby('Customer_Number').sum()['amount'].\
                    sort_values(ascending = False).
index[0]
largest_cust
```

This should return the ID of the target customer, as shown here:

```
35549
```

5. Find and list out the row for this customer in the table containing the customers:

```
largest_cust_info = \
    pd.read_sql(("select * from Customers WHERE Customer_
Number = " +
                 str(largest_cust)),
                sqlite3.connect("../datasets/supply_
company.db"))
largest_cust_info
```

This produces the following output:

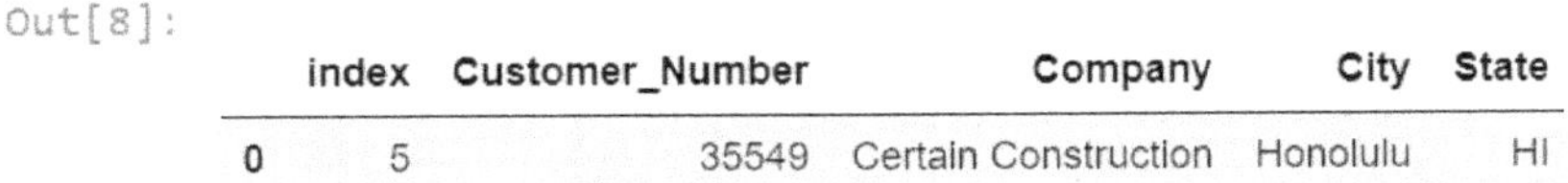

Out[8]:

	index	Customer_Number	Company	City	State
0	5	35549	Certain Construction	Honolulu	HI

Figure 15.9 – The customer with the most expenses in Q4 2020 was Certain Construction

You can now quickly get back to sales with the fact that Certain Construction in Honolulu was the largest customer in Q4 2020. In this activity, you've been able to use pandas SQL methods to access a database and extract information for a sales inquiry. pandas makes working with several data types relatively easy, and you are now equipped to handle business or technical questions such as this using pandas.

Solution 4.1

Perform the following steps to complete the activity:

1. Open a new Jupyter notebook and select the `Pandas_Workshop` kernel.
2. Import the `pandas` package:

```
import pandas as pd
```

3. Load the CSV file as a DataFrame:

```
file_url = 'https://raw.githubusercontent.com/
PacktWorkshops/The-Pandas-Workshop/master/Chapter04/Data/
car.csv'
data_frame = pd.read_csv(file_url)
```

4. Display the first `10` rows of the DataFrame:

```
data_frame.head(10)
```

The output will be as follows:

	buying	maint	doors	persons	lug_boot	safety	class
0	vhigh	vhigh	2	2.0	small	low	unacc
1	vhigh	vhigh	2	2.0	small	med	unacc
2	vhigh	vhigh	2	NaN	small	high	unacc
3	vhigh	vhigh	2	2.0	med	low	unacc
4	vhigh	vhigh	2	2.0	med	med	unacc
5	NaN	vhigh	2	2.0	med	high	NaN
6	vhigh	vhigh	2	2.0	big	low	unacc
7	vhigh	vhigh	2	2.0	big	NaN	unacc
8	vhigh	vhigh	2	2.0	big	high	unacc
9	vhigh	NaN	2	4.0	small	low	unacc

Figure 15.10 – Displaying the top 10 rows of the DataFrame

You can see some missing data (`NaN`) in a couple of columns. Displaying the DataFrame details with the `info()` function should help us to confirm this.

5. Display the data types of each column in the DataFrame using the `info()` method:

```
data_frame.info()
```

The output will be as follows:

```
<class 'pandas.core.frame.DataFrame'>
RangeIndex: 1728 entries, 0 to 1727
Data columns (total 7 columns):
 #   Column    Non-Null Count  Dtype
---  ------    --------------  -----
 0   buying    1727 non-null   object
 1   maint     1727 non-null   object
 2   doors     1728 non-null   int64
 3   persons   1151 non-null   float64
 4   lug_boot  1728 non-null   object
 5   safety    1727 non-null   object
 6   class     1727 non-null   object
dtypes: float64(1), int64(1), object(5)
memory usage: 94.6+ KB
```

Figure 15.11 – Displaying the full details of the DataFrame

As suspected, most columns have missing data. It will be a good idea to replace them with an appropriate value. Also, you can see that most of the columns are of the `object` type. You can convert them into `category` and see how this helps to optimize `memory usage`.

6. First, handle missing values by replacing them and display the top 10 rows of the DataFrame:

```
data_frame.fillna(value={'buying': 'Unknown',\
                         'maint': 'Unknown',\
                         'doors': round(data_frame.doors.
mean()),\
                         'persons':\
                         round(data_frame.persons.
mean()),\
                         'lug_boot': 'Unknown',\
                         'safety': 'Unknown',\
                         'class': 'Unknown'},\
                  inplace=True)

data_frame.head(10)
```

The output will be as follows:

	buying	maint	doors	persons	lug_boot	safety	class
0	vhigh	vhigh	2	2.0	small	low	unacc
1	vhigh	vhigh	2	2.0	small	med	unacc
2	vhigh	vhigh	2	3.0	small	high	unacc
3	vhigh	vhigh	2	2.0	med	low	unacc
4	vhigh	vhigh	2	2.0	med	med	unacc
5	Unknown	vhigh	2	2.0	med	high	Unknown
6	vhigh	vhigh	2	2.0	big	low	unacc
7	vhigh	vhigh	2	2.0	big	Unknown	unacc
8	vhigh	vhigh	2	2.0	big	high	unacc
9	vhigh	Unknown	2	4.0	small	low	unacc

Figure 15.12 – Top 10 rows of the DataFrame

Here, you replaced missing data in columns with the `object` data type with the value `Unknown` and you replaced missing data in the numeric (`float64` or `int64`) columns with the column-wise mean value.

7. Count the number of distinct unique values for `buying`, `maint`, `doors`, `persons`, `lug_boot`, `safety`, and `class`. Start with the `buying` column:

```
data_frame['buying'].nunique()
```

The output will be as follows:

```
5
```

`buying` contains `5` unique values out of `1728` rows, so you can convert it to `category` later.

Repeat this step for `maint`:

```
data_frame['maint'].nunique()
```

The output will be as follows:

```
5
```

`maint` contains `5` unique values out of `1728` rows, so you can convert this to the `category` data type later as well.

Count the unique values in the `doors` column:

```
data_frame['doors'].nunique()
```

The output will be as follows:

```
4
```

`doors` contains `4` unique values out of `1728` rows, so you can leave it as an `int64` data type.

Count the unique values for the `persons` column:

```
data_frame['persons'].nunique()
```

The output will be as follows:

```
3
```

`persons` contains `3` unique values out of `1728` rows; you can convert it to the `int64` data type later.

Count the unique values for the `lug_boot` column:

```
data_frame['lug_boot '].nunique()
```

The output will be as follows:

```
3
```

`lug_boot` contains `3` unique values, and you can convert it to the `category` data type later.

Repeat the step for the `safety` column:

```
data_frame['safety'].nunique()
```

The output will be as follows:

```
4
```

`safety` contains `4` unique values. You can convert it to the `category` data type later.

Finally, count the unique values for the `class` column:

```
data_frame['class'].nunique()
```

The output will be as follows:

```
5
```

`class` contains `5` unique values. You can convert it to the `category` data type later.

8. Convert the `object` columns into `category` columns and the `float64` columns into `int64` columns, where appropriate, and display them:

```
data_frame = data_frame.astype({'buying': 'category',\
                                'maint': 'category',\
                                'persons': 'int',\
                                'lug_boot': 'category',\
                                'safety': 'category',\
                                'class': 'category'})
```

9. Display the data types of each column in the DataFrame:

```
data_frame.info()
```

The output will be as follows:

```
<class 'pandas.core.frame.DataFrame'>
RangeIndex: 1728 entries, 0 to 1727
Data columns (total 7 columns):
 #   Column    Non-Null Count  Dtype
---  ------    --------------  -----
 0   buying    1728 non-null   category
 1   maint     1728 non-null   category
 2   doors     1728 non-null   int64
 3   persons   1728 non-null   int32
 4   lug_boot  1728 non-null   category
 5   safety    1728 non-null   category
 6   class     1728 non-null   category
dtypes: category(5), int32(1), int64(1)
memory usage: 29.8 KB
```

Figure 15.13 – Displaying the complete details of the DataFrame

Now that you have finished handling missing data and converting the data into appropriate types, you can see that `memory usage` has decreased by 60% from `94.6 KB` to `29.8 KB`. This is one of the many reasons that it is important to be certain to have the appropriate data type in your datasets.

Solution 5.1

Perform the following steps to complete the activity:

1. For this activity, all you will need is the `pandas` library. Load it in the first cell of the notebook:

```
import pandas as pd
```

2. Read in the `mushroom.csv` data from the `Datasets` directory and list the first five rows using `.head()`:

```
mushroom = pd.read_csv('../Datasets/mushroom.csv')
mushroom.head()
```

This produces the following output:

Out[3]:

	class	cap-shape	cap-surface	cap-color	bruises	odor	gill-attachment	gill-spacing	gill-size	gill-color	...	stalk-surface-below-ring	stalk-color-above-ring	stalk-color-below-ring	veil-type	veil-color	ring numbe
0	p	x	s	n	t	p	f	c	n	k	...	s	w	w	p	w	(
1	e	x	s	y	t	a	f	c	b	k	...	s	w	w	p	w	(
2	e	b	s	w	t	l	f	c	b	n	...	s	w	w	p	w	(
3	p	x	y	w	t	p	f	c	n	n	...	s	w	w	p	w	(
4	e	x	s	g	f	n	f	w	b	k	...	s	w	w	p	w	(

5 rows × 23 columns

Figure 15.14 – The mushroom data

> **Note**
>
> Please change the path of the dataset file (highlighted) based on where you have downloaded it on your system.

3. You see the `class` column and many visible attributes. List out all the columns to see what else there is to work with:

```
mushroom.columns
```

This produces the following output:

```
Out[12]: Index(['class', 'cap-shape', 'cap-surface', 'cap-color', 'bruises', 'odor',
               'gill-attachment', 'gill-spacing', 'gill-size', 'gill-color',
               'stalk-shape', 'stalk-root', 'stalk-surface-above-ring',
               'stalk-surface-below-ring', 'stalk-color-above-ring',
               'stalk-color-below-ring', 'veil-type', 'veil-color', 'ring-number',
               'ring-type', 'spore-print-color', 'population', 'habitat'],
              dtype='object')
```

Figure 15.15 – The columns of the mushroom DataFrame

4. In addition to `class`, you see `population` and `habitat`, which are not visible attributes. You decide to create a **multi-index** using `class`, `population`, and `habitat`:

```
my_index = pd.MultiIndex.from_frame(mushroom[['class',\

'population',\
                                              'habitat']])
```

5. Now, drop the columns that are in the index and set the `DataFrame` index to the multi-index. Be sure to drop the existing default index:

```
mushroom.drop(columns = ['class', 'population',
'habitat'],\
              inplace = True)
mushroom.set_index(my_index, inplace = True)
mushroom.head(10)
```

This produces the following output:

Out[19]:

class	population	habitat	cap-shape	cap-surface	cap-color	bruises	odor	gill-attachment	gill-spacing	gill-size	gill-color	stalk-shape	stalk-root	stalk-surface-above-ring	stalk-surface-below-ring	stalk-color-above-ring	stalk-color-below-ring	veil-type
p	s	u	x	s	n	t	p	f	c	n	k	e	e	s	s	w	w	p
e	n	g	x	s	y	t	a	f	c	b	k	e	c	s	s	w	w	p
		m	b	s	w	t	l	f	c	b	n	e	c	s	s	w	w	p
p	s	u	x	y	w	t	p	f	c	n	n	e	e	s	s	w	w	p
e	a	g	x	s	g	f	n	f	w	b	k	t	e	s	s	w	w	p
	n	g	x	y	y	t	a	f	c	b	n	e	c	s	s	w	w	p
		m	b	s	w	t	a	f	c	b	g	e	c	s	s	w	w	p
	s	m	b	y	w	t	l	f	c	b	n	e	c	s	s	w	w	p
p	v	g	x	y	w	t	p	f	c	n	p	e	e	s	s	w	w	p
e	s	m	b	s	y	t	a	f	c	b	g	e	c	s	s	w	w	p

Figure 15.16 – The mushroom DataFrame, with the multi-index

6. Using the `.loc[]` notation you learned, list out just the edible data:

```
mushroom.loc['e']
```

This produces the following output:

Out[20]:

population	habitat	cap-shape	cap-surface	cap-color	bruises	odor	gill-attachment	gill-spacing	gill-size	gill-color	stalk-shape	stalk-root	stalk-surface-above-ring	stalk-surface-below-ring	stalk-color-above-ring	stalk-color-below-ring	veil-type	veil-color
n	g	x	s	y	t	a	f	c	b	k	e	c	s	s	w	w	p	w
	m	b	s	w	t	l	f	c	b	n	e	c	s	s	w	w	p	w
a	g	x	s	g	f	n	f	w	b	k	t	e	s	s	w	w	p	w
n	g	x	y	y	t	a	f	c	b	n	e	c	s	s	w	w	p	w
	m	b	s	w	t	a	f	c	b	g	e	c	s	s	w	w	p	w
...	...	...	...	...	...	...	...	...	...	...	...	...	...	...	...	...	...	...
v	l	x	s	n	f	n	a	c	b	y	e	?	s	s	o	o	p	o
c	l	k	s	n	f	n	a	c	b	y	e	?	s	s	o	o	p	o
v	l	x	s	n	f	n	a	c	b	y	e	?	s	s	o	o	p	n
c	l	f	s	n	f	n	a	c	b	n	e	?	s	s	o	o	p	o
	l	x	s	n	f	n	a	c	b	y	e	?	s	s	o	o	p	o

4208 rows × 20 columns

Figure 15.17 – Mushroom data for edible mushrooms only

Solution 6.1

In this activity, you will read in some US population data for large cities for the years 2010 and 2019 and perform some analysis:

1. For this activity, all you will need is the `pandas` library. Load it in the first cell of the notebook:

```
import pandas as pd
```

2. Read a `pandas` Series from the `US_Census_SUB-IP-EST2019-ANNRNK_top_20_2010.csv` file. The city names are in the first column; read them in such that they are used as the index. List out the resulting Series:

```
populations_2010 = \
pd.read_csv('..//Datasets//US_Census_SUB-IP-EST2019-
ANNRNK_top_20_2010.csv',
                index_col = [0],
                squeeze = True)
populations_2010
```

The result should be as follows:

```
Out[13]: City
         New York                8190209
         Los Angeles             3795512
         Chicago                 2697477
         Houston                 2100280
         Phoenix                 1449038
         Philadelphia            1528283
         San Antonio             1332299
         San Diego               1305906
         Dallas                  1200350
         San Jose                 954940
         Austin                   806164
         Jacksonville             823114
         Fort Worth               748441
         Columbus                 790943
         Charlotte                738444
         San Francisco            805505
         Indianapolis             821579
         Seattle                  610630
         Denver                   603359
         District of Columbia     605226
         Name: 2010, dtype: int64
```

Figure 15.18 – Populations_2010 Series

Note that in addition to specifying `index_col = [1]`, we use `squeeze = True` to store the result in a Series, instead of a DataFrame.

3. Calculate the total population of the three largest cities in the 2010 Series (New York, Los Angeles, and Chicago) and save the result in a variable:

```
top_3_2010 = sum(populations_2010[['New York', 'Los
Angeles', 'Chicago']])
```

4. Read in the corresponding data for 2019 from the `US_Census_SUB-IP-EST2019-ANNRNK_top_20_2019.csv` file, again using the first column as the index and reading the data into a Series:

```
populations_2019 = \
pd.read_csv('..//Datasets//US_Census_SUB-IP-EST2019-
ANNRNK_top_20_2019.csv',
            index_col = [0],
            squeeze = True)
populations_2019
```

The output should look as follows:

```
Out[20]: City
         New York                  8336817
         Los Angeles               3979576
         Chicago                   2693976
         Houston                   2320268
         Phoenix                   1680992
         Philadelphia              1584064
         San Antonio               1547253
         San Diego                 1423851
         Dallas                    1343573
         San Jose                  1021795
         Austin                     978908
         Jacksonville               911507
         Fort Worth                 909585
         Columbus                   898553
         Charlotte                  885708
         San Francisco              881549
         Indianapolis               876384
         Seattle                    753675
         Denver                     727211
         District of Columbia       705749
         Name: 2019, dtype: int64
```

Figure 15.19 – The populations_2019 Series, with the updated values for each city from the 2010 Series

5. Calculate the total population for the same three cities in the 2019 Series and save the result in a variable:

```
top_3_2019 = sum(populations_2019[['New York', 'Los
Angeles', 'Chicago']])
```

6. You have been asked to consider whether there has been net migration out of the three largest cities. Using the saved values, calculate the percent change from 2010 to 2019 for the three cities. Also, calculate the percent change for all cities. Print out a comparison of the changes for the three cities versus all cities:

```
top_3_change = 100 * (top_3_2019 - top_3_2010) /
top_3_2010
all_change = 100 * sum(populations_2019 -
populations_2010) / sum(populations_2010)
#
print('top 3 changed', str(round(top_3_change, 1)),
'%\n',
      'vs. all changed', str(round(all_change, 1)), '%')
```

The output should be as follows:

```
top 3 changed 2.2 %
 vs. all changed 8.0 %
```

Figure 15.20 – Percent population change for the 3 largest cities in 2010 compared to the percent change for the top 20 largest cities from 2010

You will note that although there is still net growth in the largest cities, it is much smaller than the growth for the top 20 cities. Also, note that you were able to subtract the two Series, and then sum their values to get the total change, because the Series have the same index. Recall from earlier that pandas usually aligns the index when doing operations on entire Series or DataFrames. If we had some different cities in the Series, there would have been `NaN` values in the results.

Solution 6.2

In this activity, you are analyzing data from this year's survey of abalone oysters for the National Marine Fisheries Service. In particular, you want to get some summary values for the dimensions of male and female samples in the data, depending on the number of rings in the shells. The ring count is a measure of age, and reviewing this data provides comparisons to previous years to understand the health of the population. The data contains a number of observations, including sex, length, diameter, weight, shell weight, and the number of rings:

1. For this activity, all you will need is the `pandas` library. Load it in the first cell of the notebook:

```
import pandas as pd
```

2. Read the `abalone.csv` file into a DataFrame called `abalone` and view the first five rows:

```
abalone = pd.read_csv('..//Datasets//abalone.csv')
abalone.head()
```

The output should look as follows:

Out[4]:

	Sex	Length	Diameter	Height	Whole weight	.Shucked weight	Viscera weight	Shell weight	Rings
0	M	0.455	0.365	0.095	0.5140	0.2245	0.1010	0.150	15
1	M	0.350	0.265	0.090	0.2255	0.0995	0.0485	0.070	7
2	F	0.530	0.420	0.135	0.6770	0.2565	0.1415	0.210	9
3	M	0.440	0.365	0.125	0.5160	0.2155	0.1140	0.155	10
4	I	0.330	0.255	0.080	0.2050	0.0895	0.0395	0.055	7

Figure 15.21 – The first five rows of the abalone oyster dataset

3. Create a `MultiIndex` from the `Sex` and `Rings` columns, since these are the variables under which you want to summarize the data. Be sure to drop the `Sex` and `Rings` columns once the index is created:

```
my_index = pd.MultiIndex.from_frame(abalone[['Sex',
'Rings']])
abalone.drop(columns = ['Sex', 'Rings'], inplace = True)
abalone.set_index(my_index, inplace = True)
abalone.head(10)
```

The result should look like the following:

Out[5]:

		Length	Diameter	Height	Whole weight	.Shucked weight	Viscera weight	Shell weight
Sex	**Rings**							
M	15	0.455	0.365	0.095	0.5140	0.2245	0.1010	0.150
	7	0.350	0.265	0.090	0.2255	0.0995	0.0485	0.070
F	9	0.530	0.420	0.135	0.6770	0.2565	0.1415	0.210
M	10	0.440	0.365	0.125	0.5160	0.2155	0.1140	0.155
I	7	0.330	0.255	0.080	0.2050	0.0895	0.0395	0.055
	8	0.425	0.300	0.095	0.3515	0.1410	0.0775	0.120
F	20	0.530	0.415	0.150	0.7775	0.2370	0.1415	0.330
	16	0.545	0.425	0.125	0.7680	0.2940	0.1495	0.260
M	9	0.475	0.370	0.125	0.5095	0.2165	0.1125	0.165
F	19	0.550	0.440	0.150	0.8945	0.3145	0.1510	0.320

Figure 15.22 – The abalone DataFrame with a multi-index using the Sex and Rings columns

4. You plan to focus on oysters with more than 15 rings. Since you want statistics for each sex, you need to know the values of `Rings` in the data for each sex. Use `abalone.loc['sex'].index` to get a list of all the values for each sex (that is, replace `sex` with `M` and then with `F`). This works well because you have a two-level index, so by passing a value to filter on `Sex`, you then get the relevant items in the next level index, which is `Rings`.

   To filter the data, you want a list of the unique values of the rings. Python provides the `set()` method, which conveniently gives a set of unique values, so you can apply it as `set(abalone.loc['sex'].index)` to store the unique values of rings for each sex in a variable:

   ```
   min_rings = 16
   all_rings_M = set(abalone.loc['M'].index)
   all_rings_F = set(abalone.loc['F'].index)
   ```

5. You also need the maximum number of rings for each sex. You can get that with `max(abalone.loc['sex'].index)`, which works in the same way as getting all the values. Store this value for each sex in a variable:

   ```
   max_rings_M = max(abalone.loc['M'].index)
   max_rings_F = max(abalone.loc['F'].index)
   ```

6. You now need to find the values for each sex that are greater than 15 and are in the unique values for that sex. You can use a list comprehension to iterate over the possible values and keep only those belonging to a given sex. This looks like `[i for i in range(min_rings, max_rings + 1)  if i in all_rings]`, where `all_rings` is the list of unique values for the sex, `min_rings` is 16 (one more than 15), and `max_rings` is the maximum value for the sex. Do this operation and save the result for each sex:

   ```
   rings_M = [i for i in range(min_rings, max_rings_M + 1)
              if i in all_rings_M]
   rings_F = [i for i in range(min_rings, max_rings_F + 1)
              if i in all_rings_F]
   ```

7. You now want to select the data for each sex, for the `.Shucked weight`, `Length`, `Diameter`, and `Height` columns. For each column, you want the mean value. Thanks to your multi-index, you can do this using the following steps:

   - `abalone.loc['sex']` selects one sex (`M` or `F`).
   - `.loc[rings]` selects just the values of rings you obtained in the list comprehension.

- Bracket notation with a list of the columns selects the columns, in other words, `[['Length', 'Diameter', 'Height', '.Shucked weight']]`.
- Adding the `.mean(axis = 0)` method tells pandas to take the column means.

The entire operation for each sex looks like this:

```
abalone.loc['sex'][rings][[ 'Length', 'Diameter',
'Height', '.Shucked weight']].mean(axis = 0)
```

Perform this operation for each sex, using the correct list of rings, and save the result in a variable for each.

8. Print out a comparison of the values between the two sexes:

```
print('for oysters with', min_rings, 'or more rings\n')
print('males weigh', round(males['.Shucked weight'], 3),
      'vs. females weigh', round(females['.Shucked
weight'], 3))
print('males are', round(males['Length'], 3), 'long ',
      'vs. females are', round(females['Length'], 3),
'long')
print('males are', round(males['Diameter'], 3), 'in
diameter ',
      'vs. females are', round(females['Diameter'], 3),
'in diameter')
print('males are', round(males['Height'], 3), 'in height
',
      'vs. females are', round(females['Height'], 3), 'in
height')
```

The result should look something like the following:

```
for oysters with 16 or more rings

males weigh 0.458 vs. females weigh 0.449
males are 0.603 long  vs. females are 0.603 long
males are 0.478 in diameter  vs. females are 0.479 in diameter
males are 0.176 in height  vs. females are 0.174 in height
```

Figure 15.23 – The comparison of summary size statistics for abalone oysters with 16 or more rings

Solution 7.1

Please use the following steps to complete the activity:

1. Open a Jupyter notebook.
2. Import the pandas package:

```
import pandas as pd=
```

Load the CSV file as a DataFrame:

```
file_url = 'https://raw.githubusercontent.com/
PacktWorkshops/The-Pandas-Workshop/master/Chapter07/Data/
student-mat.csv'
data_frame = pd.read_csv(file_url, delimiter=';')
```

Note that CSV uses ; as a delimiter. So, we have used the `delimiter` option with `pd.read_csv()` to explicitly specify the correct delimiter to be used in order to read the dataset.

3. Modify the DataFrame to contain only these columns: `school`, `sex`, `age`, `address`, `health`, `absences`, `G1`, `G2`, and `G3`:

```
data_frame = data_frame[[
    'school', 'sex', 'age', 'address', 'health',
'absences', 'G1', 'G2', 'G3'
]]
```

4. Display the first 10 rows of the DataFrame:

```
data_frame.head(10)
```

The output will be as follows:

	school	sex	age	address	health	absences	G1	G2	G3
0	GP	F	18	U	3	6	5	6	6
1	GP	F	17	U	3	4	5	5	6
2	GP	F	15	U	3	10	7	8	10
3	GP	F	15	U	5	2	15	14	15
4	GP	F	16	U	5	4	6	10	10
5	GP	M	16	U	5	10	15	15	15
6	GP	M	16	U	3	0	12	12	11
7	GP	F	17	U	1	6	6	5	6
8	GP	M	15	U	1	0	16	18	19
9	GP	M	15	U	5	0	14	15	15

Figure 15.24 – Placeholder

5. Build a pivot table indexed on `'school'`:

```
pd.pivot_table(data_frame,index=['school'])
```

The output will be as follows:

school	G1	G2	G3	absences	age	health
GP	10.939828	10.782235	10.489971	5.965616	16.521490	3.575931
MS	10.673913	10.195652	9.847826	3.760870	18.021739	3.391304

Figure 15.25 – Placeholder

By default, the pivot table uses mean aggregation. The main insight you can derive from this pivot table is that school `GP` has higher absences (~5.9) than school `MS` (~3.7).

6. Build a pivot table indexed on `'school'` and `'age'`:

```
pd.pivot_table(data_frame, index=['school', 'age'])
```

The output will be as follows:

school	age	G1	G2	G3	absences	health
GP	15	11.231707	11.365854	11.256098	3.341463	3.585366
	16	10.942308	11.182692	11.028846	5.451923	3.701923
	17	10.802326	10.383721	10.232558	6.709302	3.639535
	18	10.614035	9.964912	9.157895	7.333333	3.350877
	19	11.222222	10.055556	9.055556	12.777778	3.277778
	20	17.000000	18.000000	18.000000	0.000000	5.000000
	22	6.000000	8.000000	8.000000	16.000000	1.000000
MS	17	11.583333	11.166667	10.583333	4.666667	2.500000
	18	10.960000	10.520000	10.440000	3.120000	3.640000
	19	7.333333	6.833333	5.666667	3.500000	4.166667
	20	12.000000	11.500000	12.000000	7.500000	3.500000
	21	10.000000	8.000000	7.000000	3.000000	3.000000

Figure 15.26 – Placeholder

It seems that, for school `GP`, ages 19 and 22 have the highest absences (12.7 and 16, respectively). For school `MS`, age group 20 has the highest absences (7.5).

7. Build a pivot table indexed on `'school'`, `'sex'`, and `'age'` with the mean and sum aggregation on the `'absences'` column:

```
pd.pivot_table(data_frame, index=['school', 'sex',
'age'], values = 'absences', aggfunc={'mean', 'sum'})
```

The output will be as follows:

school	sex	age	mean	sum
GP	F	15	3.894737	148.0
		16	5.888889	318.0
		17	7.120000	356.0
		18	8.137931	236.0
		19	13.083333	157.0
	M	15	2.863636	126.0
		16	4.980000	249.0
		17	6.138889	221.0
		18	6.500000	182.0
		19	12.166667	73.0
		20	0.000000	0.0
		22	16.000000	16.0
MS	F	17	5.625000	45.0
		18	1.785714	25.0
		19	2.000000	4.0
		20	4.000000	4.0
	M	17	2.750000	11.0
		18	4.818182	53.0
		19	4.250000	17.0
		20	11.000000	11.0
		21	3.000000	3.0

Figure 15.27 – Placeholder

You can clearly see that, for school `GP`, females of age group 19 and males of age groups 19 and 22 have more than 10 absences. For school `MS`, only males in the age group 20 have more than 10 absences.

Solution 8.1

Please use the following steps to complete the activity:

1. Open a new Jupyter notebook.
2. Import the pandas, numpy, and matplotlib packages:

```
import pandas as pd
import numpy as np
import matplotlib.pyplot as plt
```

3. Load the CSV file as a DataFrame:

```
file_url = 'PUF2020final_v1coll.csv'
data_frame = pd.read_csv(file_url)
```

4. Import the pandas, numpy, and matplotlib packages:

```
data_frame = data_frame[["REGION", "SQFT", "BEDROOMS",
"PRICE"]]
```

5. Display the first 10 rows of the DataFrame:

```
data_frame.head(10)
```

The output will be as follows:

	REGION	SQFT	BEDROOMS	PRICE
0	3	960	1	52000
1	3	1300	3	39900
2	4	1200	3	60000
3	4	730	1	9
4	4	500	1	87000
5	4	1100	3	56000
6	1	1000	1	9
7	3	700	1	42600
8	3	700	1	46300
9	3	1200	3	61000

Figure 15.28 – Placeholder

6. Plot a histogram chart for PRICE:

```
data_frame.PRICE.plot(kind = 'hist');
```

The output will be as follows:

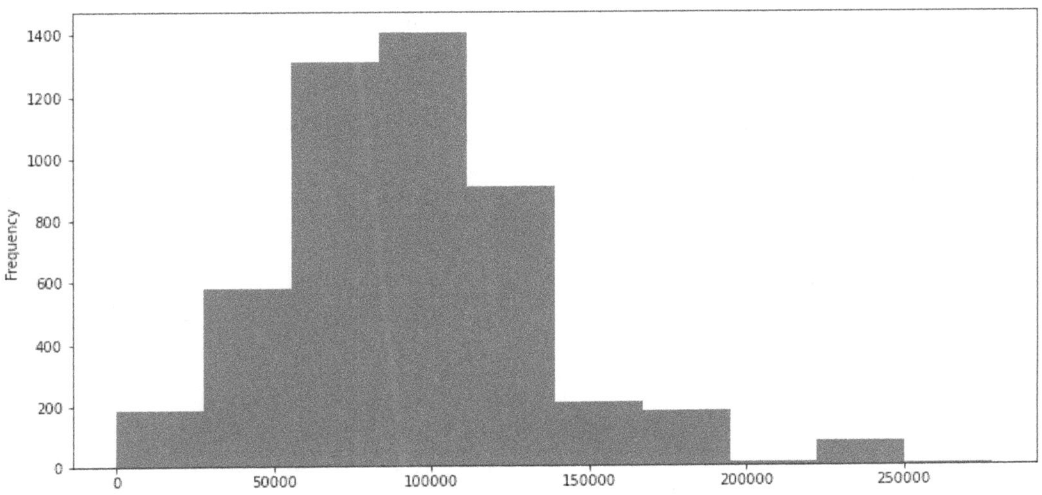

Figure 15.29 – Placeholder

It seems that most of the properties have a sale price centered around 50k-150k. There are also a few outliers with a high sale price of over 200k.

7. Plot a histogram chart for SQFT:

```
data_frame.SQFT.plot(kind = 'hist');
```

The output will be as follows:

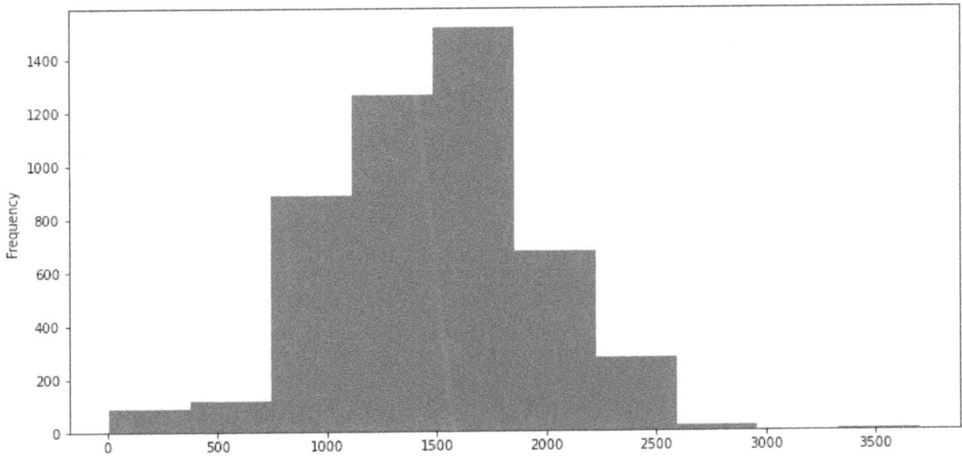

Figure 15.30 – Placeholder

SQFT has a completely different distribution as compared to PRICE. Most of the observations lie between 700 and 2,500. The rest of the observations represent a small portion of the dataset. We can also notice some extreme outliers over 3,000.

8. Plot a scatter plot chart for PRICE and SQFT:

```
data_frame.plot(kind='scatter', y = 'PRICE', x = 'SQFT');
```

The output will be as follows:

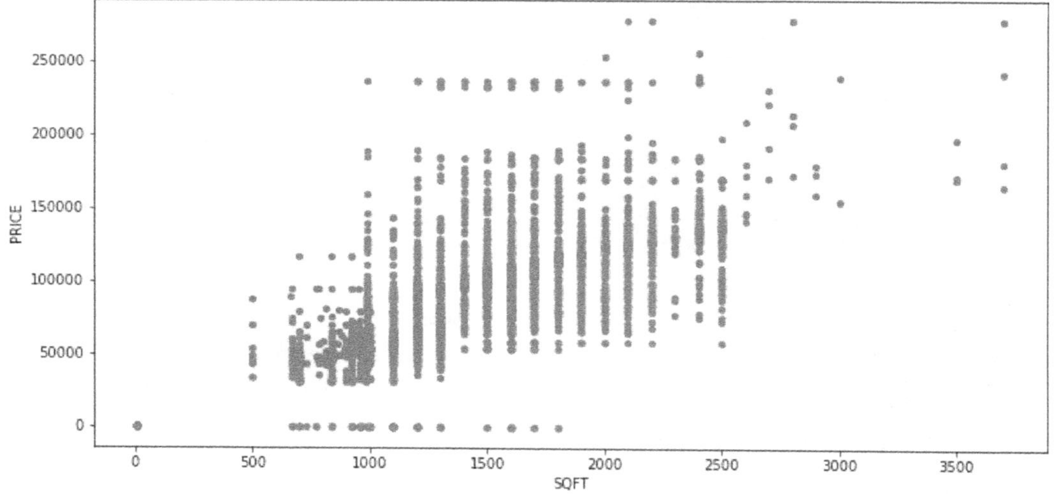

Figure 15.31 – Placeholder

There is clearly a positive correlation between the size of the property and the price. When SQFT is under 1,000, PRICE remains below 150,000. The properties get more and more expensive as they get beyond 1,000 SQFT.

9. Plot a boxplot chart for BEDROOMS and PRICE:

```
data_frame.boxplot(by='BEDROOMS', column='PRICE');
```

The output will be as follows:

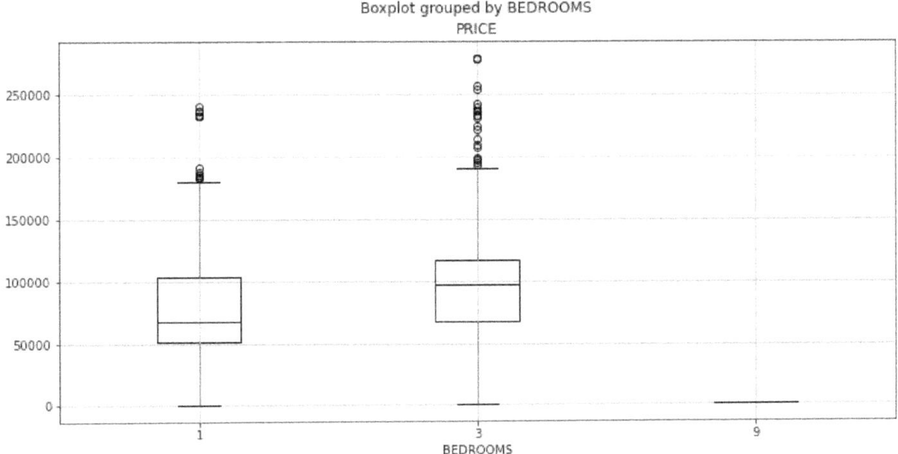

Figure 15.32 – Placeholder

It seems that the BEDROOMS variable has a positive correlation with PRICE. The median price for houses with one bedroom is closer to 60,000, whereas for three-bedroom houses, the median is close to 100,000. Notice that there might be some outlier cases with nine bedrooms.

10. Plot a scatter chart for BEDROOMS and PRICE:

```
data_frame.plot(kind='scatter', x = 'BEDROOMS', y =
'PRICE');
```

The output will be as follows:

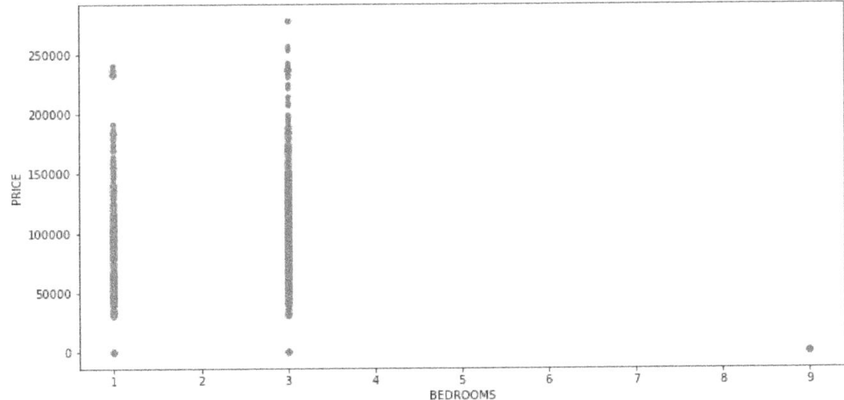

Figure 15.33 – Placeholder

Overall, this scatter plot confirms our previous findings.

11. Plot a scatter chart for `REGION` and `PRICE`:

```
data_frame.boxplot(by='REGION', column='PRICE');
```

The output will be as follows:

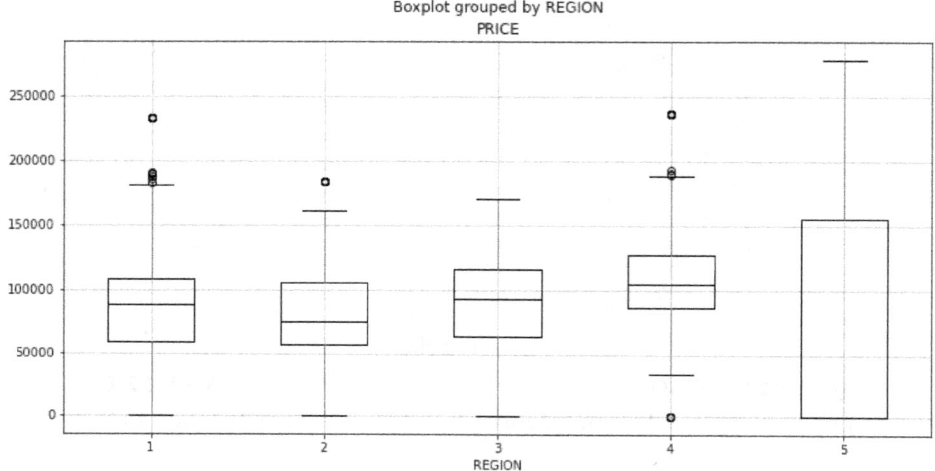

Figure 15.34 – Placeholder

We can see how the `REGION` variable corresponds to different price distributions. That is, `PRICE` may vary depending on the region where the house is located. Following the preceding plot, we can assess that houses in the region denoted as `4` generally boast higher prices than the other regions.

12. Plot a scatter chart for `REGION` on `PRICE`:

```
data_frame.plot(kind='scatter', x = 'REGION', y =
'PRICE')
plt.show()
```

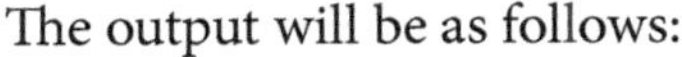

The output will be as follows:

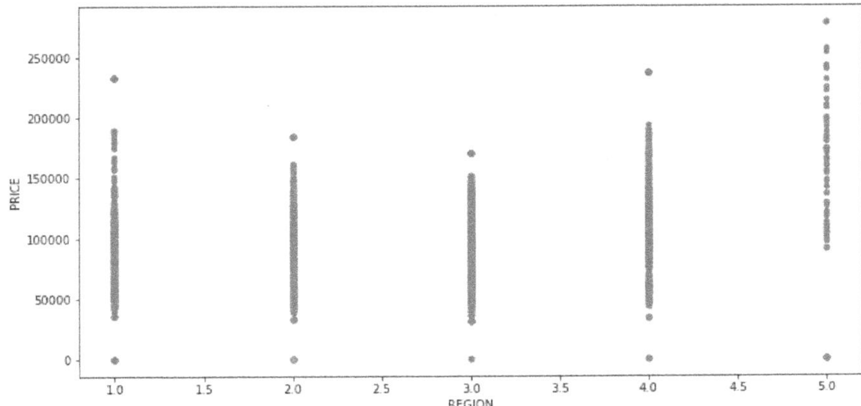

Figure 15.35 – Placeholder

This scatter plot confirms the previous finding. We can also add that the region denoted by the number `5` contains fewer houses (look at the density of dots) than the other regions and although the price generally seems higher in this region, the relative lack of data points might suggest that these cases are either outliers or this is a newly developed region.

13. Plot a horizontal bar chart for `REGION` and `PRICE`:

```
pd.pivot_table(data_frame,index=['REGION']).PRICE.
plot(kind='barh');
```

The output will be as follows:

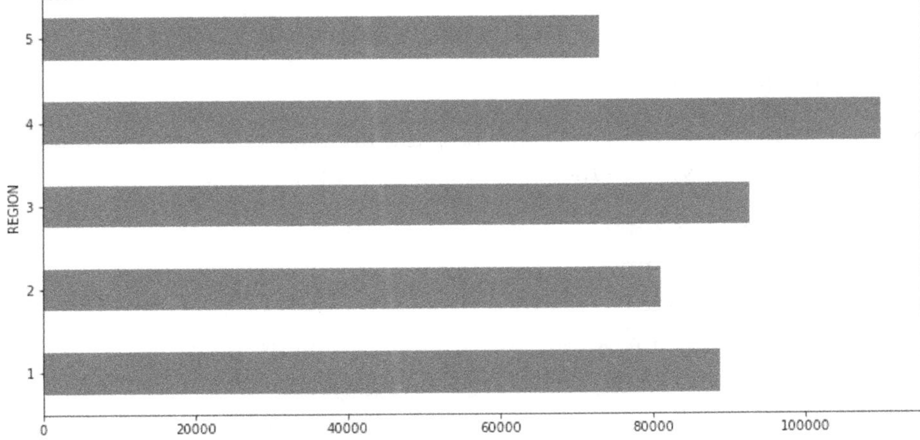

Figure 15.36 – Placeholder

By looking at this chart, we can find out which region has the highest average price, which happens to be 4 in this case.

14. Plot a line chart for REGION and PRICE:

```
pd.pivot_table(data_frame,index=['REGION']).PRICE.plot();
```

The output will be as follows:

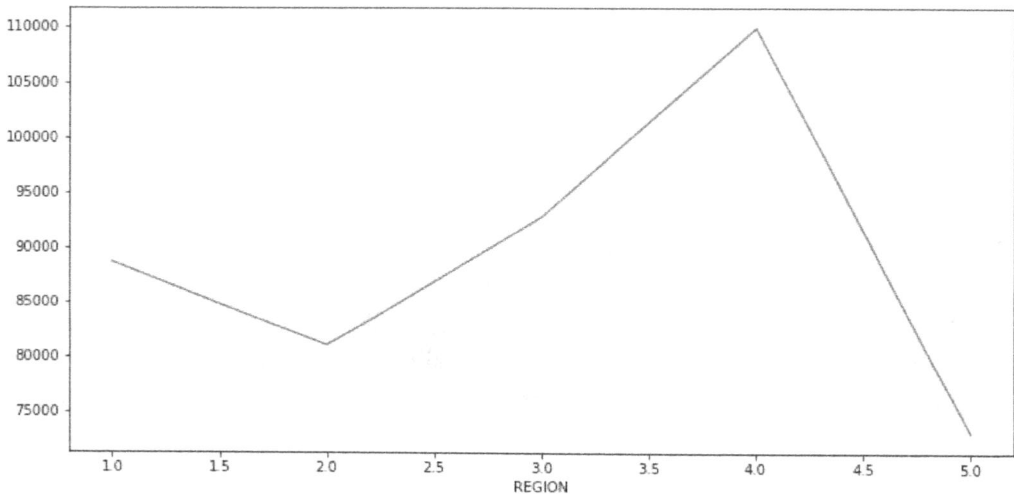

Figure 15.37 – Placeholder

This confirms our previous chart where we showed that region 4 boasts the highest average price among all regions.

Solution 9.1

Perform the following steps to complete the activity:

1. For this activity, all you will need is the pandas library, some modules from sklearn, and numpy. Load them in the first cell of the notebook:

```
import pandas as pd
import numpy as np
from sklearn.model_selection import train_test_split
from sklearn.preprocessing import StandardScaler
from sklearn.linear_model import LinearRegression as OLS
from sklearn.metrics import mean_squared_error
```

2. Use the `power_plant.csv` dataset: `'Datasets\\power_plant.csv'`. Read the data into a pandas DataFrame, print out the shape, and list the first five rows.

   The independent variables are as follows:

   - AT – ambient temperature
   - V – exhaust vacuum level
   - AP – ambient pressure
   - RH – relative humidity

   The dependent variable is the following:

   - EP – electrical power produced

```
power_data = pd.read_csv('Datasets\\power_plant.csv')
print(power_data.shape)
power_data.head()
```

Out[5]:

	AT	V	AP	RH	EP
0	8.34	40.77	1010.84	90.01	480.48
1	23.64	58.49	1011.40	74.20	445.75
2	29.74	56.90	1007.15	41.91	438.76
3	19.07	49.69	1007.22	76.79	453.09
4	11.80	40.66	1017.13	97.20	464.43

Figure 15.38 – The power plant data

3. Split the data into `train`, `val`, and `test` sets with fractions of 0.8, 0.1, and 0.1, using Python and pandas but not `sklearn` methods. You use 0.8 for the training split because there is a large number of rows, so the validation and test splits will still have enough rows:

```
np.random.seed(42)
train_rows = \
    pd.Series(np.random.choice(list(power_data.index),
                               int(0.8 * power_data.
shape[0]),
                               replace = False))
val_rows = \
    pd.Series(np.random.choice(list(power_data.
drop(train_rows,
```

```
                                                      axis
= 0).index),
                                int(0.1 * power_data.
shape[0]),
                                replace = False))
test_rows = \
    pd.Series(power_data.drop(pd.concat([train_rows,
                                          val_rows]),
                               axis = 0).index)train_data
= power_data.iloc[train_rows, :]
val_data = power_data.iloc[val_rows, :]
test_data = power_data.iloc[test_rows, :]
print('train is ', train_data.shape, ' rows, cols\n',
      'val is ', val_data.shape, ' rows, cols\n',
      'test is ', test_data.shape, 'rows, cols')#
```

You should see the following output:

```
train is  (7654, 5)  rows, cols
 val is  (956, 5)  rows, cols
 test is  (958, 5) rows, cols
```

Figure 15.39 – Manually splitting the data

4. Repeat the split in *step 3* but use `train_test_split`. Call it once to split the `train` data, and then call it again to split the remainder into `val` and `test`:

```
train_data_2, val_data_2 = \
    train_test_split(power_data, train_size = 0.8,
random_state = 42)
val_data_2, test_data_2 = \
    train_test_split(val_data_2, test_size = 0.5, random_
state = 42)
print('train is ', train_data_2.shape, ' rows, cols\
n','val is ', val_data_2.shape, ' rows, cols\n', 'test is
', test_data_2.shape, 'rows, cols')
```

Running this code will lead to the following output:

```
train is  (7654, 5)  rows, cols
 val is  (957, 5)  rows, cols
 test is  (957, 5) rows, cols
```

Figure 15.40 – Using train_test_split() twice to generate the splits

5. Show that the row counts are correct in all cases.

   Note that in the first case, `val` and `test` differ by two rows, one having one more and one having one less. This is the result of sampling the `val` set as `int(0.1 * power_data.shape[0])`, which rounds down one value.

6. Fit `.StandardScaler()` to the `train` data from *step 3*, and then transform `train`, `validation`, and `test X`. Do not transform the `EP` column, as it is the target:

```
scaler = StandardScaler()
scaler.fit(train_data.iloc[:, :-1])
train_X = scaler.transform(train_data.iloc[:, :-1])
train_y = train_data['EP']
val_X = scaler.transform(val_data.iloc[:, :-1])
val_y = val_data['EP']
test_X = scaler.transform(test_data.iloc[:, :-1])
test_y = test_data['EP']
```

7. Fit a `.LinearRegression()` model to the scaled `train` data, using the `X` variables to predict `y` (the `EP` column):

```
linear_model = OLS()
linear_model.fit(train_X, train_y)
```

8. Print the R2 score and the RMSE of the model on the `train`, `validation`, and `test` datasets:

```
print('train score: ', linear_model.score(train_X,
train_y),
      '\nvalidation score: ', linear_model.score(val_X,
val_y),
      '\ntest score: ', linear_model.score(test_X,
test_y))
print('train RMSE: ',
      mean_squared_error(linear_model.predict(train_X),
train_y),
```

```
        '\nvalidation RMSE: ',
        mean_squared_error(linear_model.predict(val_X),
val_y),
        '\ntest RMSE: ',
        mean_squared_error(linear_model.predict(test_X),
test_y))
```

This produces the following output:

```
train score:  0.9287072840354756
validation score:  0.9238845251967255
test score:  0.9333918854821254
train RMSE:  20.732519659228675
validation RMSE:  22.82059184376622
test RMSE:  19.0233909525747
```

Figure 15.41 – Results of predicting EP from the other variables

You see that the linear model predicts nearly 93% of the variation in the generated power and that the results on the validation and test splits are nearly identical. These results provide a baseline—it's best practice in modeling to create a simple baseline model early in the workflow so that as you make changes in more complex models, you have a reference for improvement.

Solution 10.1

Suppose you are an analyst in a financial advisory firm. Your manager has given three stock symbols to you and requested your input on how they may be correlated in their price behavior. You are provided with a data file called `stocks.csv`, which contains the symbols, the closing prices, the trading volumes, and a sentiment indicator (some view of the stocks' quality, but you are not told the exact definition). Your initial goal here is to determine whether all three stocks show similar market characteristics, and if any or all of them do, make an initial visualization using smoothing. The long-term goal is to try to build some predictive models, so you will split the data into training and test sets. As it is a time series, it's important to split on time, not randomly. For this activity, all you will need is the `pandas` library, a scaling module from `sklearn`, and `matplotlib`.

Perform the following steps to complete the activity:

1. Load the required libraries:

```
import pandas as pd
from sklearn.preprocessing import StandardScaler
import matplotlib.pyplot as plt
```

2. Load the `stocks.csv` file and store it in a DataFrame called `my_data`. Use the `.head()` method to display the first five rows:

```
my_data = pd.read_csv('Datasets\\stocks.csv')
my_data.head()
```

The output should be as follows:

Out[5]:

	Date	Close	Volume	symbol	sentiment
0	2017-04-17	20636.919922	229240000	S1	NEUTRAL
1	2017-04-17	20.000000	88300	S2	NEUTRAL
2	2017-04-17	5400.000000	0	S3	NEUTRAL
3	2017-04-18	20523.279297	263180000	S1	NEUTRAL
4	2017-04-18	20.150000	60500	S2	NEUTRAL

Figure 15.42 – Stocks data

3. Inspect `.dtypes` and convert the dates to pandas `datetime` if needed:

```
my_data.dtypes
```

The result is as follows:

```
Out[6]: Date          object
        Close        float64
        Volume         int64
        symbol        object
        sentiment     object
        dtype: object
```

Figure 15.43 – The object types in my_data

4. Since the date is of the `object` type (as it was read as a string), use `pd.to_datetime()` to convert it to datetime, and use `.describe()` to inspect the result:

```
my_data['Date']  = pd.to_datetime(my_data['Date'])
my_data['Date'].describe()
```

The date range should appear as follows:

```
Out[7]: count                       753
        unique                      251
        top       2017-08-29 00:00:00
        freq                          3
        first     2017-04-17 00:00:00
        last      2018-04-13 00:00:00
        Name: Date, dtype: object
```

Figure 15.44 – The dates after converting to datetime

5. Split the data into `train` and `test` based on the date, keeping the last 3 months as the `test` set. Note that you are not splitting randomly, as the eventual goal is to predict future values, so the test data needs to be the most recent and is all in the future compared to the training data. The choice of period is somewhat arbitrary but in this case, a 90-day prediction window has been requested by management:

```
train_end = '2018-01-13'
train = my_data.loc[my_data['Date'] <= train_end, :]
test = my_data.loc[my_data['Date'] > train_end, :]
```

6. Generate a scatter plot that shows the prices over time of different stock symbols and identifies the train and test splits. For this, add the following code. Note that the `.groupby()` method is used without any aggregation and creates an **iterable** object that makes the visualization convenient (you can understand that object if desired by running, for example, `list(symbols_train)` and inspecting the output):

```
figure, ax = plt.subplots(figsize = (11, 8))
symbols_train = train.groupby('symbol')
symbols_test = test.groupby('symbol')
for train_name, symbol_train in symbols_train:
    ax.scatter(symbol_train.Date,
               symbol_train.Close,
               label = 'closing ' + train_name + '
(train_set)')
for test_name, symbol_test in symbols_test:
    ax.scatter(symbol_test.Date,
               symbol_test.Close,
               label = 'closing ' + test_name + ' (test
set)')
ax.legend(fontsize = 12)
ax.set_ylabel('Closing Price', fontsize = 14)
ax.tick_params(labelsize = 12)
ax.set_title('Comparison of three stocks ca 2017-2018',
             fontsize = 16)
plt.show()
```

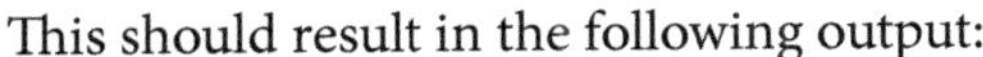

This should result in the following output:

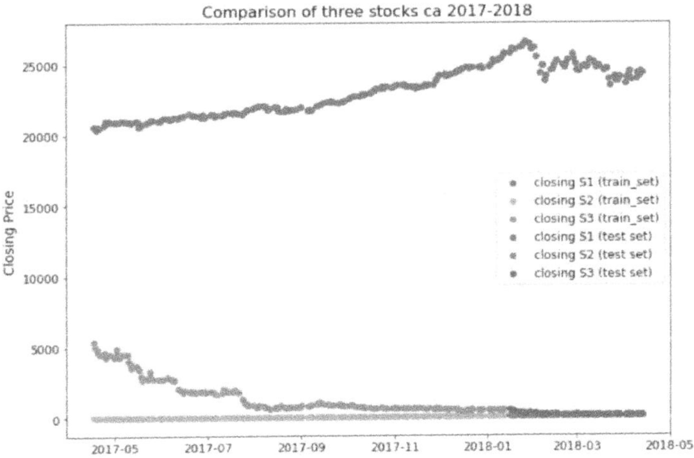

Figure 15.45 – Initial scatter plot of the stock data

You see that since S2 pricing is so much different, we can't tell anything about S2 in this chart. Also, S1 and S3 differ by a factor of 4x, so both are compressed on the *y* axis.

7. The initial scatter plot isn't very informative because the different symbols have very different pricing, so some are compressed at the bottom and top of the *y* axis. Plot a histogram of the price distribution for each symbol separately, and use enough bins to see the detail (in this case, we've used 50):

```
symbols = my_data.symbol.unique()
for i in range(len(symbols)):
    fig, ax = plt.subplots(figsize = (5.5, 4.5))
    ax.hist(my_data.groupby('symbol').get_
group(symbols[i])['Close'],
           bins = 50)
    ax.set_title('Closing Price Distribution\nSymbol ' +
symbols[i])
    plt.show()
```

The three histograms should look as follows:

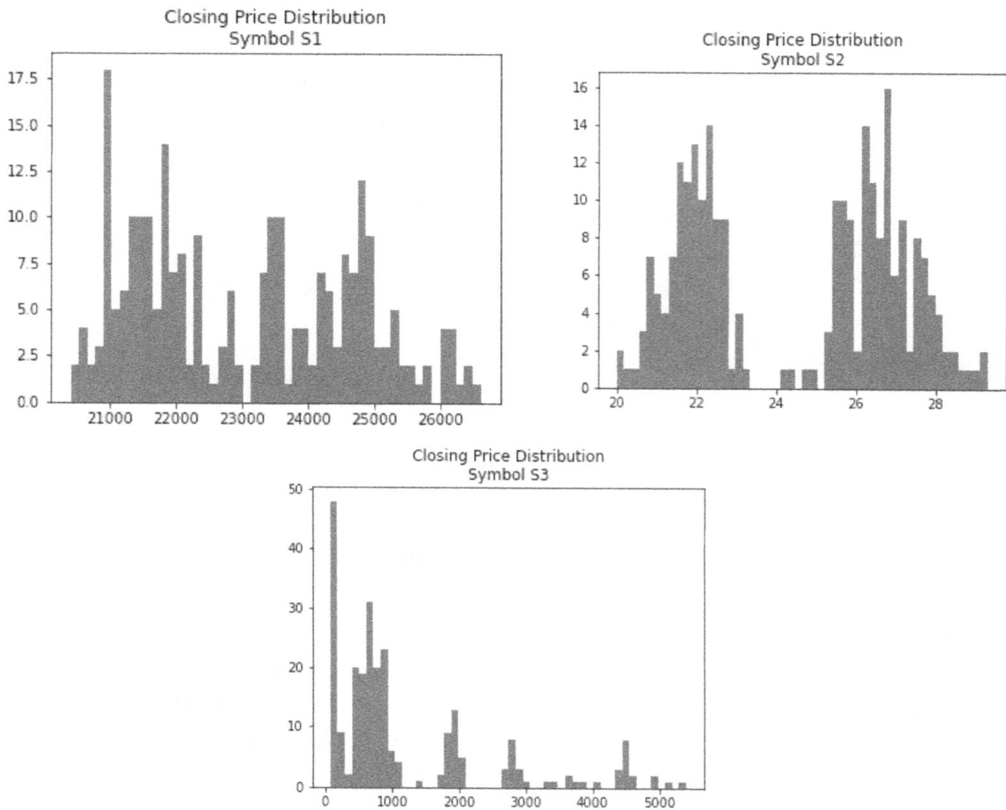

Figure 15.46 – Histograms of the raw prices of the three stock symbols

8. We have an indication from the distributions that `S1` and `S2` are behaving more similarly. Since we see the prices have very different ranges, we need to scale each symbol separately. Use `sklearn` to scale the original price and volume data by symbol, storing each symbol as a new DataFrame in a list, with the scalers as another list:

```
scale_cols = ['Close', 'Volume'] # to which columns we'll
apply a scaler
scalers = [] # keep a record of the scaler method used
for each symbol
scaled_data = [] # store scaled data in a list
```

```
for this_symbol in range(len(symbols)):
    scalers.append(StandardScaler())
    (scaled_data.append(my_data.groupby('symbol').
                          get_group(symbols[this_symbol]).
copy()))
    scaled_data[this_symbol].loc[:, scale_cols] = \
        (scalers[this_symbol].
         fit_transform(scaled_data[this_symbol].loc[:,
scale_cols]))
[data.head() for data in scaled_data]
```

This produces the following output. Note that since `scaled_data` is a list, with each element of the list being a DataFrame containing the data with only one symbol, the **list comprehension** iterates through `scaled_data` and displays the `.head()` of each DataFrame in turn:

```
Out[41]: [         Date     Close    Volume symbol sentiment
          0  2017-04-17 -1.469506 -1.175399     S1   NEUTRAL
          3  2017-04-18 -1.538998 -0.840327     S1   NEUTRAL
          6  2017-04-19 -1.611638 -0.528257     S1   NEUTRAL
          9  2017-04-20 -1.505101 -0.354008     S1       POS
          14 2017-04-21 -1.524028  0.210303     S1   NEUTRAL,
                   Date     Close    Volume symbol sentiment
          1  2017-04-17 -1.757829  0.198494     S2   NEUTRAL
          4  2017-04-18 -1.699092 -0.359611     S2   NEUTRAL
          7  2017-04-19 -1.640355  0.351069     S2   NEUTRAL
          10 2017-04-20 -1.424984 -0.443929     S2       POS
          12 2017-04-21 -1.483721 -0.259233     S2   NEUTRAL,
                   Date     Close    Volume symbol sentiment
          2  2017-04-17  3.342186 -0.211226     S3   NEUTRAL
          5  2017-04-18  3.104449 -0.211226     S3   NEUTRAL
          8  2017-04-19  2.985580 -0.211226     S3   NEUTRAL
          11 2017-04-20  2.747843 -0.211226     S3       NEG
          13 2017-04-21  2.628974 -0.211226     S3   NEUTRAL]
```

Figure 15.47 – The scaled stock data by symbol

9. Plot the training/test data as before, using a loop over the symbols. You can check out the code on GitHub: `https://github.com/PacktWorkshops/The-Pandas-Workshop/blob/master/Chapter10/Activity10.01.ipynb`.

   The chart should look as follows:

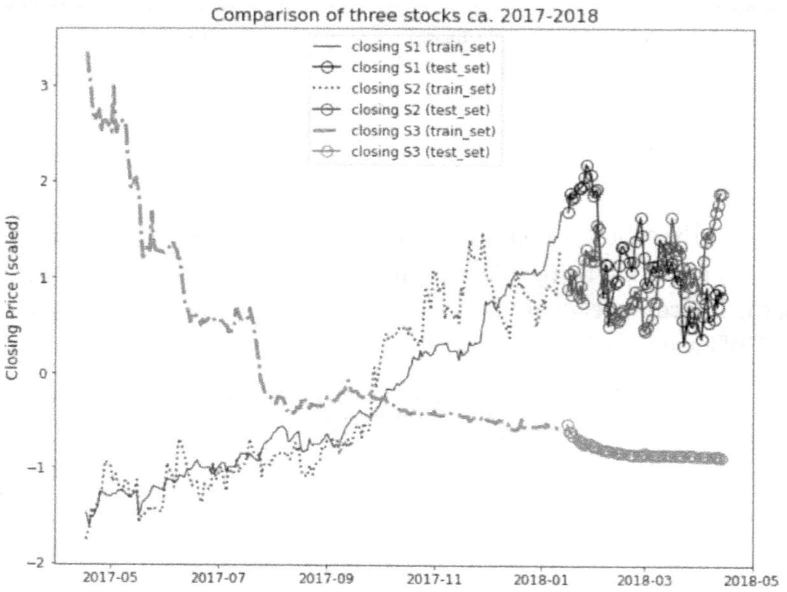

Figure 15.48 – Plotting the three symbols, each scaled separately

10. *Figure 15.48* gives a very different picture than *Figure 15.47*. You can now see that `S1` and `S2` are trending quite similarly, and `S3` is very different. Replot `S1` and `S2`, but apply a smoothing of 14 days and compare whether the two stocks are behaving the same way over the period from **2017-09** onward. You can find the code on GitHub here: `https://github.com/PacktWorkshops/The-Pandas-Workshop/blob/master/Chapter10/Activity10.01.ipynb`.

    The plot should be as follows:

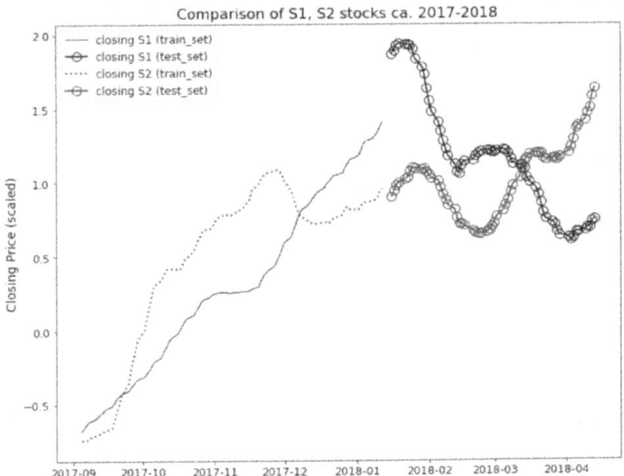

Figure 15.49 – Smoothed data for the two similar stocks

You note that in the latter part of the plot, there seem to be some similarities in the peaks and valleys of the two stocks, although not perfectly aligned. With this observation, you decide to do some more market research to see whether there are events or other factors that might be causing both stocks to behave similarly in this time period.

Solution 11.1

As part of a research effort to improve metallic-oxide semiconductor sensors for the toxic gas CO (carbon monoxide), you are asked to investigate models of the sensor response for an array of sensors. You will review the data, perform some feature engineering for non-linear features, and then compare a baseline linear regression approach to a random forest model.

Perform the following steps to complete the activity:

1. For this exercise, you will need the `pandas` and `numpy` libraries, and three modules from `sklearn`, `matplotlib`, and `seaborn`. Load them in the first cell of the notebook:

```
import pandas as pd
import numpy as np
from sklearn.linear_model import LinearRegression as OLS
from sklearn.ensemble import RandomForestRegressor
from sklearn.preprocessing import StandardScaler
import matplotlib.pyplot as plt
import seaborn as sns
```

2. As we have done before, create a utility function to plot a grid of histograms given the data, which variables to plot, the rows and columns of the grid, and how many bins. Similarly, create a utility function that allows plotting a list of variables as scatter plots against a given *x* variable, also given the rows and columns of the grid. You can find the code here: `https://github.com/PacktWorkshops/The-Pandas-Workshop/blob/master/Chapter11/Activity11_01/Activity11.01.ipynb`.
3. Now, load the `CO_sensors.csv` file into a DataFrame called `my_data`:

```
my_data = pd.read_csv('Datasets\\CO_sensors.csv')
my_data.head()
```

This should produce the following:

	Time (s)	CO (ppm)	Humidity (%r.h.)	Temperature (C)	Flow rate (mL/min)	Heater voltage (V)	R1 (MOhm)	R2 (MOhm)	R3 (MOhm)	R4 (MOhm)	R5 (MOhm)	R6 (MOhm)	R7 (MOhm)	R8 (MOhm)	R9 (MOhm)	R10 (MOhm)	F (MOh
0	0.000	0.0	49.21	26.38	247.2771	0.1994	0.5114	0.5863	0.5716	1.9386	1.1669	0.7103	0.5541	51.0146	40.8079	47.8748	4.6(
1	0.311	0.0	49.21	26.38	243.3618	0.7158	0.0626	0.1586	0.1161	0.1347	0.1385	0.1545	0.1307	0.1935	0.1341	0.1773	0.1
2	0.620	0.0	49.21	26.38	242.4944	0.8840	0.0654	0.1496	0.1075	0.1076	0.1131	0.1363	0.1188	0.1195	0.1049	0.1289	0.1
3	0.930	0.0	49.21	26.38	241.6242	0.8932	0.0722	0.1444	0.1074	0.1032	0.1106	0.1306	0.1190	0.1125	0.1014	0.1232	0.1
4	1.238	0.0	49.21	26.38	240.8151	0.8974	0.0767	0.1417	0.1098	0.1025	0.1116	0.1284	0.1208	0.1111	0.1008	0.1226	0.1

Figure 15.50 – The CO sensor data

4. Use `.describe().T` to inspect the data further:

```
my_data.describe().T
```

This produces the following output:

	count	mean	std	min	25%	50%	75%	max
Time (s)	295700.0	45435.140266	26245.705362	0.0000	22696.21350	45430.5430	68165.08150	90901.7260
CO (ppm)	295700.0	9.900266	6.426957	0.0000	4.44000	8.8900	15.56000	20.0000
Humidity (%r.h.)	295700.0	45.607506	12.445601	16.4300	36.14000	46.7000	55.37000	72.9800
Temperature (C)	295700.0	26.720057	0.418020	25.3800	26.38000	26.6600	27.06000	27.4200
Flow rate (mL/min)	295700.0	239.943680	1.697848	0.0000	239.90420	239.9716	240.03660	262.3167
Heater voltage (V)	295700.0	0.355212	0.288572	0.1990	0.20000	0.2000	0.20700	0.9010
R1 (MOhm)	295700.0	15.198374	22.583110	0.0324	0.40480	1.7121	25.85040	119.5851
R2 (MOhm)	295700.0	17.440031	26.665302	0.0555	0.48140	1.3664	29.05830	142.5199
R3 (MOhm)	295700.0	22.151461	28.585001	0.0541	0.57940	4.0667	44.88580	127.2483
R4 (MOhm)	295700.0	19.759571	16.412620	0.0394	1.94360	19.9434	31.75500	78.4601
R5 (MOhm)	295700.0	31.360319	27.068315	0.0480	1.72010	32.3170	51.48750	194.6753
R6 (MOhm)	295700.0	28.601243	27.198270	0.0493	1.50860	22.5929	49.60550	122.0913
R7 (MOhm)	295700.0	31.640992	27.612186	0.0517	1.80335	31.2996	52.41740	177.9975
R8 (MOhm)	295700.0	26.658295	19.523869	0.0334	11.69870	26.4721	40.41290	93.4149
R9 (MOhm)	295700.0	23.000006	17.919762	0.0291	8.44600	21.5685	35.50410	109.1693
R10 (MOhm)	295700.0	25.417975	20.410103	0.0368	7.56070	23.1211	39.88530	92.5828
R11 (MOhm)	295700.0	27.205435	20.348773	0.0309	10.29880	26.6826	41.73510	105.0967
R12 (MOhm)	295700.0	25.201259	18.560530	0.0327	9.45670	25.2860	38.99700	129.9261
R13 (MOhm)	295700.0	22.026591	17.036098	0.0331	7.59640	20.8730	34.05870	74.7083
R14 (MOhm)	295700.0	28.258380	21.982871	0.0316	9.47520	26.3557	44.15375	92.5210

Figure 15.51 – Details of the sensor data

5. Use the histogram grid utility function to plot histograms of all rows except `Time (s)`:

```
plot_histogram_grid(my_data, my_data.columns[1:], 7, 3,
25)
```

The output should appear as follows:

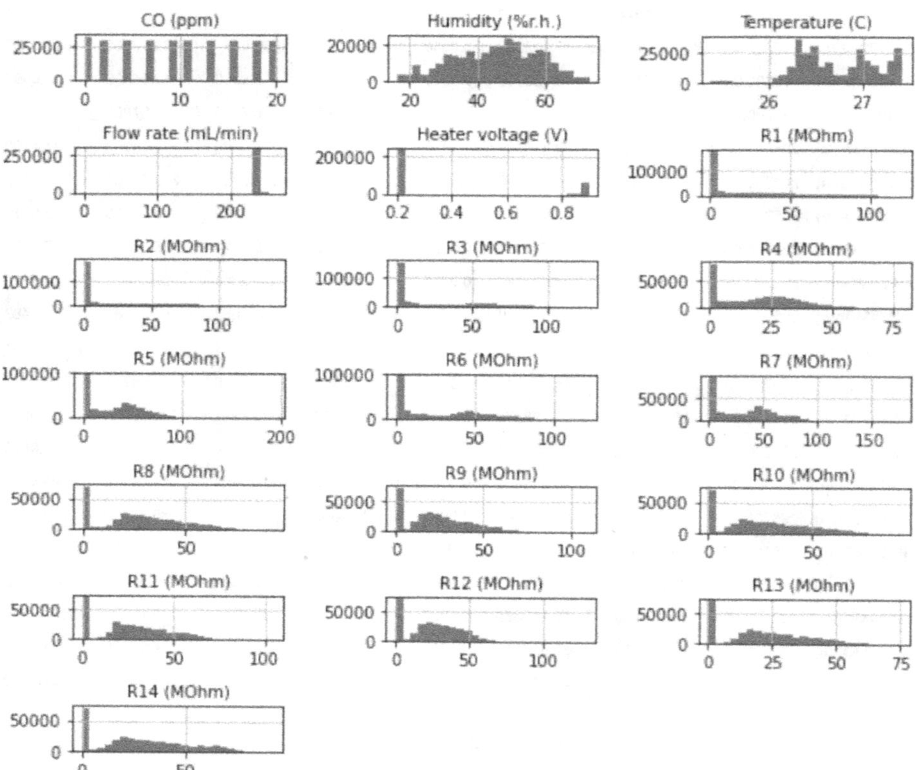

Figure 15.52 – Histograms of the sensor data

6. Use `seaborn` to generate a pairplot of the first five columns (excluding the sensor readings):

```
sns.pairplot(my_data.iloc[:, :5],
             height = 1.5, aspect = 1)
```

The plot should look as follows (depending on how you defined your function and options):

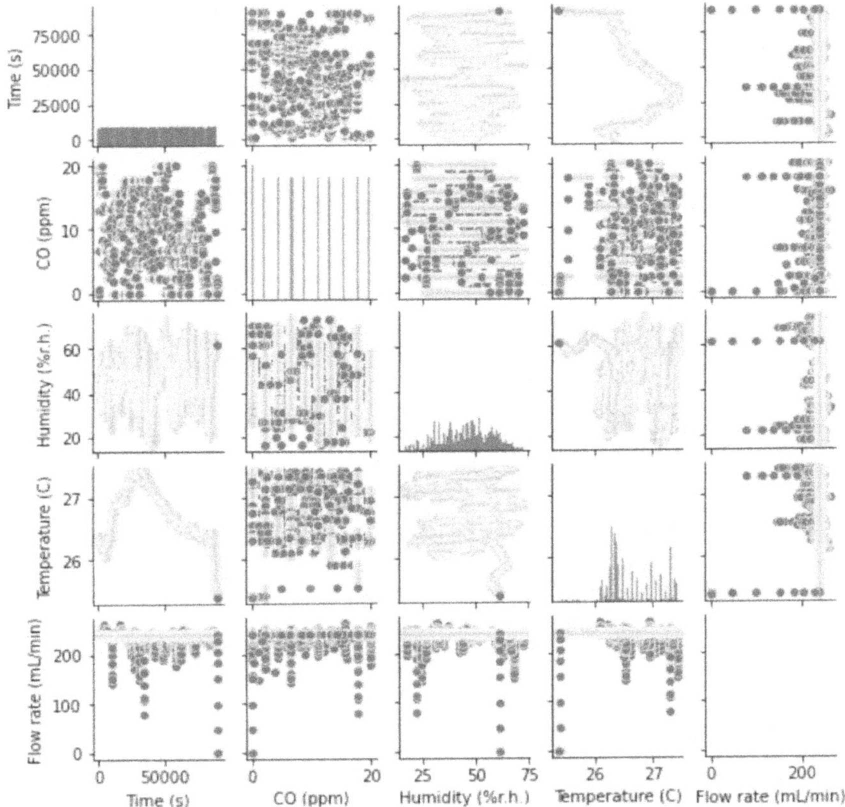

Figure 15.53 – Pairplot of the sensor data

7. Use the scatter plot grid utility function to plot all the sensor data versus time:

```
plot_cols = list(my_data.loc[:, 'R1 (MOhm)': ].columns)
plot_scatter_grid(my_data, plot_cols, 'Time (s)', 5, 4)
```

The result should look as follows (depending on how you defined your function and options):

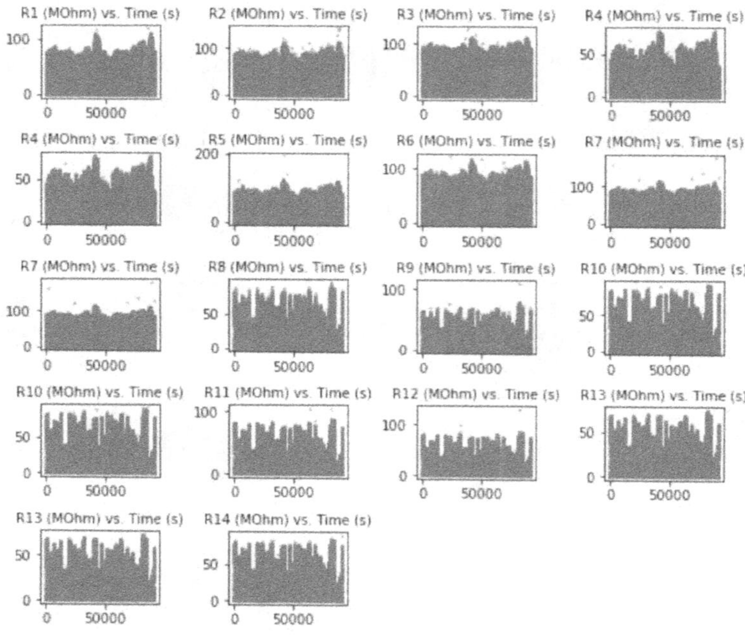

Figure 15.54 – The sensor traces versus time

8. It's difficult to tell whether there is a time dependency or a periodic component. Zoom in on `R13` over the time from 40,000 to 45,000 seconds:

```
fig, ax = plt.subplots(figsize = (11, 8))
ax.scatter(my_data.loc[(my_data['Time (s)'] > 40000) &
                        (my_data['Time (s)'] < 45000),
'Time (s)'],
           my_data.loc[(my_data['Time (s)'] > 40000) &
                        (my_data['Time (s)'] < 45000), 'R13
(MOhm)'])
ax.set_title('R13 vs. time')
plt.show()
```

This detail should look as follows:

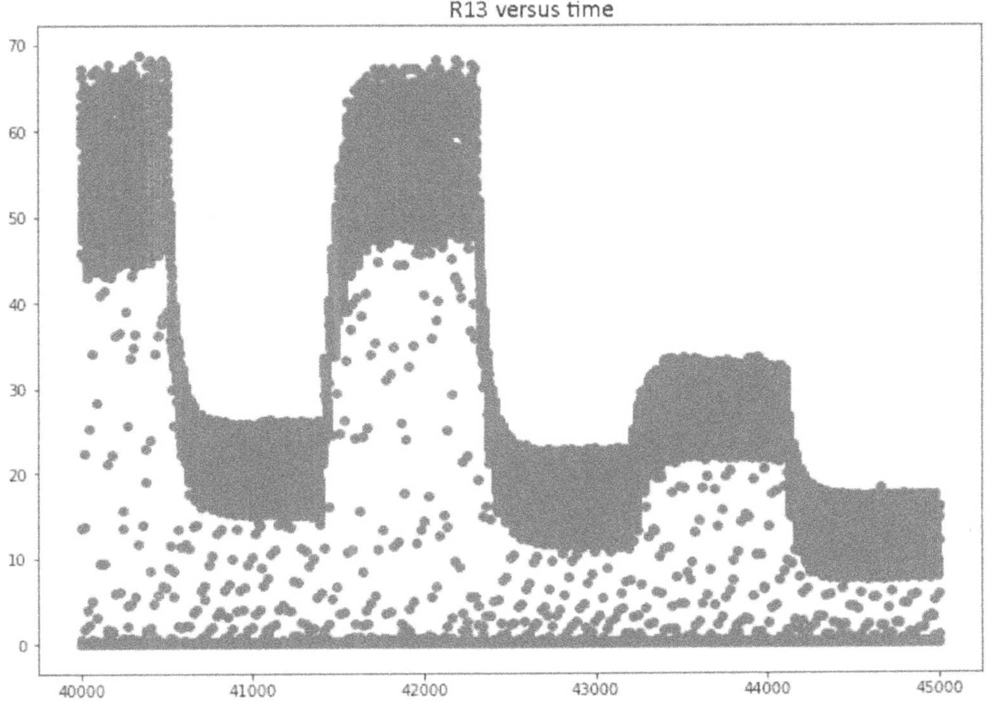

Figure 15.55 – Detail of one of the sensor traces

You can now see that the tests appear to comprise step functions of various sizes. This shows that the time variable is arbitrary and not useful for modeling the CO response. We also see that there is a significant number of values that deviate from the steps, which may be due to the humidity variations or could be measurement errors or some other issues. These may limit how well we can model the results.

9. Investigate the relationship of the changes in `R13` to the CO and humidity during one step change—for example, from about 41,250 to 42,500. Plot the `R13` values using the `.plot()` method in `matplotlib`, and overlay the CO and humidity values as line plots on the same plot:

```
fig, ax = plt.subplots(figsize = (15, 8))
ax.plot(my_data['Time (s)'],
        my_data['R13 (MOhm)'],
        color = 'red', lw = 0.5,
        label = 'R13')
```

```
ax.plot(my_data['Time (s)'],
        my_data['CO (ppm)'],
        label = 'CO')
ax.plot(my_data['Time (s)'],
        my_data['Humidity (%r.h.)'],
        label = 'Humidity')
ax.set_title('R13 vs. time')
ax.legend(loc = 'upper left', markerscale = 2)
ax.set_xlim((41250, 42500))
plt.show()
```

The resulting plot should look something like the following (depending on your choices):

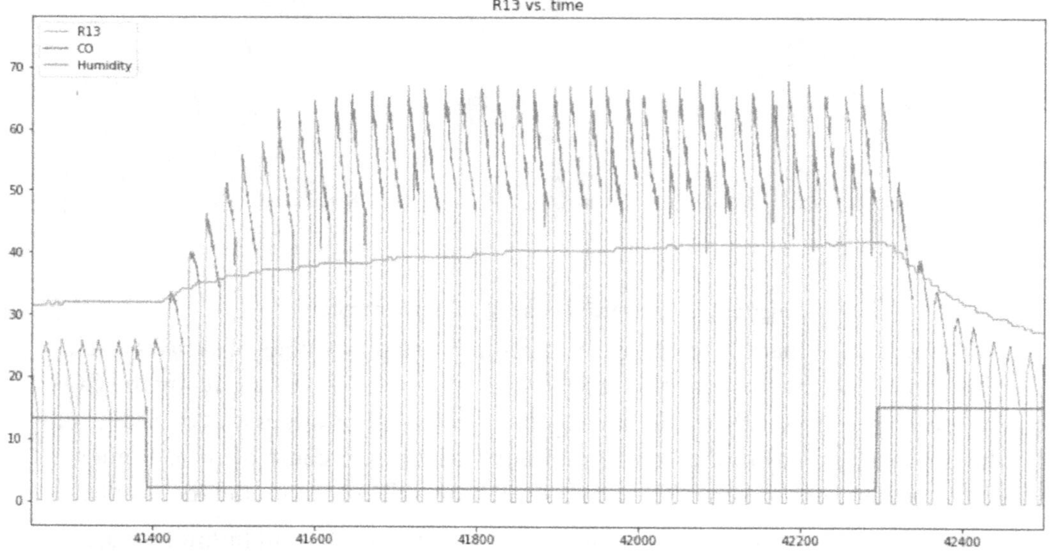

Figure 15.56 – R13, CO, and humidity versus time

As you saw in the graph, there are a series of step changes for both CO and humidity, resulting in changes in the resistance values. However, there are evident time lags involved, as shown by the curved traces of humidity. In addition, R13 seems to spike and then fall and has intervening periods where the value appears to be 0 and at a steady state. Perhaps this is a function of the electronics, but would require further investigation to be sure.

10. Now, use `seaborn` to plot a correlation heatmap for the sensor columns:

```
plt.figure(figsize = (10, 8))
sns.heatmap(my_data.loc[:, 'R1 (MOhm)':].corr())
```

The heatmap should look as follows:

```
Out[9]: <matplotlib.axes._subplots.AxesSubplot at 0x18d58f87908>
```

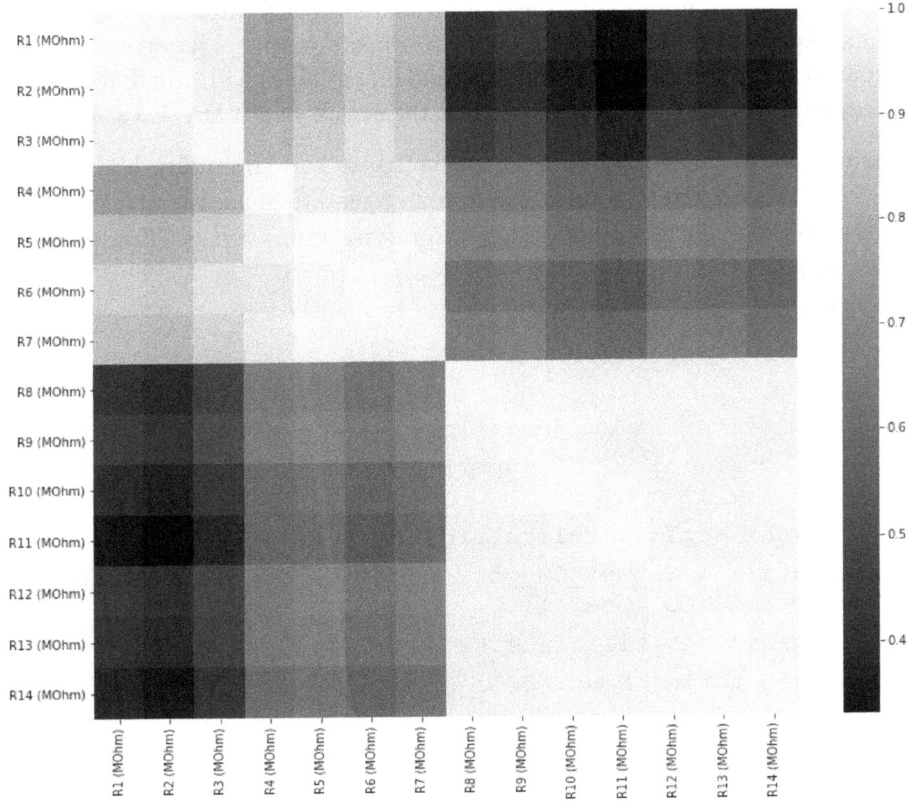

Figure 15.57 – Correlations among the sensors

Notice that there are two or three groups in the plot. The last seven sensors are all highly correlated with one another. The first three are as well, as are the next four.

11. The data description for this data (and the provided text file in the `Datasets` directory) says there are two kinds of sensors: "Figaro Engineering (7 units of TGS 3870-A04) and FIS (7 units of SB-500-12)". Now, it is apparent that R1 to R7 are one kind of sensor and R8 to R14 are the other kind. The data was collected to evaluate the performance of the sensors measuring CO at various conditions of temperature and humidity. In particular, the humidity is taken to be an "uncontrolled variable," and during the tests, random levels of humidity were imposed. In the field, the humidity would not be controlled or measured, which impacts the interpretation of data, especially for low levels of CO. The sensors' output is reported as the resistance in MOhms, which are the main independent variables with which to predict CO. Temperature and the voltage applied to the sensor heater are also available.

    Investigate the behavior of the sensors versus CO and humidity. Use pandas `.corr()` to generate the correlation matrix, and then use the first two rows of the result to make a bar plot of the sensor correlations versus CO and humidity, respectively:

```
Sensor_CO_corr = \
    (pd.concat([my_data.loc[:, ['CO (ppm)',
                                 'Humidity (%r.h.)']],
                my_data.loc[:, 'R1 (MOhm)':]], axis = 1).
     corr().loc['CO (ppm)':'Humidity (%r.h.)', 'R1
(MOhm)':])
fig, ax = plt.subplots(figsize = (9, 8))
ax.bar(x = Sensor_CO_corr.columns, height = Sensor_CO_
corr.loc['CO (ppm)'])
ax.xaxis.set_ticks_position('top')
ax.set_title('R vs. CO')
plt.xticks(rotation = 90)
plt.show()
fig, ax = plt.subplots(figsize = (9, 8))
ax.bar(x = Sensor_CO_corr.columns, height = Sensor_CO_
corr.loc['Humidity (%r.h.)'])
ax.xaxis.set_ticks_position('top')
ax.set_title('R vs. Humidity')
plt.xticks(rotation = 90)
plt.show()
```

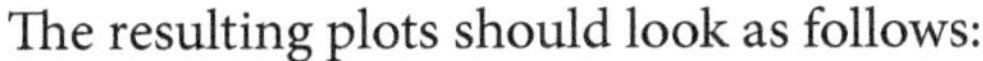

The resulting plots should look as follows:

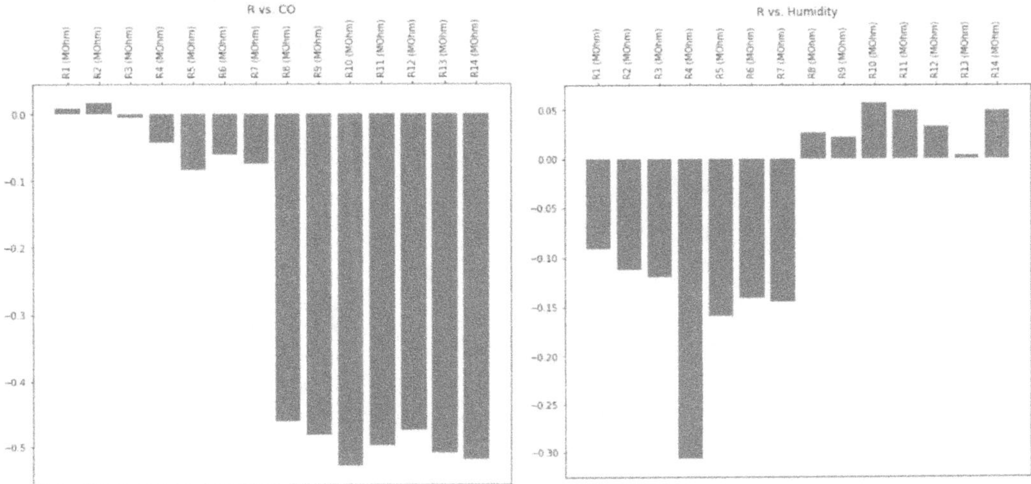

Figure 15.58 – Correlation of the sensor outputs to CO (top) and humidity (bottom)

You see that while all the sensors are to measure the CO, they have markedly different behavior depending on which of the two types we are measuring. From the problem description, it is evident that the sensors are impacted by humidity, but in the application, humidity is an uncontrolled and possibly unknown value. Hopefully, the different sensor behaviors can provide humidity information to a model and enable good predictions.

12. Apply a `sqrt()` transform to each of the sensor columns (since there are 0 or near-zero values, a log transform would not be appropriate) and add the columns to the dataset:

```
sensor_cols = list(my_data.loc[:, 'R1 (MOhm)': ].columns)
for i in range(len(sensor_cols)):
    my_data['sqrt_' + sensor_cols[i]] = np.sqrt(my_
data[sensor_cols[i]])
```

13. For the initial model, drop `Time`, `Humidity`, and `CO` from the `X` data. Use `CO` as the `y` data. Use `LinearRegression` to fit a model and plot the residuals as well as the predicted versus actual values:

```
model_X = pd.concat([my_data.loc[:, 'Temperature
(C)':'Heater voltage (V)'],
                     my_data.loc[:, 'sqrt_R1 (MOhm)':]],
axis = 1)
model_y = my_data.loc[:, 'CO (ppm)']
my_model = OLS()
my_model.fit(model_X, model_y)
preds = my_model.predict(model_X)
residuals = preds - model_y
fig, ax = plt.subplots(figsize = (9, 8))
ax.hist(residuals, bins = 50)
plt.show()
fig, ax = plt.subplots(figsize = (9, 8))
ax.scatter(model_y, preds)
ax.plot([0, 20], [0, 20], color = 'black', lw = 1)
ax.set_xlim(0, 20)
ax.set_ylim(0, 20)
ax.set_ylabel('predicted CO (ppm)')
ax.set_xlabel('actual CO (ppm)')
plt.show()
```

The outputs should look as follows:

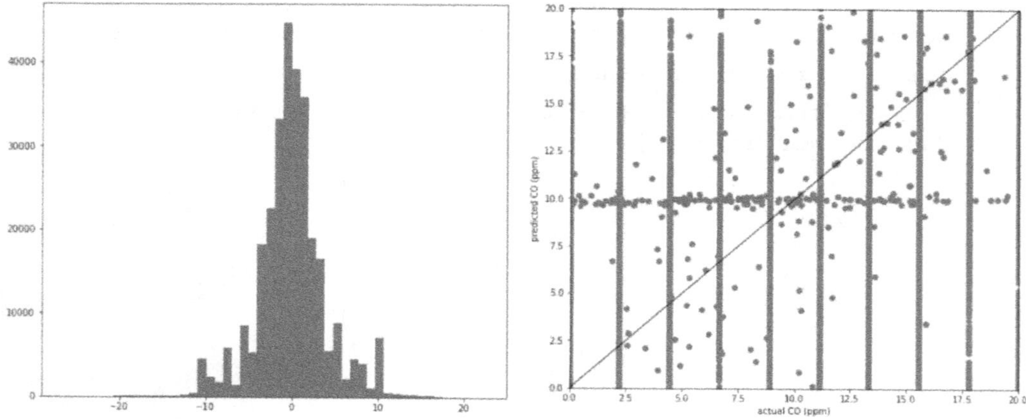

Figure 15.59 – The initial linear model. Residuals (top), predicted versus actual (bottom)

This model produces unbiased results, as shown by the residuals centered around 0, but from the second plot, we can see that there are multiple issues. There are groups of incorrect predictions at various levels, along with a clump near the middle of predicted CO readings of 10 ppm. This result clearly is not acceptable.

14. Scale the data with `StandardScaler()`, and then fit `RandomForestRegressor()` to the model. Plot the residuals and the predicted versus actual values:

```
scaler = StandardScaler()
model_X = scaler.fit_transform(model_X)
RF_model = RandomForestRegressor(n_estimators = 100)
RF_model.fit(model_X, model_y)
preds = RF_model.predict(model_X)
residuals = preds - model_y
fig, ax = plt.subplots(figsize = (9, 8))
ax.hist(residuals, bins = 50)
plt.show()
fig, ax = plt.subplots(figsize = (9, 8))
ax.scatter(model_y, preds)
ax.plot([0, 20], [0, 20], color = 'black', lw = 1)
ax.set_xlim(0, 20)
ax.set_ylim(0, 20)
ax.set_ylabel('predicted CO (ppm)')
ax.set_xlabel('actual CO (ppm)')
plt.show()
```

The plots should look as follows:

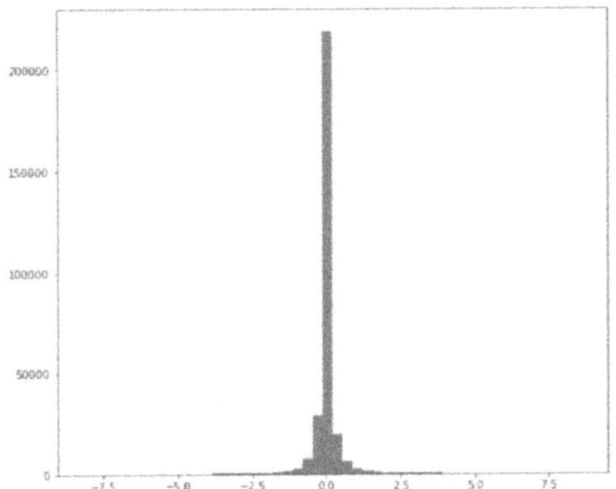

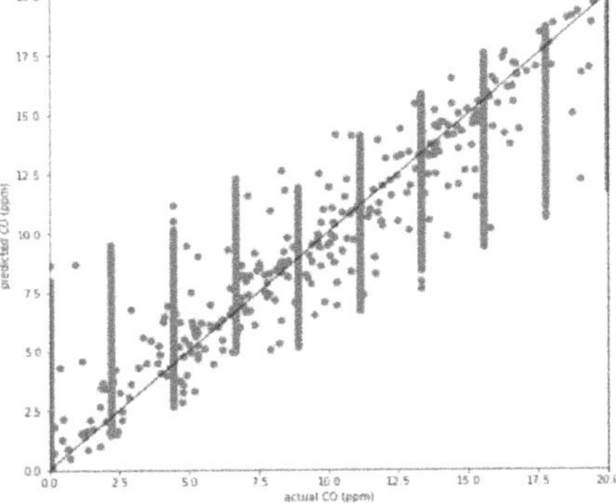

Figure 15.60 – Results of the RandomForestRegressor() fit

Although it is evident that the Random Forest model has reduced the residuals, the vertical groupings are still present and this is not a satisfactory result. Reviewing *Figure 15.56*, note that although the CO values are shown and are nearly constant, there is a time lag in the humidity and sensor resistance values. A possible approach would be to average over readings. A simple test of this idea is to group by the CO values and take the mean sensor values, and then model with those. In addition, it seems reasonable to filter out the regions where the resistance values drop to low values, as those seem anomalous.

15. Create a dataset by filtering out all the rows where a sensor resistance value drops to a low value, say 0.1. Then, group by `'CO (ppm)'` and aggregate as the mean values. Build a Random Forest model using the sensor mean resistances and the CO group values. Also, refit a linear regression model to this data. Plot the predicted values versus the actual values for both results.

    You can find the code here: `https://github.com/PacktWorkshops/The-Pandas-Workshop/blob/master/Chapter11/Activity11_01/Activity11.01.ipynb`.

    The resulting plot should look something like the following:

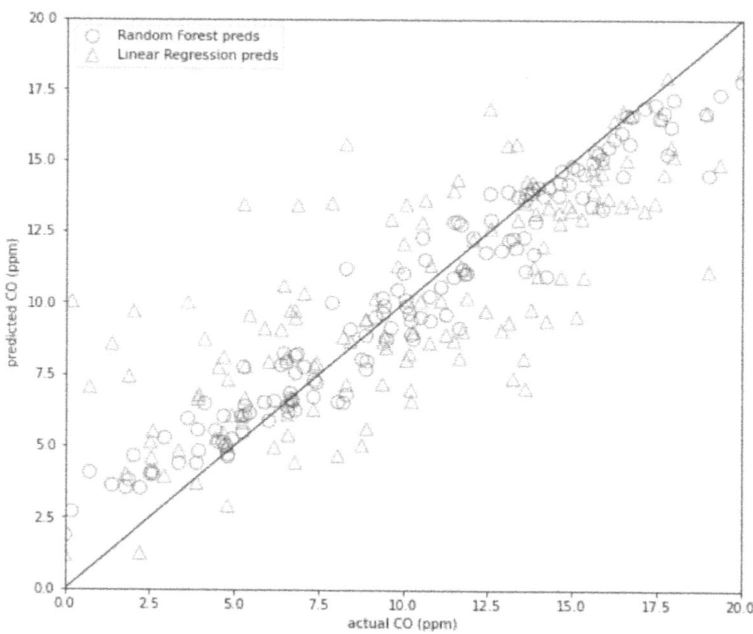

Figure 15.61 – Results using grouped CO values and mean sensor values

These results are much better. It would require more discussion with expert stakeholders to confirm this approach, but it is a promising direction to obtain a calibration for the sensors. Note that the linear regression model cannot fit a lot of the data nearly as well. Also note that the vertical scatter in the Random Forest predictions might be an indicator of noise caused by the random humidity values. This could be investigated further by building another model and including humidity as an independent variable.

You have now tied together the ideas of preprocessing, feature engineering, and linear and non-linear models for data modeling.

Solution 12.1

Perform the following steps to complete the activity:

1. For this activity, you will need the `pandas` and `numpy` libraries. Load them in the first cell of the notebook:

```
import pandas as pd
import numpy as np
```

2. Read in the `household_power_consumption.csv` data from the `Datasets` directory and list the first few rows:

```
data_fn = 'household_power_consumption.csv'
household_electricity = \
    pd.read_csv('../datasets/' + data_fn,
    sep = ';',
    low_memory = False)
household_electricity.head()
```

This generates the following:

Out[5]:

	Date	Time	Global_active_power	Global_reactive_power	Voltage	Global_intensity	Sub_metering_1	Sub_metering_2	Sub_metering_3
0	1/8/2008	00:00:00	0.500	0.226	239.750	2.400	0.000	0.000	1.0
1	1/8/2008	00:01:00	0.482	0.224	240.340	2.200	0.000	0.000	1.0
2	1/8/2008	00:02:00	0.502	0.234	241.680	2.400	0.000	0.000	0.0
3	1/8/2008	00:03:00	0.556	0.228	241.750	2.600	0.000	0.000	1.0
4	1/8/2008	00:04:00	0.854	0.342	241.550	4.000	0.000	1.000	7.0

Figure 15.62 – The household_power_consumption.csv data

3. You should inspect the data types of the columns and further investigate whether there are non-numeric values. If so, correct them by converting them to `NA` values and then filling them by interpolation:

```
Household_electricity.dtypes
```

This generates the following output:

```
Out[4]: Date                      object
        Time                      object
        Global_active_power       object
        Global_reactive_power     object
        Voltage                   object
        Global_intensity          object
        Sub_metering_1            object
        Sub_metering_2            object
        Sub_metering_3           float64
        dtype: object
```

Figure 15.63 – The data types of the electricity data

4. The columns from `Global_active_power` to `Sub_metering_3` should be numeric, but some rows contain a placeholder for missing data, `?`. You can see that by using `.describe()` on each column:

```
for col in household_electricity.columns:
    print('information for column ' +
          col +
          ':\n',
          household_electricity[col].describe())
```

This generates the following (we have limited *Figure 15.64* to a range of columns with missing values):

```
information for column Global_active_power:
 count     1049760
unique       3852
top             ?
freq         9570
Name: Global_active_power, dtype: object
information for column Global_reactive_power:
 count     1049760
unique        510
top         0.000
freq       230359
Name: Global_reactive_power, dtype: object
information for column Voltage:
 count     1049760
unique       2738
top             ?
freq         9570
Name: Voltage, dtype: object
```

Figure 15.64 – Two columns containing ? values

Notice the `?` character in the `Global_active_power` and `Voltage` columns.

5. Replace `?` by first using pandas `.replace()` to replace `?` with `np.nan` values, and then pandas `.interpolate()` to fill the `NAN` values by simple interpolation between the previous and next values:

```
home_elec.replace('?', np.nan, inplace = True)
home_elec.interpolate(inplace = True)
for col in home_elec.columns[2:]:
    home_elec[col] = home_elec[col].astype(float)
```

6. Make a quick visualization to understand the timeframe of the data. Your plan is to identify a year with complete data and focus on that:

```
(home_elec[['Date',
            'Sub_metering_1']].
            plot(x = 'Date',
                 y = 'Sub_metering_1'))
```

This generates the following output:

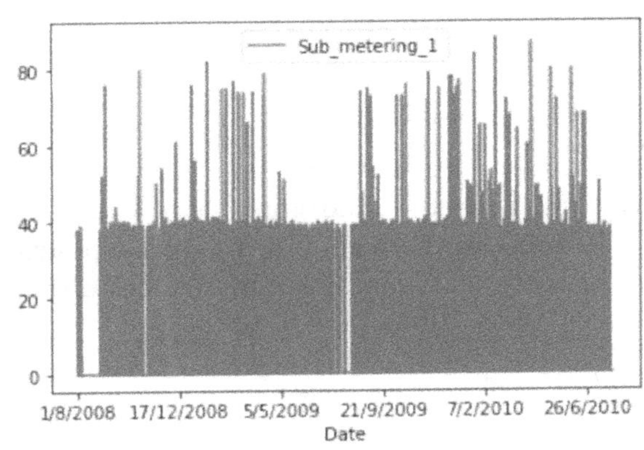

Figure 15.65 – Sub_metering_1 data versus date. The year 2009 is a full year

7. Using the year you identified, create a new DataFrame with `Date`, `Time`, and `Kitchen_power_use`. `Sub_metering_1` is the kitchen:

```
kitchen_elec = home_elec[['Date',
                          'Time',
                          'Sub_metering_1']]
(kitchen_elec = \
     kitchen_elec.loc[kitchen_elec['Date'].
                 str.contains('2009'), :])
kitchen_elec.columns = ['Date',
                        'Time',
                        'Kitchen_power_use']
kitchen_elec.head()
```

Out[13]:

	Date	Time	Kitchen_power_use
1074636	1/1/2009	00:00:00	0.0
1074637	1/1/2009	00:01:00	0.0
1074638	1/1/2009	00:02:00	0.0
1074639	1/1/2009	00:03:00	0.0
1074640	1/1/2009	00:04:00	0.0

Figure 15.66 – The new DataFrame containing kitchen power use

8. `Date` and `Time` are strings; combine them on each row, then convert the combined string into a datetime and store that in a new column called `timestamp`. Keep in mind the European format of the original date strings. To address that, we pass the `dayfirst = True` argument to the `.to_datetime()` method:

```
kitchen_elec.loc[:, 'timestamp'] = \
    pd.to_datetime(kitchen_elec.loc[:, 'Date'] + ' '
                   + kitchen_elec.loc[:, 'Time'],
                  dayfirst = True)
kitchen_elec.sort_values('timestamp',
                         inplace = True)
kitchen_elec.head()
```

Out[12]:

	Date	Time	Kitchen_power_use	timestamp
1074636	1/1/2009	00:00:00	0.0	2009-01-01 00:00:00
1074637	1/1/2009	00:01:00	0.0	2009-01-01 00:01:00
1074638	1/1/2009	00:02:00	0.0	2009-01-01 00:02:00
1074639	1/1/2009	00:03:00	0.0	2009-01-01 00:03:00
1074640	1/1/2009	00:04:00	0.0	2009-01-01 00:04:00

Figure 15.67 – Creation of a timestamp

9. Create an `hour` and `date` column using methods on the `timestamp` column to represent the hour of the day and the date in a standard format:

```
kitchen_elec['hour'] = \
    kitchen_elec['timestamp'].dt.hour
kitchen_elec['date'] = \
    kitchen_elec['timestamp'].dt.date
kitchen_elec.head()
```

Out[34]:

	Date	Time	Kitchen_power_use	timestamp	hour	date
1074636	1/1/2009	00:00:00	0.0	2009-01-01 00:00:00	0	2009-01-01
1074637	1/1/2009	00:01:00	0.0	2009-01-01 00:01:00	0	2009-01-01
1074638	1/1/2009	00:02:00	0.0	2009-01-01 00:02:00	0	2009-01-01
1074639	1/1/2009	00:03:00	0.0	2009-01-01 00:03:00	0	2009-01-01
1074640	1/1/2009	00:04:00	0.0	2009-01-01 00:04:00	0	2009-01-01

Figure 15.68 – Addition of the hour and new date column

10. Group the data by `date` and `hour`, aggregating `Kitchen_power_use`:

```
kitchen_elec = \
    (kitchen_elec[['date',
                   'hour',
                   'Kitchen_power_use']].
                   groupby(['date',
                            'hour']).sum())
kitchen_elec.reset_index(inplace = True)
kitchen_elec.iloc[20:28, :]
```

Out[55]:

	date	hour	Kitchen_power_use
20	2009-01-01	20	0.0
21	2009-01-01	21	0.0
22	2009-01-01	22	0.0
23	2009-01-01	23	0.0
24	2009-01-02	0	0.0
25	2009-01-02	1	0.0
26	2009-01-02	2	0.0
27	2009-01-02	3	0.0

Figure 15.69 – Kitchen power use by hour for each day

11. For January, aggregate the data by hour and make a bar plot by hour of kitchen energy use:

```
(kitchen_elec.loc[((kitchen_elec['date'] >=
                    pd.to_datetime('2009-01-01')) &
                   (kitchen_elec['date'] <
                    pd.to_datetime('2009-02-01'))),
                  ['hour',
                   'Kitchen_power_use']].
                  groupby('hour').mean().plot(kind =
                                                'bar'))
```

```
Out[50]: <AxesSubplot:xlabel='hour'>
```

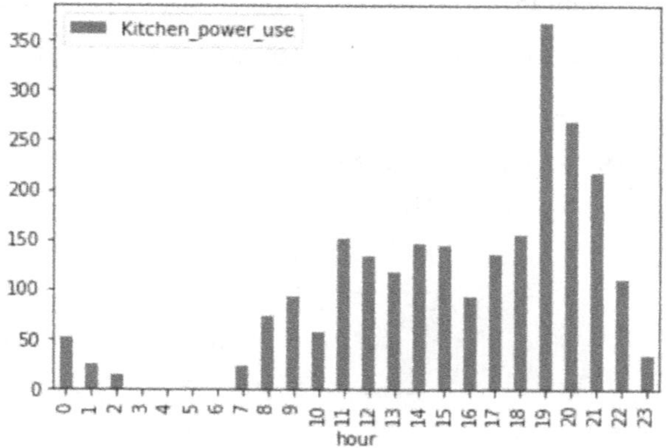

Figure 15.70 – Kitchen energy use by hour for January 2009

12. You see that usage seems to begin around breakfast time, continues throughout the day, with a peak at dinner time, and then trails off. Make a similar plot for the entire year to compare:

```
(kitchen_elec.loc[:,
                  ['hour',
                   'Kitchen_power_use']].
                   groupby('hour').mean().plot(kind =
                                               'bar'))
```

```
Out[56]: <AxesSubplot:xlabel='hour'>
```

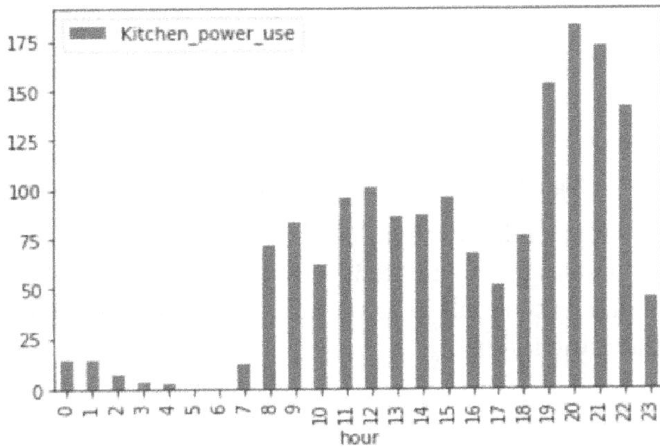

Figure 15.71 – Kitchen energy use by hour for all of 2009

You see that the two charts are similar, although there are some differences to be explored.

Solution 13.1

Perform the following steps to complete the activity:

1. For this activity, you will need the `pandas` library, the `matplotlib.pyplot` library, and the `sklearn.linear_model.LinearRegression` module. Load them in the first cell of the notebook:

```
import pandas as pd
import matplotlib.pyplot as plt
from sklearn.linear_model import LinearRegression
```

2. Read in the `bike_share.csv` data from the `Datasets` directory and list the first five rows using `.head()`:

```
rental_data = pd.read_csv('../Datasets/bike_share.csv')
rental_data.head()
```

This produces the following:

Out[91]:

	date	hour	rentals
0	1/1/2011	0	16
1	1/1/2011	1	40
2	1/1/2011	2	32
3	1/1/2011	3	13
4	1/1/2011	4	1

Figure 15.72 – The first five rows of the bike_share data

3. You need to create a datetime index. Begin by creating a column that combines the date and hour strings into a datetime-like string, and then convert that string to a datetime, storing the result in a column. Finally, set the index to the new column, so there is a datetime index:

```
rental_data['date_time'] = \
[(rental_data.date[i] +
  ' ' +
  '{:02}'.format(rental_data.hour[i]) +
  ':00:00')
 for i in rental_data.index]
rental_data.set_index(pd.to_datetime(rental_data['date_
time']),
                      inplace = True,
                      drop = True)
rental_data
```

This provides the following output:

Out[137]:

	date	hour	rentals	date_time
date_time				
2011-01-01 00:00:00	1/1/2011	0	16	1/1/2011 00:00:00
2011-01-01 01:00:00	1/1/2011	1	40	1/1/2011 01:00:00
2011-01-01 02:00:00	1/1/2011	2	32	1/1/2011 02:00:00
2011-01-01 03:00:00	1/1/2011	3	13	1/1/2011 03:00:00
2011-01-01 04:00:00	1/1/2011	4	1	1/1/2011 04:00:00
...	...	...	...	...
2012-12-31 19:00:00	12/31/2012	19	119	12/31/2012 19:00:00
2012-12-31 20:00:00	12/31/2012	20	89	12/31/2012 20:00:00
2012-12-31 21:00:00	12/31/2012	21	90	12/31/2012 21:00:00
2012-12-31 22:00:00	12/31/2012	22	61	12/31/2012 22:00:00
2012-12-31 23:00:00	12/31/2012	23	49	12/31/2012 23:00:00

17379 rows × 4 columns

Figure 15.73 – The rental_data DataFrame with the added date_time column and index

4. Generate a simple plot of the first 240 hours of data:

```
rental_data['rentals'][:240].plot()
```

This should generate a plot as follows:

Out[94]: <matplotlib.axes._subplots.AxesSubplot at 0x25969241208>

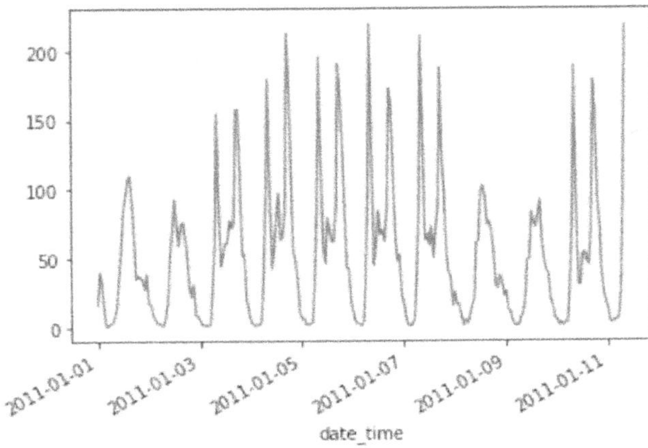

Figure 15.74 – Initial plot of the data

5. Using the index and the `rentals` column, downsample the data to 1-day intervals. You want total rentals per day, so choose the appropriate aggregation function:

```
rental_data = pd.DataFrame(rental_data['rentals'].
resample('1d').sum())
rental_data.head(14)
```

The data should look as follows:

Out[95]:

date_time	rentals
2011-01-01	985
2011-01-02	801
2011-01-03	1349
2011-01-04	1562
2011-01-05	1600
2011-01-06	1606
2011-01-07	1510
2011-01-08	959
2011-01-09	822
2011-01-10	1321
2011-01-11	1263
2011-01-12	1162
2011-01-13	1406
2011-01-14	1421

Figure 15.75 – rental_data resampled to 1-day intervals

6. Generate a simple plot of the first 8 weeks (56 days) of the resampled data:

```
rental_data[:56].plot()
```

The plot should be as follows:

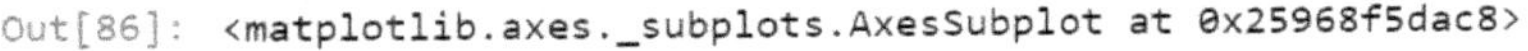
Out[86]: <matplotlib.axes._subplots.AxesSubplot at 0x25968f5dac8>

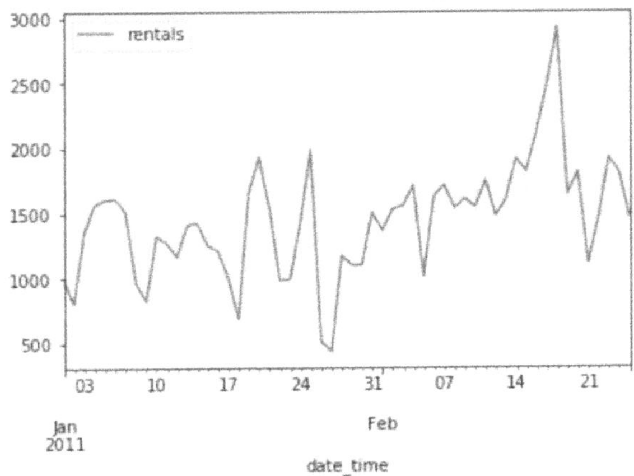

Figure 15.76 – The resampled rental_data

7. You should notice that there are ups and downs that seem to be on a 7-day cycle. Make a plot of the rentals on a given day versus the rentals 7 days in the past to explore this idea:

```
fig, ax = plt.subplots()
ax.scatter(rental_data['rentals'][:(rental_data.shape[0]
- 7)],
           rental_data['rentals'][7:])
ax.set_title('Rentals vs. rentals 7 days ago')
plt.show()
```

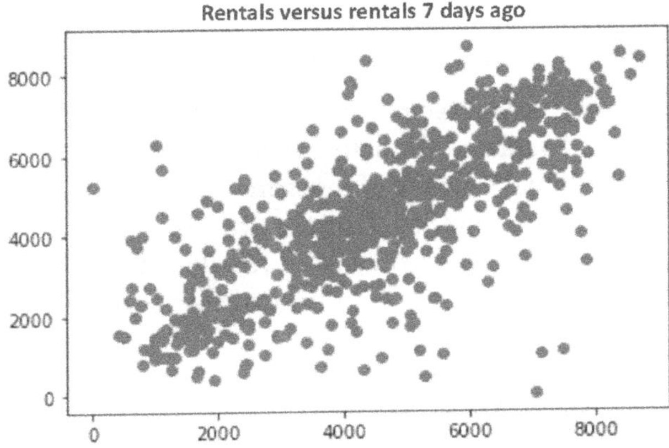

Figure 15.77 – Plotting the daily rentals versus the same data 7 days in the past

8. Since there appears to be a strong correlation on a 7-day period, you want to generate a new column in the data with the rentals from 7 days in the past for each day. Note that the first 7 days don't have past data, so those will be `NaN`:

```
lagged_rentals =\
    rental_data['rentals'][:(rental_data.shape[0] - 7)]
lagged_rentals.index = rental_data.index[7:]
rental_data['lagged_rentals'] = lagged_rentals
rental_data
```

Out[109]:

date_time	rentals	lagged_rentals
2011-01-01	985	NaN
2011-01-02	801	NaN
2011-01-03	1349	NaN
2011-01-04	1562	NaN
2011-01-05	1600	NaN
...	...	...
2012-12-27	2114	4128.0
2012-12-28	3095	3623.0
2012-12-29	1341	1749.0
2012-12-30	1796	1787.0
2012-12-31	2729	920.0

731 rows × 2 columns

Figure 15.78 – rental_data with an added column, containing the rental totals 7 days in the past

9. Now, you will use the `LinearRegression` module to fit a linear model, using the lagged data as the `X` data and the actual rentals as the `Y` data. You need to drop the `NaN` values, so make a copy of the `rental_data` DataFrame in a new DataFrame, `model_data`, and then drop the `NaN` rows. By using `model_data = rental_data.copy()` first, you ensure you don't make any changes to the original data. Create an instance of `LinearRegression()`, fit the model, and get the R2 score:

```
model_data = rental_data.copy()[rental_data['lagged_
rentals'].isna() == False]
lagged_model = LinearRegression()
```

```
lagged_model.fit(model_data['lagged_rentals'].values.
reshape(-1, 1),
                 model_data['rentals'].values.reshape(-1,
1))
model_data['predicted'] =\
    lagged_model.predict(model_data['lagged_rentals'].
values.reshape(-1, 1))
R2 = lagged_model.score(model_data['rentals'].values.
reshape(-1, 1),
                        model_data['predicted'].values.
reshape(-1, 1))
print('R2 is ', R2, ' using:')
print(model_data[['rentals', 'lagged_rentals']].head())
```

Don't worry too much about the details of the model calls. The `.values.reshape(-1, 1)` method calls are due to the fact that `sklearn` uses `numpy` arrays, and we need to reshape the pandas Series into `numpy` arrays. You see that once you create an instance of `LinearRegression` (as `lagged_model`), it has additional attributes and methods, such as `.score()` and `.predict()`. We'll see more about those in the next chapter!

The output should be as follows:

```
R2 is  0.5145071365683822  using:
            rentals  lagged_rentals
date_time
2011-01-08      959           985.0
2011-01-09      822           801.0
2011-01-10     1321          1349.0
2011-01-11     1263          1562.0
2011-01-12     1162          1600.0
```

Figure 15.79 – The R2 score and the original data for the rental model

10. Plot the predicted values versus the actual values. Make the *x* and *y* scales the same and add a diagonal line. The idea is that a perfect prediction would be along the diagonal, so you can compare your results to that:

```
fig, ax = plt.subplots()
ax.scatter(model_data['rentals'],
           model_data['predicted'])
xlim = (0, max(pd.concat([model_data['predicted'],
                          model_data['rentals']])))
```

```
ylim = xlim
ax.set_xlim(xlim)
ax.set_ylim(ylim)
ax.plot([xlim[0], xlim[1]],
        [ylim[0], ylim[1]],
       color = 'red')
ax.set_title('Predicted vs. Actual Rentals\n' +
             'R2 = ' + str(round(R2, 2)))
ax.set_xlabel('Actual Rentals per day')
ax.set_ylabel('Predicted Rentals per day')
plt.show()
```

The plot should be as follows:

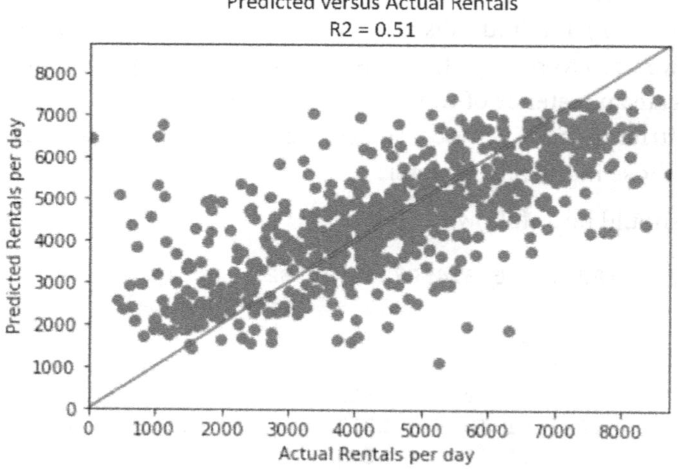

Figure 15.80 – The predictions of the simple linear model versus the actual data

You see that the fit isn't great, but with a few simple steps, you have accounted for just over half the variation in the data! Some of the variation may be inherently unpredictable, caused by myriad random factors combining to lead to a certain value of rentals on any given day. However, you might improve the model by adding features to account for weekdays, weekends, holidays, and weather, as just a few examples. In time series modeling, we use the past to predict the future. In such a case, we don't always have every explanatory variable, such as weather, in the future. Therefore, sometimes we need to forecast something such as weather in order to use it in another model. This is a specific aspect that makes time series modeling more challenging than simple regression.

Solution 14.1

Perform the following steps to complete the activity:

1. Open a new Jupyter notebook file. Import `pandas` and `dateutil.parser` into your notebook:

```
import pandas as pd
from dateutil.parser import parse
```

2. Define the path for the dataset and read the data:

```
# Defining the paths of the files

filePath = '/content/drive/MyDrive/Packt_Colab/pandas_
chapter11/chapter11/AirQualityUCI.csv'



# Reading the text files
data = pd.read_csv(filePath,delimiter=";")
data.head()
```

You should get the following output:

PT08.S1(CO)	NMHC(GT)	C6H6(GT)	PT08.S2(NMHC)	NOx(GT)	PT08.S3(NOx)	NO2(GT)	PT08.S4(NO2)	PT08.S5(O3)	T	RH	AH	Unnamed: 15	Unnamed: 16
1360.0	150.0	11,9	1046.0	166.0	1056.0	113.0	1692.0	1268.0	13,6	48,9	0,7578	NaN	NaN
1292.0	112.0	9,4	955.0	103.0	1174.0	92.0	1559.0	972.0	13,3	47,7	0,7255	NaN	NaN
1402.0	88.0	9,0	939.0	131.0	1140.0	114.0	1555.0	1074.0	11,9	54,0	0,7502	NaN	NaN
1376.0	80.0	9,2	948.0	172.0	1092.0	122.0	1584.0	1203.0	11,0	60,0	0,7867	NaN	NaN
1272.0	51.0	6,5	836.0	131.0	1205.0	116.0	1490.0	1110.0	11,2	59,6	0,7888	NaN	NaN

Figure 15.81 – Placeholder

3. Drop the unwanted columns, as follows:

```
data = data.drop(['Unnamed: 15','Unnamed: 16'],axis=1)
data.head()
```

You should get an output as follows:

Date	Time	CO(GT)	PT08.S1(CO)	NMHC(GT)	C6H6(GT)	PT08.S2(NMHC)	NOx(GT)	PT08.S3(NOx)	NO2(GT)	PT08.S4(NO2)	PT08.S5(O3)	T	RH	AH
10/03/2004	18.00.00	2,6	1360.0	150.0	11,9	1046.0	166.0	1056.0	113.0	1692.0	1268.0	13,6	48,9	0,7578
10/03/2004	19.00.00	2	1292.0	112.0	9,4	955.0	103.0	1174.0	92.0	1559.0	972.0	13,3	47,7	0,7255
10/03/2004	20.00.00	2,2	1402.0	88.0	9,0	939.0	131.0	1140.0	114.0	1555.0	1074.0	11,9	54,0	0,7502
10/03/2004	21.00.00	2,2	1376.0	80.0	9,2	948.0	172.0	1092.0	122.0	1584.0	1203.0	11,0	60,0	0,7867
10/03/2004	22.00.00	1,6	1272.0	51.0	6,5	836.0	131.0	1205.0	116.0	1490.0	1110.0	11,2	59,6	0,7888

Figure 15.82 – Placeholder

```
data.shape
```

You should get the following output:

```
(9471, 15)
```

4. Remove the NA values:

```
data = data.dropna()
data.shape
```

You should get the following output:

```
(9357, 15)
```

5. Check whether there are any data points with NA:

```
data.info()
```

You should get the following output:

```
Int64Index: 9357 entries, 0 to 9356
Data columns (total 15 columns):
Date             9357 non-null object
Time             9357 non-null object
CO(GT)           9357 non-null object
PT08.S1(CO)      9357 non-null float64
NMHC(GT)         9357 non-null float64
C6H6(GT)         9357 non-null object
PT08.S2(NMHC)    9357 non-null float64
NOx(GT)          9357 non-null float64
PT08.S3(NOx)     9357 non-null float64
NO2(GT)          9357 non-null float64
PT08.S4(NO2)     9357 non-null float64
PT08.S5(O3)      9357 non-null float64
T                9357 non-null object
RH               9357 non-null object
AH               9357 non-null object
dtypes: float64(8), object(7)
```

Figure 15.83 – Placeholder

From the output, you can see that there are no NA values in the dataset.

6. Parse the date column and extract the weekday, day, and month, as follows:

```
# Parsing the date
data['Parse_date'] = data['Date'].apply(lambda x:
parse(x))
# Parsing the weekdaty
data['Weekday'] = data['Parse_date'].apply(lambda x:
```

```
x.weekday())
# Parsing the Day
data['Day'] = data['Parse_date']\
.apply(lambda x: x.strftime("%A"))
# Parsing the Month
data['Month'] = data['Parse_date']\
.apply(lambda x: x.strftime("%B"))
data.head()
```

You should see the following output:

MHC(GT)	C6H6(GT)	PT08.S2(NMHC)	NOx(GT)	PT08.S3(NOx)	NO2(GT)	PT08.S4(NO2)	PT08.S5(O3)	T	RH	AH	Parse_date	Weekday	Day	Month
150.0	11,9	1046.0	166.0	1056.0	113.0	1692.0	1268.0	13,6	48,9	0,7578	2004-10-03	6	Sunday	October
112.0	9,4	955.0	103.0	1174.0	92.0	1559.0	972.0	13,3	47,7	0,7255	2004-10-03	6	Sunday	October
88.0	9,0	939.0	131.0	1140.0	114.0	1555.0	1074.0	11,9	54,0	0,7502	2004-10-03	6	Sunday	October
80.0	9,2	948.0	172.0	1092.0	122.0	1584.0	1203.0	11,0	60,0	0,7867	2004-10-03	6	Sunday	October
51.0	6,5	836.0	131.0	1205.0	116.0	1490.0	1110.0	11,2	59,6	0,7888	2004-10-03	6	Sunday	October

Figure 15.84 – Placeholder

7. **Question 1: Which day of the week has the highest NO2(GT) emissions?**

   Aggregate **NO2(GT)** based on **Day** and find the mean emissions for the days:

```
# Aggregate NO2(GT) based on Day and find the mean
emissions for the days
Q1_1 = pd.DataFrame(data.groupby(['Day'])['NO2(GT)'].
agg('mean'))
Q1_1
```

   You should get the following values:

Day	NO2(GT)
Friday	70.924851
Monday	65.771155
Saturday	58.110340
Sunday	50.844978
Thursday	52.760417
Tuesday	48.289394
Wednesday	60.279545

Figure 15.85 – Placeholder

8. Make a bar plot with the data:

```
# Make a bar plot with the data
DayPlot = Q1_1.plot.bar(y='NO2(GT)',rot=90,\
title = 'Emissions on week days')

DayPlot.set_xlabel("Week Days")
DayPlot.set_ylabel("Average Emissions")
```

You should get the following output:

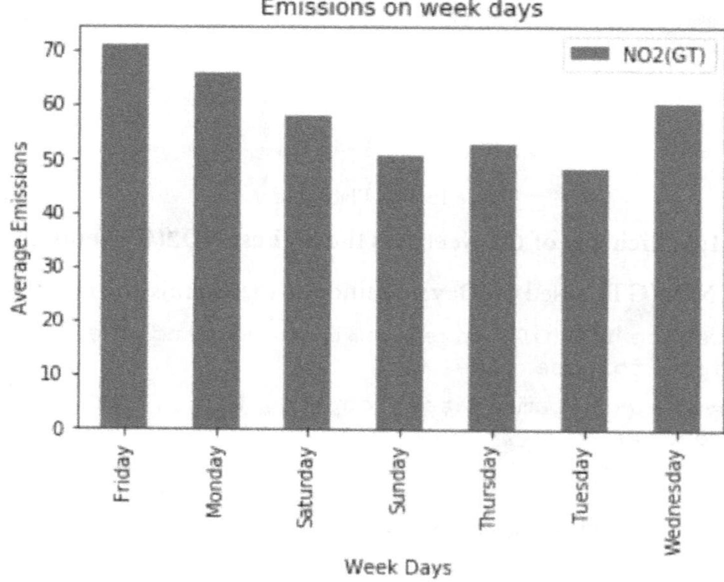

Figure 15.86 – Placeholder

From the output, you can see that Fridays have the highest NO2 emissions.

9. **Solution for question 2: At what time of the day are NMHC(GT) emissions highest?**

   Aggregate the data based on the time column and find the mean of 'NMHC(GT)':

```
# Aggregate the data based on the time column and find
the mean of 'S
Q2_1 = pd.DataFrame(data.groupby(['Time'])['NMHC(GT)'].
agg('mean'))
```

10. Plot a horizontal bar plot on the data:

```
# Horizontal bar plot
timePlot = Q2_1.plot.barh(y= 'NMHC(GT)',\
title = 'Emissions at hours of day')

timePlot.set_xlabel("Hours")
timePlot.set_ylabel("Average Emissions")
```

You should get a plot similar to the following:

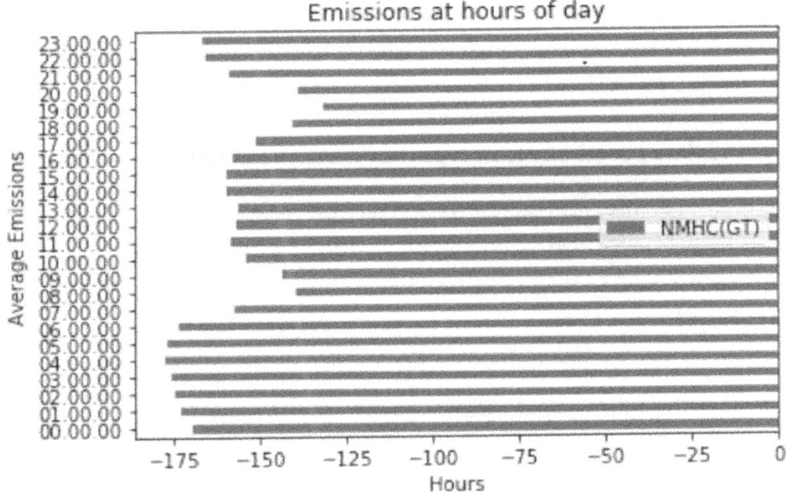

Figure 15.87 – Placeholder

From the plot, you can see that the early morning hours have the highest NMHC(GT) emissions.

11. **Solution for question 3: Which month has the highest CO(GT) emissions?**

Now, you need to aggregate based on the month column and find the mean of the `CO(GT)` column. There is a bit of cleaning to be done on the `CO(GT)` column as the data has a `,` between the figures. You can introduce a decimal and then convert it into numeric data. Create a function to do this, and then use the `lambda` function for the conversion of data:

```
# Creating the function to clean
def cleanFeat(x):
    return pd.to_numeric(".".join(x.split(',')))
```

Use the `lambda` function to clean the column:

```
# Cleaning up the CO format
data['CO(GT)'] = data['CO(GT)'].apply(lambda x:
cleanFeat(x))
data.head()
```

You should see the following output:

	Date	Time	CO(GT)	PT08.S1(CO)	NMHC(GT)	C6H6(GT)	PT08.S2(NMHC)	NOx(GT)	PT08.S3(NOx)	NO2(GT)	PT08.S4(NO2)	PT08.S5(O3)	T
0	10/03/2004	18.00.00	2.6	1360.0	150.0	11,9	1046.0	166.0	1056.0	113.0	1692.0	1268.0	13,6
1	10/03/2004	19.00.00	2.0	1292.0	112.0	9,4	955.0	103.0	1174.0	92.0	1559.0	972.0	13,3
2	10/03/2004	20.00.00	2.2	1402.0	88.0	9,0	939.0	131.0	1140.0	114.0	1555.0	1074.0	11,9
3	10/03/2004	21.00.00	2.2	1376.0	80.0	9,2	948.0	172.0	1092.0	122.0	1584.0	1203.0	11,0
4	10/03/2004	22.00.00	1.6	1272.0	51.0	6,5	836.0	131.0	1205.0	116.0	1490.0	1110.0	11,2

Figure 15.88 – Placeholder

12. Aggregate the data based on `Month` and find the mean of the `CO(GT)` column:

```
# Aggregate the data based on the Month and find the mean
of 'CO(GT)' column
Q3_1 = pd.DataFrame(data.groupby(['Month'])['CO(GT)']\
.agg('mean'))
Q3_1
```

You should get the following output:

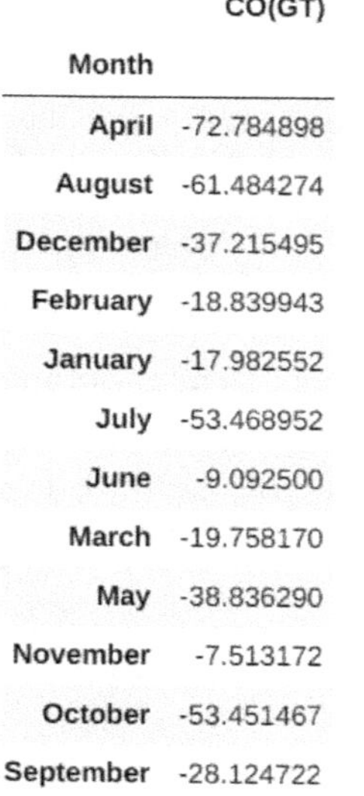

	CO(GT)
Month	
April	-72.784898
August	-61.484274
December	-37.215495
February	-18.839943
January	-17.982552
July	-53.468952
June	-9.092500
March	-19.758170
May	-38.836290
November	-7.513172
October	-53.451467
September	-28.124722

Figure 15.89 – Aggregated DataFrame

13. Plot the data using a horizontal bar chart:

```
# Plot the data
monthPlot = Q3_1.plot.barh(y= 'CO(GT)',\
title = 'Monthly average emissions')

monthPlot.set_xlabel("Months")
monthPlot.set_ylabel("Average Emissions")
```

You should get the following output:

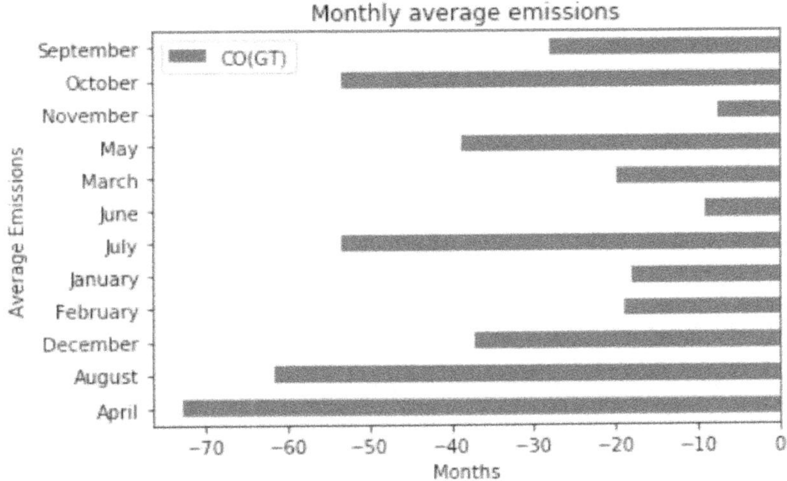

Figure 15.90 – Final output of average emissions

From the output, you can see that the month of November has the lowest emissions.

Index

D

E

F

G

H

I

J

L

T

U

V

W

X

Z

Other Books You May Enjoy

If you enjoyed this book, you may be interested in these other books by Packt:

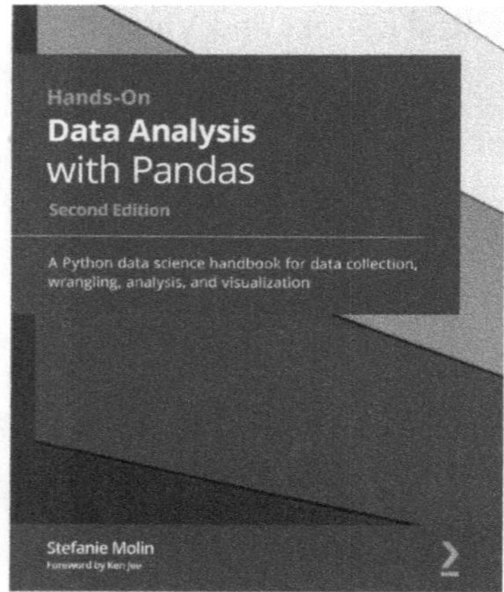

Data Analysis with Pandas, Second Edition

Stefanie Molin

ISBN: 9781800563452

- Understand how data analysts and scientists gather and analyze data
- Perform data analysis and data wrangling using Python
- Combine, group, and aggregate data from multiple sources
- Create data visualizations with pandas, matplotlib, and seaborn
- Apply machine learning algorithms to identify patterns and make predictions
- Use Python data science libraries to analyze real-world datasets
- Solve common data representation and analysis problems using pandas
- Build Python scripts, modules, and packages for reusable analysis code

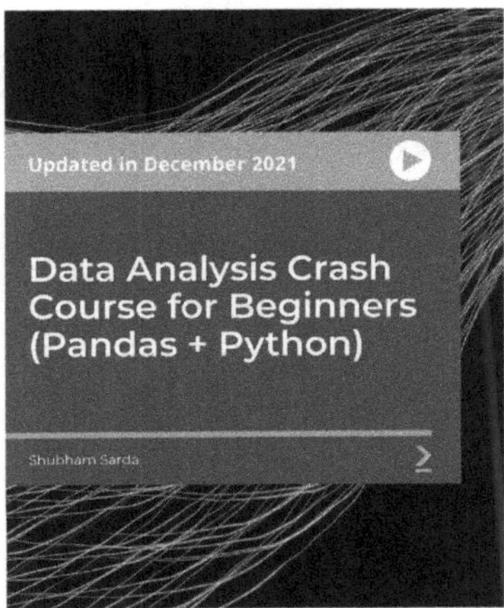

Data Analysis Crash Course for Beginners

Shubham Sarda

ISBN: 9781803242354

- Learn fundamentals of data analysis
- Understand important Jupyter Notebook commands
- Work with DataFrames (indexing, slicing, adding and deleting)
- Work with Pandas, iPython, Jupyter Notebook
- Work with CSV, Excel, TXT, JSON Files, and API Responses
- Add, delete, and update rows and columns

Packt is searching for authors like you

If you're interested in becoming an author for Packt, please visit `authors.packtpub.com` and apply today. We have worked with thousands of developers and tech professionals, just like you, to help them share their insight with the global tech community. You can make a general application, apply for a specific hot topic that we are recruiting an author for, or submit your own idea.

Share Your Thoughts

Now you've finished *The Pandas Workshop*, we'd love to hear your thoughts! Scan the QR code below to go straight to the Amazon review page for this book and share your feedback or leave a review on the site that you purchased it from.

`https://packt.link/r/1-800-20893-6`

Your review is important to us and the tech community and will help us make sure we're delivering excellent quality content.

www.ingramcontent.com/pod-product-compliance
Lightning Source LLC
Chambersburg PA
CBHW081447050326
40690CB00015B/2713
* 9 7 8 1 8 0 0 2 0 8 9 3 3 *